Advances in Particle Physics

Advances in Particle Physics

Edited by
Joy Moody

Willford Press
www.willfordpress.com

Published by Willford Press,
118-35 Queens Blvd., Suite 400,
Forest Hills, NY 11375, USA

ISBN: 978-1-68285-645-1

Cataloging-in-Publication Data

Advances in particle physics / edited by Joy Moody.
p. cm.
Includes bibliographical references and index.
ISBN 978-1-68285-645-1
1. Particles (Nuclear physics). 2. Nuclear physics. 3. Physics.
I. Moody, Joy.
QC793.2 .A28 2019
539.72--dc23

For information on all Willford Press publications
visit our website at www.willfordpress.com

Willford Press

Contents

Preface

Fundamental particles that constitute matter and radiation are studied in the domain of particle physics. Particle interaction and behavior is governed by the laws of quantum mechanics and described by a quantum field theory model called the Standard Model. Particle interactions can be in the form of strong, weak and electromagnetic interactions. Such interactions are mediated by the gauge bosons. Research in modern particle physics focuses on subatomic particles such as electrons, neutrons and protons as well as a wide range of exotic particles. This book unfolds the innovative aspects of particle physics which will be crucial for the progress of this field in the future. It aims to present researches that have transformed this discipline and aided its advancement. Particle physicists, phenomenologists, lattice theorists, researchers and students involved in this area of study will benefit alike from this book.

This book is the end result of constructive efforts and intensive research done by experts in this field. The aim of this book is to enlighten the readers with recent information in this area of research. The information provided in this profound book would serve as a valuable reference to students and researchers in this field.

At the end, I would like to thank all the authors for devoting their precious time and providing their valuable contribution to this book. I would also like to express my gratitude to my fellow colleagues who encouraged me throughout the process.

Editor

1

A framework for second-order parton showers

Hai Tao Li, Peter Skands *

ARC Centre of Excellence for Particle Physics at the Terascale, School of Physics and Astronomy, Monash University, VIC-3800, Australia

ARTICLE INFO

Editor: L. Rolandi

Keywords:
Parton showers
Monte Carlo event generators
QCD resummation

ABSTRACT

A framework is presented for including second-order perturbative corrections to the radiation patterns of parton showers. The formalism allows to combine $\mathcal{O}(\alpha_s^2)$-corrected iterated $2 \to 3$ kernels for "ordered" gluon emissions with tree-level $2 \to 4$ kernels for "unordered" ones. The combined Sudakov evolution kernel is thus accurate to $\mathcal{O}(\alpha_s^2)$. As a first step towards a full-fledged implementation of these ideas, we develop an explicit implementation of $2 \to 4$ shower branchings in this letter.

1. Introduction

Recent decades have seen tremendous improvements in our ability to combine fixed-order and resummed calculations in QCD. In the context of Monte Carlo event generators, the state of the art is now that several next-to-leading order (NLO) matrix elements can be combined with parton showers, with some progress even at the NNLO level. A new generation of shower algorithms has also been developed [1–11], most of which are based on colour dipoles or antennae, however the formal accuracy of these showers remains governed by leading-order splitting functions [12–15].

One option for improving the accuracy of the resummed part of the calculation is to match the shower evolution to higher-order analytical resummation [16]. In this letter, we instead propose to include higher-order corrections directly in the shower evolution, truncated in such a way that the full evolution kernels are accurate to $\mathcal{O}(\alpha_s^2)$. We construct a framework to include NLO corrections into the Sudakov form factor for final-state showers, which implies that more subleading logarithmic terms will be resummed. We do this by writing the Sudakov factor as a product of $2 \to 3$ splittings, which are responsible for the ordinary strongly-ordered shower evolution (with one-loop corrections and tree-level $2 \to 4$ corrections applied to products of $2 \to 3$ branchings), and direct $2 \to 4$ ones which access parts of phase space that would look "unordered" from the iterated $2 \to 3$ perspective and hence would not be reached by such branchings. A main stumbling block to achieving this in the past, which we address in this letter, is how to avoid double-counting between iterated $2 \to 3$ branchings and direct $2 \to 4$ ones. Our solution to this problem is to order the two branching types in a common measure of $p_\perp$, allowing a clean phase-space separation between them. We also ensure that there is a smooth transition at the boundary between the ordered and unordered regions.

We work in the framework of dipole-antenna showers which combine dipole showers [1] with the antenna subtraction formalism [17–19], as embodied in the VINCIA shower code[1] [20,21]. VINCIA was initially developed for final-state showers [6,22,23,13] and was recently extended to hadronic collisions [24,20]. The aim of this letter is to demonstrate the basic formalism for second-order shower kernels (at leading colour) and provide a concrete proof-of-concept implementation of $2 \to 4$ showers with two-gluon emission. We leave implementations of $g \to q\bar{q}$ splittings, one-loop corrections to $2 \to 3$ showers, and a discussion of initial-state antennae to forthcoming work.

This letter is organised as follows. In Section 2 we discuss the Sudakov factor and partition it into a product of $2 \to 3$ and $2 \to 4$ ones. Section 3 presents the method for implementing $2 \to 4$ branchings using the veto algorithm. In Section 4 we describe the $2 \to 4$ antenna functions and compare them with corresponding matrix elements. In Section 5 we discuss numerical results and collect our conclusions in Section 6.

* Corresponding author.
E-mail addresses: haitao.li@monash.edu (H.T. Li), peter.skands@monash.edu (P. Skands).

[1] http://vincia.hepforge.org/.

2. Shower framework

Within the existing antenna-shower formalism for a shower evolved in a generic measure of jet resolution Q, the LO subtraction term (antenna function) corresponding to a specific colour-connected pair of partons, call it a_3^0 [19], is exponentiated to define an all-orders Sudakov factor, $\Delta(Q_1^2, Q_2^2)$, which represents the no-branching probability for that parton pair between scales Q_1 and Q_2. As such, the differential branching probability per phase-space element is given by the derivative of the Sudakov factor,

$$\frac{d}{dQ^2}\left(1-\Delta(Q_0^2,Q^2)\right) = -\int \frac{d\Phi_3}{d\Phi_2}\,\delta(Q^2-Q^2(\Phi_3))\,a_3^0\,\Delta(Q_0^2,Q^2)\,, \quad (1)$$

where the δ function projects out a contour of constant Q^2 in the $2\to 3$ antenna phase space and we leave colour and coupling factors implicit in a_3^0. Typically, the phase space is then rewritten explicitly in terms of Q and two complementary phase-space variables, which we denote ζ and ϕ:

$$\frac{d\ln\Delta(Q_0^2,Q^2)}{dQ^2} = \int_{\zeta_-(Q)}^{\zeta_+(Q)} d\zeta \int_0^{2\pi} \frac{d\phi}{2\pi}\,\frac{|J|a_3^0}{16\pi^2 m^2}, \quad (2)$$

with m the invariant mass of the mother (2-parton) antenna. The Jacobian factor $|J|$ arises from the transformation to the (Q,ζ) variables and the $\zeta_\pm$ phase-space boundaries are defined by the specific choice of Q and ζ, see e.g. [13]. It is now straightforward to apply more derivatives, in ζ and ϕ, to obtain the fully differential branching probability in terms of the shower variables.

The essential point is that, for a_3^0 to be the proper subtraction term for NLO calculations, it must contain all relevant poles corresponding to single-unresolved limits of QCD matrix elements. Thus, a shower based on a_3^0 is guaranteed to produce the same LL structure as DGLAP ones in the collinear limit [25,26], while simultaneously respecting the dipole coherence embodied by the eikonal formula in the soft limit; the latter *without* a need to average over azimuthal angles (as required for the angular-ordered approach to coherence, see e.g. [27]).

Generalising this formalism to use NNLO subtraction terms requires the introduction of the one-loop correction to a_3^0, call it a_3^1, as well as the tree-level double-emission antenna function, a_4^0. Explicit forms for all second-order antennae in QCD can be found in [19], including their pole structure and factorisation properties in all single- and double-unresolved limits.[2] Note that a_3^1 contains explicit singularities which appear as poles in ϵ in dimensional regularisation. These are cancelled by the poles in a_4^0 upon integration of one unresolved parton (while logarithms beyond those generated at LL will in general remain).

By analogy with eq. (1), we define the differential branching probability as

$$\frac{d}{dQ^2}\Delta(Q_0^2,Q^2) = \int \frac{d\Phi_3}{d\Phi_2}\,\delta(Q^2-Q^2(\Phi_3))\left(a_3^0+a_3^1\right)\Delta(Q_0^2,Q^2) + \int \frac{d\Phi_4}{d\Phi_2}\,\delta(Q^2-Q^2(\Phi_4))\,a_4^0\,\Delta(Q_0^2,Q^2)\,, \quad (3)$$

where $Q^2(\Phi_4)$ denotes the hardest clustering scale in Φ_4, with the softer one being integrated over. Specifically, for a double clustering of $4\to 3\to 2$ partons, we define $Q(\Phi_4)\equiv \max(Q_4,Q_3)$; for an ordinary strongly ordered history, it is thus equal to the resolution scale of the clustered 3-parton configuration, Q_3, while for an unordered sequence, it is the 4-parton resolution scale, Q_4.

We now come to the central part of our proposal: how to reorganise eq. (3) in terms of finite branching probabilities (as mentioned above, the a_3^1 term and the integral over a_4^0 are separately divergent), expressed in shower variables and allowing iterated $2\to 3$ splittings and direct $2\to 4$ ones to coexist with the correct limiting behaviours (and no double counting) for both single- and double-unresolved emissions.

We first partition the a_4^0 function into two terms, one for each of the possible iterated $2\to 3$ histories, which we label a and b respectively. Suppressing the zero superscripts to avoid clutter, we define a $2\to 4$ correction factor in close analogy with the matrix-element-correction factors defined in [22],

$$R_{2\to 4} = \frac{a_4}{a_3 a_3' + b_3 b_3'}\,, \quad (4)$$

where a_3 and b_3 (a_3' and b_3') denote the antenna functions for the first (second) $2\to 3$ splittings in the a and b histories, respectively. E.g., for $1_q 2_{\bar q}\to 3_q 4_g 5_g 6_{\bar q}$, the a history is produced by the product of $a_3'(3,4,5)$ and $a_3(\widehat{34},\widehat{45},6)$, with the (on-shell) momenta of the intermediate 3-parton state, $\widehat{34}$ and $\widehat{45}$, defined by the phase-space map of the shower / clustering algorithm. The b history is produced by the product of $b_3'(4,5,6)$ and $b_3(3,\widehat{45},\widehat{56})$. We emphasise that the denominator of eq. (4) is nothing but the incoherent sum of the a and b antenna patterns (modulo the ordering variable), as would be obtained from the uncorrected (LL) antenna shower, while the numerator is the full (coherent) $2\to 4$ radiation pattern. Among other things, the factor $R_{2\to 4}$ therefore contains precisely the modulations that account for coherence between colour-neighbouring antennae.

We use the definition of $R_{2\to 4}$, eq. (4), to partition a_4 into two terms, $a_4 = R_{2\to 4}\,(a_3 a_3' + b_3 b_3')$, each of which isolates a specific (colour-ordered) single-unresolved limit, corresponding to either g_4 or g_5 becoming soft, respectively. For each term we iterate the exact antenna phase-space factorisation [19],

$$d\Phi_{m+1}(p_1,\ldots,p_{m+1}) = d\Phi_m(p_1,\ldots,p_I,p_K,\ldots,p_{m+1})\times d\Phi_{\rm ant}(i,j,k)\,, \quad (5)$$

with all momenta on shell and $p_i+p_j+p_k = p_I+p_K$, to write

$$\frac{d\Phi_4(3,4,5,6)}{d\Phi_2(1,2)} = \begin{cases} \text{path a:} & d\Phi_{\rm ant}(\widehat{34},\widehat{45},6)\,d\Phi_{\rm ant}(3,4,5) \\ \text{path b:} & d\Phi_{\rm ant}(3,\widehat{45},\widehat{56})\,d\Phi_{\rm ant}(4,5,6) \end{cases}, \quad (6)$$

where we have chosen the nesting of the antenna phase spaces such that the soft parton in the given history is always the one clustered first. We also divide up each of the resulting 4-parton integrals into ordered and unordered clustering sequences, for which $Q(\Phi_4)=Q_3$ and $Q(\Phi_4)=Q_4$, respectively (see above). The result is

$$\frac{d\Delta(Q_0^2,Q^2)}{dQ^2} = \int d\Phi_{\rm ant}\Bigg[\delta(Q^2-Q^2(\Phi_3))\,a_3^0 \times\left(1+\frac{a_3^1}{a_3^0}+\sum_{s\in a,b_{\rm ord}}\int d\Phi_{\rm ant}^s\,R_{2\to 4}\,s_3'\right)\Delta(Q_0^2,Q^2)$$

[2] Note that, for the 4-parton antenna functions, [19] only provides explicit formulae summed over permutations of identical gluons. These must then subsequently be partitioned into individual (sub-antenna) contributions from each permutation separately.

$$+\sum_{s\in a,b_{\rm unord}} \int d\Phi^s_{\rm ant}\delta(Q^2-Q^2(\Phi_4))R_{2\to4s_3s'_3}\Delta(Q_0^2,Q^2)\Bigg] \quad (7)$$

where the sums in the last two lines run over the clustering sectors (= histories), a and b.

We may now interpret the first two lines as an effective second-order probability density for $2\to3$ branchings, while the last line represents a contribution from direct $2\to4$ branchings. The solution of eq. (7) can be written as the product of $2\to3$ and $2\to4$ Sudakov form factors

$$\Delta(Q_0^2,Q^2)=\Delta_{2\to3}(Q_0^2,Q^2)\Delta_{2\to4}(Q_0^2,Q^2)\,. \quad (8)$$

Using the same notation as in eq. (2) and with Q_3 denoting a 3-parton resolution scale, the second-order $2\to3$ Sudakov factor is:

$$\Delta_{2\to3}(Q_0^2,Q^2)=\exp\Bigg[-\int_{Q^2}^{Q_0^2} dQ_3^2\int_{\zeta_-(Q_3)}^{\zeta_+(Q_3)} d\zeta \times\frac{|J|}{16\pi^2m^2}a_3^0\left(1+\frac{a_3^1}{a_3^0}+\sum_{s\in a,b_{\rm ord}}\int d\Phi^s_{\rm ant}\,R_{2\to4}\,s'_3 +\int_{Q_3^2}^{Q_0^2} d\tilde{Q}_3^2\int_{\zeta_-(\tilde{Q}_3)}^{\zeta_+(\tilde{Q}_3)} d\tilde{\zeta}\,\frac{|\tilde{J}|}{16\pi^2m^2}\,a_3^0\right)\Bigg], \quad (9)$$

where the integral over $a_3^0\equiv a_3^0(\tilde{Q}_3,\tilde{\zeta})$ is generated by the $\Delta(Q_0^2,Q^2)$ term in the second line of eq. (7), and $|\tilde{J}|\equiv|J(\tilde{Q},\tilde{\zeta})|$. The functional form of $\tilde{Q}$ must be the same as that of Q while the form of $\tilde{\zeta}$ can in principle be chosen independently of that of ζ.

The $2\to4$ Sudakov factor is defined by the last term in eq. (7). However since the $\delta(Q^2-Q^2(\Phi_4))$ function projects out the 4-parton resolution scale in this case, we interchange the order of the nested phase-space integrations, utilising that

$$\int_0^{Q_0^2} dQ_3^2\int_{Q^2}^{Q_0^2} dQ_4^2\,\Theta(Q_4^2-Q_3^2)\,f(Q_3^2,Q_4^2)= \int_{Q^2}^{Q_0^2} dQ_4^2\int_0^{Q_4^2} dQ_3^2\,f(Q_3^2,Q_4^2)\,, \quad (10)$$

for a generic integrand, f, with the result:

$$\Delta_{2\to4}(Q_0^2,Q^2)=\exp\Bigg[-\sum_{s\in a,b}\int_{Q^2}^{Q_0^2} dQ_4^2\int_0^{Q_4^2} dQ_3^2 \int_{\zeta_{4-}}^{\zeta_{4+}} d\zeta_4\int_{\zeta_{3-}}^{\zeta_{3+}} d\zeta_3\,\frac{|J_3J_4|}{(16\pi^2)^2m^2m_s^2}\int_0^{2\pi}\frac{d\phi_4}{2\pi}R_{2\to4s_3s'_3}\Bigg], \quad (11)$$

where the nested antenna phase spaces of eq. (7), $d\Phi_{\rm ant}\,d\Phi^s_{\rm ant}$ have now been expressed in terms of shower variables, with an associated combined Jacobian $|J_3J_4|$. In section 3, we show how to construct an explicit shower algorithm based on eq. (11) while we refer to [13] for a proof of concept of an NLO-corrected $2\to3$ shower based on a formula that only differs from eq. (9) by finite terms.

Let us now turn our attention to whether the integrands in each of the Sudakov form factors, eqs. (9) and (11), are well-defined and finite. For $\Delta_{2\to3}$, this amounts to showing whether the singularities present in the a_3^1 term are fully cancelled by those coming from the integral over $R_{2\to4}s'_3$. We start from the observation that the single-unresolved limits of the 4-parton antenna functions are fully captured by the LL $2\to3$ ones (up to angular terms which cancel upon integration over the unresolved region [19]), hence

$$a_4\;\to\;a_3a'_3\;+\;b_3b'_3\;+\;\text{ang.}\,, \quad (12)$$

which in turn implies that $R_{2\to4}\to1$ in any single-unresolved limit (modulo the angular terms), hence the pole structure of the $R_{2\to4}s'_3$ integrals is the same as that of the unmodified antenna functions,

$$\text{Poles}\left\{\int_{\rm ord} d\Phi^s_{\rm ant}\,R_{2\to4}\,s'_3\right\}=\text{Poles}\left\{\int d\Phi^s_{\rm ant}\,s'_3\right\}, \quad (13)$$

where the integration region can be extended to all of phase space since the ordered region by definition includes all single-unresolved limits,[3] and use of the angular-averaged $R_{2\to4}$ is justified since s'_3 itself does not depend on the azimuth angle. The sum of two sub-antenna integrals like the ones on the right-hand side of eq. (13) precisely cancels the singularities of the corresponding one-loop antenna functions, a_3^1 [19], thus establishing that the integrand in eq. (9) is free of poles in ϵ.

In the unordered part of phase space, singularities only occur when both $Q_4\to0$ & $Q_3\to0$ which corresponds to part of the double-unresolved contribution. In the shower context, these singularities are controlled via the assumption of unitarity. Thus, the $2\to4$ Sudakov factor is also well defined. Since the NLO $2\to3$ and $2\to4$ contributions are therefore both free of explicit poles in ϵ, and since they generate corrections in different parts of phase space, they may be developed as separate algorithms, provided they use the same set of antenna functions. (Full second-order precision is of course only achieved when both components are included.) Given that a proof-of-concept study of NLO corrections to $\Delta_{2\to3}$ already exists [13], we focus in the following sections on the previously missing piece: explicit construction of the $2\to4$ component.

We round off the discussion of the Sudakov form factors by illustrating the scale evolutions for $2\to3$ and $2\to4$ showers in Fig. 1. An ordered sequence of $2\to3$ branchings is represented by path $A\to C\to D$ and the corresponding combined Sudakov factor is $\Delta_{2\to3}(Q_A^2,Q_C^2)\Delta_{3\to4}(Q_C^2,Q_D^2)$. The $2\to4$ shower explores more phase space by including path $A\to B$ which lives in unordered phase space compared with the ordinary strongly-ordered shower. Path $A\to C\to B$ shows the possible branching in "smoothly-ordered showers" [22] which can also access unordered phase space. However, for smooth ordering the combined Sudakov factor $\Delta_{2\to3}(Q_A^2,Q_C^2)\Delta_{3\to4}(Q_C'^2,Q_B^2)$ is used where $Q'_C>Q_B$ represents the restart scale of the smooth-ordering shower. As pointed out in [13], the $\Delta_{2\to3}(Q_A^2,Q_C^2)$ factor implies an LL sensitivity to the intermediate scale Q_C; an undesired byproduct of the use of iterated on-shell $2\to3$ phase-space factorisations. The direct $2\to4$ shower avoids this by using the exact Sudakov factor $\Delta_{2\to4}(Q_A^2,Q_B^2)$ in which Q_C only appears implicitly as an auxiliary integration variable.

[3] This is true for all evolution variables considered in Vincia and, more generally, for any evolution variable that defines an infrared safe observable. Without this property, an explicit regularisation has to be introduced, see e.g., the case of energy ordering considered in [13].

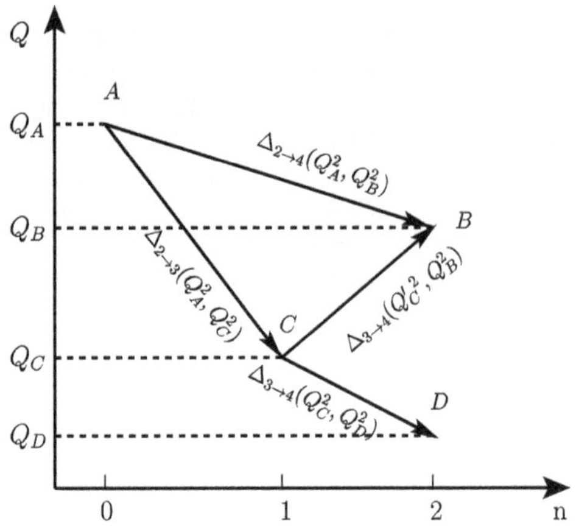

Fig. 1. Illustration of scales and Sudakov factors in strongly ordered (ACD), smoothly (un)ordered (ACB), and direct $2 \to 4$ (AB) branching processes, as a function of the number of emitted partons, n.

Finally, let us consider what happens in the vicinity of the boundary between what we label as ordered and unordered emissions, i.e., when there is no "strong" ordering between two successive (colour-connected) emissions. This is particularly relevant for the double-unresolved limits characterised by a single unresolved scale. The boundary can be approached either from the unordered region, or from the ordered one, and in general both regions will contribute to the double-unresolved limits. In the unordered region, the $2 \to 4$ antenna functions are used directly, capturing both the single- and double-unresolved (soft and collinear) limits of QCD [19]. They are also in our formalism intrinsically characterised by a single scale, as discussed above. In the ordered region, the product of $2 \to 3$ antennae is modulated by the correction factors $R_{2\to4}$, to reproduce the full $2 \to 4$ functions, and the two separate scales coincide as we approach the boundary, interpolating smoothly between the single-unresolved (iterated, strongly ordered) and double-unresolved (single-scale) limits.

3. Explicit construction of the $2 \to 4$ shower

For a branching $1\,2 \to 3\,4\,5\,6$ we define the resolution scale as $Q_4 = 2\min(p_\perp^{345}, p_\perp^{456})$, with $(p_\perp^{ijk})^2 = s_{ij}s_{jk}/s_{ijk}$. We let the direct $2 \to 4$ shower populate all configurations for which the clustering corresponding to Q_4 is unordered. (Conversely, iterated $2 \to 3$ splittings populate those configurations for which the clustering corresponding to Q_4 is ordered, with the correction factor $R_{2\to4}$ reducing to $R_{2\to4} \to a_4/(a_3 a_3')$ when there is only a single ordered path, and, for gluon neighbours, the neighbour with the smaller resolution scale used to define a_4.)

We partition the direct $2 \to 4$ phase space into two sectors: sector A with condition $p_\perp^{345} < p_\perp^{456}$ and sector B with $p_\perp^{345} > p_\perp^{456}$. For each sector, branching scales for $2 \to 4$ emissions are generated from a uniformly distributed random number $R \in [0, 1]$ by solving the following equation for Q^2:

$$R = \Delta_{2\to4}(Q_0^2, Q^2) = \exp[-\mathcal{A}(Q_0^2, Q^2)]\,. \tag{14}$$

This is done by means of the veto algorithm, which allows us to replace the (complicated) a_4 by a simple overestimate of it. We construct an appropriate "trial function" from the product of two eikonal functions $a_{\rm trial}^{2\to3} = 2g_s^2 C_A/p_\perp^2$, with an improvement factor $P_{\rm imp}$ from smooth-ordering showers [22] which improves the approximation in the unordered region of phase space, and an overall factor 2 ensuring that the overestimate remains valid in the region $p_\perp^{345} \sim p_\perp^{456}$. Thus,

$$\frac{1}{(16\pi^2)^2} a_{\rm trial}^{2\to4} = \frac{2}{(16\pi^2)^2} a_{\rm trial}^{2\to3}(Q_3^2) P_{\rm imp} a_{\rm trial}^{2\to3}(Q_4^2) = \mathcal{C}\left(\frac{\alpha_s}{4\pi}\right)^2 \frac{128}{(Q_3^2+Q_4^2)Q_4^2}\,, \tag{15}$$

where $\mathcal{C}$ is the colour factor for the double branching, normalised so that $\mathcal{C} \to C_A^2$ at leading colour. In particular, the trial function for sector A (B) is independent of momentum p_6 (p_3) which makes it easy to translate the $2 \to 4$ phase spaces defined in eq. (6) to shower variables. Technically, we generate these phase spaces by oversampling, vetoing configurations which do not fall in the appropriate sector.

For a fixed trial coupling $\hat\alpha_s$, integration yields

$$\mathcal{A}_{2\to4}^{\rm trial}(Q_0^2, Q^2) = \mathcal{C}\, I_\zeta \frac{\ln(2)\hat\alpha_s^2}{8\pi^2} \ln\frac{Q_0^2}{Q^2} \ln\frac{m^4}{Q_0^2 Q^2}\,, \tag{16}$$

where I_ζ is the ζ integral pertaining to the 4-parton phase space, defined as for $p_\perp$-ordering in ref. [22]. The solution for Q^2 in eq. (14) is thus

$$Q^2 = m^2 \exp\left(-\sqrt{\ln^2(Q_0^2/m^2) + 2f_R/\hat\alpha_s^2}\right) \tag{17}$$

where $f_R = -4\pi^2 \ln R/(\ln(2)\mathcal{C} I_\zeta)$.

The trial generator can be made more efficient by including the leading effect of scaling violation, specifically the first-order running of α_s,

$$\hat\alpha_s(k_\mu^2 p_\perp^2) = \frac{1}{b_0 \ln(k_\mu^2 p_\perp^2/\Lambda^2)}\,, \tag{18}$$

where $b_0 = (11C_A - 4n_f T_R)/(12\pi)$ and k_μ allows to apply a user-definable pre-factor. In the following equations we replace $k_\mu p_\perp$ by $k_\mu Q/2$. The trial integral then becomes

$$\mathcal{A}_{2\to4}^{\rm trial}(Q_0^2, Q^2) = \mathcal{C}\, I_\zeta \frac{\ln(2)}{4\pi^2 b_0^2}\left[\frac{\ln(m^2/Q^2)}{\ln(k_\mu^2 Q^2/4\Lambda^2)} - \frac{\ln(m^2/Q_0^2)}{\ln(k_\mu^2 Q_0^2/4\Lambda^2)} - \ln\left(\frac{\ln(k_\mu^2 Q_0^2/4\Lambda^2)}{\ln(k_\mu^2 Q^2/4\Lambda^2)}\right)\right], \tag{19}$$

and the solution to eq. (14) is

$$Q^2 = \frac{4\Lambda^2}{k_\mu^2}\left(\frac{k_\mu^2 m^2}{4\Lambda^2}\right)^{-1/W_{-1}(-y)}\,, \tag{20}$$

where

$$y = \frac{\ln k_\mu^2 m^2/4\Lambda^2}{\ln k_\mu^2 Q_0^2/4\Lambda^2} \exp\left[-f_R b_0^2 - \frac{\ln k_\mu^2 m^2/4\Lambda^2}{\ln k_\mu^2 Q_0^2/4\Lambda^2}\right]\,, \tag{21}$$

and $W_{-1}(z)$ is the Lambert W function (solving $z = we^w$ for w when $w \le -1$) for which we use the numerical implementation of [28,29].

With a trial scale Q having been generated, the remaining 4 kinematic variables (up to a global orientation) are generated according to the trial phase space integral in eq. (11), allowing to construct explicit four-momenta. The sector veto is then applied and, if the sector is accepted, the trial is accepted with a probability

$$P_{\rm trial}^{2\to4} = \frac{\alpha_s^2}{\hat\alpha_s^2}\frac{a_4}{a_{\rm trial}^{2\to4}}\,, \tag{22}$$

where higher-order running effects can be included via the α_s ratio. Note that the final orientation of the post-branching system will also depend on the specific choice of kinematics map, see [6, 30].

The last piece required for the construction of the $2 \to 4$ shower is the set of antenna functions, a_4, for $q\bar{q}$, qg, and gg parent antennae. These are defined in the following section.

4. Antenna functions

For a branching $1\ 2 \to 3\ 4\ 5\ 6$ we consider partons 1 and 2 (3 and 6) as the hard radiators (recoilers) and partons 4 and 5 as the radiated soft and/or collinear partons. (This is equivalent to the treatment of $2 \to 3$ branchings.) These partons are colour-ordered and hence the antenna function for 3 4 5 6 is not identical to that for 3 5 4 6. This is referred to as sub-antenna functions in the antenna-subtraction literature [19]. Since the shower framework is probabilistic, we also require that the antenna functions should be positive definite[4] (and bounded by the trial functions). For a $q\bar{q}$ parent antenna, the sub-antenna functions are equal to the full ones and we use a_4^0 from [19].

For qg and gg parent antennae, the full leading-colour antenna functions in [19] contain several sub-antenna configurations corresponding to any quark-gluon or gluon-gluon pair as hard partons. Moreover, some of them include terms representing two colour-unconnected emissions, for which the definition of the hard radiators and recoilers is ambiguous. The general problem of partitioning the full antenna functions into sub-antennae for colour-connected and colour-unconnected double emissions, with all singularities correct, is nontrivial. Therefore, rather than using the full antenna functions, we construct new sub-antennae based on a_4^0 by applying explicit modification factors to it, so that the unresolved limits agree with those of the relevant full antenna functions.

Specifically, for a qg parent antenna, we define the sub-antenna d_4^0:

$$d_4^0 = \begin{cases} a_4^0(3,4,5,6)\dfrac{d_3^0(\widehat{34},\widehat{45},6)}{a_3^0(\widehat{34},\widehat{45},6)} & \text{Sec. A} \\ a_4^0(3,4,5,6)\dfrac{d_3^0(3,\widehat{45},\widehat{56})}{a_3^0(3,\widehat{45},\widehat{56})}\dfrac{f_3^0(4,5,6)}{d_3^0(4,5,6)} & \text{Sec. B} \end{cases}, \tag{23}$$

where a_3^0, d_3^0, and f_3^0 are the single-emission $q\bar{q}$ and qg (sub-)antenna functions defined in [19], truncated so that only their singular terms are kept. For a gg parent antenna, we define the double-emission sub-antenna function as

$$f_4^0 = \begin{cases} a_4^0(3,4,5,6)\dfrac{f_3^0(\widehat{34},\widehat{45},6)}{a_3^0(\widehat{34},\widehat{45},6)}\dfrac{f_3^0(3,4,5)}{d_3^0(3,4,5)} & \text{Sec. A} \\ a_4^0(3,4,5,6)\dfrac{f_3^0(3,\widehat{45},\widehat{56})}{a_3^0(3,\widehat{45},\widehat{56})}\dfrac{f_3^0(4,5,6)}{d_3^0(4,5,6)} & \text{Sec. B} \end{cases}. \tag{24}$$

In all but the triple-collinear limits, it can be shown analytically that our constructions of the sub-antenna functions, d_4^0 and f_4^0, exhibit the correct infrared singularities. In the limit in which three final-state partons are collinear, we have compared numerically with matrix elements and find good agreement.

Finally, we note that the original parton pair, 1 2, is assumed to be on shell in the antenna formalism. For strongly ordered branchings, this is an excellent approximation, but for high-scale branchings, it was found in [22] that by including an effective off-shellness term for the original parton pair, products of $2 \to 3$ branchings could be brought into much better agreement with the full $2 \to 4$ ones. Analogously, we may define an effective $2 \to 5$ improvement factor, $Q_2^2/(Q_2^2 + Q_4^2)$, which can be applied to the definitions of the 4-parton antenna functions whenever they appear in the context of 5- or higher-parton configurations, with Q_2 defined as the $3 \to 2$ clustering scale of the parent partons together with their nearest colour neighbour.

As a further numerical validation of our 4-parton antenna functions, away from the singular limits, we compare the leading-colour matrix element squared for $Z \to q_1 g_2 g_3 g_4 \bar{q}_5$ with the approximation obtained using our sub-antenna function d_4^0, by defining the quantity

$$\begin{aligned} R_5 = {} & \frac{|M(Z \to q\bar{q})|^2}{|M(1,2,3,4,5)|^2} \times \\ & \Big(a_3^0(\widehat{123},\widehat{234},5)\, d_4^0(1,2,3,4) \\ & + a_3^0(1,\widehat{234},\widehat{345})\, d_4^0(5,4,3,2) \\ & + a_3^0(\widehat{12},\widehat{234},\widehat{45})\, d_3^0(1,2,\widehat{34})\, d_3^0(5,4,3) \\ & + a_3^0(\widehat{12},\widehat{234},\widehat{45})\, d_3^0(5,4,\widehat{23})\, d_3^0(1,2,3) \Big), \end{aligned} \tag{25}$$

where the second and third lines represent the two possible $2 \to 4$ branchings ($qg\bar{q} \to qggg\bar{q}$). Note that, in the sub-antenna functions d_4^0 only one sector, A or B, contributes at a time, with the phase-space factorisations given by eq. (6). The fourth and fifth lines correspond to two colour-unconnected $2 \to 3$ branchings, correlations between which could only be taken into account properly at the $2 \to 5$ level. For a gg parent antenna, we compare the leading-colour matrix element for $H \to g_1 g_2 g_3 g_4$ to the result using our f_4^0 sub-antenna function, via the quantity

$$\begin{aligned} R_4 = {} & \frac{|M(h \to gg)|^2}{|M(1,2,3,4)|^2} \\ & \times \Big(f_4^0(1,2,3,4) + f_4^0(2,3,4,1) \\ & + f_4^0(3,4,1,2) + f_4^0(4,1,2,3) \\ & + f_3^0(\widehat{23},1,\widehat{34})\, f_3^0(2,3,4) + f_3^0(\widehat{34},2,\widehat{41})\, f_3^0(3,4,1) \\ & + f_3^0(\widehat{41},3,\widehat{12})\, f_3^0(4,1,2) + f_3^0(\widehat{12},4,\widehat{23})\, f_3^0(1,2,3) \Big). \end{aligned} \tag{26}$$

Because the matrix element squared is symmetric under cyclic interchanges of the momenta, there are cases for which there is no way to define parent partons as radiators for $2 \to 4$ branchings, represented by the fourth and fifth lines in eq. (26).

Results for R_5 and R_4 are shown in the left- and right-hand panes of Fig. 2, respectively. We use a smoothly ordered $2 \to 3$ shower to generate phase space, and show the average values of R_i, differentially in the resolution scales for the last two branchings (averaging over the other phase-space variables of 5- and 4-parton phase space, respectively). As can be seen the effective sub-antennae d_4^0 and f_4^0 agree well with the matrix elements in both the ordered and unordered regions of phase space. Note that, at the very "top" of phase-space there is no overall scale hierarchy and hence no logarithmic enhancements. (The result in this region depends on a non-singular term which could be fixed by matching to the relevant fixed-order matrix elements.)

5. Numerical results

We restrict ourselves to strong ordering in the resolution scale, defined as the scale of the last radiation. That is, a $2 \to 4$ branching can have an "unordered" intermediate step, but the resolution scale of the resulting 4-parton configuration will still be required to be lower than that at which the parent 2-parton antenna was

[4] We note that negative ones could in principle be treated using the formalism presented in [31–33].

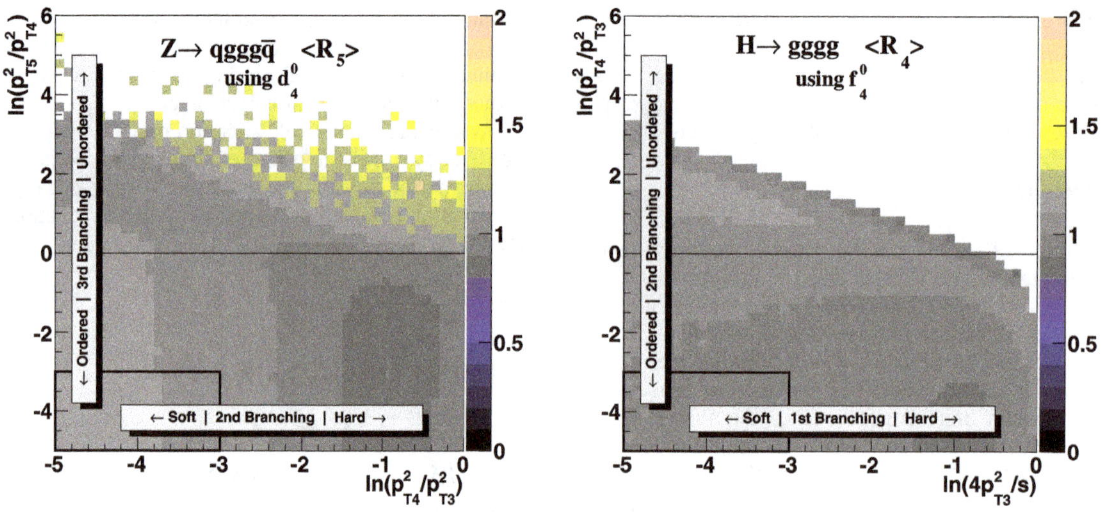

Fig. 2. Mean value of R_5 (left) and R_4 (right) differentially over phase space, with n-parton clustering scales $p_{Tn} \equiv p_T(\Phi_n)$. Note that the "top edges" of the phase spaces correspond to hard configurations that are not logarithmically enhanced.

created. Notice that, due to momentum recoil effects, the local definition of the evolution scale (inside a branching antenna) might not be the smallest global scale in a given event. (In general, the local and global resolution scale is only guaranteed to agree for so-called sector showers [23].)

The number of direct $2 \to 4$ branchings is suppressed by the small volume of unordered phase space. For a $q\bar{q}$ parent antenna, the probability of a direct $2 \to 4$ branching is around 5%. The effect of the $2 \to 4$ shower on inclusive distributions is therefore small. Since we moreover only have partial $\mathcal{O}(\alpha_s)$ corrections in the Sudakov factor, we focus on theory-level distributions to further validate the $2 \to 4$ implementation, postponing a more detailed comparison with physical observables to future work.

To compare $2 \to 3$ and $2 \to 4$ showers, we truncate the shower such that every event has at most four partons in the final state and we select the events with two gluon emissions for original colour dipole. Using shower history and colour information we identify the two radiated gluons and define the smaller transverse momentum associated with these two gluons emissions as $Q_4/2$. For example in branching $1\ 2 \to 3\ 4\ 5\ 6$ where 4 and 5 are radiations $Q_4 = 2\min(p_\perp^{345}, p_\perp^{456})$. For the case of 4 and 6 as emissions which corresponds to two iterated colour-unconnected $2 \to 3$ branchings, $Q_4 = 2\min(p_\perp^{345}, p_\perp^{365})$. The intermediate states are obtained by clustering three partons to two intermediate ones according to the choice of Q_4. And Q_3 is the scale of the other emitted parton, in the intermediate (3-parton) state. With this definition, all direct $2 \to 4$ branchings satisfy the condition $Q_4 > Q_3$ and most $2 \to 3$ branchings satisfy the complementary condition $Q_4 < Q_3$. In order to make the theory prediction consistent on the boundary $Q_4 = Q_3$, we modify the last $2 \to 3$ branching by using the strong coupling at the scale of Q_4, thus making the product $\alpha_s^2(Q_4)$ the same for branchings on either side of the ordering threshold.

Fig. 3 shows the distribution of the ratio of Q_4/Q_3 for Z decaying into $qgg\bar{q}$ (top row) and Higgs decaying into $gggg$ (bottom row). It can be seen that, due to momentum recoil effects, the pure $2 \to 3$ shower *can* generate some (highly suppressed) contributions in the phase-space region $Q_4 > Q_3$. In the $2 \to 4$ shower, the iterated $2 \to 3$ branchings are matched by sub-antenna functions a_4^0 or f_4^0, as discussed above, and the direct $2 \to 4$ branches are used to populate the unordered phase space. For Z boson decay, the iterated $2 \to 3$ and $2 \to 4$ branchings are effectively generated by the same function a_4^0. However, for Higgs boson decay, the $2 \to 4$ functions only include colour-connected double emissions, while the presence of two antennae in the Born configuration means that the iterated $2 \to 3$ branchings can also generate two colour-unconnected emissions, which are not matched to $2 \to 4$. In order to clarify the correctness of the $2 \to 4$ implementation in Fig. 3 we do not include the contribution from colour-unconnected branching sequences in the $2 \to 3$ contributions for the case with $2 \to 4$ shower. As shown in the right-hand plots, the $2 \to 4$ shower fills in the unordered phase space, and, in the limit $Q_4 \sim Q_3$, consistently matches onto the $2 \to 3$ result.

6. Conclusion and outlook

We have presented a framework for deriving corrections at the NLO level to Sudakov form-factor integrands, which generates $2 \to 3$ and $2 \to 4$ strongly-ordered showers. Compared to matching and merging methods for each branching, our corrections are generated directly by the Sudakov form factor and are present throughout the shower evolution. We hope that this framework may serve as a useful conceptual step towards the resummation of further (subleading) logarithmic terms in parton showers.

A proof-of-concept implementation of NLO corrections to a single gluon emission from a $q\bar{q}$ antenna was presented in [13]. A crucial new ingredient developed here for the first time are direct $2 \to 4$ branchings, for which we have presented an explicit Sudakov-type phase-space generator, in which the resolution scale of the 4-parton state is used as the shower evolution measure. We applied this to the case of a colour antenna radiating two (non-strongly-ordered) gluons, via a decomposition of the phase space into two sectors. For each sector we construct a trial function and trial integral based on iterated $2 \to 3$ ones, with the scale of the intermediate 3-parton state integrated over. (I.e., the intermediate 3-parton resolution scale is only used to separate what is ordered — and hence accessible by the iterated $2 \to 3$ evolution — from what is unordered.) We also define sub-antenna functions for dipole-antennae in which one or both of the parent partons are gluons, starting from the antenna function for quark-antiquark pairs, which is a good first approximation to the amplitude squared. As a validation, we compare $2 \to 4$ and $2 \to 3$ branchings in Fig. 3. As expected, the $2 \to 4$ branchings extend the phase-space population into the unordered region. Importantly, the $2 \to 4$ and $2 \to 3$ branchings produce consistent results on the boundary $Q_4 = Q_3$.

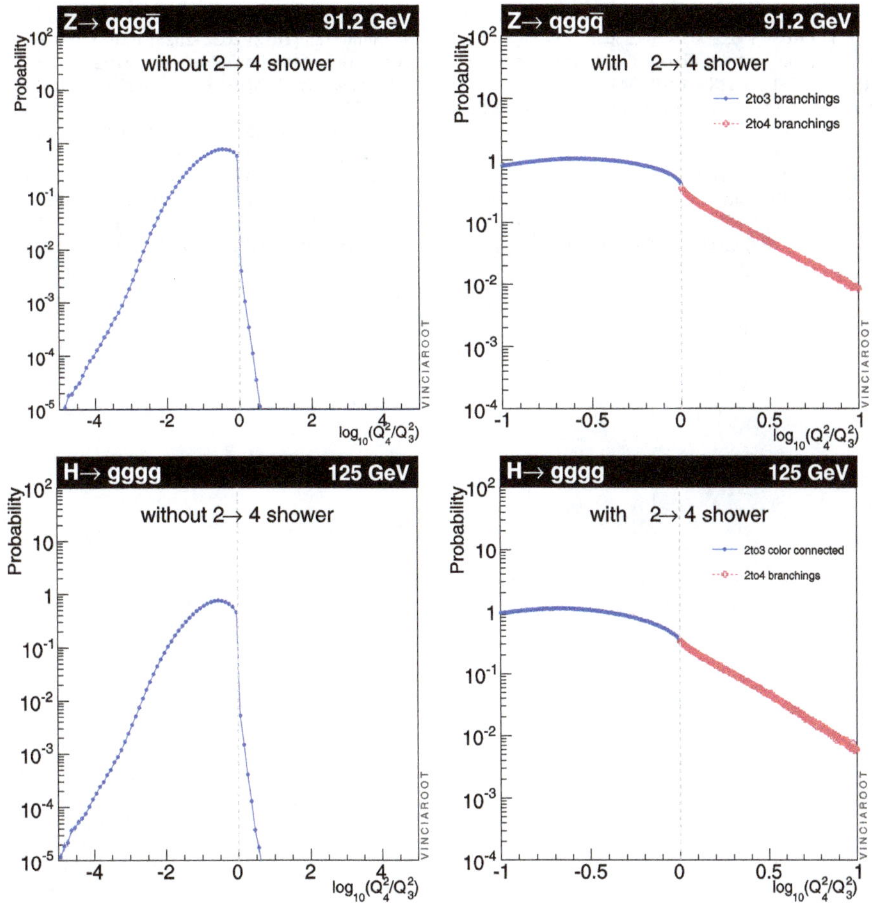

Fig. 3. Top left: the ratio of sequential clustering scales Q_4/Q_3 for a strongly ordered $2 \to 3$ shower, for $Z \to qgg\bar{q}$ (on log-log axes). Top right: closeup of the region around $Q_4/Q_3 \sim 1$, with $2 \to 4$ branchings included. Bottom row: the same for $H \to gggg$.

In the near future we will extend the $2 \to 4$ shower formalism to include $g \to q\bar{q}$ splittings. We also expect to include the second-order correction to the $2 \to 3$ Sudakov form factor defined in eq. (9). Finally, in the longer term we plan to turn our attention to the initial state, extending the formalism to the case of hadron collisions.

Acknowledgements

We are grateful to S. Prestel for comments on the manuscript. This work was supported in part by the ARC Centre of Excellence for Particle Physics at the Tera-Scale (CoEPP), CE1101004. PS is the recipient of an ARC Future Fellowship, FT130100744.

References

[1] G. Gustafson, U. Pettersson, Dipole formulation of QCD cascades, Nucl. Phys. B 306 (1988) 746–758.
[2] L. Lönnblad, ARIADNE version 4: a program for simulation of QCD cascades implementing the color dipole model, Comput. Phys. Commun. 71 (1992) 15–31.
[3] T. Sjöstrand, P.Z. Skands, Transverse-momentum-ordered showers and interleaved multiple interactions, Eur. Phys. J. C 39 (2005) 129–154, arXiv:hep-ph/0408302.
[4] Z. Nagy, D.E. Soper, Matching parton showers to NLO computations, J. High Energy Phys. 10 (2005) 024, arXiv:hep-ph/0503053.
[5] Z. Nagy, D.E. Soper, Parton showers with quantum interference, J. High Energy Phys. 09 (2007) 114, arXiv:0706.0017.
[6] W.T. Giele, D.A. Kosower, P.Z. Skands, A simple shower and matching algorithm, Phys. Rev. D 78 (2008) 014026, arXiv:0707.3652.
[7] M. Dinsdale, M. Ternick, S. Weinzierl, Parton showers from the dipole formalism, Phys. Rev. D 76 (2007) 094003, arXiv:0709.1026.
[8] S. Schumann, F. Krauss, A parton shower algorithm based on Catani–Seymour dipole factorisation, J. High Energy Phys. 03 (2008) 038, arXiv:0709.1027.
[9] J.-C. Winter, F. Krauss, Initial-state showering based on colour dipoles connected to incoming parton lines, J. High Energy Phys. 07 (2008) 040, arXiv:0712.3913.
[10] S. Plätzer, S. Gieseke, Coherent parton showers with local recoils, J. High Energy Phys. 01 (2011) 024, arXiv:0909.5593.
[11] S. Höche, S. Prestel, The midpoint between dipole and parton showers, Eur. Phys. J. C 75 (9) (2015) 461, arXiv:1506.0505.
[12] S. Jadach, A. Kusina, W. Placzek, M. Skrzypek, M. Slawinska, Inclusion of the QCD next-to-leading order corrections in the quark-gluon Monte Carlo shower, Phys. Rev. D 87 (3) (2013) 034029, arXiv:1103.5015.
[13] L. Hartgring, E. Laenen, P. Skands, Antenna showers with one-loop matrix elements, J. High Energy Phys. 10 (2013) 127, arXiv:1303.4974.
[14] Z. Nagy, D.E. Soper, A parton shower based on factorization of the quantum density matrix, J. High Energy Phys. 06 (2014) 097, arXiv:1401.6364.
[15] S. Jadach, A. Kusina, W. Placzek, M. Skrzypek, On the dependence of QCD splitting functions on the choice of the evolution variable, J. High Energy Phys. 08 (2016) 092, arXiv:1606.0123.
[16] S. Alioli, C.W. Bauer, C.J. Berggren, A. Hornig, F.J. Tackmann, C.K. Vermilion, J.R. Walsh, S. Zuberi, Combining higher-order resummation with multiple NLO calculations and parton showers in Geneva, J. High Energy Phys. 09 (2013) 120, arXiv:1211.7049.
[17] Y.I. Azimov, Y.L. Dokshitzer, V.A. Khoze, S.I. Troian, The string effect and QCD coherence, Phys. Lett. B 165 (1985) 147–150.
[18] D.A. Kosower, Antenna factorization of gauge theory amplitudes, Phys. Rev. D 57 (1998) 5410–5416, arXiv:hep-ph/9710213.
[19] A. Gehrmann-De Ridder, T. Gehrmann, E.W.N. Glover, Antenna subtraction at NNLO, J. High Energy Phys. 09 (2005) 056, arXiv:hep-ph/0505111.
[20] N. Fischer, S. Prestel, M. Ritzmann, P. Skands, Vincia for hadron colliders, Eur. Phys. J. C 76 (11) (2016) 589, arXiv:1605.0614.
[21] T. Sjöstrand, S. Ask, J.R. Christiansen, R. Corke, N. Desai, P. Ilten, S. Mrenna, S. Prestel, C.O. Rasmussen, P.Z. Skands, An introduction to PYTHIA 8.2, Comput. Phys. Commun. 191 (2015) 159–177, arXiv:1410.3012.
[22] W.T. Giele, D.A. Kosower, P.Z. Skands, Higher-order corrections to timelike jets, Phys. Rev. D 84 (2011) 054003, arXiv:1102.2126.

[23] J.J. Lopez-Villarejo, P.Z. Skands, Efficient matrix-element matching with sector showers, J. High Energy Phys. 11 (2011) 150, arXiv:1109.3608.
[24] M. Ritzmann, D.A. Kosower, P. Skands, Antenna showers with hadronic initial states, Phys. Lett. B 718 (2013) 1345–1350, arXiv:1210.6345.
[25] Z. Nagy, D.E. Soper, Final state dipole showers and the DGLAP equation, J. High Energy Phys. 05 (2009) 088, arXiv:0901.3587.
[26] P.Z. Skands, S. Weinzierl, Some remarks on dipole showers and the DGLAP equation, Phys. Rev. D 79 (2009) 074021, arXiv:0903.2150.
[27] R.K. Ellis, W.J. Stirling, B.R. Webber, QCD and collider physics, Camb. Monogr. Part. Phys. Nucl. Phys. Cosmol. 8 (1996) 1–435.
[28] D. Veberic, Having fun with Lambert W(x) function, arXiv:1003.1628.
[29] D. Veberic, Lambert W function for applications in physics, Comput. Phys. Commun. 183 (2012) 2622–2628, arXiv:1209.0735.
[30] A. Gehrmann-De Ridder, M. Ritzmann, P.Z. Skands, Timelike dipole-antenna showers with massive fermions, Phys. Rev. D 85 (2012) 014013, arXiv:1108.6172.
[31] S. Plätzer, M. Sjödahl, The Sudakov veto algorithm reloaded, Eur. Phys. J. Plus 127 (2012) 26, arXiv:1108.6180.
[32] S. Hoeche, F. Krauss, M. Schönherr, F. Siegert, A critical appraisal of NLO+PS matching methods, J. High Energy Phys. 09 (2012) 049, arXiv:1111.1220.
[33] L. Lönnblad, Fooling around with the Sudakov veto algorithm, Eur. Phys. J. C 73 (3) (2013) 2350, arXiv:1211.7204.

2

Effect of CP violation in the singlet-doublet dark matter model

Tomohiro Abe [a,b,*]

[a] Institute for Advanced Research, Nagoya University, Furo-cho Chikusa-ku, Nagoya, Aichi, 464-8602, Japan
[b] Kobayashi-Maskawa Institute for the Origin of Particles and the Universe, Nagoya University, Furo-cho Chikusa-ku, Nagoya, Aichi, 464-8602, Japan

ARTICLE INFO

Editor: J. Hisano

ABSTRACT

We revisit the singlet-doublet dark matter model with a special emphasis on the effect of CP violation on the dark matter phenomenology. The CP violation in the dark sector induces a pseudoscalar interaction of a fermionic dark matter candidate with the SM Higgs boson. The pseudoscalar interaction helps the dark matter candidate evade the strong constraints from the dark matter direct detection experiments. We show that the model can explain the measured value of the dark matter density even if dark matter direct detection experiments do not observe any signal. We also show that the electron electric dipole moment is an important complement to the direct detection for testing this model. Its value is smaller than the current upper bound but within the reach of future experiments.

1. Introduction

Dark matter (DM) is a leading candidate for physics beyond the standard model of particle physics. Models based on the WIMP paradigm are popular and have been widely studied. On the other hand, the recent dark matter direct detection experiments [1,2] give severe constraints on the models.

It is possible to evade the constraints from the direct detection experiments if a DM candidate is a fermion and interacts with the standard model (SM) sector through pseudoscalar interactions [3, 4]. In that case, the cross section for the direct detection is suppressed by the velocity of the DM and thus is negligible. On the other hand, the same interactions are relevant to DM annihilation processes and we can obtain the DM thermal relic abundance that matches the measured value of the DM density.

There are two simple ways to introduce the pseudoscalar interactions. One way is to add CP-odd scalar mediators that couple both to a DM candidate and the SM particles [5–16]. The other way is to introduce CP-violation into the dark sector. In the latter case, the SM Higgs boson can be a mediator, and we do not need CP-odd scalar mediators. We focus on the latter possibility.

We consider the singlet-doublet model [17–19]. This model is one of the minimal setups in simplified dark matter models with a fermionic dark matter candidate and has been widely studied [20–33]. The stability of dark matter is guaranteed by a Z_2 symmetry. All the SM particles are Z_2-even. There are a gauge singlet Weyl fermion and an SU(2) doublet Dirac fermion. They are Z_2-odd. The singlet, the doublet, and the Higgs boson form Yukawa interactions. There is a CP phase in the dark sector, and thus the model naturally contains a pseudoscalar interaction. The effect of this CP violation on the electron electric dipole moment (EDM) was discussed in Ref. [17,18]. On the other hand, the CP violation effect on the DM phenomenology has hardly ever been discussed.[1]

In this paper, we examine the effect of the CP violation in the model. In particular, we focus on its effect on the cross section for the direct detection. We show that the pseudoscalar interaction generated by the CP violation helps to evade the strong constraints from the direct detection experiments. We also investigate the electron EDM with emphasis on the relation to the DM phenomenology.

The rest of this paper is organized as follows. In section 2, we briefly review the singlet-doublet dark matter model. In section 3, we discuss the current constraints and prospects of this model from the viewpoint of DM direct searches and the electron EDM. We devote section 4 to our conclusion.

2. Model

In this section, we briefly review the model. We introduce a gauge singlet Majorana fermion (ω) and an $SU(2)_L$ doublet Dirac fermion with hypercharge $Y = 1/2$ which is composed of a left-

* Correspondence to: Institute for Advanced Research, Nagoya University, Furo-cho Chikusa-ku, Nagoya, Aichi, 464-8602, Japan.
E-mail address: abetomo@kmi.nagoya-u.ac.jp.

[1] The analysis of DM phenomenology with CP phase for the light DM mass region, $m_{DM} \lesssim 100$ GeV, was discussed in [23].

handed Weyl fermion ($\eta = (\eta^+, \eta^0)^T$) and a right-handed Weyl fermion ($\xi^\dagger = ((\xi^-)^\dagger, \xi^{0\dagger})^T$). We impose a Z_2 symmetry on the model. Under the Z_2 symmetry, all the SM particles are even, and all the fermions we introduced in the above are odd. The lightest neutral Z_2-odd fermion is a DM candidate.

The mass and Yukawa interaction terms for the Z_2 odd particles are given by

$$\mathcal{L}_{int.} = -\frac{M_1}{2}\omega\omega - M_2 e^{-i\phi}\xi\eta - y\omega H^\dagger\eta - y'\xi H\omega + (h.c.). \tag{2.1}$$

All the parameters have a CP violating phase in general, but we can eliminate three of them by the redefinition of the fermion fields. We work in the basis where only the Dirac mass term has a phase, and we explicitly write down the phase as $M_2 e^{-i\phi}$. In this basis, all the parameters except ϕ are positive. After the Higgs field develops a vacuum expectation value (VEV), we find the following mass terms

$$\mathcal{L}_{mass} = -\frac{1}{2}\begin{pmatrix}\omega & \eta^0 & \xi^0\end{pmatrix}\begin{pmatrix} M_1 & \frac{v}{\sqrt{2}}y & \frac{v}{\sqrt{2}}y' \\ \frac{v}{\sqrt{2}}y & 0 & M_2 e^{-i\phi} \\ \frac{v}{\sqrt{2}}y' & M_2 e^{-i\phi} & 0 \end{pmatrix}\begin{pmatrix}\omega \\ \eta^0 \\ \xi^0\end{pmatrix} - M_2\xi^-\eta^+ + (h.c.), \tag{2.2}$$

where v is the VEV of the Higgs boson, $v \simeq 246$ GeV. We introduce λ, θ, and r for later convenience,

$$y = \lambda\sin\theta, \quad y' = \lambda\cos\theta, \quad r = \frac{y}{y'} = \tan\theta. \tag{2.3}$$

The mass of the charged Dirac fermion is M_2. After we diagonalize the mass matrix, we obtain three neutral Weyl fermions $(\chi_1^0, \chi_2^0, \chi_3^0)$ that are related to (ω, η^0, ξ^0) by a unitary matrix V as follows.

$$\begin{pmatrix}\omega \\ \eta^0 \\ \xi^0\end{pmatrix} = \begin{pmatrix} V_{11} & V_{12} & V_{13} \\ V_{21} & V_{22} & V_{23} \\ V_{31} & V_{32} & V_{33}\end{pmatrix}\begin{pmatrix}\chi_1^0 \\ \chi_2^0 \\ \chi_3^0\end{pmatrix}, \tag{2.4}$$

where χ_1^0 is the lightest neutral Z_2-odd field and thus is the DM candidate in this model.

It is sufficient to study ϕ in the range $0 \le \phi \le \pi$, although $-\pi < \phi \le \pi$ in general. Physics in the negative ϕ regime is related to physics in the positive ϕ regime by the complex conjugate of the mass matrix, and thus of the mixing matrix. If observables respect the CP symmetry, they do not depend on the sign of ϕ. All the processes for the relic abundance and the direct detection are independent of the sign of ϕ because they respect the CP symmetry. While the CP violating EDM depends on the sign of ϕ, the sign of ϕ merely changes the sign of the EDM, so we focus only on the positive ϕ region.

The ratio of the two Yukawa couplings is important as we will see below. We denote it by $r = y/y'$ and focus on $0 < r \le 1$. Since both Yukawa couplings are positive, r is positive definite. For $r = 0$, we can eliminate ϕ by the redefinition of the fermion fields. We do not discuss that situation because we aim to examine the CP violation effect in the dark sector. Physics for $1 < r < \infty$ is equivalent to physics for $0 < r < 1$ by renaming η^0 and ξ^0 as ξ^0 and η^0 respectively as can be seen from the mass matrix. Therefore it is sufficient to discuss for $0 < r \le 1$.

2.1. couplings

The Z_2-odd fermions couple to the Higgs boson and the gauge bosons. We can obtain the interaction terms by diagonalizing the mass matrix and going to the mass eigenbasis. Four component notation is useful in calculations. The mass eigenstates of the charged and neutral Z_2 odd particles in four component notation are given by

$$\Psi_+ = \begin{pmatrix}\eta^+ \\ (\xi^-)^\dagger\end{pmatrix}, \quad \Psi_j = \begin{pmatrix}\chi_j^0 \\ \chi_j^{0\dagger}\end{pmatrix}. \tag{2.5}$$

The interaction terms including Z_2-odd particles are[2]

$$\begin{aligned}\mathcal{L}_{int.} \supset & -\sum_j \overline{\Psi_+}\gamma^\mu(P_L c^L_{\chi_j} + P_R c^R_{\chi_j})\Psi_j W^+_\mu \\ & -\sum_j \overline{\Psi_j}\gamma^\mu(P_L(c^L_{\chi_j})^* + P_R(c^R_{\chi_j})^*)\Psi_+ W^-_\mu \\ & -\left(\frac{e}{2s_W c_W}(c_W^2 - s_W^2)Z_\mu + eA_\mu\right)\overline{\Psi_+}\gamma^\mu\Psi_+ \\ & -\frac{1}{2}\sum_{j,k} c_{Z\chi_j\chi_k}\overline{\Psi_j}\gamma^\mu\gamma^5\Psi_k Z_\mu + \frac{1}{2}\sum_{j,k} c^p_{Z\chi_j\chi_k}\overline{\Psi_j}i\gamma^\mu\Psi_k Z_\mu \\ & -\frac{1}{2}\sum_{j,k} c_{h\chi_j\chi_k}\overline{\Psi_j}\Psi_k h + \frac{1}{2}\sum_{j,k} c^p_{h\chi_j\chi_k}\overline{\Psi_j}i\gamma_5\Psi_k h,\end{aligned} \tag{2.6}$$

where

$$c^L_{\chi_j} = \frac{e}{\sqrt{2}s_W}V_{2j}, \tag{2.7}$$

$$c^R_{\chi_j} = \frac{e}{\sqrt{2}s_W}V^*_{3j}, \tag{2.8}$$

$$c_{Z\chi_j\chi_k} = \frac{e}{2s_W c_W}\mathrm{Re}(V^*_{2j}V_{2k} - V^*_{3j}V_{3k}), \tag{2.9}$$

$$c^p_{Z\chi_j\chi_k} = \frac{e}{2s_W c_W}\mathrm{Im}(V^*_{2j}V_{2k} - V^*_{3j}V_{3k}), \tag{2.10}$$

$$c_{h\chi_j\chi_k} = \sqrt{2}\mathrm{Re}(yV_{1j}V_{2k} + y'V_{1j}V_{3k}), \tag{2.11}$$

$$c^p_{h\chi_j\chi_k} = \sqrt{2}\mathrm{Im}(yV_{1j}V_{2k} + y'V_{1j}V_{3k}). \tag{2.12}$$

Among these terms, the following interactions are particularly important for our discussion.

$$-\frac{1}{2}c_{Z\chi_1\chi_1}\bar{\Psi}_1\gamma^\mu\gamma^5\Psi_1 Z_\mu - \frac{1}{2}c_{h\chi_1\chi_1}\bar{\Psi}_1\Psi_1 h + \frac{1}{2}c^p_{h\chi_1\chi_1}\bar{\Psi}_1 i\gamma^5\Psi_1 h. \tag{2.13}$$

All these three terms contribute to DM annihilation processes. On the other hand, they play different roles in elastic scattering processes of the dark matter with nucleon. The scalar interaction ($c_{h\chi_1\chi_1}$) contributes to the spin-independent cross section (σ_{SI}), the Z interaction ($c_{Z\chi_1\chi_1}$) contributes to the spin-dependent cross section (σ_{SD}), and the pseudoscalar interaction ($c^p_{h\chi_1\chi_1}$) does not contribute to the scattering process due to the velocity suppression. Therefore, if the $c_{h\chi_1\chi_1} = c_{Z\chi_1\chi_1} = 0$, then DM can completely evade the current strong constraints from the direct detection experiments. Meanwhile, a nonzero value of $c^p_{h\chi\chi}$ can ensure the necessary annihilation cross section.

There are symmetries that may force $c^p_{h\chi_j\chi_k}$ and $c_{Z\chi_j\chi_k}$ to be zero. In the CP invariant situations, $\phi = 0$ or π, then $c^p_{h\chi_j\chi_k} = 0$ because the pseudoscalar interaction originates from the CP violation. If $y = y'$, then $c_{Z\chi_j\chi_k} = 0$ because the model becomes symmetric under the exchange of χ^0 and ξ^0 as can be seen from Eq. (2.2).

---

[2] We have checked these interaction terms are consistent with Ref. [23,24] up to conventions of the fields and the gauge couplings.

This symmetry implies $V_{2i} = V_{3i}$, and thus $c_{Z\chi_j\chi_k} = 0$ as can be seen from Eq. (2.9).

The scalar coupling $c_{h\chi_j\chi_k}$ can be zero as well. However, in contrast to $c^p_{h\chi_j\chi_k}$ and $c_{Z\chi_j\chi_k}$, it becomes zero accidentally rather than by symmetries. We discuss the condition on the parameters for vanishing $c_{h\chi_1\chi_1}$ in the next subsection.

2.2. Blind spot

There is a condition that the scalar interaction of the DM vanishes, $c_{h\chi_1\chi_1} = 0$. The condition is called the blind spot [20–22]. In this subsection, we discuss the condition for the blind spot.

For the purpose of finding the condition that $c_{h\chi_1\chi_1} = 0$, the expression of the scalar coupling given in Eq. (2.11) is not convenient. We can also obtain the scalar coupling from the derivative of the dark matter mass with respect to the VEV of the Higgs boson, $c_{h\chi_1\chi_1} = \partial m_{\rm DM}/\partial v$. The expression obtained in this way is useful to find the blind spot. The DM mass satisfies the characteristic equation given by

$$\begin{aligned}0 =& m_{\rm DM}^6 - m_{\rm DM}^4\left(M_1^2 + 2M_2^2 + v^2\lambda^2\right)\\ &+ m_{\rm DM}^2\left(2M_1^2M_2^2 + \left(M_2^2 + \frac{v^2\lambda^2}{2}\right)^2\right.\\ &\left. - M_1M_2v^2\lambda^2\sin 2\theta\cos\phi\right)\\ &- M_2^2\left(M_1^2M_2^2 - M_1M_2v^2\lambda^2\sin 2\theta\cos\phi + \frac{1}{4}v^4\lambda^4\sin^2 2\theta\right),\end{aligned} \tag{2.14}$$

where $\tan\theta = y/y' = r$. By differentiating this equation with respect to v and setting $\partial m_{\rm DM}/\partial v = 0$, we find

$$\begin{aligned}0 =& m_{\rm DM}^4 - m_{\rm DM}^2\left(M_2^2 + \frac{v^2\lambda^2}{2} - M_1M_2\sin 2\theta\cos\phi\right)\\ &- M_2^2\sin 2\theta\left(M_1M_2\cos\phi - \frac{v^2}{2}\lambda^2\sin 2\theta\right).\end{aligned} \tag{2.15}$$

Using Eqs. (2.14) and (2.15), we can obtain two relations. For example, we can solve for $m_{\rm DM}$ and λ,

$$m_{\rm DM} = \left(\frac{M_1^2M_2^2\sin^2 2\theta\sin^2\phi}{M_1^2 + M_2^2\sin^2 2\theta + 2M_1M_2\sin 2\theta\cos\phi}\right)^{1/2}, \tag{2.16}$$

$$\lambda = \sqrt{\frac{2(M_2^2 - m_{\rm DM}^2)(m_{\rm DM}^2 + M_1M_2\sin 2\theta\cos\phi)}{v^2(M_2^2\sin^2 2\theta - m_{\rm DM}^2)}}. \tag{2.17}$$

This is the blind spot condition where $c_{h\chi_1\chi_1} = 0$. Eq. (2.16) is given in Ref. [23]

For $\phi = 0$, there is no blind spot because it requires $m_{\rm DM} = 0$. For $\phi = \pi$, $m_{\rm DM}$ is non-zero if the denominator of Eq. (2.16) is zero and we find the following blind spot condition for $\phi = \pi$,

$$m_{DM} = M_1 = M_2\sin 2\theta. \tag{2.18}$$

This is the same as the blind spot condition given in Refs. [22,24].[3]

[3] Note that $m_{\rm DM} = M_2\sin 2\theta = -M_2\cos\pi\sin 2\theta$, and $\sin 2\theta$ in Refs. [22,24] corresponds to $\cos\phi\sin 2\theta$ in our notation.

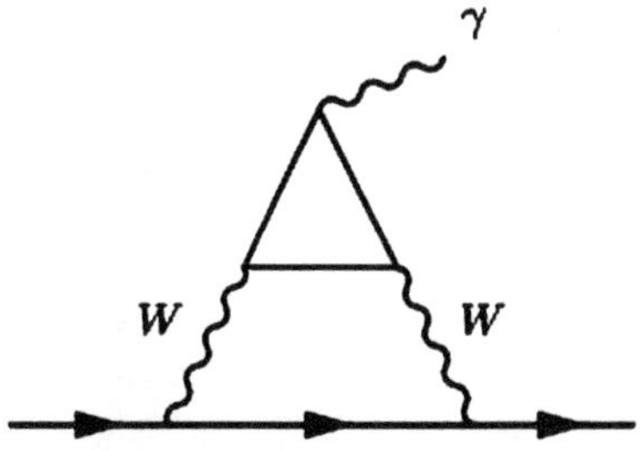

Fig. 1. Barr–Zee type contributions to the EDM. The Z_2-odd charged and neutral fermions run in the triangle part.

2.3. EDM

This model predicts a new contribution to the EDM because of the new source of CP violation ϕ. The electron EDM is an important complement to direct detection experiments as we will see in Sec. 3. We summarize the formulae here. The leading contribution comes from the Barr–Zee type diagram shown in Fig. 1. The electron EDM is defined through

$$\mathcal{H}_{eff} = i\frac{d_e}{2}\bar{\psi}_e\sigma_{\mu\nu}\gamma_5\psi_e F^{\mu\nu}. \tag{2.19}$$

We find [23].

$$\frac{d_e}{e} = -\frac{2\alpha}{(4\pi)^3 s_W^2}\sqrt{2}G_F m_{\chi^\pm} m_e \sum_{j=1}^{3}{\rm Im}(V_{2j}V_{3j}) m_{\chi_j^0}\mathcal{I}_j, \tag{2.20}$$

where

$$\mathcal{I}_j = \int_0^1 dz \frac{1-z}{m_{\chi^\pm}^2(1-z) + m_{\chi_j^0}^2 z - m_W^2 z(1-z)} \times \ln\frac{m_{\chi^\pm}^2(1-z) + m_{\chi_j^0}^2 z}{m_W^2 z(1-z)}. \tag{2.21}$$

3. Current and future prospects on direct detection and EDM

In this section, we discuss the current constraints from the dark matter direct searches [1,2] and prospects for the direct searches [34]. We assume the thermal relic scenario. We have calculated the relic abundance and the scattering cross section of dark matter with nucleon by using `micrOMEGAs v4.3.1` [35]. We choose the value of λ so as to reproduce the measured dark matter density, $\Omega h^2 = 0.1198 \pm 0.0015$ [36].

The current constraints from the dark matter direct searches and future prospects are shown in Fig. 2. Here we take $M_2 = 1000$ GeV. The red regions are excluded by the constraint on $\sigma_{\rm SI}$ from LUX experiment [1], and the orange regions are excluded by the constraint on $\sigma_{\rm SD}$ from PandaX-II experiment [2]. We use the projection for LZ experiment in Ref. [34] for the future prospects of $\sigma_{\rm SI}$ and $\sigma_{\rm SD}$. We find that the constraint on $\sigma_{\rm SI}$ already excludes a large region of the parameter space. However, the spin-independent scattering process cannot cover some of the regions due to the existence of the blind spot. In those regions, the spin-dependent scattering process is helpful. The constraint on $\sigma_{\rm SD}$ is currently much weaker than the constraint from $\sigma_{\rm SI}$. However, it plays an important role in future as we can see from the lower panels in the figure. The spin-dependent scattering process can cover the parameter space where the spin-independent scattering process cannot. In $r \sim 1$ regime, $\sigma_{\rm SD}$ becomes small because it depends on $c_{Z\chi_1\chi_1}$ that becomes zero for $r = 1$ as we discussed in the previous section.

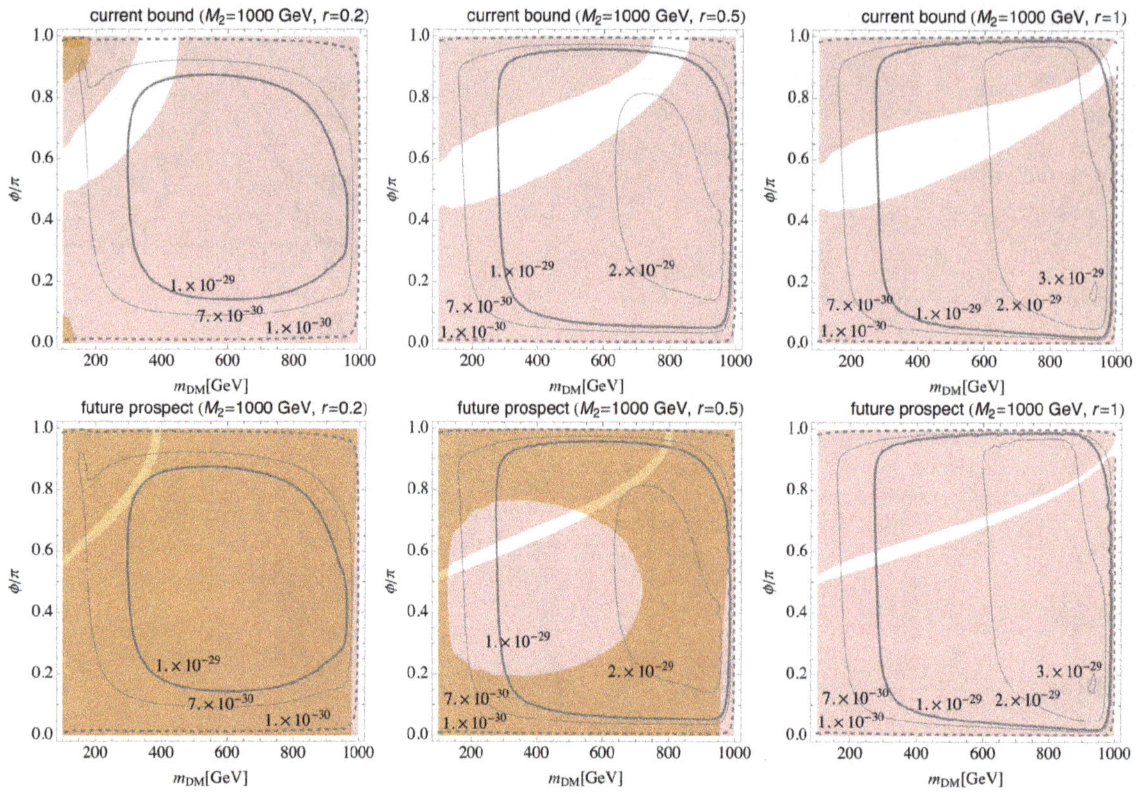

Fig. 2. The current bound and prospects. The red regions in the upper panels are excluded by the constraint on the $\sigma_{\rm SI}$ from LUX experiment [1]. The orange regions in the upper panels are excluded by the constraint on $\sigma_{\rm SD}$ from PandaX-II experiment [2]. The red and orange regions in the lower panels show the prospects for $\sigma_{\rm SI}$ and $\sigma_{\rm SD}$, respectively. We use the projection for LZ experiment in Ref. [34] for the future prospects. The contour lines show the absolute values of the model prediction of the electron EDM. (For interpretation of the references to color in this figure legend, the reader is referred to the web version of this article.)

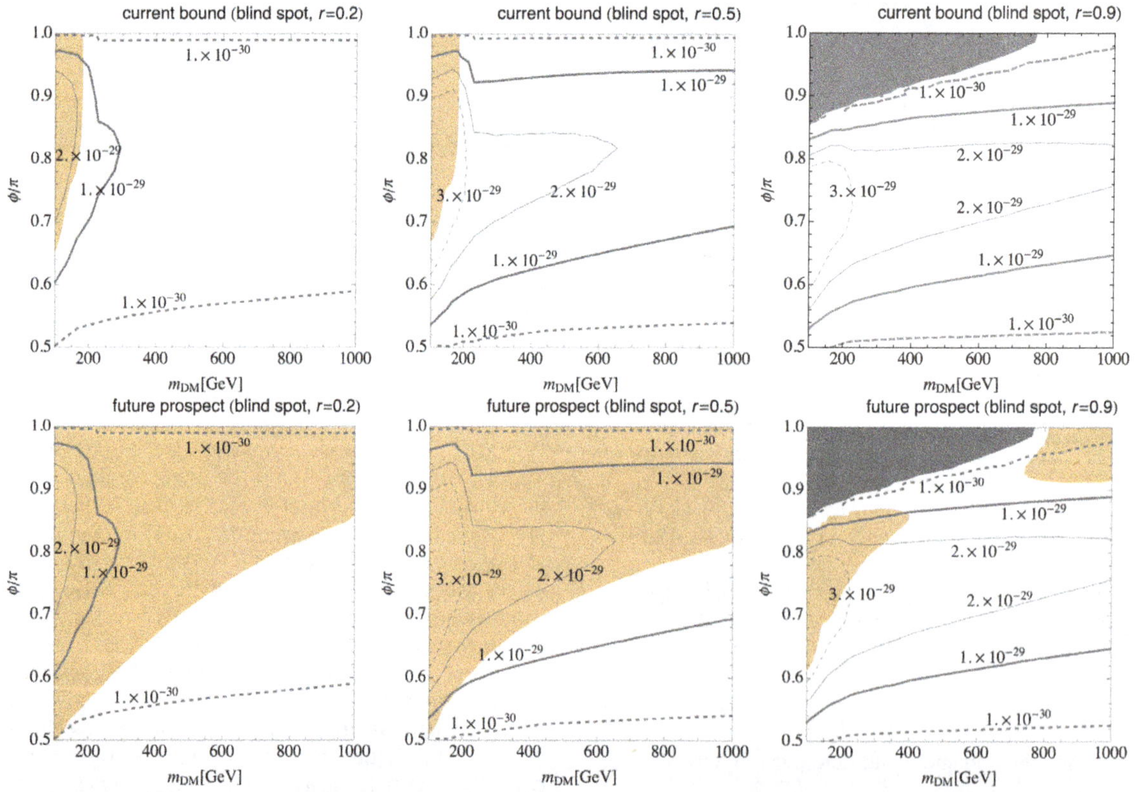

Fig. 3. The current constraints and prospects in the blind spot. In the gray regions, we cannot obtain the DM thermal relic abundance that matched the measured value of DM density. The notations of the other colors and the contours are the same as in Fig. 2. (For interpretation of the references to color in this figure legend, the reader is referred to the web version of this article.)

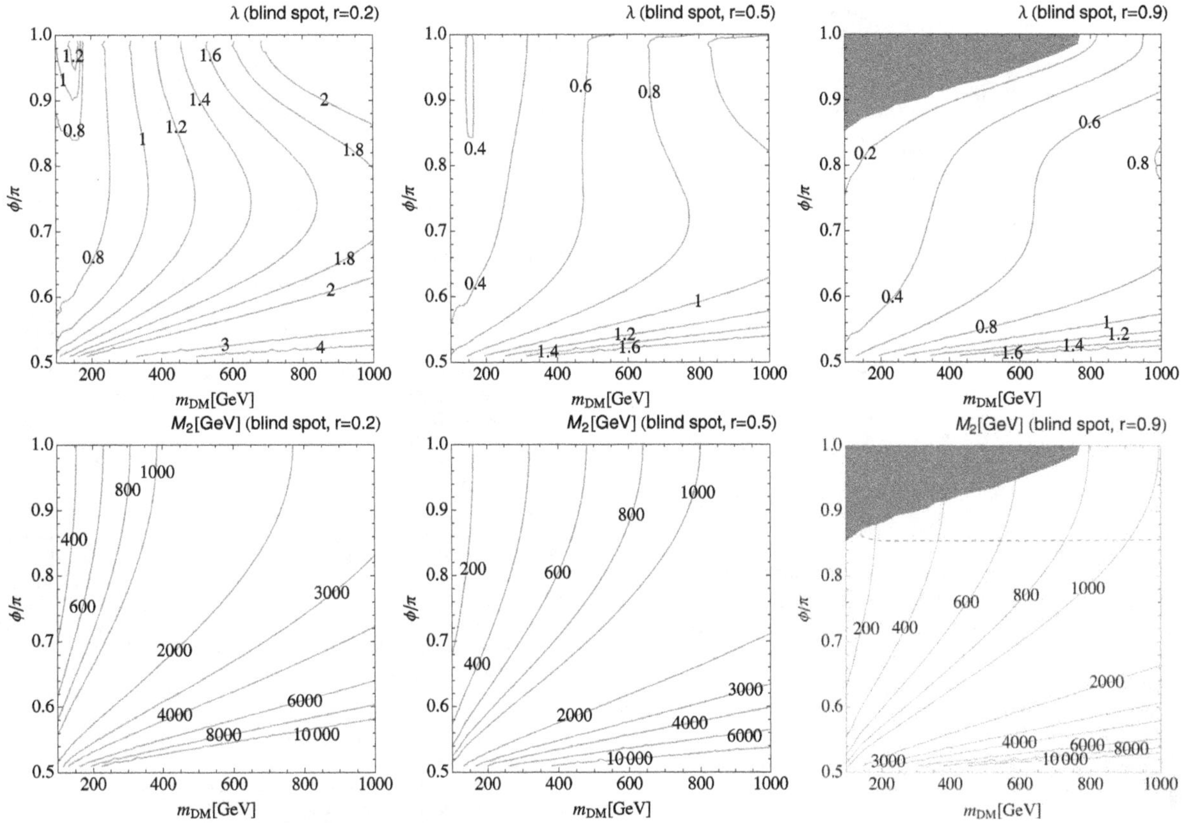

Fig. 4. The values of λ and M_2 in the blind spot. The mass of the charged Z_2-odd particle is M_2. The other masses, m_{χ_1} and m_{χ_2}, are almost same as M_2. In the region above the blue dashed curve, $M_2 - m_{\rm DM} < 0.1 m_{\rm DM}$, and co-annihilation processes for the relic abundance are efficient and cannot be ignored.

We also show the absolute value of the electron EDM by the contours in the figure. The current bound is $|d_e| \leq 8.7 \times 10^{-29}$ e cm by the ACME experiment [37]. We find $|d_e| \lesssim 3 \times 10^{-29}$ e cm in the figure. Therefore there is no constraint from the current upper bound on the electron EDM. We can expect the regions where $|d_e| \gtrsim \mathcal{O}(10^{-30})$ e cm is detectable in future [38–41]. This model predicts $|d_e| \gtrsim 10^{-30}$ e cm in nearly all of the parameter space, and thus the electron EDM is expected to be probed in future.

We next focus on the blind spot where $\sigma_{\rm SI} = 0$. Here we determine the value of M_2 so that the blind spot condition is satisfied. We show the current bound and prospects of $\sigma_{\rm SD}$ with the electron EDM in Fig. 3. In the gray region and for $0 \leq \phi \lesssim \pi/2$, the DM thermal relic abundance does not match the measured value of DM density while satisfying the blind spot condition. We find that the spin-dependent scattering process is a powerful tool to detecting DM in future. Although some regions of the parameter space cannot be covered by the spin-dependent scattering process, the electron EDM is within the reach of the future experiments in the large parts of those regions. Therefore the electron EDM is an important complement to the direct detection in the blind spot.

Fig. 4 shows that the values of λ and M_2 in the blind spot. We can see that $\lambda \lesssim 1$ in the wide regions of the parameter space. The mass of the charged Z_2-odd particle is M_2, and the masses of heavier Z_2-odd neutral particles are almost same as M_2 in Fig. 3 and 4. Therefore the lower panels in Fig. 4 shows that the masses of the Z_2-odd particles other than the dark matter candidate. In the region above the blue dashed curve in the lower-right panel, the mass differences between the dark matter and the other Z_2-odd particles are within the 10% of the dark matter mass, $M_2 - m_{\rm DM} < 0.1 m_{\rm DM}$. In that region, we find that co-annihilation processes for the relic abundance are efficient and cannot be ignored. For r = 0.1 and 0.5, we have checked the co-annihilation processes are negligible. We also have checked the constraint from the ST parameters [42,43], and found that the model is consistent with the current electroweak precision measurements [44] in all the regions of the parameter space shown here.

Finally, we discuss the blind spot with $r = 1$. This is a special parameter choice where the model can completely evade the constraints from the direct detection, namely $\sigma_{\rm SI} = \sigma_{\rm SD} = 0$. In this case, both $c_{h\chi_1\chi_1}$ and $c_{Z\chi_1\chi_1}$ vanish, and thus we have to rely on the pseudoscalar coupling $c^p_{h\chi\chi}$ to realize the thermal relic scenario. In the white region in Fig. 5, we can obtain the DM thermal relic abundance that matches the measured value of DM density. Therefore this model is viable even if direct detection experiments do not find any signal. We also show the absolute value of the electron EDM in the figure. The value is within the reach of the future experiments in most of the parameter space, and thus we can confirm the validity of this model by the observation of the electron EDM.

4. Conclusion

The recent progress of the dark matter direct detection experiments gives stringent constraint on the scattering cross section of dark matter with nucleon. Dark matter models in the thermal relic scenario have to evade this constraint. One simple way to evade the constraint is to rely on pseudoscalar interactions of a fermionic dark matter candidate.

We have studied the singlet-doublet dark matter model with a special emphasis on the CP violation effect. The model contains pseudoscalar interactions that originate from the CP violation in the dark sector. Even if dark matter direct detection experiments do not observe any signal, the model can explain the dark matter thermal relic abundance that matches the measured value of

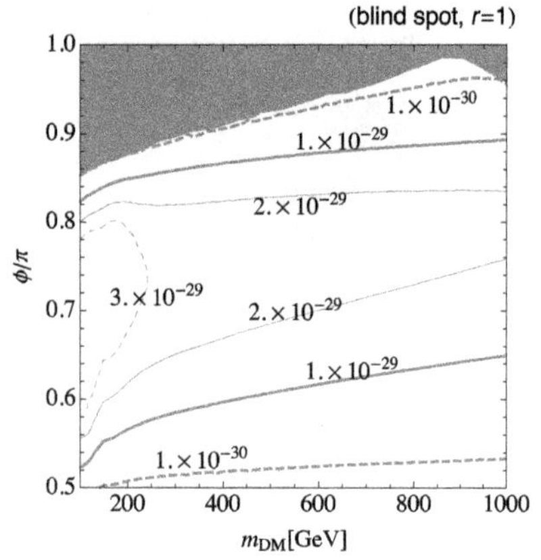

Fig. 5. The contours show the electron EDM in a special case where dark matter completely evade the direct detection via scattering with nucleon, namely $\sigma_{SI} = \sigma_{SD} = 0$. In the gray regions, we cannot obtain the DM thermal relic abundance that matched the measured value of DM density.

DM density thanks to the CP violation in the dark sector. The CP violation in the dark sector induces the electron EDM. We have shown that its value is larger than 10^{-30} e cm in a large region of the parameter space where we can obtain the measured value of DM density. This value is within the reach of the future experiments, and the electron EDM is thus strongly expected to be observed. Therefore, the electron EDM measurement is an important complement to the direct dark matter detection experiments for testing this model.

References

[1] D.S. Akerib, et al., LUX Collaboration, Phys. Rev. Lett. 118 (2) (2017) 021303, http://dx.doi.org/10.1103/PhysRevLett.118.021303, arXiv:1608.07648 [astro-ph.CO].

[2] C. Fu, et al., PandaX-II Collaboration, Phys. Rev. Lett. 118 (7) (2017) 071301, http://dx.doi.org/10.1103/PhysRevLett.118.071301, arXiv:1611.06553 [hep-ex].

[3] M. Escudero, A. Berlin, D. Hooper, M.X. Lin, J. Cosmol. Astropart. Phys. 1612 (2016) 029, http://dx.doi.org/10.1088/1475-7516/2016/12/029, arXiv:1609.09079 [hep-ph].

[4] M. Escudero, D. Hooper, S.J. Witte, J. Cosmol. Astropart. Phys. 1702 (02) (2017) 038, http://dx.doi.org/10.1088/1475-7516/2017/02/038, arXiv:1612.06462 [hep-ph].

[5] C. Boehm, M.J. Dolan, C. McCabe, M. Spannowsky, C.J. Wallace, J. Cosmol. Astropart. Phys. 1405 (2014) 009, http://dx.doi.org/10.1088/1475-7516/2014/05/009, arXiv:1401.6458 [hep-ph].

[6] A. Hektor, L. Marzola, Phys. Rev. D 90 (5) (2014) 053007, http://dx.doi.org/10.1103/PhysRevD.90.053007, arXiv:1403.3401 [hep-ph].

[7] K. Ghorbani, J. Cosmol. Astropart. Phys. 1501 (2015) 015, http://dx.doi.org/10.1088/1475-7516/2015/01/015, arXiv:1408.4929 [hep-ph].

[8] J. Kozaczuk, T.A.W. Martin, J. High Energy Phys. 1504 (2015) 046, http://dx.doi.org/10.1007/JHEP04(2015)046, arXiv:1501.07275 [hep-ph].

[9] A. Berlin, S. Gori, T. Lin, L.T. Wang, Phys. Rev. D 92 (2015) 015005, http://dx.doi.org/10.1103/PhysRevD.92.015005, arXiv:1502.06000 [hep-ph].

[10] J. Abdallah, et al., Phys. Dark Universe 9–10 (2015) 8, http://dx.doi.org/10.1016/j.dark.2015.08.001, arXiv:1506.03116 [hep-ph].

[11] J.M. No, Phys. Rev. D 93 (3) (2016) 031701, http://dx.doi.org/10.1103/PhysRevD.93.031701, arXiv:1509.01110 [hep-ph].

[12] M. Escudero, A. Berlin, D. Hooper, M.X. Lin, J. Cosmol. Astropart. Phys. 1612 (2016) 029, http://dx.doi.org/10.1088/1475-7516/2016/12/029, arXiv:1609.09079 [hep-ph].

[13] D. Goncalves, P.A.N. Machado, J.M. No, arXiv:1611.04593 [hep-ph].

[14] M. Escudero, D. Hooper, S.J. Witte, J. Cosmol. Astropart. Phys. 1702 (02) (2017) 038, http://dx.doi.org/10.1088/1475-7516/2017/02/038, arXiv:1612.06462 [hep-ph].

[15] S. Baek, P. Ko, J. Li, arXiv:1701.04131 [hep-ph].

[16] M. Bauer, U. Haisch, F. Kahlhoefer, arXiv:1701.07427 [hep-ph].

[17] R. Mahbubani, L. Senatore, Phys. Rev. D 73 (2006) 043510, http://dx.doi.org/10.1103/PhysRevD.73.043510, arXiv:hep-ph/0510064.

[18] F. D'Eramo, Phys. Rev. D 76 (2007) 083522, http://dx.doi.org/10.1103/PhysRevD.76.083522, arXiv:0705.4493 [hep-ph].

[19] R. Enberg, P.J. Fox, L.J. Hall, A.Y. Papaioannou, M. Papucci, J. High Energy Phys. 0711 (2007) 014, http://dx.doi.org/10.1088/1126-6708/2007/11/014, arXiv:0706.0918 [hep-ph].

[20] T. Cohen, J. Kearney, A. Pierce, D. Tucker-Smith, Phys. Rev. D 85 (2012) 075003, http://dx.doi.org/10.1103/PhysRevD.85.075003, arXiv:1109.2604 [hep-ph].

[21] C. Cheung, L.J. Hall, D. Pinner, J.T. Ruderman, J. High Energy Phys. 1305 (2013) 100, http://dx.doi.org/10.1007/JHEP05(2013)100, arXiv:1211.4873 [hep-ph].

[22] C. Cheung, D. Sanford, J. Cosmol. Astropart. Phys. 1402 (2014) 011, http://dx.doi.org/10.1088/1475-7516/2014/02/011, arXiv:1311.5896 [hep-ph].

[23] T. Abe, R. Kitano, R. Sato, Phys. Rev. D 91 (9) (2015) 095004, http://dx.doi.org/10.1103/PhysRevD.91.095004, arXiv:1411.1335 [hep-ph].

[24] L. Calibbi, A. Mariotti, P. Tziveloglou, J. High Energy Phys. 1510 (2015) 116, http://dx.doi.org/10.1007/JHEP10(2015)116, arXiv:1505.03867 [hep-ph].

[25] A. Freitas, S. Westhoff, J. Zupan, J. High Energy Phys. 1509 (2015) 015, http://dx.doi.org/10.1007/JHEP09(2015)015, arXiv:1506.04149 [hep-ph].

[26] G. Cynolter, J. Kovács, E. Lendvai, Mod. Phys. Lett. A 31 (01) (2016) 1650013, http://dx.doi.org/10.1142/S0217732316500139, arXiv:1509.05323 [hep-ph].

[27] K. Hamaguchi, K. Ishikawa, Phys. Rev. D 93 (5) (2016) 055009, http://dx.doi.org/10.1103/PhysRevD.93.055009, arXiv:1510.05378 [hep-ph].

[28] M. Badziak, M. Olechowski, P. Szczerbiak, J. High Energy Phys. 1603 (2016) 179, http://dx.doi.org/10.1007/JHEP03(2016)179, arXiv:1512.02472 [hep-ph].

[29] S. Horiuchi, O. Macias, D. Restrepo, A. Rivera, O. Zapata, H. Silverwood, J. Cosmol. Astropart. Phys. 1603 (03) (2016) 048, http://dx.doi.org/10.1088/1475-7516/2016/03/048, arXiv:1602.04788 [hep-ph].

[30] S. Banerjee, S. Matsumoto, K. Mukaida, Y.L.S. Tsai, J. High Energy Phys. 1611 (2016) 070, http://dx.doi.org/10.1007/JHEP11(2016)070, arXiv:1603.07387 [hep-ph].

[31] J. Kearney, N. Orlofsky, A. Pierce, Phys. Rev. D 95 (3) (2017) 035020, http://dx.doi.org/10.1103/PhysRevD.95.035020, arXiv:1611.05048 [hep-ph].

[32] P. Huang, R.A. Roglans, D.D. Spiegel, Y. Sun, C.E.M. Wagner, arXiv:1701.02737 [hep-ph].

[33] M. Badziak, M. Olechowski, P. Szczerbiak, arXiv:1701.05869 [hep-ph].

[34] P. Cushman, et al., arXiv:1310.8327 [hep-ex].

[35] G. Bélanger, F. Boudjema, A. Pukhov, A. Semenov, Comput. Phys. Commun. 192 (2015) 322, http://dx.doi.org/10.1016/j.cpc.2015.03.003, arXiv:1407.6129 [hep-ph].

[36] P.A.R. Ade, et al., Planck Collaboration, Astron. Astrophys. 594 (2016) A13, http://dx.doi.org/10.1051/0004-6361/201525830, arXiv:1502.01589 [astro-ph.CO].

[37] J. Baron, et al., ACME Collaboration, Science 343 (2014) 269, http://dx.doi.org/10.1126/science.1248213, arXiv:1310.7534 [physics.atom-ph].

[38] Y. Sakemi, et al., J. Phys. Conf. Ser. 302 (2011) 012051, http://dx.doi.org/10.1088/1742-6596/302/1/012051.

[39] D.M. Kara, I.J. Smallman, J.J. Hudson, B.E. Sauer, M.R. Tarbutt, E.A. Hinds, New J. Phys. 14 (2012) 103051, http://dx.doi.org/10.1088/1367-2630/14/10/103051, arXiv:1208.4507 [physics.atom-ph].

[40] D. Kawall, J. Phys. Conf. Ser. 295 (2011) 012031, http://dx.doi.org/10.1088/1742-6596/295/1/012031.

[41] V. Anastassopoulos, et al., Rev. Sci. Instrum. 87 (11) (2016) 115116, http://dx.doi.org/10.1063/1.4967465, arXiv:1502.04317 [physics.acc-ph].

[42] M.E. Peskin, T. Takeuchi, Phys. Rev. Lett. 65 (1990) 964, http://dx.doi.org/10.1103/PhysRevLett.65.964.

[43] M.E. Peskin, T. Takeuchi, Phys. Rev. D 46 (1992) 381, http://dx.doi.org/10.1103/PhysRevD.46.381.

[44] C. Patrignani, et al., Particle Data Group, Chin. Phys. C 40 (10) (2016) 100001, http://dx.doi.org/10.1088/1674-1137/40/10/100001.

3

Elliptic flow in small systems due to elliptic gluon distributions?

Yoshikazu Hagiwara [a], Yoshitaka Hatta [b,*], Bo-Wen Xiao [c], Feng Yuan [d]

[a] *Department of Physics, Kyoto University, Kyoto 606-8502, Japan*
[b] *Yukawa Institute for Theoretical Physics, Kyoto University, Kyoto 606-8502, Japan*
[c] *Key Laboratory of Quark and Lepton Physics (MOE) and Institute of Particle Physics, Central China Normal University, Wuhan 430079, China*
[d] *Nuclear Science Division, Lawrence Berkeley National Laboratory, Berkeley, CA 94720, USA*

ARTICLE INFO

Editor: J.-P. Blaizot

ABSTRACT

We investigate the contributions from the so-called elliptic gluon Wigner distributions to the rapidity and azimuthal correlations of particles produced in high energy pp and pA collisions by applying the double parton scattering mechanism. We compute the 'elliptic flow' parameter v_2 as a function of the transverse momentum and rapidity, and find qualitative agreement with experimental observations. This shall encourage further developments with more rigorous studies of the elliptic gluon distributions and their applications in hard scattering processes in pp and pA collisions.

1. Introduction

One of the interesting experimental observations from the proton–proton and proton–nucleus collisions at the Large Hadron Collider (LHC) and Relativistic Heavy Ion Collider (RHIC) is the long range rapidity and azimuthal angle correlations between hadrons [1–9], see, e.g., a recent review in Ref. [10]. These intriguing observations have generated great theoretical investigations, and many models have been proposed to explain the experimental results, including (but not limited to) hydrodynamics [11–14], QCD motivated models [15–18], and in particular, the multi-gluon correlations calculated in the Color Glass Condensate (CGC) framework [19,20,22–33].

In this paper, we investigate the contribution from the double parton scattering (DPS) [34,35] coupled with the so-called elliptic gluon Wigner distribution [36–38]. In high energy collisions, we expect the DPS, or in general, the multi-parton scattering, is the dominant source for multi-particle productions. A unique feature of DPS is that its contribution is not strongly suppressed for near-side particle productions with large rapidity separation as compared to the single parton scattering (SPS) contribution. Therefore, DPS may well be the dominant source for long range correlations among produced hadrons.

It was first pointed out in Ref. [39] that the DPS plays an important role in two particle production in forward pA and dA collisions at RHIC. This idea was followed up in the saturation formalism in Ref. [40] to estimate the so-called pedestal contribution in the correlation measurements. Further study in Ref. [41] also confirmed the importance of these contributions in the two particle production in pA collisions. However, all these studies assumed that the two hard scatterings are essentially uncorrelated. In the following, we will extend the DPS mechanism to include the impact parameter dependence which naturally encodes the correlation between the two scatterings. If we average over the impact parameter space, this will reduce to the previous applications of the DPS mechanism in the CGC framework. However, the unintegrated gluon distribution involved in these scatterings depends on the impact parameter. In particular, there is a nonzero $\cos(2\phi)$ azimuthal correlation between the transverse momentum $k_\perp$ and the impact parameter $b_\perp$, which was referred to as the elliptic gluon Wigner distribution in Ref. [36]. Since the impact parameters for the two hard scatterings are correlated due to the DPS mechanism, we expect the transverse momenta from the two hard scatterings are correlated as well. This will naturally give rise to the $\cos(2\phi)$ two-particle correlation in the final state.

In Ref. [36], the elliptic gluon Wigner distribution has been shown to be measurable in diffractive dijet production in lepton-nucleon collisions at the future electron-ion collider (EIC). The present study suggests that the same distribution can affect various observables in different types of collisions.

The rest of this paper is organized as follows. In Sec. 2, we study the DPS contributions to the two particle production in the dilute-dense collisions and derive a formula for the 'elliptic flow' parameter v_2. The result is relevant to pp and pA experiments at RHIC and the LHC. In Sec. 3, we numerically evaluate v_2 in a model which incorporates the saturation effect in the target. We point out some generic features of the DPS contributions which

* Corresponding author.
E-mail address: hatta@yukawa.kyoto-u.ac.jp (Y. Hatta).

can be compared to the experimental observations. We summarize our paper in Sec. 4.

2. Double parton scattering contributions in the dilute-dense collisions

In order to describe the near-side two particle correlations in pp and pA collisions, we introduce the impact parameter dependence in the DPS framework. Similarly to the derivation of DPS in Refs. [34,35], we write down the generic expression for the differential cross section of two parton production as

$$\left.\frac{d\sigma}{dy_1 d^2k_{1\perp} dy_2 d^2k_{2\perp}}\right|_{DPS} = \int d^2x_\perp d^2y_\perp d^2b_{1\perp} d^2b_{2\perp} e^{ik_{1\perp}\cdot x_\perp} e^{ik_{2\perp}\cdot y_\perp} F_A(x_p, x_p'; z_\perp) \times F_B(x_A, x_A'; \vec{b}_{1\perp}, \vec{b}_{2\perp}; \vec{x}_\perp, \vec{y}_\perp), \tag{1}$$

where $z_\perp = |\vec{b}_{1\perp} - \vec{b}_{2\perp}|$, and $\vec{b}_{1\perp}$ and $\vec{b}_{2\perp}$ denote the two hard scattering positions with respect to the center of the target. The 'dipole sizes' $x_\perp$ and $y_\perp$ are Fourier-conjugate variables to the partons' outgoing transverse momentum $k_{1\perp}$ and $k_{2\perp}$, respectively. The longitudinal momentum fractions x_p, x_p', x_A, and x_A' are determined by the final state kinematics. The physics picture is that two partons from the incoming proton encounter multiple scattering off the target, and fragment into two final state particles. The multiple scattering is described in the CGC framework or in the color-dipole model. For a large nucleus, we can assume a factorized form

$$F_B \approx S_{x_A}\left(\vec{b}_{1\perp}, \vec{x}_\perp\right) S_{x_A'}\left(\vec{b}_{2\perp}, \vec{y}_\perp\right), \tag{2}$$

where S is the dipole S-matrix which may be in the fundamental or adjoint representation depending on the partonic channels involved in the DPS. The terms neglected in (2) are of order $1/N_c^2$. It has been argued [33] that these color-suppressed, but 'connected' contributions can give rise to nonvanishing v_2 in pp and pA collisions. Moreover, if the target is small, as in pp collisions, factorization (2) is violated even in the large-N_c limit due to the small-x evolution in the target. (In the case of a dipole target, this can be shown analytically [42,43].) Such factorization breaking effects have been considered as another source of v_2 in small systems [19–21].

Here we show that, even if the factorization (2) holds strictly, there exist non-trivial angular correlations between the two outgoing particles due to the angular correlation between $\vec{b}_\perp$ and $\vec{x}_\perp$ in the S-matrix. It should be mentioned that the idea that the correlation between impact parameter and dipole orientation generates anisotropy in the final state has been previously studied in the context of single [15,16] (see also, [45]) and double [31,32] parton scattering. Thus, the approach here is essentially the same as in [31,32]. Yet, our formulation is considerably more concise and clearly establishes the connection to the elliptic gluon Wigner distribution which is a fundamental object in the tomographic study of the nucleon/nucleus.

For this purpose, let us write (1) as

$$\left.\frac{d\sigma}{dy_1 d^2k_{1\perp} dy_2 d^2k_{2\perp}}\right|_{DPS} = \int d^2b_{1\perp} d^2b_{2\perp} F_A(x_p, x_p'; z_\perp) G_{x_A}(\vec{b}_{1\perp}, \vec{k}_{1\perp}) G_{x_A'}(\vec{b}_{2\perp}, \vec{k}_{2\perp}), \tag{3}$$

where $G(\vec{b}_\perp, \vec{k}_\perp)$ is the Fourier transform of $S(\vec{b}_\perp, \vec{x}_\perp)$ and we assumed (2). The angular correlation between $\vec{b}_\perp$ and $\vec{x}_\perp$ is transformed into the one between $\vec{b}_\perp$ and $\vec{k}_\perp$. At small-x, this correlation is dominantly *elliptic* [36,37], namely,

$$G(\vec{b}_\perp, \vec{k}_\perp) = G^0(b_\perp, k_\perp) + 2\cos 2(\phi_b - \phi_k)\widetilde{G}(b_\perp, k_\perp) + \cdots. \tag{4}$$

The angular integrals in (3) then lead to an elliptic angular correlation of the form $\cos 2(\phi_{k_1} - \phi_{k_2})$.

This can be seen most clearly and model-independently at large impact parameter where it is convenient to write $\vec{b}_{1,2\perp} = \vec{b}_\perp \pm \vec{z}_\perp/2$, so that $d^2b_{1\perp}d^2b_{2\perp} = d^2z_\perp d^2b_\perp$. Since the two partons are confined in the proton, the $z_\perp$ integral is limited within the confinement radius $z_\perp \lesssim 1/\Lambda$. When $b_\perp \gg 1/\Lambda \sim z_\perp$, we can approximately integrate over $z_\perp$ to obtain the collinear double parton distribution of the proton,

$$\int d^2z_\perp F_A(x_p, x_p'; z_\perp) = \mathcal{D}_p(x_p, x_p'), \tag{5}$$

which can be further simplified as $\mathcal{D}_p(x_p, x_p') = \mathcal{C}(x_p, x_p') f(x_p) \times f(x_p')$ with $\mathcal{C} \approx 1$. With this approximation, we can write down the differential cross section as

$$\left.\frac{d\sigma}{dy_1 d^2k_{1\perp} dy_2 d^2k_{2\perp}}\right|_{DPS} \sim \int_{1/\Lambda} d^2b_\perp f(x_p) f(x_p') G_{x_A}(\vec{b}_\perp, \vec{k}_{1\perp}) G_{x_A'}(\vec{b}_\perp, \vec{k}_{2\perp}) \tag{6}$$

$$\propto \pi \int_{1/\Lambda} db_\perp^2 \Big[G^0_{x_A}(b_\perp, k_{1\perp}) G^0_{x_A'}(b_\perp, k_{2\perp}) + 2\cos 2(\phi_{k_{1\perp}} - \phi_{k_{2\perp}}) \widetilde{G}_{x_A}(b_\perp, k_{1\perp}) \widetilde{G}_{x_A'}(b_\perp, k_{2\perp}) \Big].$$

As expected, we recognize the $\cos 2(\phi_{k_1} - \phi_{k_2})$ correlation proportional to the elliptic part $\widetilde{G}$ squared.

We now turn to the small impact parameter region $b_\perp \sim z_\perp \sim 1/\Lambda$. To proceed, we introduce a Gaussian model $F_A(z_\perp) \propto e^{-z_\perp^2 \Lambda^2}$. The angular integrals can then be performed as

$$\int^{1/\Lambda} d^2b_{1\perp} d^2b_{2\perp} e^{-\Lambda^2 |\vec{b}_{1\perp} - \vec{b}_{2\perp}|^2} G_{x_A}(b_{1\perp}, k_{1\perp}) G_{x_A'}(b_{2\perp}, k_{2\perp})$$

$$= 4\pi^2 \int_0^{1/\Lambda} b_{1\perp} db_{1\perp} b_{2\perp} db_{2\perp} e^{-\Lambda^2 (b_{1\perp}^2 + b_{2\perp}^2)} \times \Big[I_0(2\Lambda^2 b_{1\perp} b_{2\perp}) G^0_{x_A}(b_{1\perp}, k_{1\perp}) G^0_{x_A'}(b_{2\perp}, k_{2\perp}) + 2\cos 2(\phi_{k_{1\perp}} - \phi_{k_{2\perp}}) I_2(2\Lambda^2 b_{1\perp} b_{2\perp}) \times \widetilde{G}_{x_A}(b_{1\perp}, k_{1\perp}) \widetilde{G}_{x_A'}(b_{2\perp}, k_{2\perp}) \Big]. \tag{7}$$

We again find the elliptic correlation $\cos 2(\phi_{k_1} - \phi_{k_2})$. Other models of F_A will also give rise to this correlation, as long as F_A depends on the angle between $\vec{b}_{1\perp}$ and $\vec{b}_{2\perp}$ via $z_\perp = |\vec{b}_{1\perp} - \vec{b}_{2\perp}|$.

Noting that the upper limit of the $b_{1,2\perp}$-integrations in (7) can actually be extended to some value $R_{cut} > 1/\Lambda$, we define

$$V_2(k_{1\perp}, k_{2\perp}) \equiv \frac{\int_0^{R_{cut}} b_{1\perp} db_{1\perp} b_{2\perp} db_{2\perp} e^{-\Lambda^2 (b_{1\perp}^2 + b_{2\perp}^2)} I_2(2\Lambda^2 b_{1\perp} b_{2\perp}) \widetilde{G}_{x_A}(b_{1\perp}, k_{1\perp}) \widetilde{G}_{x_A'}(b_{2\perp}, k_{2\perp})}{\int_0^{R_{cut}} b_{1\perp} db_{1\perp} b_{2\perp} db_{2\perp} e^{-\Lambda^2 (b_{1\perp}^2 + b_{2\perp}^2)} I_0(2\Lambda^2 b_{1\perp} b_{2\perp}) G^0_{x_A}(b_{1\perp}, k_{1\perp}) G^0_{x_A'}(b_{2\perp}, k_{2\perp})}, \tag{8}$$

This is related to the experimentally measured v_2 via

$$v_2(k_\perp, k_\perp^{ref}) \equiv \frac{V_2(k_\perp, k_\perp^{ref})}{\sqrt{V_2(k_\perp^{ref}, k_\perp^{ref})}}, \tag{9}$$

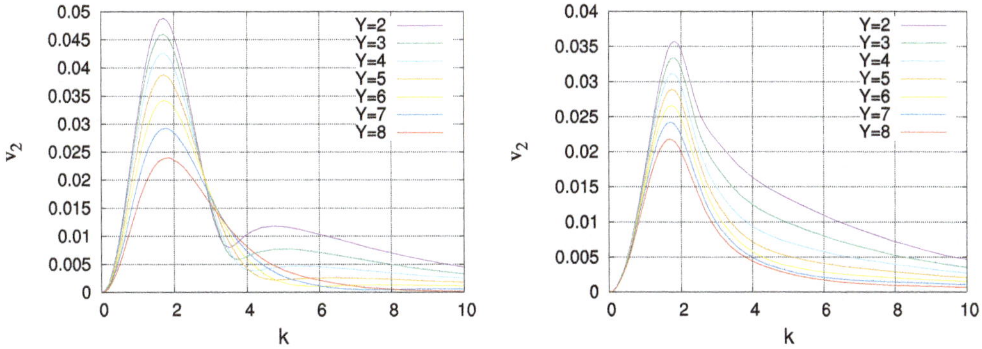

Fig. 1. v_2 as a function of $k_\perp$. The cutoff in the $b_\perp$-integration in (8) is $R_{cut} = 2$ (left) and $R_{cut} = 5$ (right). (For interpretation of the references to color in this figure legend, the reader is referred to the web version of this article.)

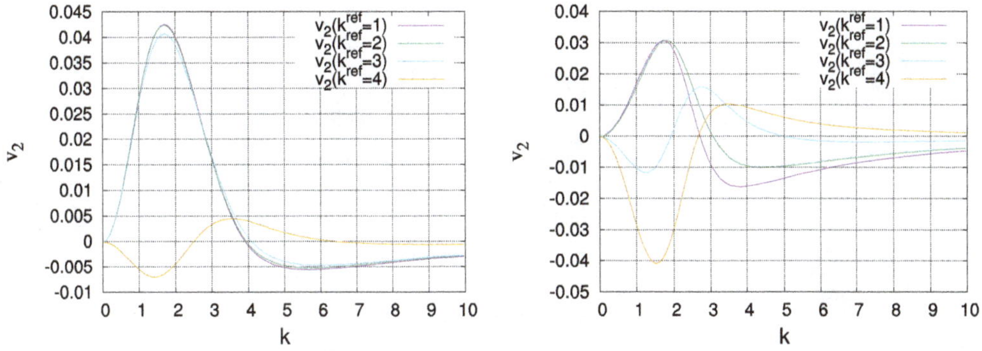

Fig. 2. $k_\perp^{ref}$-dependence at $Y = 4$. Left: $R_{cut} = 2$, Right: $R_{cut} = 5$. (For interpretation of the references to color in this figure legend, the reader is referred to the web version of this article.)

where $k_\perp^{ref}$ denotes some reference momentum. Experimentally, it has been observed that v_2 is roughly independent of $k_\perp^{ref}$. This implies that the two-particle correlation V_2 factorizes $V_2(k_{1\perp}, k_{2\perp}) \approx v_2(k_{1\perp})v_2(k_{2\perp})$, and this is consistent with the hydrodynamic interpretation of v_2. In our case (8), the integrand approximately factorizes at small $b_{1,2\perp}$. However, the result after the full $b_{1,2\perp}$-integrations does not factorize in general.

3. Model calculation

To illustrate the DPS contribution discussed above, we evaluate (8) in a model that incorporates the gluon saturation effect at small-x. For definiteness, we consider the pp collisions partly because realistic models for pA with both the $b_\perp$-dependence and the small-x evolution are not available to us. The angular independent part G^0 and the elliptic part $\tilde{G}$ are computed from the solution of the impact parameter dependent Balitsky–Kovchegov (BK) equation in the same way as explained in [37]. The only difference is that here we use a different initial condition to be slightly more realistic ($e^{-d^2} \to e^{-cd^2}$ with $c = 6$ in Eq. (7) of [37]). For small dipole sizes $r_\perp \to 0$ and at small impact parameter $b_\perp \approx 0$, this gives the initial condition $S_{Y=0}(b_\perp, r_\perp) \approx e^{-cr_\perp^2/R^2}$ where R is roughly the size of the target. For the proton, we use $R = 1$ fm so that the initial condition is $\sim e^{-r_\perp^2/(0.4\text{ fm})^2}$ which is the same as the original GBW model [44]. Below we use R as the unit of length, so for example $R_{cut} = 2$ means $R_{cut} = 2R$. The current numerical result is intended to be viewed as an illustrative example which shows that this mechanism due to the elliptic Wigner distribution can generate sizeable elliptic flow in small systems. Of course, a more realistic numerical model calculation should be carried out in the future in order to compare with the experimental data measured for small systems created in pp and pA collisions.

The results for $k_\perp = k_\perp^{ref}$ and $x_A = x_{A'} = e^{-Y}$ are shown in Fig. 1 for different values of Y with $\Lambda = 1$. In the two figures, we used different values of the cutoff R_{cut} in the $b_\perp$-integral. Actually, the elliptic part $\tilde{G}$ becomes negative in the large-$k_\perp$ region. This means that the squared function $(\tilde{G})^2$ in the numerator has two peaks in the $k_\perp$ direction at fixed $b_{1\perp} \approx b_{2\perp}$. However, the zero of $\tilde{G}$ is $b_\perp$-dependent, so that after integrating over $b_\perp$ we obtain the single-peak structure as in the right figure. On the other hand, if the cutoff is small as in the left figure, $b_\perp$-averaging is insufficient and we see a remnant of the double-peak structure. Note that the peak position of v_2 is almost independent of Y, and does *not* coincide with the saturation momentum $k \sim Q_s(Y)$ which is a rapidly increasing function of Y. This is a characteristic property of the elliptic part $\tilde{G}$ observed in [37]. Interestingly, Ref. [21] argued that the peak position of v_2 from the present mechanism should occur at the inverse correlation length of Q_s in the transverse plane which is much smaller than Q_s. It remains to be seen whether the observation in [37] can be physically interpreted as such. We also note that the height of the peak decreases with increasing Y largely because $\tilde{G}$ has the same property. Compared with the experimental data, the magnitude of v_2 in our model is somewhat smaller even for the smallest values of Y.

Next we test the degree of factorization. We choose $k_\perp \neq k_\perp^{ref}$ and check the $k_\perp^{ref}$-dependence of the result. The integrand of (8) factorizes at small $b_\perp$, because $I_2(2\Lambda^2 b_{1\perp}b_{2\perp}) \sim (b_{1\perp}b_{2\perp})^2$. Thus, factorization is good if R_{cut} is small and this is clearly seen in Fig. 2 ($Y = 4$) and Fig. 3 ($Y = 8$). We also see that the factorization holds better for larger Y values. Note that v_2 is no longer positive definite once we allow $k_\perp \neq k_\perp^{ref}$ because $\tilde{G}$ is negative for large $k_\perp^{ref}$. When Y becomes large, the negative region of $\tilde{G}$ is pushed to a larger k-region and v_2 tends to become positive.

Finally we study the rapidity correlation between two particles. We take x_A and $x_{A'}$ in (8) to be different and compute $v_2(k_\perp =$

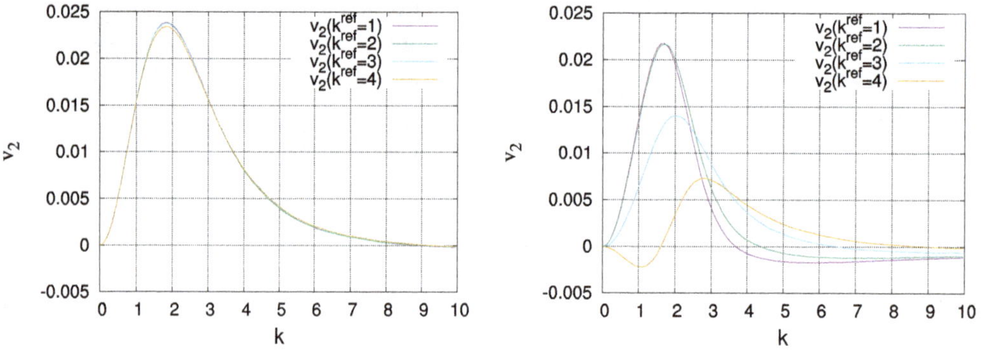

Fig. 3. $k_\perp^{ref}$-dependence at $Y = 8$. Left: $R_{cut} = 2$, Right: $R_{cut} = 5$. (For interpretation of the references to color in this figure legend, the reader is referred to the web version of this article.)

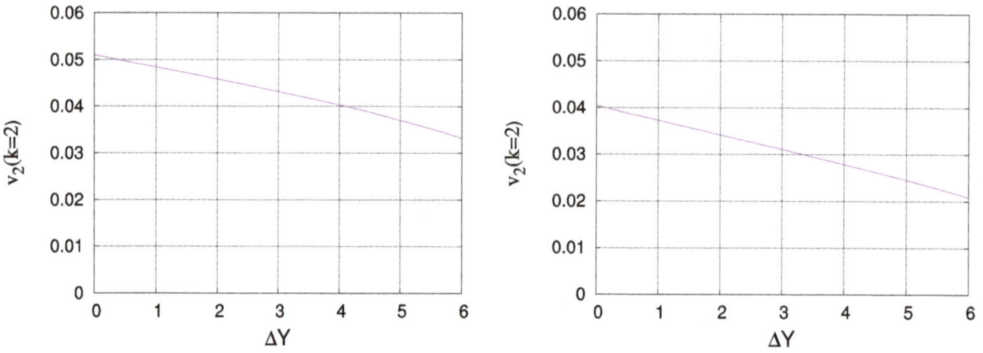

Fig. 4. $v_2(k_\perp = k_\perp^{ref} = 2)$ as a function of the rapidity difference ΔY between two particles. Left: $R_{cut} = 2$, Right: $R_{cut} = 5$.

$k_\perp^{ref} = 2)$ from (9) as a function of $\Delta Y = Y - Y^{ref} = \ln\frac{1}{x} - \ln\frac{1}{x^{ref}}$. (In practice we set $Y^{ref} = 0$.) The result is shown in Fig. 4. We recognize a certain degree of 'long-range rapidity correlation', namely, v_2 decreases only slowly with increasing ΔY. This is largely due to approximate factorization of the double scattering amplitude (2). We should mention that when ΔY becomes too large such that $\alpha_s \Delta Y \sim 1$, one has to consider the small-x evolution between the two rapidities [25]. This is however beyond the scope of this work.

4. Conclusions and discussions

To summarize, we have explored the two-particle productions in small systems created in high energy pp and pA collisions from the double parton scattering mechanism, where the elliptic gluon Wigner distributions give rise to the desired $\cos(2\phi)$ azimuthal angular correlations. By applying the DPS idea, we derived a formula for the two particle correlation in the small-x saturation formalism, where two partons from the incoming nucleon scatter with the target nucleon/nucleus and produce two particles in the final state. The DPS mechanism imposes the impact parameter correlation between the two hard partonic scattering processes, which results in a correlation between the transverse momenta of the final state particles. Due to the unique feature of the DPS mechanism, this correlation will not decrease dramatically with the increase of the rapidity difference between the two particles.

We have also applied a recent result of the elliptic gluon distributions from the BK equation [37] to illustrate their contributions to the $\cos(2\phi)$ correlation between two particles produced in high energy pp and pA collisions. The size of the elliptic flow parameter v_2 was found in a similar range as the experimental observations at RHIC and the LHC. This is an encouraging message, and demonstrates that the long range correlation of v_2 may have significant contributions from the elliptic gluon distributions in the target. However, we would like to emphasize that we have limited knowledge on the elliptic gluon distribution as compared to the usual dipole gluon distribution, and more studies are needed to compare to the experimental data. This will also help to pin down the underlying mechanism for the novel "flow" phenomena in high energy pp and pA collisions.

Finally, we would like to point out that the elliptic gluon Wigner distribution represents a nontrivial tomography structure of gluons inside the nucleon/nucleus. Its dependence on x, $b_\perp$ and $k_\perp$ will provide not only the imaging of parton distributions at small-x, but also the unique opportunity to explore the QCD dynamics associated with the small-x evolution [37]. Since the elliptic gluon distribution can be well studied in hard diffractive dijet production at the EIC [36], the comparison between the two particle elliptic flow in pp and pA collisions and further observables in eA collisions will be crucial to understand the gluon dynamics under extreme conditions. We hope to come back to this issue soon.

Acknowledgements

We thank Edmond Iancu and Amir Rezaeian for correspondence about their related work [45]. This material is based upon work supported by the U.S. Department of Energy, Office of Science, Office of Nuclear Physics, under contract number DE-AC02-05CH11231 and by the Natural Science Foundation of China (NSFC) under Grant No. 11575070.

References

[1] V. Khachatryan, et al., CMS Collaboration, J. High Energy Phys. 1009 (2010) 091, http://dx.doi.org/10.1007/JHEP09(2010)091, arXiv:1009.4122 [hep-ex].
[2] B. Abelev, et al., ALICE Collaboration, Phys. Lett. B 719 (2013) 29, http://dx.doi.org/10.1016/j.physletb.2013.01.012, arXiv:1212.2001 [nucl-ex].
[3] G. Aad, et al., ATLAS Collaboration, Phys. Rev. Lett. 110 (18) (2013) 182302, http://dx.doi.org/10.1103/PhysRevLett.110.182302, arXiv:1212.5198 [hep-ex].

[4] S. Chatrchyan, et al., CMS Collaboration, Phys. Lett. B 718 (2013) 795, http://dx.doi.org/10.1016/j.physletb.2012.11.025, arXiv:1210.5482 [nucl-ex].
[5] A. Adare, et al., PHENIX Collaboration, Phys. Rev. Lett. 114 (19) (2015) 192301, http://dx.doi.org/10.1103/PhysRevLett.114.192301, arXiv:1404.7461 [nucl-ex].
[6] A. Adare, et al., PHENIX Collaboration, Phys. Rev. Lett. 115 (14) (2015) 142301, http://dx.doi.org/10.1103/PhysRevLett.115.142301, arXiv:1507.06273 [nucl-ex].
[7] G. Aad, et al., ATLAS Collaboration, Phys. Rev. C 90 (4) (2014) 044906, http://dx.doi.org/10.1103/PhysRevC.90.044906, arXiv:1409.1792 [hep-ex].
[8] G. Aad, et al., ATLAS Collaboration, Phys. Rev. Lett. 116 (17) (2016) 172301, http://dx.doi.org/10.1103/PhysRevLett.116.172301, arXiv:1509.04776 [hep-ex].
[9] M. Aaboud, et al., ATLAS Collaboration, arXiv:1609.06213 [nucl-ex].
[10] K. Dusling, W. Li, B. Schenke, Int. J. Mod. Phys. E 25 (01) (2016) 1630002, http://dx.doi.org/10.1142/S0218301316300022, arXiv:1509.07939 [nucl-ex].
[11] J. Casalderrey-Solana, U.A. Wiedemann, Phys. Rev. Lett. 104 (2010) 102301, http://dx.doi.org/10.1103/PhysRevLett.104.102301, arXiv:0911.4400 [hep-ph].
[12] E. Avsar, C. Flensburg, Y. Hatta, J.Y. Ollitrault, T. Ueda, Phys. Lett. B 702 (2011) 394, http://dx.doi.org/10.1016/j.physletb.2011.07.031, arXiv:1009.5643 [hep-ph].
[13] B. Schenke, R. Venugopalan, Phys. Rev. Lett. 113 (2014) 102301, http://dx.doi.org/10.1103/PhysRevLett.113.102301, arXiv:1405.3605 [nucl-th].
[14] L. Yan, J.Y. Ollitrault, Phys. Rev. Lett. 112 (2014) 082301, http://dx.doi.org/10.1103/PhysRevLett.112.082301, arXiv:1312.6555 [nucl-th].
[15] B.Z. Kopeliovich, H.J. Pirner, A.H. Rezaeian, I. Schmidt, Phys. Rev. D 77 (2008) 034011, http://dx.doi.org/10.1103/PhysRevD.77.034011, arXiv:0711.3010 [hep-ph].
[16] B.Z. Kopeliovich, A.H. Rezaeian, I. Schmidt, Phys. Rev. D 78 (2008) 114009, http://dx.doi.org/10.1103/PhysRevD.78.114009, arXiv:0809.4327 [hep-ph].
[17] M. Gyulassy, P. Levai, I. Vitev, T.S. Biro, Phys. Rev. D 90 (5) (2014) 054025, http://dx.doi.org/10.1103/PhysRevD.90.054025, arXiv:1405.7825 [hep-ph].
[18] G. Gambini, G. Torrieri, arXiv:1606.07865 [nucl-th].
[19] A. Kovner, M. Lublinsky, Phys. Rev. D 83 (2011) 034017, http://dx.doi.org/10.1103/PhysRevD.83.034017, arXiv:1012.3398 [hep-ph].
[20] A. Kovner, M. Lublinsky, Phys. Rev. D 84 (2011) 094011, http://dx.doi.org/10.1103/PhysRevD.84.094011, arXiv:1109.0347 [hep-ph].
[21] A. Kovner, M. Lublinsky, Int. J. Mod. Phys. E 22 (2013) 1330001, http://dx.doi.org/10.1142/S0218301313300014, arXiv:1211.1928 [hep-ph].
[22] K. Dusling, R. Venugopalan, Phys. Rev. D 87 (5) (2013) 054014, http://dx.doi.org/10.1103/PhysRevD.87.054014, arXiv:1211.3701 [hep-ph].
[23] K. Dusling, R. Venugopalan, Phys. Rev. D 87 (5) (2013) 051502, http://dx.doi.org/10.1103/PhysRevD.87.051502, arXiv:1210.3890 [hep-ph].
[24] Y.V. Kovchegov, D.E. Wertepny, Nucl. Phys. A 906 (2013) 50, http://dx.doi.org/10.1016/j.nuclphysa.2013.03.006, arXiv:1212.1195 [hep-ph].
[25] E. Iancu, D.N. Triantafyllopoulos, J. High Energy Phys. 1311 (2013) 067, http://dx.doi.org/10.1007/JHEP11(2013)067, arXiv:1307.1559 [hep-ph].
[26] B. Schenke, S. Schlichting, R. Venugopalan, Phys. Lett. B 747 (2015) 76, http://dx.doi.org/10.1016/j.physletb.2015.05.051, arXiv:1502.01331 [hep-ph].
[27] A. Dumitru, A.V. Giannini, Nucl. Phys. A 933 (2015) 212, http://dx.doi.org/10.1016/j.nuclphysa.2014.10.037, arXiv:1406.5781 [hep-ph].
[28] A. Dumitru, J. Jalilian-Marian, Phys. Rev. D 81 (2010) 094015, http://dx.doi.org/10.1103/PhysRevD.81.094015, arXiv:1001.4820 [hep-ph].
[29] A. Dumitru, L. McLerran, V. Skokov, Phys. Lett. B 743 (2015) 134, http://dx.doi.org/10.1016/j.physletb.2015.02.046, arXiv:1410.4844 [hep-ph].
[30] T. Lappi, Phys. Lett. B 744 (2015) 315, http://dx.doi.org/10.1016/j.physletb.2015.04.015, arXiv:1501.05505 [hep-ph].
[31] E. Levin, A.H. Rezaeian, Phys. Rev. D 84 (2011) 034031, http://dx.doi.org/10.1103/PhysRevD.84.034031, arXiv:1105.3275 [hep-ph].
[32] E. Gotsman, E. Levin, U. Maor, S. Tapia, Phys. Rev. D 93 (7) (2016) 074029, http://dx.doi.org/10.1103/PhysRevD.93.074029, arXiv:1603.02143 [hep-ph].
[33] T. Lappi, B. Schenke, S. Schlichting, R. Venugopalan, J. High Energy Phys. 1601 (2016) 061, http://dx.doi.org/10.1007/JHEP01(2016)061, arXiv:1509.03499 [hep-ph].
[34] M. Diehl, D. Ostermeier, A. Schafer, J. High Energy Phys. 1203 (2012) 089, http://dx.doi.org/10.1007/JHEP03(2012)089, arXiv:1111.0910 [hep-ph].
[35] A.V. Manohar, W.J. Waalewijn, Phys. Lett. B 713 (2012) 196, http://dx.doi.org/10.1016/j.physletb.2012.05.044, arXiv:1202.5034 [hep-ph].
[36] Y. Hatta, B.W. Xiao, F. Yuan, Phys. Rev. Lett. 116 (20) (2016) 202301, http://dx.doi.org/10.1103/PhysRevLett.116.202301, arXiv:1601.01585 [hep-ph].
[37] Y. Hagiwara, Y. Hatta, T. Ueda, Phys. Rev. D 94 (9) (2016) 094036, http://dx.doi.org/10.1103/PhysRevD.94.094036, arXiv:1609.05773 [hep-ph].
[38] J. Zhou, arXiv:1611.02397 [hep-ph].
[39] M. Strikman, W. Vogelsang, Phys. Rev. D 83 (2011) 034029, http://dx.doi.org/10.1103/PhysRevD.83.034029, arXiv:1009.6123 [hep-ph].
[40] A. Stasto, B.W. Xiao, F. Yuan, Phys. Lett. B 716 (2012) 430, http://dx.doi.org/10.1016/j.physletb.2012.08.044, arXiv:1109.1817 [hep-ph].
[41] T. Lappi, H. Mantysaari, Nucl. Phys. A 908 (2013) 51, http://dx.doi.org/10.1016/j.nuclphysa.2013.03.017, arXiv:1209.2853 [hep-ph].
[42] Y. Hatta, A.H. Mueller, Nucl. Phys. A 789 (2007) 285, http://dx.doi.org/10.1016/j.nuclphysa.2007.03.003, arXiv:hep-ph/0702023v2.
[43] E. Avsar, Y. Hatta, J. High Energy Phys. 0809 (2008) 102, http://dx.doi.org/10.1088/1126-6708/2008/09/102, arXiv:0805.0710 [hep-ph].
[44] K.J. Golec-Biernat, M. Wusthoff, Phys. Rev. D 59 (1998) 014017, http://dx.doi.org/10.1103/PhysRevD.59.014017, arXiv:hep-ph/9807513.
[45] E. Iancu, A.H. Rezaeian, arXiv:1702.03943 [hep-ph].

Extending the LHC reach for new physics with sub-millimeter displaced vertices

Hayato Ito [a,*], Osamu Jinnouchi [b], Takeo Moroi [a], Natsumi Nagata [a], Hidetoshi Otono [c]

[a] *Department of Physics, University of Tokyo, Tokyo 113-0033, Japan*
[b] *Department of Physics, Tokyo Institute of Technology, Tokyo 152-8551, Japan*
[c] *Research Center for Advanced Particle Physics, Kyushu University, Fukuoka 819-0395, Japan*

ARTICLE INFO

Editor: J. Hisano

ABSTRACT

Particles with a sub-millimeter decay length appear in many models of physics beyond the Standard Model. However, their longevity has been often ignored in their LHC searches and they have been regarded as promptly-decaying particles. In this letter, we show that, by requiring displaced vertices on top of the event selection criteria used in the ordinary search strategies for promptly-decaying particles, we can considerably extend the LHC reach for particles with a decay length of $\gtrsim 100$ μm. We discuss a way of reconstructing sub-millimeter displaced vertices by exploiting the same technique used for the primary vertex reconstruction on the assumption that the metastable particles are always pair-produced and their decay products contain high-p_T jets. We show that, by applying a cut based on displaced vertices on top of standard kinematical cuts for the search of new particles, the LHC reach can be significantly extended if the decay length is $\gtrsim 100$ μm. In addition, we may measure the lifetime of the target particle through the reconstruction of displaced vertices, which plays an important role in understanding the new physics behind the metastable particles.

1. Introduction

New metastable massive particles are predicted in a variety of extensions of the Standard Model (SM) [1], and have been explored at colliders such as the LHC. If these particles have a decay length (*i.e.*, the product of the lifetime τ and the speed of light c) of $\mathcal{O}(1)$ m or shorter, then their decay can occur within tracking detectors and thus it is in principle possible to directly observe their decay points, which are away from the production point. In fact, such attempts have been made in the LHC experiments. For example, the ATLAS Collaboration has searched for displaced vertices (DVs) that originate from decay of long-lived particles by investigating charged tracks with a transverse impact parameter, d_0, of 2 mm $< |d_0| <$ 300 mm, requiring that the transverse distance between DVs and any of the primary vertices be longer than 4 mm [2]. This search is therefore sensitive to metastable particles whose decay length is $c\tau \sim \mathcal{O}(1 - 1000)$ mm. The disappearing-track searches [3] can also probe a long-lived charged particle when it decays into a neutral particle which is degenerate with the charged particle in mass [4–7]; the target of these searches is $c\tau \gtrsim 10$ cm.

On the contrary, particles with a sub-millimeter decay length have been beyond the reach of these searches. Such rather short-lived particles have been often regarded as promptly-decaying particles and probed without relying on their longevity. Exceptionally, recently, Ref. [8] considered R-parity violating supersymmetric (SUSY) model to which "ordinary" search strategies does not apply, and showed that DV-based cuts may be useful for the LHC study of such a model if the decay length of the lightest SUSY particle is longer than $\mathcal{O}(100)$ μm. From the point of view of physics beyond the SM, however, there are a variety of well-motivated new particles with $c\tau \sim$ sub-millimeter besides the above case; although LHC constraints on some of those have been already stringent even with the analysis assuming that they decay promptly, inclusion of DV-based cuts upon it significantly extends the reach of those. One of the important examples for such particles is metastable gluino in SUSY models with heavy squarks [5,9,10]. In particular, if the squark masses are as heavy as the PeV scale, the decay length of the gluino can be $c\tau \sim \mathcal{O}(100)$ μm (assuming that the gluino mass is around TeV) [11]. Metastable SUSY particles are also found in the gauge-mediation models [12,13], where the decay length of the next-to-lightest SUSY particle decaying into gravitino can be order sub-millimeter, as well as in R-parity violating SUSY models [14,15]. In addition, theories of Neutral Naturalness [16,17],

* Corresponding author.
E-mail address: ito@hep-th.phys.s.u-tokyo.ac.jp (H. Ito).

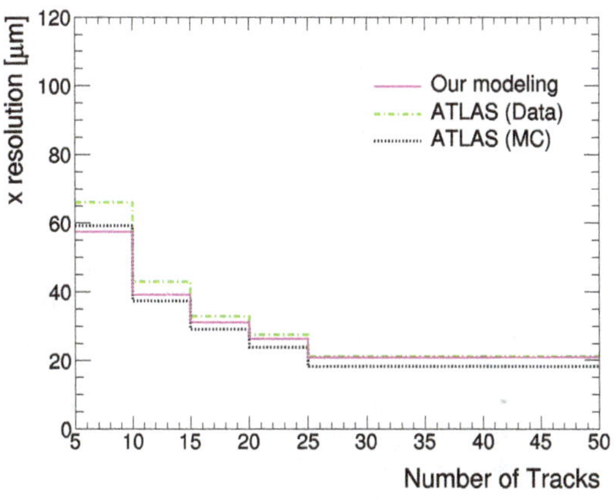

(a) x-direction

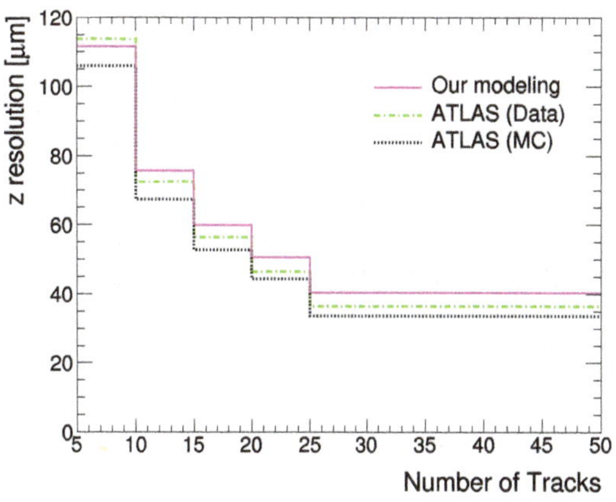

(b) z-direction

Fig. 1. The resolutions of reconstructed primary vertex position as a function of the number of tracks associated with the primary vertex. The resolutions obtained with our modeling are shown in purple lines while those provided by the ATLAS Collaboration [22] are shown in green dot-dashed (derived from data) and black dotted (derived from MC samples) ones. (For interpretation of the references to color in this figure legend, the reader is referred to the web version of this article.)

hidden-valley models [18], composite Higgs models [19], dark matter models [20], and models with sterile neutrinos [21] predict metastable particles with an $\mathcal{O}(100)$ μm decay length.

In this letter, we discuss a method of searching for metastable particles with DVs that is sensitive to $c\tau \lesssim 1$ mm as well. Here, we focus on the cases where the target metastable particles are always pair-produced, which is assured if the new particles have a conserved quantum number; for instance, in R-parity conserving SUSY models, SUSY particles, being R-parity odd, are always produced in pairs. In these cases, there are two decay points in each event, which are separated from each other by order of the decay length of these particles. We reconstruct these decay vertices in a similar manner that is used for the primary vertex reconstruction. As shown below, using this method, we can distinguish the decay vertices if these two are separated by $\gtrsim 100$ μm. This method therefore enables us to search for sub-millimeter DVs. By requiring DVs in addition to the event selection criteria used in the promptly-decaying particle searches, we can go beyond the reach of these searches if the target particle has a decay length of $c\tau \gtrsim 100$ μm. Moreover, when they are discovered, it is also possible to measure the typical distance of the two decay points and thus to estimate the decay length of the particles, which can provide important information about the nature of the new physics behind the new particles. To be specific, in this letter, we consider metastable gluinos as an example and demonstrate that the study of sub-millimeter DVs can significantly enlarge the parameter region covered by the LHC.

The organization of this letter is as follows. We first summarize how the vertex is reconstructed in Sec. 2. In Sec. 3, we describe our method to search for DVs. Then, we apply our method to the metastable gluino search and show that the LHC reach can be extended with the study of the DVs. We also point out that it is possible to measure the decay length of the metastable particle by means of DV reconstruction. Sec. 4 is devoted to conclusions and discussion.

2. Vertex reconstruction

First, let us briefly summarize how vertices are reconstructed at the LHC experiment. In order to make the argument clear, we use the performance of the ATLAS detector. In this letter, we concentrate on the case where a number of charged particles are emitted from vertices, which is the case when the production and the decay of metastable colored particles, like gluino, occur. Then, with the precise tracking of the charged particles by inner tracking detectors, the decay vertex of the parent particle may be reconstructed.

A similar analysis, *i.e.*, track-based reconstruction of primary vertex in proton-proton collision, has been already performed by the ATLAS [22,23] and CMS [24] Collaborations from which we can estimate the accuracy of the determination of the vertex position at the LHC. In Ref. [22], charged tracks with $p_T > 400$ MeV were used to reconstruct the primary vertex. In Fig. 1, we show the vertex resolutions to x- and z-directions (corresponding to the directions perpendicular and parallel to the beam axis) provided by the ATLAS Collaboration [22]; the green (dot-dashed) and black (dotted) lines show the data and Monte Carlo (MC) results. Thus, if a sizable number of charged tracks are associated with the vertex, we expect that the vertex position is reconstructed with the accuracy of $\mathcal{O}(10)$ μm. This fact indicates that, if the distance between two decay vertices is longer than $\mathcal{O}(10)$ μm in the pair production process of metastable particles, it may be possible to distinguish two vertices. Existence of two distinct DVs can be used to reduce SM backgrounds, as we discuss below.

In the following, we quantitatively study how well we can improve the discovery reach for the new particles with the reconstruction of DVs. For this purpose, in our MC analysis, we implement an algorithm to reconstruct DVs using charged tracks. As we mentioned, we mainly focus on DVs which are away from the interaction point by $\lesssim 1$ mm, though our method can be used for more displaced cases as well. Our strategy to reconstruct DVs relies on tracking performance of charged-particle tracks in the inner detector.

The tracking performance of the ATLAS inner detector for $\sqrt{s} = 13$ TeV is given in Ref. [25].[1] In our MC analysis, to take account of the performance of track reconstruction, we shift each track obtained from the MC-truth information in parallel by impact parameters; the shifted track is regarded as reconstructed one. We neglect the effect of the curvature of the tracks in this procedure since we focus on DVs which are very close to the interaction point. We also neglect the track parameter resolutions regarding its direction, *i.e.*, the azimuthal angle ϕ and the polar angle θ, as their resolutions are sufficiently small: $\sigma_\phi \sim 100\,\mu$rad and $\sigma_{\cot\theta} \sim 10^{-3}$ [27]. Thus, we only consider the resolutions of the transverse and longitudinal impact parameters, d_0 and $z_0 \sin\theta$,

[1] Before the LHC Run-II started, the insertable B-layer (IBL) [26] was installed, which improves the performance of track and vertex reconstruction.

respectively. Effects of these are taken into account by random parallel shift of each track. The resolutions of the impact parameters depend on the transverse momentum p_T and the pseudorapidity η of the track. In the processes we consider in this letter, jets have relatively high p_T and do not have any preference for the small polar angle regions. In addition, it is found that the η dependences of the resolutions of the impact parameters become sufficiently small for $p_T \gtrsim$ a few GeV [27,28]. For these reasons, we neglect the η dependences of the resolutions in this analysis. Following Ref. [27], we parametrize the p_T dependence of the track impact parameter resolutions as

$$\sigma_X(p_T) = \sigma_X(\infty)(1 \oplus p_X/p_T) , \tag{1}$$

(with $X = d_0$ and $z_0 \sin\theta$) where $\sigma_X(\infty)$ and p_X are track-resolution parameters. We determine the values of $\sigma_X(\infty)$ and p_X by fitting this expression onto the p_T dependence of the track impact parameter resolutions measured by the ATLAS Collaboration [25].

Next, let us describe the procedure of the vertex reconstruction used in our analysis, which gives the best-fit point of the vertex for a given set of charged tracks. We follow the prescription given in Refs. [23]. In this prescription, the adaptive vertex fitting algorithm [29], which we briefly review in Appendix A, is exploited to determine the vertex position. At the outset of this algorithm, for a given set of charged tracks, a vertex seed is found from the crossing points of the reconstructed tracks by means of a method called the fraction of sample mode with weights (FSMW) [30]. Once the initial vertex is fixed, we assign a weight, which is given in Eq. (A.3), to each track such that tracks far from the vertex point are down-weighted. We then determine another vertex position at which an objective function, which corresponds to the vertex χ^2 multiplied by the above weights, is minimized. We iterate this χ^2 fitting steps with varying a parameter for the weight assignment until the vertex position converges within 1 µm. The parameters in this algorithm are set to be the default values given in Ref. [29] and references therein, though the results are rather insensitive to these parameters.

To validate our modeling of impact-parameter resolutions and the vertex reconstruction, we reconstruct the position of primary vertices in minimum-bias events using our procedure. We generate 47,000 minimum-bias event MC samples with `PYTHIA v8.2` [31]. Here, we use only tracks with $p_T > 400$ MeV and $|\eta| < 2.5$ in accordance with the ATLAS study [22]. (For this choice of minimal p_T, the best-fit values of $\sigma_{d_0}(\infty)$ ($\sigma_{z_0 \sin\theta}(\infty)$) and p_{d_0} ($p_{z_0 \sin\theta}$) in Eq. (1) are 30 µm (90 µm) and 2.1 GeV (1.0 GeV), respectively.) We then evaluate the resolutions of primary vertices as a function of the number of tracks. The results are also shown in Fig. 1. As can be seen from this figure, our result is in good agreement with the ATLAS results [22]. In the following analysis, we will use the procedure explained in the preceding paragraph to determine the best-fit points of the decay vertices of pair-produced new particles; in the calculation of the χ^2 variable given in Eq. (A.2), we will adopt the track-resolution parameters given in [22].

3. Extending the reach with DVs

Now, we discuss how and how well the reach for the new physics can be extended by using the information about the DVs. We are interested in the case where

(a) the metastable particles are pair produced, and
(b) the metastable particles decay into SM colored particles (*i.e.*, quarks and/or gluons) as well as possibly other particles.

In the pair production processes of new metastable particles, no hard particles are produced at the interaction point (assuming that the new particles decay after flying sizable amount of distance) except those from initial state radiation. For this reason, we do not try to determine the position of the interaction point in each event.[2] We instead use the distance between the two reconstructed DVs, $|\boldsymbol{r}_{DV1} - \boldsymbol{r}_{DV2}|$, as a discriminating variable in our study, where $\boldsymbol{r}_{DV1}$ and $\boldsymbol{r}_{DV2}$ are positions of the reconstructed vertices. As we demonstrate below, we may extend the LHC reach for new particles by combining conventional kinematical cuts with the new cuts based on $|\boldsymbol{r}_{DV1} - \boldsymbol{r}_{DV2}|$.

Although the strategy we propose is applicable to the class of models satisfying the conditions (a) and (b) mentioned above, a quantitative study needs to be performed on a model-by-model basis. Thus, we consider metastable gluino as an example, and discuss the implication of the study of the sub-millimeter DVs.

3.1. Gluino properties

Before discussing the LHC search for the metastable gluino with DVs, we summarize gluino properties which are important for the following discussion.

A gluino decays through the exchange of squarks. If squarks $\tilde{q}$ are heavier than gluino $\tilde{g}$, and also if a neutralino $\tilde{\chi}^0$ and/or a chargino $\tilde{\chi}^\pm$ have a mass sufficiently smaller than the gluino mass, then the tree-level three-body decay processes $\tilde{g} \to \bar{q}' q \tilde{\chi}^{0,\pm}$ dominate the two-body one $\tilde{g} \to \tilde{\chi}^0 g$, which occurs at one-loop level. The decay length of gluino strongly depends on the masses of the squarks exchanged in the tree-level three-body decay processes. Assuming that the first-generation squarks are sufficiently lighter than the second- and third-generation ones, the decay length of the gluino is approximately given by [11]

$$c\tau_{\tilde{g}} \simeq 200\,\mu\text{m} \times \left(\frac{m_{\tilde{q}}}{10^3\ \text{TeV}}\right)^4 \left(\frac{2\ \text{TeV}}{m_{\tilde{g}}}\right)^5 , \tag{2}$$

where $m_{\tilde{g}}$ is the gluino mass, $m_{\tilde{q}}$ is the masses of all the first-generation squarks (which are assume to be degenerate). In addition, the masses of bino and wino are assumed to be much smaller than the gluino mass, while the higgsino mass is assumed to be larger than the gluino mass. Note that the above expression should be multiplied by a factor of $\simeq 1/3$ if squarks in all generations are degenerate in mass. Eq. (2) indicates that the gluino decay length can be as long as $\gtrsim 100$ µm for the PeV-scale squarks. Such heavy squarks, especially heavy stops, are in fact motivated by the measured value of the mass of the SM-like Higgs boson, $m_h \simeq 125$ GeV [32], as a large radiative correction from heavy stops can easily raise the Higgs-boson mass from its tree-level value [33] which is predicted to be smaller than the Z-boson mass [34] in the minimal supersymmetric SM (MSSM).

Even though the squark masses are at the PeV scale, gluino can still be around the TeV scale in a technically natural way since the gluino mass is protected by chiral symmetry. We may find a simple scenario for the mediation of SUSY-breaking to assure such a split mass spectrum [9,10]. For example, if all of the SUSY-breaking fields in the hidden sector are charged under some symmetries, then the dimension-five operators that give rise to the gaugino masses are forbidden. In this case, the gaugino masses are mainly induced by quantum effects, such as the anomaly mediation effects [35,36] and threshold corrections at the SUSY-breaking scale

[2] We however note that the reconstruction of the primary vertex is possible if hard jets or leptons are associated with the production point. It may also be possible to reconstruct the primary vertex using initial state radiation. Information about the primary vertex may also be utilized to eliminate the background.

[36,37], and are suppressed by a loop factor compared with the gravitino mass $m_{3/2}$. The squark masses are, on the other hand, generated by dimension-six Kähler-type operators. If these operators are induced at the Planck scale, the squark masses are expected to be $\mathcal{O}(m_{3/2})$, while if they are induced at a lower scale, then the resultant squark masses become heavier. Motivated by this consideration, we regard the squark masses, and thus $c\tau_{\tilde{g}}$ as well through Eq. (2), as free parameters in the following analysis.

3.2. Gluino search with DVs

Now we discuss the gluino search with DVs. In this letter, in order to demonstrate that the LHC reach for gluino can be extended with the information about DVs, we impose a DV-based cut on top of the event selection criteria used in the ordinary gluino search. Because DV-based cut may significantly reduce the SM backgrounds, one had better optimize the cut parameters to maximize the reach for new physics. Such an issue is beyond the scope of this letter, and we leave it for future study [38].

In gluino searches, we focus on events with relatively high-p_T jets. In reconstructing DVs, this allows us to tighten the track selection cuts to $p_T > 1$ GeV in order to eliminate low-p_T tracks, whose impact-parameter resolutions are rather poor as can be seen from Eq. (1). (For tracks with $p_T > 1$ GeV, we found that the best-fit values of $\sigma_{d_0}(\infty)$ ($\sigma_{z_0 \sin\theta}(\infty)$) and p_{d_0} ($p_{z_0 \sin\theta}$) in Eq. (1) are 23 μm (78 μm) and 3.1 GeV (1.6 GeV), respectively.) We also require the tracks used for DV reconstruction to satisfy $|d_0| < 10$ mm and $|z_0| < 320$ mm [39].

For DV reconstruction, we only use tracks associated with four-highest p_T jets.[3] If one of these jets contains no track satisfying the above requirements, then we add the fifth-highest p_T jet. If more than one jets among these five high p_T jets do not offer any tracks which meet the above conditions, then we suppose that DV reconstruction is not possible in such an event. Since we do not know which pair of jets originate from a common parent gluino, we study all possible patterns of pairings out of the four jets. For each paring, we find two DVs, each of which is reconstructed from tracks associated with the corresponding jet pair. Among the possible pairings, we adopt the one which minimizes an objective function that is defined by the sum of the weighted vertex χ^2 divided by the sum of the weights over the two DVs, where we use the same weight as that given in Ref. [29] (see Eq. (A.5) in Appendix A for a concrete expression). We regard the vertices reconstructed for this jet pairing as the reconstructed DVs in the following analysis.

In order to see how the variable $|\boldsymbol{r}_{DV1} - \boldsymbol{r}_{DV2}|$ distributes, we perform MC simulation for the gluino pair production processes. We first fix the mass and the decay length $c\tau_{\tilde{g}}$ of gluino (as well as other MSSM parameters). Then, event samples for the gluino pair production process are generated; `MadGraph5_aMC@NLO v2` [40] and `PYTHIA v8.2` are used for this purpose. We generate 50,000 events for each mass and lifetime of gluino. For each event, we determine the flight lengths of two gluinos using the lifetime of the gluino, and hence two decay vertices. The production point of each final-state particle is shifted by the flight length of its parent gluino. Signal event samples are normalized according to the NLL+NLO gluino pair production cross section [41]. The produced gluinos are forced to decay into first-generation quarks and a neutralino with a mass of 100 GeV; we refer to these samples as "light flavor samples." After a fast detector simulation with `DELPHES v3` [42], we select only event samples in the signal region of `Meff-4j-2600` defined in the ATLAS gluino search [43], which adopts events with $E_T^{(\rm miss)} > 250$ GeV (with $E_T^{(\rm miss)}$ being the missing energy), $p_T(j_1) > 200$ GeV (with $p_T(j_i)$ being the transverse momentum of i-th jet), $p_T(j_4) > 150$ GeV, $\Delta\phi(j_{1,2,3,4}, E_T^{(\rm miss)})_{\rm min} > 0.4$ (with $\Delta\phi$ being the azimuthal angle between the jet and the missing energy), aplanarity larger than 0.04, $E_T^{(\rm miss)}/m_{\rm eff}(4) > 0.2$ (with $m_{\rm eff}(4)$ being the scalar sum of $E_T^{(\rm miss)}$ and the transverse momenta of leading 4-jets), and $m_{\rm eff}(\rm incl.) > 2600$ GeV (with $m_{\rm eff}(\rm incl.)$ being the scalar sum of $E_T^{(\rm miss)}$ and the transverse momenta of jets with $p_T > 50$ GeV).[4]

In order to discuss the discovery reach, we should also consider backgrounds. As we have mentioned, we require the presence of DVs on top of the event selection conditions used in the ordinary gluino searches. The latter conditions drop most of the SM background, and thus most of the fake DV events are also expected to be eliminated. Since we impose relatively tight kinematical selection cuts (*i.e.*, `Meff-4j-2600`), we expect the properties of the SM background events relevant to tracking and DV reconstruction, such as the multiplicity and p_T of charged tracks, to resemble those of the signal events after applying the kinematical selection cuts. With this expectation, we approximate the background event samples which pass the kinematical selection cuts by the signal event samples with $c\tau_{\tilde{g}} = 0$. However, one possible difference between these two is that the SM background may contain b quarks while our signal event samples called "light flavor samples" only include the first-generation quarks. b quarks tend to be long-lived and thus may contribute to background considerably. To take into account this possibility, we generate event samples called "heavy flavor samples," where the produced gluinos are forced to decay into b quarks, and use them as background samples to be conservative. We normalize the cross section of the background events to be 0.20 fb as observed in the ATLAS gluino search [43]. In addition, since we mainly consider DVs inside the beam pipe, we neglect background vertices from hadronic interactions in the detector materials and only consider background vertices which are mis-reconstructed as displaced ones due to the resolution of track impact parameters. With this simplification, we reject an event with a DV whose reconstructed position is inside the detector materials: *i.e.*, 22 mm $\leq |(\boldsymbol{r}_{DV})_T| \leq$ 25 mm, 29 mm $\leq |(\boldsymbol{r}_{DV})_T| \leq$ 38 mm, 46 mm $\leq |(\boldsymbol{r}_{DV})_T| \leq$ 73 mm, 84 mm $\leq |(\boldsymbol{r}_{DV})_T| \leq$ 111 mm, or $|(\boldsymbol{r}_{DV})_T| \geq$ 120 mm [26,44–46].

In Fig. 2a, we show the $|\boldsymbol{r}_{DV1} - \boldsymbol{r}_{DV2}|$ distribution in the signal region `Meff-4j-2600`. In addition, in Fig. 2b we plot the fraction of events passing the selection cut of $|\boldsymbol{r}_{DV1} - \boldsymbol{r}_{DV2}| > r_{\rm cut}$ as a function of $r_{\rm cut}$. Note that the background distribution deviates from the signal distribution with $c\tau_{\tilde{g}} = 0$ because the background contains b hadrons in jets. These figures show that if we set $r_{\rm cut}$ to be $\gtrsim 100$ μm, then a significant fraction of SM background fails to pass the selection cut while a sizable number of signal events for $c\tau_{\tilde{g}} \gtrsim 100$ μm remain after the cut. This observation indicates that this cut may be useful to probe a gluino with a decay length of $c\tau_{\tilde{g}} \gtrsim 100$ μm.

To demonstrate the performance of the new selection cut based on DVs, we show how far we can extend the discovery reach and exclusion limit of the gluino searches. We apply the cut to both signal and background events in the signal region `Meff-4j-2600` and estimate the expected exclusion and discovery reaches for gluino. We vary the cut parameter $r_{\rm cut}$ from 0 μm to 1000 μm by 20 μm, and determine the highest value of the gluino mass

[3] This reflects the event topology under consideration; gluinos are always pair-produced and each of them decays into two quarks and a neutralino.

[4] In our analysis, we consider the signal region `Meff-4j-2600` which is relevant for the case with the lightest neutralino mass of 100 GeV. Notice that the improvement of the discovery reach using the DVs is expected in other signal regions as far as the splitting of the masses of gluino and the lightest neutralino is sizable.

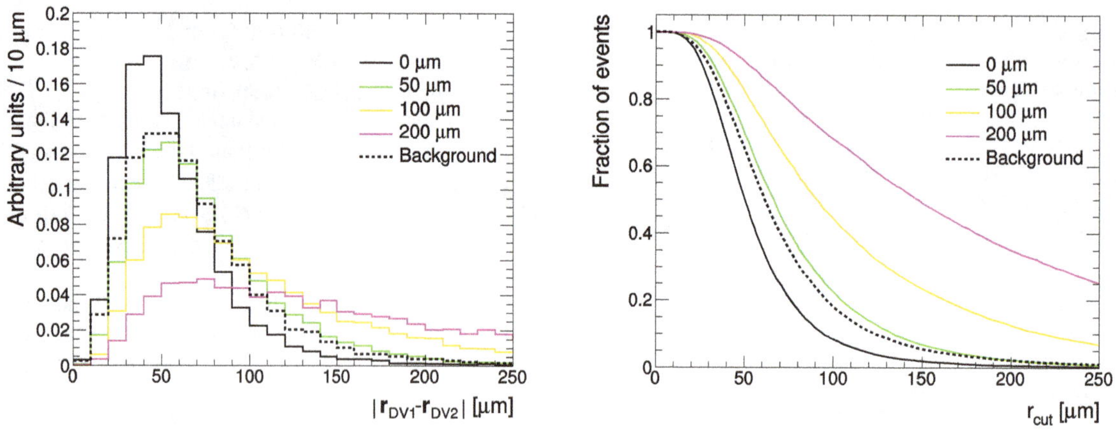

(a) $|\boldsymbol{r}_{\mathrm{DV1}} - \boldsymbol{r}_{\mathrm{DV2}}|$ distribution (b) Efficiency

Fig. 2. (a) $|\mathbf{r}_{\mathrm{DV1}} - \mathbf{r}_{\mathrm{DV2}}|$ distribution in the signal region `Meff-4j-2600` for various $c\tau_{\tilde{g}}$. (b) the fraction of events passing the selection cut of $|\mathbf{r}_{\mathrm{DV1}} - \mathbf{r}_{\mathrm{DV2}}| > r_{\mathrm{cut}}$ as a function of r_{cut}. Here we set $m_{\tilde{g}} = 2.2$ TeV in both figures.

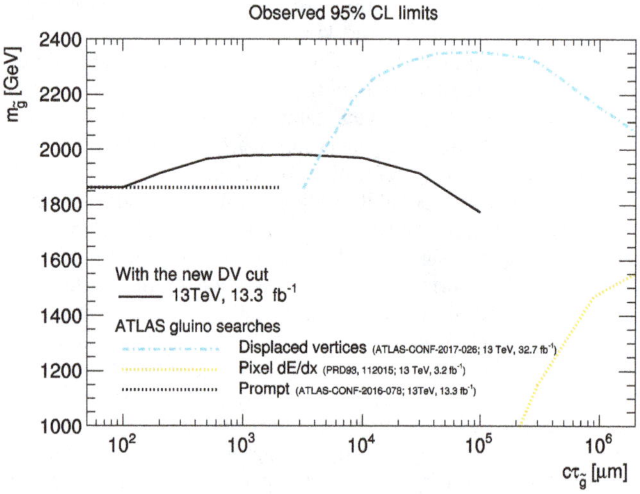

Fig. 3. The 95% CL expected exclusion limits on the gluino mass with $\mathcal{L} = 13.3$ fb^{-1} at the 13 TeV LHC run as a function of $c\tau_{\tilde{g}}$ (black solid line). For comparison, we also show the 95% CL exclusion limits given by the ATLAS prompt-decay gluino search (red dotted line) [43], the ATLAS DV search (blue dot-dashed line) [2], and the ATLAS search of large ionization energy loss in the Pixel detector (orange dotted line) [47]. (For interpretation of the references to color in this figure legend, the reader is referred to the web version of this article.)

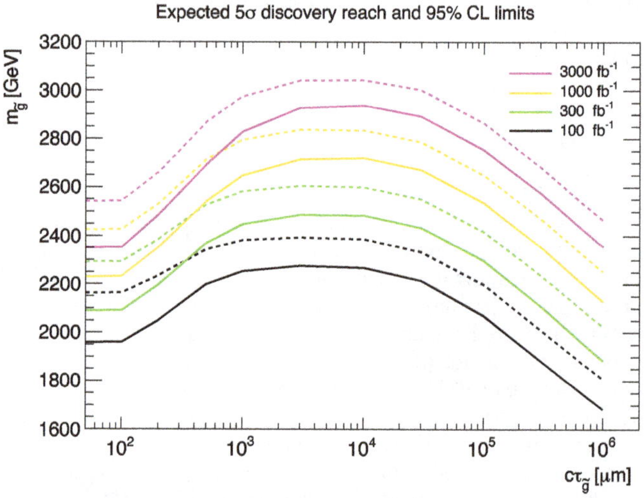

Fig. 4. The expected 95% CL exclusion limits (dotted) and 5σ discovery reaches (solid) as functions of $c\tau_{\tilde{g}}$ for different values of integrated luminosity at the 13 TeV LHC run.

as a gluino mass reach for each $c\tau_{\tilde{g}}$. For the integrated luminosity of $\mathcal{L} = 100$ fb^{-1} (1000 fb^{-1}) at the 13 TeV LHC, we find that $r_{\mathrm{cut}} \sim 200$ μm (400 μm) yields the best discovery and exclusion performance for a gluino with $c\tau_{\tilde{g}} \gtrsim 200$ μm.

For exclusion limits, we compute the expected 95% confidence level (CL) limits on the gluino mass using the CL_s prescription [48]. In Fig. 3, we show the expected limit on the gluino mass as a function of $c\tau_{\tilde{g}}$ based on the currently available luminosity of 13.3 fb^{-1} at the 13 TeV LHC. We can see that, even with the current data, the exclusion limit can be improved by about 80 and 120 GeV for $c\tau_{\tilde{g}} = 0.3$ and 1 mm, respectively. To compare the result with the current sensitivities of other gluino searches, we also show the 95% CL exclusion limits given by the ATLAS prompt-decay gluino search with the 13 TeV 13.3 fb^{-1} data (red dotted line) [43], the ATLAS DV search with the 13 TeV 32.7 fb^{-1} data (blue dot-dashed line) [2], and the ATLAS search of large ionization energy loss in the Pixel detector with the 13 TeV 3.2 fb^{-1} data (orange dotted line) [47]. Note that we extend the red dotted line for the ATLAS prompt-decay gluino search up to $c\tau_{\tilde{g}} \sim \mathcal{O}(1)$ mm just for comparison; the reach of the prompt-decay gluino search is expected to become worse when $c\tau_{\tilde{g}} \gtrsim \mathcal{O}(1)$ mm [49]. The existing metastable gluino searches are insensitive to a gluino with $c\tau_{\tilde{g}} \lesssim 1$ mm, as shown in Fig. 3 (blue dot-dashed and orange dotted lines), to which searches with the new DV cut may offer a good sensitivity. In this sense, this new search strategy plays a complementary role in probing metastable gluinos.

To see the future prospect, we also derive the expected 95% CL exclusion limits as well as 5σ discovery reach with larger luminosity. The expected discovery reach is determined by calculating the expected significance of discovery Z_0 [50]:

$$Z_0 = \sqrt{2\{(S+B)\log(1+S/B) - S\}}\,, \tag{3}$$

where S (B) is the expected number of signal (background) events. We then require Z_0 to be larger than 5 for discovery. In Fig. 4, we show the expected 95% CL exclusion limits and 5σ discovery reaches for gluino as functions of $c\tau_{\tilde{g}}$ for different values of integrated luminosity at the 13 TeV LHC run. Notice that the expected reaches for an extremely small $c\tau_{\tilde{g}}$ should correspond to those for the prompt-decay gluino with the same data set since the new DV cut plays no role in this case. As can be seen from this figure, the

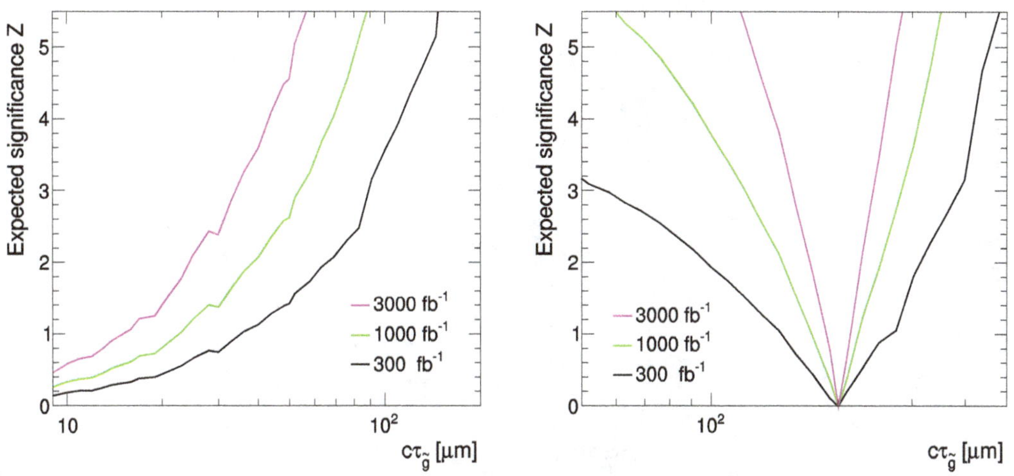

(a) $c\tau_{\tilde{g}}^{(\text{hypo})} = 0\ \mu\text{m}$ (b) $c\tau_{\tilde{g}}^{(\text{hypo})} = 200\ \mu\text{m}$

Fig. 5. The expected significance of rejection $\langle Z_{c\tau_{\tilde{g}}^{(\text{hypo})}} \rangle_{c\tau_{\tilde{g}}}$ as a function of $c\tau_{\tilde{g}}$ for an integrated luminosity of 300, 1000, and 3000 fb^{-1} in the black, green, and purple lines, respectively. Here, we set $m_{\tilde{g}} = 2.2$ TeV. See text for the definition of $\langle Z_{c\tau_{\tilde{g}}^{(\text{hypo})}} \rangle_{c\tau_{\tilde{g}}}$. (For interpretation of the references to color in this figure legend, the reader is referred to the web version of this article.)

reach for the gluino can be extended with the help of the additional DV selection cut for $c\tau_{\tilde{g}} \gtrsim 100$ μm; for instance, for a gluino with $c\tau_{\tilde{g}} \sim \mathcal{O}(1\text{–}10)$ mm, the expected discovery reach for the gluino mass can be extended by as large as ~ 300 GeV (500 GeV) with an integrated luminosity of $\mathcal{L} = 100$ fb^{-1} (1000 fb^{-1}). Because charged tracks with $|d_0| > 10$ mm are not included in the analysis, and also because we reject all events with a DV whose reconstructed position radius is larger than 120 mm, the expected exclusion limits decrease for $c\tau_{\tilde{g}} \gtrsim 100$ mm. Such a larger $c\tau_{\tilde{g}}$ region can however be covered by other long-lived gluino searches. (Remember that these numbers are based on the events in the signal region `Meff-4j-2600`. For more accurate estimation of the improvement, one should carefully optimize the selection criteria, with which we may have better reach.)

3.3. *Lifetime measurement*

If a new metastable particle is discovered at the LHC, measurement of its lifetime is of crucial importance to understand the nature of new physics behind this metastable particle. In this subsection, we discuss the prospect of the lifetime measurement by means of the DV reconstruction method we have discussed.

To see the prospect of the lifetime measurement, we study the expected significance of rejection of a hypothesis that the gluino decay length is $c\tau_{\tilde{g}}^{(\text{hypo})}$ for gluino samples with a decay length of $c\tau_{\tilde{g}}$. Event samples are binned according to the DV distance $|\boldsymbol{r}_{\text{DV1}} - \boldsymbol{r}_{\text{DV2}}|$ of the events. Then the expected significance $\langle Z_{c\tau_{\tilde{g}}^{(\text{hypo})}} \rangle_{c\tau_{\tilde{g}}}$ is determined as $\langle Z_{c\tau_{\tilde{g}}^{(\text{hypo})}} \rangle_{c\tau_{\tilde{g}}} \equiv \sqrt{\Delta\chi^2(c\tau_{\tilde{g}}^{(\text{hypo})}, c\tau_{\tilde{g}})}$, where

$$\Delta\chi^2(c\tau_{\tilde{g}}^{(\text{hypo})}, c\tau_{\tilde{g}}) = \sum_{\text{bin } i} \frac{\left\{S_i(c\tau_{\tilde{g}}^{(\text{hypo})}) - S_i(c\tau_{\tilde{g}})\right\}^2}{S_i(c\tau_{\tilde{g}}^{(\text{hypo})}) + B_i}. \tag{4}$$

Here, $S_i(c\tau)$ is the expected number of signal events in the bin i on the assumption that gluinos have a decay length of $c\tau$, while B_i is the number of SM background. We show the expected significance for $c\tau_{\tilde{g}}^{(\text{hypo})} = 0$ and 200 μm as a function of the gluino decay length $c\tau_{\tilde{g}}$ used to generate the data sample in Figs. 5a and 5b, respectively, for a gluino with a mass of 2.2 TeV. Here we use the visible cross-section $\epsilon\sigma$ of 8.8×10^{-2} fb for signal events, which is defined by the product of the production cross-section σ and the fraction of signal events in the signal region `Meff-4j-2600` estimated from our fast detector simulation, ϵ. In Figs. 6a and 6b, we also show the expected upper and lower bounds on the decay length as a function of $c\tau_{\tilde{g}}$. From the figures, we find that a metastable gluino with $c\tau_{\tilde{g}} \gtrsim 25$ (50) μm can be distinguished from a promptly decaying one with the significance of 2σ (5σ) with an integrated luminosity of 3000 fb^{-1}. Moreover, Fig. 5b shows that the decay length of a gluino with $c\tau_{\tilde{g}} \sim \mathcal{O}(100)$ μm can be measured with an $\mathcal{O}(1)$ accuracy at the high-luminosity LHC. With such a measurement, we may probe the squark mass scale $m_{\tilde{q}}$ via Eq. (2) even though squarks are inaccessible at the LHC, which sheds light on the SUSY mass spectrum as well as the mediation mechanism of SUSY-breaking effects.

4. Conclusions and discussion

In this letter, we have discussed a method of reconstructing DVs that originate from decay of metastable particles on the assumption that these metastable particles are always pair-produced and their decay products contain high-p_T jets. We especially consider gluinos in the SUSY models as an example, which tend to be metastable when squarks have masses much larger than the TeV scale. It is found that this method can separate out DVs if the gluino decay length is $\gtrsim 100$ μm. Then, we have seen that an event selection cut based on this DV reconstruction may be utilized to improve the potential of the gluino searches for a gluino with $c\tau_{\tilde{g}} \gtrsim 100$ μm. In particular, if $c\tau_{\tilde{g}} \sim \mathcal{O}(1\text{–}10)$ mm, then the exclusion and discovery reaches for the gluino mass can be extended by about 370 GeV and 500 GeV, respectively, with an integrated luminosity of 1000 fb^{-1} at the 13 TeV LHC. Furthermore, with an integrated luminosity of 3000 fb^{-1}, it is possible to measure the gluino decay length with an $\mathcal{O}(1)$ accuracy for a gluino with $c\tau_{\tilde{g}} \sim \mathcal{O}(100)$ μm and $m_{\tilde{g}} = 2.2$ TeV, which may allow us to probe the PeV-scale squarks indirectly. Although we have concentrated on metastable gluinos in SUSY models, a similar technique may be used to probe DV signatures from other metastable particles. A more extensive study will be done elsewhere [38].

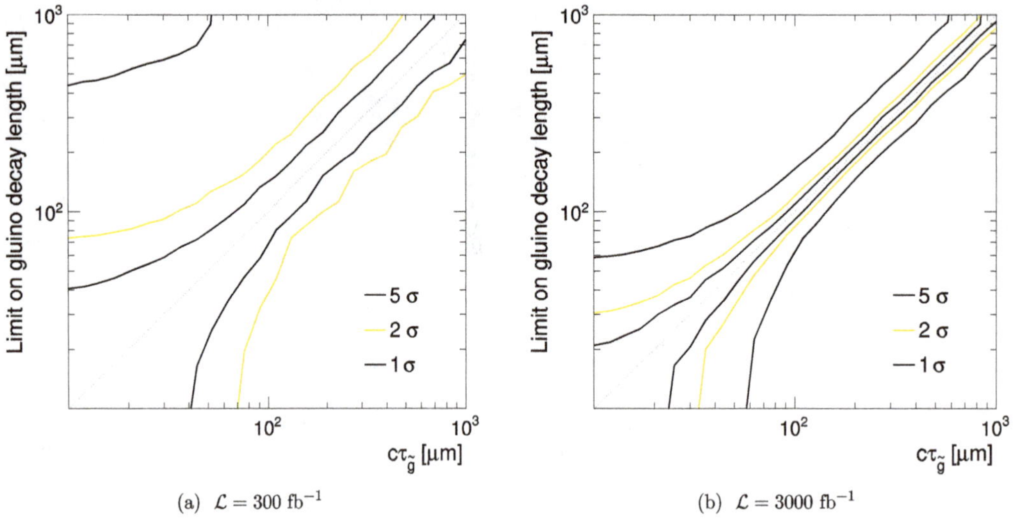

Fig. 6. The expected upper and lower bounds decay length of gluino as a function of the underlying value of $c\tau_{\tilde{g}}$. Here, we set $m_{\tilde{g}} = 2.2$ TeV.

Acknowledgements

The authors thank T. Yamanaka and S. Adachi for useful discussion. This work was supported by the Grant-in-Aid for Scientific Research on Scientific Research C (No. 26400239 [TM]), Young Scientists B (No. 15K17653 [HO]), and Innovative Areas (No. 16H06489 [OJ], No. 16H06490 [TM]).

Appendix A. Vertexing method

Here, we give a brief review on the vertexing method exploited in our analysis, as well as a concrete expression for the objective function used to determine the reconstructed DVs.

Our vertexing method is based on the adaptive vertex fitting algorithm [29]. In this algorithm, an initial vertex position is found using the FSMW method [30] for a pair of jets in question. This method first defines a crossing point for a pair of the two tracks chosen from each jet as the closest midpoint of these tracks. We then assign a weight to this crossing point,

$$w \equiv (d + d_{\rm min})^{-\frac{1}{2}} \,, \tag{A.1}$$

where d is the distance between the two tracks, and we set $d_{\rm min} = 10$ μm following Ref. [29]. This weight gets larger if the distance between the two tracks associated with the crossing point is smaller. Next, for a spatial coordinate, say, the x-coordinate, we consider a distribution of the crossing points and define a weighted interval for the distribution as the length of the interval divided by the sum of the weights of the points in the interval. We then find the smallest weighted interval that covers at least 40% of all the points. This process is recursively performed for the obtained smallest weighted interval until the interval contains only two points, and eventually the midpoint of the remaining two points is defined as the x-coordinate of the initial vertex position. We perform this procedure for each spatial direction.

For the vertex position $\boldsymbol{v}$ determined above, we define

$$\chi_i^2(\boldsymbol{v}) \equiv \frac{d_i^2(\boldsymbol{v})}{\sigma_{d_0}^2 + \sigma_{z_0 \sin\theta}^2} \,, \tag{A.2}$$

for each track i, where $d_i(\boldsymbol{v})$ denotes its distance from the vertex $\boldsymbol{v}$. We further assign a weight w_i to each track that is defined by

$$w_i(\chi_i^2) \equiv \frac{\exp\left(-\chi_i^2/2T\right)}{\exp\left(-\chi_i^2/2T\right) + \exp\left(-\chi_c^2/2T\right)} \,, \tag{A.3}$$

where we use $\chi_c = 3$ [29] and T is a parameter that we choose in the following. As can be seen from this expression, if a track is far away from the vertex $\boldsymbol{v}$, a fairly small weight is assigned to the track. Then, we determine a new vertex position by solving

$$\sum_i w_i\left(\chi_i^2(\boldsymbol{v})\right) \chi_i(\boldsymbol{v}_{\rm new}) \frac{\partial \chi_i(\boldsymbol{v}_{\rm new})}{\partial \boldsymbol{v}} = 0 \,, \tag{A.4}$$

with respect to $\boldsymbol{v}_{\rm new}$. This new vertex position ($\boldsymbol{v}_{\rm new}$) is then used as an initial vertex position to repeat this process. We iterate the above process with varying the parameter T as $T = 256 \to 64 \to 16 \to 4 \to 1 \to 1 \to \dots$ until $T = 1$ and the vertex position converges within 1 μm.

As mentioned in Sec. 3.2, the weight w_i defined in Eq. (A.3) is also used to determine the jet pairing for the reconstruction of DVs out of four jets. Among the three possible pairings of the four jets, we choose the one which minimizes

$$\chi^2 \equiv \frac{\sum_{i\in{\rm trk}(\boldsymbol{v}_1)} w_i\left(\chi_i^2(\boldsymbol{v}_1)\right)\chi_i^2(\boldsymbol{v}_1) + \sum_{j\in{\rm trk}(\boldsymbol{v}_2)} w_j\left(\chi_j^2(\boldsymbol{v}_2)\right)\chi_j^2(\boldsymbol{v}_2)}{\sum_{i\in{\rm trk}(\boldsymbol{v}_1)} w_i\left(\chi_i^2(\boldsymbol{v}_1)\right) + \sum_{j\in{\rm trk}(\boldsymbol{v}_2)} w_j\left(\chi_j^2(\boldsymbol{v}_2)\right)} \,, \tag{A.5}$$

where ${\rm trk}(\boldsymbol{v}_{1,2})$ denotes the set of tracks associated with the DV $\boldsymbol{v}_{1,2}$ reconstructed for each pair of jets, and we take $T = 1$ and $\chi_c = 3$ in the weights. We define the reconstructed DVs by $\boldsymbol{r}_{{\rm DV}1,2} \equiv \boldsymbol{v}_{1,2}$ for the jet pairing that minimizes this χ^2, and use them in our analysis.

References

[1] M. Fairbairn, A.C. Kraan, D.A. Milstead, T. Sjostrand, P.Z. Skands, T. Sloan, Phys. Rep. 438 (2007) 1, arXiv:hep-ph/0611040.
[2] Search for long-lived, massive particles in events with displaced vertices and missing transverse momentum in 13 TeV pp collisions with the ATLAS detector, Tech. Rep. ATLAS-CONF-2017-026, CERN, Geneva, 2017.
[3] G. Aad, et al., ATLAS, Phys. Rev. D 88 (2013) 112006, arXiv:1310.3675 [hep-ex].
[4] J.L. Feng, T. Moroi, L. Randall, M. Strassler, S.-f. Su, Phys. Rev. Lett. 83 (1999) 1731, arXiv:hep-ph/9904250.
[5] M. Ibe, T. Moroi, T.T. Yanagida, Phys. Lett. B 644 (2007) 355, arXiv:hep-ph/0610277.
[6] S. Asai, T. Moroi, K. Nishihara, T.T. Yanagida, Phys. Lett. B 653 (2007) 81, arXiv:0705.3086 [hep-ph];
S. Asai, T. Moroi, T.T. Yanagida, Phys. Lett. B 664 (2008) 185, arXiv:0802.3725 [hep-ph];
S. Asai, Y. Azuma, O. Jinnouchi, T. Moroi, S. Shirai, T.T. Yanagida, Phys. Lett. B 672 (2009) 339, arXiv:0807.4987 [hep-ph].
[7] N. Nagata, H. Otono, S. Shirai, arXiv:1701.07664 [hep-ph], 2017.

[8] V. Khachatryan, et al., CMS, Phys. Rev. D (2016), http://dx.doi.org/10.1103/PhysRevD.95.012009, Phys. Rev. D 95 (2017) 012009, arXiv:1610.05133 [hep-ex].
[9] J.D. Wells, arXiv:hep-ph/0306127, 2003;
N. Arkani-Hamed, S. Dimopoulos, J. High Energy Phys. 0506 (2005) 073, arXiv:hep-th/0405159;
G. Giudice, A. Romanino, Nucl. Phys. B 699 (2004) 65, arXiv:hep-ph/0406088;
N. Arkani-Hamed, S. Dimopoulos, G. Giudice, A. Romanino, Nucl. Phys. B 709 (2005) 3, arXiv:hep-ph/0409232;
J.D. Wells, Phys. Rev. D 71 (2005) 015013, arXiv:hep-ph/0411041.
[10] L.J. Hall, Y. Nomura, J. High Energy Phys. 1201 (2012) 082, arXiv:1111.4519 [hep-ph];
L.J. Hall, Y. Nomura, S. Shirai, J. High Energy Phys. 1301 (2013) 036, arXiv:1210.2395 [hep-ph];
M. Ibe, T.T. Yanagida, Phys. Lett. B 709 (2012) 374, arXiv:1112.2462 [hep-ph];
M. Ibe, S. Matsumoto, T.T. Yanagida, Phys. Rev. D 85 (2012) 095011, arXiv:1202.2253 [hep-ph];
A. Arvanitaki, N. Craig, S. Dimopoulos, G. Villadoro, J. High Energy Phys. 1302 (2013) 126, arXiv:1210.0555 [hep-ph];
N. Arkani-Hamed, A. Gupta, D.E. Kaplan, N. Weiner, T. Zorawski, arXiv:1212.6971 [hep-ph], 2012;
J.L. Evans, M. Ibe, K.A. Olive, T.T. Yanagida, Eur. Phys. J. C 73 (2013) 2468, arXiv:1302.5346 [hep-ph];
J.L. Evans, K.A. Olive, M. Ibe, T.T. Yanagida, Eur. Phys. J. C 73 (2013) 2611, arXiv:1305.7461 [hep-ph].
[11] M. Toharia, J.D. Wells, J. High Energy Phys. 02 (2006) 015, arXiv:hep-ph/0503175;
P. Gambino, G.F. Giudice, P. Slavich, Nucl. Phys. B 726 (2005) 35, arXiv:hep-ph/0506214;
R. Sato, S. Shirai, K. Tobioka, J. High Energy Phys. 11 (2012) 041, arXiv:1207.3608 [hep-ph].
[12] G.F. Giudice, R. Rattazzi, Phys. Rep. 322 (1999) 419, arXiv:hep-ph/9801271.
[13] P. Draper, P. Meade, M. Reece, D. Shih, Phys. Rev. D 85 (2012) 095007, arXiv:1112.3068 [hep-ph];
J.A. Evans, J. Shelton, J. High Energy Phys. 04 (2016) 056, arXiv:1601.01326 [hep-ph];
B.C. Allanach, M. Badziak, G. Cottin, N. Desai, C. Hugonie, R. Ziegler, Eur. Phys. J. C 76 (2016) 482, arXiv:1606.03099 [hep-ph].
[14] R. Barbier, et al., Phys. Rep. 420 (2005) 1, arXiv:hep-ph/0406039.
[15] P.W. Graham, D.E. Kaplan, S. Rajendran, P. Saraswat, J. High Energy Phys. 07 (2012) 149, arXiv:1204.6038 [hep-ph];
K. Barry, P.W. Graham, S. Rajendran, Phys. Rev. D 89 (2014) 054003, arXiv:1310.3853 [hep-ph];
C. Csaki, E. Kuflik, T. Volansky, Phys. Rev. Lett. 112 (2014) 131801, arXiv:1309.5957 [hep-ph];
C. Csaki, E. Kuflik, S. Lombardo, O. Slone, T. Volansky, J. High Energy Phys. 08 (2015) 016, arXiv:1505.00784 [hep-ph].
[16] Z. Chacko, H.-S. Goh, R. Harnik, Phys. Rev. Lett. 96 (2006) 231802, arXiv:hep-ph/0506256;
G. Burdman, Z. Chacko, H.-S. Goh, R. Harnik, J. High Energy Phys. 02 (2007) 009, arXiv:hep-ph/0609152;
H. Cai, H.-C. Cheng, J. Terning, J. High Energy Phys. 05 (2009) 045, arXiv:0812.0843 [hep-ph].
[17] Z. Chacko, D. Curtin, C.B. Verhaaren, Phys. Rev. D 94 (2016) 011504, arXiv:1512.05782 [hep-ph].
[18] M.J. Strassler, K.M. Zurek, Phys. Lett. B 651 (2007) 374, arXiv:hep-ph/0604261.
[19] J. Barnard, P. Cox, T. Gherghetta, A. Spray, J. High Energy Phys. 03 (2016) 003, arXiv:1510.06405 [hep-ph].
[20] S. Chang, M.A. Luty, arXiv:0906.5013 [hep-ph], 2009;
R.T. Co, F. D'Eramo, L.J. Hall, D. Pappadopulo, J. Cosmol. Astropart. Phys. 1512 (2015) 024, arXiv:1506.07532 [hep-ph].
[21] L. Basso, A. Belyaev, S. Moretti, C.H. Shepherd-Themistocleous, Phys. Rev. D 80 (2009) 055030, arXiv:0812.4313 [hep-ph];
J.C. Helo, M. Hirsch, S. Kovalenko, Phys. Rev. D 89 (2014) 073005, arXiv:1312.2900 [hep-ph], Phys. Rev. D 93 (9) (2016) 099902 (Erratum);
E. Izaguirre, B. Shuve, Phys. Rev. D 91 (2015) 093010, arXiv:1504.02470 [hep-ph];
S. Antusch, E. Cazzato, O. Fischer, J. High Energy Phys. 12 (2016) 007, arXiv:1604.02420 [hep-ph];
S. Antusch, E. Cazzato, O. Fischer, arXiv:1612.02728 [hep-ph], 2016;
E. Accomando, L. Delle Rose, S. Moretti, E. Olaiya, C. Shepherd-Themistocleous, arXiv:1612.05977 [hep-ph], 2016;
P.S.B. Dev, R.N. Mohapatra, Y. Zhang, arXiv:1612.09587 [hep-ph], 2016;
A. Maiezza, M. Nemevšek, F. Nesti, Phys. Rev. Lett. 115 (2015) 081802, arXiv:1503.06834 [hep-ph];
M. Nemevšek, F. Nesti, J.C. Vasquez, arXiv:1612.06840 [hep-ph], 2016.
[22] Vertex Reconstruction Performance of the ATLAS Detector at $\sqrt{s} = 13$ TeV, Tech. Rep. ATL-PHYS-PUB-2015-026, CERN, Geneva, 2015.
[23] Performance of primary vertex reconstruction in proton-proton collisions at $\sqrt{s}=7$ TeV in the ATLAS experiment, Tech. Rep. ATLAS-CONF-2010-069, CERN, Geneva, 2010;
M. Aaboud, et al., ATLAS, arXiv:1611.10235 [physics.ins-det], 2016.
[24] S. Chatrchyan, et al., CMS, J. Instrum. 9 (2014) P10009, arXiv:1405.6569 [physics.ins-det];
CMS, CMS Collaboration, Primary vertex resolution in 2016, Tech. Rep. CMS-DP-2016-041, 2016.
[25] ATLAS, Impact parameter resolution, https://atlas.web.cern.ch/Atlas/GROUPS/PHYSICS/PLOTS/IDTR-2015-007/, 2015;
Track Reconstruction Performance of the ATLAS Inner Detector at $\sqrt{s} = 13$ TeV, Tech. Rep. ATL-PHYS-PUB-2015-018, CERN, Geneva, 2015.
[26] M. Capeans, G. Darbo, K. Einsweiller, M. Elsing, T. Flick, M. Garcia-Sciveres, C. Gemme, H. Pernegger, O. Rohne, R. Vuillermet, ATLAS Insertable B-Layer Technical Design Report, Tech. Rep. CERN-LHCC-2010-013, ATLAS-TDR-19, 2010.
[27] G. Aad, et al., ATLAS, arXiv:0901.0512 [hep-ex], 2009.
[28] ATLAS, Impact parameter resolution using 2016 MB data, https://atlas.web.cern.ch/Atlas/GROUPS/PHYSICS/PLOTS/IDTR-2016-018/, 2016.
[29] R. Fruhwirth, W. Waltenberger, P. Vanlaer, J. Phys. G 34 (2007) N343.
[30] D.R. Bickel, R. Fruhwirth, Comput. Stat. Data Anal. 50 (2006) 3500, arXiv:math/0505419.
[31] T. Sjostrand, S. Mrenna, P.Z. Skands, Comput. Phys. Commun. 178 (2008) 852, arXiv:0710.3820 [hep-ph].
[32] G. Aad, et al., ATLAS, CMS, Phys. Rev. Lett. 114 (2015) 191803, arXiv:1503.07589 [hep-ex].
[33] Y. Okada, M. Yamaguchi, T. Yanagida, Prog. Theor. Phys. 85 (1991) 1;
Y. Okada, M. Yamaguchi, T. Yanagida, Phys. Lett. B 262 (1991) 54;
J.R. Ellis, G. Ridolfi, F. Zwirner, Phys. Lett. B 257 (1991) 83;
H.E. Haber, R. Hempfling, Phys. Rev. Lett. 66 (1991) 1815;
J.R. Ellis, G. Ridolfi, F. Zwirner, Phys. Lett. B 262 (1991) 477.
[34] K. Inoue, A. Kakuto, H. Komatsu, S. Takeshita, Prog. Theor. Phys. 67 (1982) 1889;
R.A. Flores, M. Sher, Ann. Phys. 148 (1983) 95.
[35] L. Randall, R. Sundrum, Nucl. Phys. B 557 (1999) 79, arXiv:hep-th/9810155.
[36] G.F. Giudice, M.A. Luty, H. Murayama, R. Rattazzi, J. High Energy Phys. 12 (1998) 027, arXiv:hep-ph/9810442.
[37] D.M. Pierce, J.A. Bagger, K.T. Matchev, R.-j. Zhang, Nucl. Phys. B 491 (1997) 3, arXiv:hep-ph/9606211.
[38] H. Ito, O. Jinnouchi, T. Moroi, N. Nagata, H. Otono, work in progress.
[39] G. Aad, et al., ATLAS, Phys. Rev. D 92 (2015) 012010, arXiv:1504.03634 [hep-ex].
[40] J. Alwall, R. Frederix, S. Frixione, V. Hirschi, F. Maltoni, O. Mattelaer, H.S. Shao, T. Stelzer, P. Torrielli, M. Zaro, J. High Energy Phys. 07 (2014) 079, arXiv:1405.0301 [hep-ph].
[41] C. Borschensky, M. Krämer, A. Kulesza, M. Mangano, S. Padhi, T. Plehn, X. Portell, Eur. Phys. J. C 74 (2014) 3174, arXiv:1407.5066 [hep-ph].
[42] J. de Favereau, et al., DELPHES 3, J. High Energy Phys. 1402 (2014) 057, arXiv:1307.6346 [hep-ex].
[43] Further searches for squarks and gluinos in final states with jets and missing transverse momentum at $\sqrt{s}$ =13 TeV with the ATLAS detector, Tech. Rep. ATLAS-CONF-2016-078, CERN, Geneva, 2016.
[44] ATLAS Collaboration, J. Instrum. 3 (2008) S08003.
[45] A. Miucci, J. Instrum. 9 (2014) C02018.
[46] G.A.M. Aaboud, et al., J. Instrum. 11 (2016) P11020.
[47] M. Aaboud, et al., ATLAS, Phys. Rev. D 93 (2016) 112015, arXiv:1604.04520 [hep-ex].
[48] A.L. Read, Advanced statistical techniques in particle physics, in: Proceedings, Conference, Durham, UK, March 18–22, 2002, J. Phys. G 28 (2002) 2693;
T. Junk, Nucl. Instrum. Methods A 434 (1999) 435, arXiv:hep-ex/9902006.
[49] Limits on metastable gluinos from ATLAS SUSY searches at 8 TeV, Tech. Rep. ATLAS-CONF-2014-037, CERN, Geneva, 2014,.
[50] G. Cowan, K. Cranmer, E. Gross, O. Vitells, Eur. Phys. J. C 71 (2011) 1554, arXiv:1007.1727 [physics.data-an], Eur. Phys. J. C 73 (2013) 2501 (Erratum).

5

Fermion number violating effects in low scale leptogenesis

Shintaro Eijima *, Mikhail Shaposhnikov

Institute of Physics, Laboratory for Particle Physics and Cosmology, École Polytechnique Fédérale de Lausanne, CH-1015 Lausanne, Switzerland

ARTICLE INFO

Editor: A. Ringwald

ABSTRACT

The existence of baryon asymmetry and dark matter in the Universe may be related to CP-violating reactions of three heavy neutral leptons (HNLs) with masses well below the Fermi scale. The dynamical description of the lepton asymmetry generation, which is the key ingredient of baryogenesis and of dark matter production, is quite complicated due to the presence of many different relaxation time scales and the necessity to include quantum-mechanical coherent effects in HNL oscillations. We derive kinetic equations accounting for fermion number violating effects missed so far and identify one of the domains of HNL masses that can potentially lead to large lepton asymmetry generation boosting the sterile neutrino dark matter production.

1. Introduction

Though the canonical Standard Model (SM) has been completed by the discovery of the Higgs boson and may be a valid effective quantum field theory all the way up to the Planck scale (for recent discussions see [1–3]) it is inconsistent with a number of observations. They include the non-zero neutrino masses, the presence of Dark Matter (DM) in the Universe, and its baryon asymmetry (BAU). Perhaps, the most minimal way to address all these problems on the same footing is to extend the SM by three right-handed neutrinos with masses below the Fermi scale [4,5]. These new fermions N_I, $I = 1, 2, 3$ (following the Particle Data Group [6] we will call them Heavy Neutral Leptons or HNLs for short) are singlets with respect to the SM gauge group and thus are allowed to have Majorana neutrino masses. The lightest of these particles, N_1, may play a role of Dark Matter [7,8]. Two others (N_2 and N_3), if (almost) degenerate, can produce the baryon asymmetry of the Universe [9,5] and explain non-zero neutrino masses and mixings at the same time. This model was dubbed the νMSM for "Neutrino Minimal Standard Model" [4]. For a number of computations of baryon asymmetry in this model see [10–20].

The most conservative scenario of the Universe evolution, which does not require any new physics beyond the νMSM, proceeds as follows. First, the Universe is inflated by the SM Higgs field [21] and heated up due to Higgs field oscillations to temperatures $T \sim 10^{14}$ GeV [22–24]. The Higgs inflation prepares the initial conditions for the Hot Big Bang [22] at $T \sim 10^{14}$ GeV: baryon and lepton numbers of the Universe are equal to zero, and the number densities of HNLs at this time are zero as well. The particles N_2 and N_3 enter into thermal equilibrium below the sphaleron freeze-out temperature $T_{sph} \simeq 130$ GeV and produce baryon asymmetry of the Universe by processes which include their coherent oscillations, transfer of lepton number from HNLs to active leptons and back [9,5], and rapid anomalous sphaleron transitions [25]. The lighter HNL – N_1 – DM sterile neutrino never equilibrates and is mainly produced at temperatures $T_{DM} \sim 100$–300 MeV by transitions from the ordinary neutrinos to N_1 [7,8,26–30]. The combination of X-ray and Lyman-α bounds on the DM sterile neutrino excludes the "non-resonant" Dodelson–Widrow mechanism [7] for their production, which operates in the cosmic plasma with small lepton asymmetries. In other words, to get enough DM particles N_1, the processes involving $N_{2,3}$ should produce [11] sufficiently large lepton asymmetry $\Delta L/L > 2 \times 10^{-3}$ which must be present at temperatures T_{DM}. This is needed to boost the production of N_1 due to the resonant mechanism proposed by Shi and Fuller in [8] and developed in a rigorous way in [27–30]. The production of this large lepton asymmetry must take place below the sphaleron temperature T_{sph}, otherwise the baryon asymmetry will be too large [11].

The estimates of the equilibration rates of $N_{2,3}$ in [11,15,16] and in more recent works [31,32] based on careful thermal field theory computations showed that for all parameter choices consistent with observed pattern of neutrino masses and oscillations the HNLs $N_{2,3}$ enter in thermal equilibrium at some temperature T_{in} exceeding tens of GeV and go out of thermal equilibrium at temperatures $T_{\text{out}} < T_{\text{in}}$ which can be as small as 1 GeV. This has led to the conclusion that the equilibrium period between T_{in} and

* Corresponding author.
E-mail addresses: Shintaro.Eijima@epfl.ch (S. Eijima), Mikhail.Shaposhnikov@epfl.ch (M. Shaposhnikov).

T_{out} erases all the lepton asymmetry which could have been generated at freeze-in temperature T_{in}, requiring that the large lepton asymmetry needed for effective dark matter production must be created at $T < T_{out}$. The analysis made in [16] demonstrated that a large lepton asymmetry can indeed be generated in the scattering processes involving $N_{2,3}$ at the freeze-out temperature T_{out} and below it in out-of-equilibrium decays of $N_{2,3}$. This asymmetry does not exceed $\Delta L/L \simeq 3 \times 10^{-2}$, leading to the conclusion that the mass of the DM sterile neutrino must lie in the interval from 1 to 50 keV, to be consistent with the Lyman-α and phase density constraints coming from observations of dwarf galaxies [33, 34]. As for $N_{2,3}$, their *physical masses* should be between 1.5 GeV and $\simeq 80$ GeV (the W-boson mass) and be extremely degenerate, $\Delta M_{phys}/M < 10^{-15}$ [11,15,16,35,36].[1] The latter condition comes from the requirement that the period of $N_2 \leftrightarrow N_3$ oscillations should be comparable with the age of the Universe at the time of lepton asymmetry production, to insure the resonance [11,35]. The minimal scenario that has been proven to work, albeit under the requirement of a *strong fine-tuning* (of the order of 10^{-4} [16]) between two different contributions to the physical mass difference: one coming from the Yukawa couplings and the Higgs condensate, and another from Majorana masses of $N_{2,3}$.

The aim of the present paper is to show that the part of the lepton asymmetry generated at T_{in} *can in fact survive until the temperatures of sterile neutrino DM production* ~ 100 MeV, in-spite of the fact that HNLs are well in thermal equilibrium between T_{in} and T_{out}. Qualitatively, this comes about because of the following reasons. In the symmetric phase of the electroweak theory the transfer of asymmetry from active to sterile sector and back occurs mainly via the processes with fermion number conservation (we attribute positive fermion number to left-handed neutrinos and to right-handed HNLs) with the rate Γ_+. The rate of fermion number non-conserving processes Γ_- is suppressed by a kinematic factor $(M/k)^2$, where M is the HNL mass, and $k \sim 3T$ is the typical momentum of fermions in the plasma.

On the contrary, in the Higgs phase, at temperatures of the order of tens GeV, the dominant reaction is induced by the mixing term between ν's and N's and has a rate Γ_- exceeding that of the Universe expansion at $T_{out} < T < T_{in}$. It proceeds with fermion number non-conservation: left handed neutrinos go into left-handed anti-HNLs and vice-versa.

In a large portion of the νMSM parameters the reactions with fermion number conservations are faster and give the main contribution to baryogenesis at the sphaleron freeze-out temperature $T \simeq 130$ GeV. However, in a specific domain of the NHL masses and couplings, the rate of these processes never exceeds the Hubble rate. Thus, the asymmetry in this almost conserved number is *protected from dilution*, in-spite of the fact that HNLs are equilibrated due to the processes with fermion number violation.[2] Moreover, at $T \simeq T_{in}$ the rate Γ_+ can be close to the Hubble rate, meaning that large asymmetry in this number *can be generated*. To understand whether it is indeed produced would require numerical solution of our integro-differential kinetic equations for many parameters of the νMSM, which is not attempted here.

The paper is organised as follows. In Section 2 we will derive kinetic equations accounting for helicity structure of HNL interactions. In Section 3 we analyse the different rates and identify the range of νMSM parameters which may potentially lead to large lepton asymmetries surviving until small temperatures where the production of DM sterile neutrino takes place. In Section 4 we summarise our results.

Table 1
Fermion numbers of creation operators.

plus, particles	minus, anti-particles
$a^\dagger_+(k)$, $b^\dagger_-(k)$	$a^\dagger_-(k)$, $b^\dagger_+(k)$

2. Kinetic equations and helicity

In this section we derive the kinetic equations describing the evolution of HNLs density matrix and lepton densities. To elucidate their structure, we will consider first the temperatures well below the electroweak scale, deeply in the Higgs phase. We also neglect for the time being the subtleties related to electric neutrality of the plasma [41–44] and consider the system with HNLs and active neutrinos only, this will be corrected towards the end of this Section. We take a pair of almost degenerate HNLs N_2 and N_3, the generalisation to the case of arbitrary number of N is straightforward. It is convenient to unify N_2 and N_3 in one Dirac spinor, as has been done in [11], and consider the νMSM Lagrangian

$$\begin{aligned}\mathcal{L} &= \mathcal{L}_{SM} + \overline{\Psi} i\partial_\mu \gamma^\mu \Psi - M\overline{\Psi}\Psi + \mathcal{L}_{int},\\ \mathcal{L}_{int} &= -\frac{\Delta M}{2}(\overline{\Psi}\Psi^c + \overline{\Psi^c}\Psi)\\ &\quad - (h_{\alpha 2}\langle\Phi\rangle\overline{\nu_{L\alpha}}\Psi + h_{\alpha 3}\langle\Phi\rangle\overline{\nu_{L\alpha}}\Psi^c + h.c.),\end{aligned} \tag{1}$$

where $\mathcal{L}_{SM}$ is the SM part, $\Psi = N_2 + N_3^c$ is the HNL field in the pseudo-Dirac basis, $M = (M_3 + M_2)/2$ and $\Delta M = (M_3 - M_2)/2$ are the common mass and Majorana mass difference of HNLs, respectively, $h_{\alpha I}$ is a matrix of Yukawa coupling constants and $\langle\Phi\rangle$ is the temperature dependent Higgs vacuum expectation value,[3] which is 174.1 GeV at zero temperature. The HNL field Ψ is given in terms of creation and annihilation operators by

$$\Psi = \int \frac{d^3k}{(2\pi)^3}\frac{1}{\sqrt{2k_0}}\sum_\sigma\left[a_\sigma(k)u_\sigma(k)e^{-ik\cdot x} + b^\dagger_\sigma(k)v_\sigma(k)e^{ik\cdot x}\right], \tag{2}$$

where $\sigma = \pm$ describes the HNL helicity. The attribution of fermion numbers to a_σ, b_σ leading to fermion number conservation in the limit $M \to 0$, $\Delta M \to 0$ is shown in Table 1.

We will consider $\mathcal{L}_{int}$ as a perturbation and work in the second order in Yukawa couplings and first order in ΔM assuming for power counting that $M\Delta M \sim h^2\langle\Phi\rangle^2$, where h is a typical value of the Yukawa couplings. The quadratic mixing term $\mathcal{L}_{int}$ leads to communication between sterile sector of HNLs and the rest of the SM, ensuring creation and destruction of HNLs, their coherent oscillations, lepton number non-conservation, and transfer of asymmetries from active flavours to sterile and back.

To construct kinetic equations, we work in the Heisenberg picture of quantum mechanics and use the ideas of [45] in what follows.[4] The derivation is performed by using four creation operators

[1] These restrictions are not required if the DM sterile neutrino is produced by some new interactions not contained in the renormalizable νMSM Lagrangian [37–39].

[2] The importance of the processes with and without fermion number violation has been already realised in [11]. Unfortunately, the estimates of the relevant rates were not done correctly and the conclusions we arrived in our present work were not achieved at that time. Some further studies of the processes with fermion number non-conservation in this context were carried out in [40].

[3] At small temperatures we are working now the dynamical character of the Higgs field is not important.

[4] The novel feature of our derivation is that we keep the evolution of rapidly oscillating products of creation and annihilation operators of the type $a^\dagger b^\dagger$, $a^\dagger a^\dagger$, ab, etc, playing a crucial role for description of the processes with fermion number conservation.

$a^\dagger_\sigma(k), b^\dagger_\sigma(k)$ for HNLs and two (for each generation α) neutrino and antineutrino operators, $a^\dagger_{\nu_\alpha}(k), b^\dagger_{\nu_\alpha}(k)$.

Let ρ be a (time-independent) density matrix of the complete system. The HNL abundances, including coherent quantum-mechanical correlations between N_2 and N_3, are given by the averages $\mathrm{Tr}[a^\dagger_\sigma(k)a_{\sigma'}(k)\rho]$, $\mathrm{Tr}[b^\dagger_\sigma(k)b_{\sigma'}(k)\rho]$, $\mathrm{Tr}[a^\dagger_\sigma(k)b_{\sigma'}(k)\rho]$, and $\mathrm{Tr}[b^\dagger_\sigma(k)a_{\sigma'}(k)\rho]$. The neutrino number densities are $a^\dagger_{\nu_\alpha}(k)a_{\nu_\beta}(k)$ and $b^\dagger_{\nu_\alpha}(k)b_{\nu_\beta}(k)$. Symbolically, all these number-density operators will be denoted by Q^0_i. With the use of our generic notation Q^0_i, the abundances are given by $q^0_i = \mathrm{Tr}[Q^0_i\rho]$.

The time derivative of q^0_i is readily found:

$$i\dot{q}^0_i = \mathrm{Tr}\left([\mathbf{H}, Q^0_i]\rho\right), \tag{3}$$

where $\mathbf{H}$ is the total Hamiltonian of the system under consideration. Since we neglected the existence of charged leptons for the moment, the commutator of Q^0_i with the SM Hamiltonian is zero, the only non-trivial contribution comes from the interaction Hamiltonian H_{int}, associated with $\mathcal{L}_{int}$ defined in (1). It is easy to see that the commutators $[H_{int}, Q^0_i]$ are the quadratic polynomials with respect to creation and annihilation operators, containing all possible terms, to list just a few: $a^\dagger_\sigma(k)b_{\nu_\beta}(k)$, $a^\dagger_\sigma(k)b^\dagger_{\nu_\beta}(k)$, $a_\sigma(k)b_{\nu_\beta}(-k)$. These operators are multiplied by the first power of Yukawa couplings or by ΔM. Denote by Q^1_i these binomials, and by $q^1_i = \mathrm{Tr}[Q^1_i\rho]$. Now we continue the chain, and write an equation for every q^1_i which appeared at the first step:

$$i\dot{q}^1_i = \mathrm{Tr}\left([\mathbf{H}, Q^1_i]\rho\right), \tag{4}$$

and then repeat the procedure again and again. In this way we get an infinite chain of kinetic equations, which includes the averages of higher and higher polynomials in creation and annihilation operators.

To truncate the system, we proceed as follows. The total Hamiltonian of the system can be written as

$$\mathbf{H} = H_2 + H_{int} + H^{SM}_{int}, \tag{5}$$

where H_2 is the quadratic part including HNLs and active neutrinos, and H^{SM}_{int} describes the SM interactions. As an example, let us take an operator $Q^1_1 = a^\dagger_\sigma(k)b_{\nu_\beta}(k)$. Its commutator with the total Hamiltonian contains 3 terms. The first one is

$$[Q^1_1, H_2] = -(E_N(k) - \epsilon_{\nu_\beta}(k))Q^1_1, \tag{6}$$

where $E_N(k) = \sqrt{k^2+M^2}$ and $\epsilon_\nu(k) = k$ are the energies of HNL and active neutrino respectively (the small neutrino mass can be safely neglected here). The second is $[Q^1_1, H_{int}] = \Sigma_i C_i Q^0_i$ where C_i are the coefficients containing Yukawa couplings and ΔM. The third one is $[Q^1_1, H^{SM}_{int}]$ and is of the order of Fermi constant G_F and of 4th order in creation and annihilation operators. It accounts for neutrino interaction in the medium. The first two terms contain the operators that have already showed up in the first kinetic equation (3), while the third one contains new operators. To find their time evolution would require next steps in the iterative procedure.

To deal with the third term, we note that the neutrino interactions in the plasma can be accounted for by modification of neutrino energy $\epsilon_\nu(k)$, replacing it with the temperature dependent dispersion relation $E_\nu(k)$ (related to the real part of neutrino propagator Σ) and by attributing to it an imaginary part $\gamma_\nu(k) > 0$ (associated with absorptive part of Σ). These considerations suggest the following modification of the commutation relations:

$$[a_{\nu_\alpha}(k), H_2] \to (E_\nu(k) + i\gamma_\nu(k)/2)a_{\nu_\alpha}(k), \tag{7}$$

$$[a^\dagger_{\nu_\alpha}(k), H_2] \to -(E_\nu(k) - i\gamma_\nu(k)/2)a^\dagger_{\nu_\alpha}(k), \tag{8}$$

where the signs in front of $\gamma_\nu(k)$ are chosen in such a way that they correspond to damping rather than an instability. The similar rules apply to antineutrinos. These substitutions effectively account for the third term which can now be removed.

The system of kinetic equations for q^0_i and q^1_i is now complete. It can be simplified even further as the active neutrinos are well in thermal equilibrium at all temperatures we are interested it, $\gamma_\nu/H \gg 1$, where H is the Hubble rate. Again, we take Q^1_1 as an example. The equation for it now reads

$$i\dot{q}^1_1 = -(E_N(k) - (E_\nu(k) + i\gamma_\nu/2))q^1_1 + \Sigma_i C_i q^0_i, \tag{9}$$

and has an approximate slow varying solution

$$q^1_1 = \frac{\Sigma_i C_i q^0_i}{(E_N(k) - (E_\nu(k) + i\gamma_\nu/2))}. \tag{10}$$

All q^1_i can be found in this way and inserted into equation (3) for q^0_i. As a result, we get the kinetic description in terms of q^0_i only.

The realisation of this program requires a straightforward but tedious computation, which we have done with the use of DiracQ Mathematica package [46]. By introducing the notations

$$\rho_{\nu_\alpha} = \mathrm{Tr}[a^\dagger_{\nu_\alpha}(k)a_{\nu_\alpha}(k)\rho] - \rho^{eq}_\nu, \tag{11}$$

$$\rho_{\bar{\nu}_\alpha} = \mathrm{Tr}[b^\dagger_{\nu_\alpha}(k)b_{\nu_\alpha}(k)\rho] - \rho^{eq}_\nu, \tag{12}$$

$$\rho_N = \begin{pmatrix} \mathrm{Tr}[a^\dagger_+(k)a_+(k)\rho] & \mathrm{Tr}[a^\dagger_+(k)b_-(k)\rho] \\ \mathrm{Tr}[b^\dagger_-(k)a_+(k)\rho] & \mathrm{Tr}[b^\dagger_-(k)b_-(k)\rho] \end{pmatrix} - \rho^{eq}_N\mathbf{1}, \tag{13}$$

$$\rho_{\bar{N}} = \begin{pmatrix} \mathrm{Tr}[a^\dagger_-(k)a_-(k)\rho] & \mathrm{Tr}[a^\dagger_-(k)b_+(k)\rho] \\ \mathrm{Tr}[b^\dagger_+(k)a_-(k)\rho] & \mathrm{Tr}[b^\dagger_+(k)b_+(k)\rho] \end{pmatrix} - \rho^{eq}_N\mathbf{1}, \tag{14}$$

where ρ^{eq}_ν and ρ^{eq}_N are equilibrium distribution functions of neutrinos and HNLs, and "**1**" is the unity matrix, we arrived to the following result[5]

$$i\frac{d\rho_{\nu_\alpha}}{dt} = -i\Gamma_{\nu_\alpha}\rho_{\nu_\alpha} + i\,\mathrm{Tr}[\tilde{\Gamma}_{\nu_\alpha}\rho_{\bar{N}}], \tag{15}$$

$$i\frac{d\rho_{\bar{\nu}_\alpha}}{dt} = -i\Gamma^*_{\nu_\alpha}\rho_{\bar{\nu}_\alpha} + i\,\mathrm{Tr}[\tilde{\Gamma}^*_{\nu_\alpha}\rho_N], \tag{16}$$

$$i\frac{d\rho_N}{dt} = [H_N, \rho_N] - \frac{i}{2}\{\Gamma_N, \rho_N\} + i\sum_\alpha \tilde{\Gamma}^\alpha_N\rho_{\bar{\nu}_\alpha}, \tag{17}$$

$$i\frac{d\rho_{\bar{N}}}{dt} = [H^*_N, \rho_{\bar{N}}] - \frac{i}{2}\{\Gamma^*_N, \rho_{\bar{N}}\} + i\sum_\alpha (\tilde{\Gamma}^\alpha_N)^*\rho_{\nu_\alpha}. \tag{18}$$

The effective Hamiltonian is

$$H_N = H_0 + H_I, \tag{19}$$

$$H_0 = -\frac{\Delta M M}{E_N}\sigma_1 \tag{20}$$

$$H_I = h_+\sum_\alpha Y^N_{+,\alpha} + h_-\sum_\alpha Y^N_{-,\alpha}, \tag{21}$$

where σ_1 is the Pauli matrix. The production rates for HNLs are

[5] We do not account here the expansion of the universe, but it can be easily accommodated.

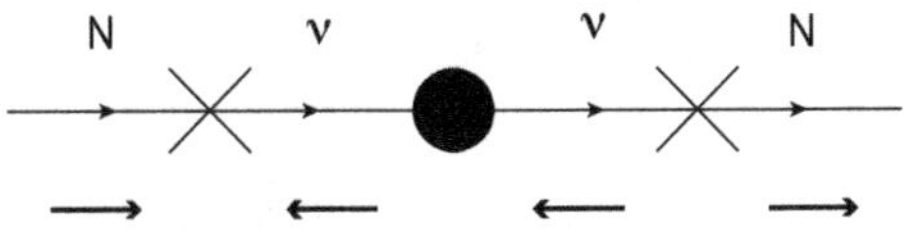

(a) Fermion number conserving process

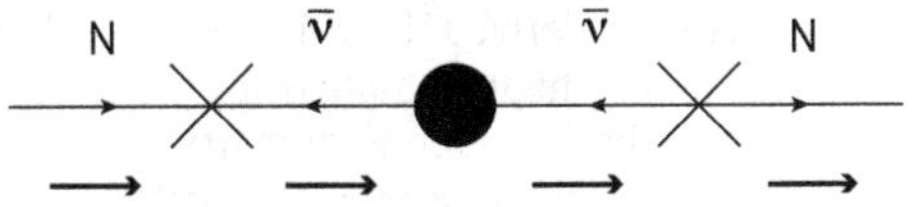

(b) Fermion number violating process

Fig. 1. Diagrammatical descriptions of fermion number conserving and violating processes. The arrows under particle lines show the direction of momentum, and the black spot indicates interactions in plasma.

$$\Gamma_N = \Gamma_+ + \Gamma_-, \tag{22}$$

$$\Gamma_+ = \gamma_+ \sum_\alpha Y^N_{+,\alpha}, \tag{23}$$

$$\Gamma_- = \gamma_- \sum_\alpha Y^N_{-,\alpha}, \tag{24}$$

$$\tilde{\Gamma}^\alpha_N = -\gamma_+ Y^N_{+,\alpha} + \gamma_- Y^N_{-,\alpha}, \tag{25}$$

and those for active neutrinos are

$$\Gamma_{\nu_\alpha} = (\gamma_+ + \gamma_-) \sum_I h_{\alpha I} h^*_{\alpha I}, \tag{26}$$

$$\tilde{\Gamma}_{\nu_\alpha} = -\gamma^\nu_{+,\alpha} Y^\nu_{+,\alpha} + \gamma^\nu_{-,\alpha} Y^\nu_{-,\alpha}. \tag{27}$$

The coefficients are given by

$$h_+ = \frac{2\langle\Phi\rangle^2 E_\nu (E_N + k)(E_N + E_\nu)}{k E_N (4(E_N + E_\nu)^2 + \gamma_\nu^2)}, \tag{28}$$

$$h_- = \frac{2\langle\Phi\rangle^2 E_\nu (E_N - k)(E_N - E_\nu)}{k E_N (4(E_N - E_\nu)^2 + \gamma_\nu^2)}, \tag{29}$$

$$\gamma_+ = \frac{2\langle\Phi\rangle^2 E_\nu (E_N + k)\gamma_\nu}{k E_N (4(E_N + E_\nu)^2 + \gamma_\nu^2)}, \tag{30}$$

$$\gamma_- = \frac{2\langle\Phi\rangle^2 E_\nu (E_N - k)\gamma_\nu}{k E_N (4(E_N - E_\nu)^2 + \gamma_\nu^2)}, \tag{31}$$

and the matrices of Yukawa coupling constants are

$$Y^N_{+,\alpha} = \begin{pmatrix} h_{\alpha 3}h^*_{\alpha 3} & -h_{\alpha 3}h^*_{\alpha 2} \\ -h_{\alpha 2}h^*_{\alpha 3} & h_{\alpha 2}h^*_{\alpha 2} \end{pmatrix}, \tag{32}$$

$$Y^N_{-,\alpha} = \begin{pmatrix} h_{\alpha 2}h^*_{\alpha 2} & -h_{\alpha 3}h^*_{\alpha 2} \\ -h_{\alpha 2}h^*_{\alpha 3} & h_{\alpha 3}h^*_{\alpha 3} \end{pmatrix}, \tag{33}$$

$$Y^\nu_{+,\alpha} = \begin{pmatrix} h_{\alpha 3}h^*_{\alpha 3} & -h_{\alpha 2}h^*_{\alpha 3} \\ -h_{\alpha 3}h^*_{\alpha 2} & h_{\alpha 2}h^*_{\alpha 2} \end{pmatrix}, \tag{34}$$

$$Y^\nu_{-,\alpha} = \begin{pmatrix} h_{\alpha 2}h^*_{\alpha 2} & -h_{\alpha 2}h^*_{\alpha 3} \\ -h_{\alpha 3}h^*_{\alpha 2} & h_{\alpha 3}h^*_{\alpha 3} \end{pmatrix}, \tag{35}$$

where $E_\nu = k - b_L$ and function b_L is often called neutrino potential in the medium. It has been computed in a number of papers in different limits [47,48]. The neutrino damping rate as well as b_L can be taken from a recent work [32].

The parts of the Hamiltonian and production rates with subscript "+" and "−" are associated with the fermion number conserving and violating operators, respectively. In Fig. 1 they are expressed diagrammatically, where the vertexes (denoted by the cross) in (a) come from the structures in H_{int} containing the product of two creation (or annihilation) operators, as for example in (36)

$$\begin{aligned} &h_{\alpha 2}\, a_+(k)\, b_\nu(-k)\, e^{-i(E_N+E_\nu)t}, \\ &h^*_{\alpha 2}\, b^\dagger_\nu(-k)\, a^\dagger_+(k)\, e^{i(E_N+E_\nu)t}, \end{aligned} \tag{36}$$

and those in (b) coming from a product of creation and annihilation operators as, for instance:

$$\begin{aligned} &h^*_{\alpha 3}\, a_+(k)\, b_\nu(k)^\dagger\, e^{-i(E_N-E_\nu)t}, \\ &h_{\alpha 3}\, b_\nu(k)\, a^\dagger_+(k)\, e^{i(E_N-E_\nu)t}. \end{aligned} \tag{37}$$

The structure of the kinetic equations is exactly the same as it was first found in [5] and elucidated in [11],[6] but now with all the terms expressed explicitly through parameters of the theory. The same set of equations (15)–(18) was used in [10,12–20], for analysis of baryon asymmetry generation in the νMSM, but with different choices of kinetic coefficients. The novel results of our work are the formulas for the Hamiltonian and different rates (28)–(33) neatly separating the effects of the processes with and without fermion number non-conservation, the goal which was attempted already in [11] but not achieved at that time.

Now, we make these equations more realistic, accounting for the presence of charged fermions in the plasma, equilibrium character of electroweak reactions, and eventually sphaleron transitions. For this end we introduce leptonic numbers ΔL_α of every generation, being a sum of asymmetries in neutrinos and charged leptons, integrated over momentum, and consider $\Delta_\alpha = \Delta L_\alpha - 1/3\Delta B$, where ΔB is the baryon asymmetry. Due to weak interactions the asymmetries in neutrinos are rapidly transferred to charge leptons and distributed among different momenta, whereas when sphalerons are operating there is a rapid transfer of lepton number to baryons. However, the rate of Δ_α change is proportional to HNL Yukawa couplings and corresponds to a slow process. A well defined procedure accounting for equilibrium character of weak reactions and electric neutrality of the plasma [41–44] allows to write kinetic equations for Δ_α instead of ρ_{ν_α} and $\rho_{\bar{\nu}_\alpha}$:

$$\begin{aligned} i\frac{d\Delta_\alpha}{dt} &= -i\left[\frac{12}{T^3}\int \frac{d^3k}{(2\pi)^3}\Gamma_{\nu_\alpha} f_f(1-f_f)\right]\omega_{\alpha\beta}\Delta_\beta \\ &\quad + i\int \frac{d^3k}{(2\pi)^3}\left[\mathrm{Tr}[\tilde{\Gamma}_{\nu_\alpha}\rho_{\bar{N}}] - \mathrm{Tr}[\tilde{\Gamma}^*_{\nu_\alpha}\rho_N]\right], \end{aligned} \tag{38}$$

$$\begin{aligned} i\frac{d\rho_N}{dt} &= [H_N, \rho_N] - \frac{i}{2}\{\Gamma_N, \rho_N\} \\ &\quad - \frac{i}{2}\sum_\alpha \tilde{\Gamma}^\alpha_N \omega_{\alpha\beta}\left[\frac{12\Delta_\beta}{T^3} f_f(1-f_f)\right], \end{aligned} \tag{39}$$

$$\begin{aligned} i\frac{d\rho_{\bar{N}}}{dt} &= [H^*_N, \rho_{\bar{N}}] - \frac{i}{2}\{\Gamma^*_N, \rho_{\bar{N}}\} \\ &\quad + \frac{i}{2}\sum_\alpha (\tilde{\Gamma}^\alpha_N)^* \omega_{\alpha\beta}\left[\frac{12\Delta_\beta}{T^3} f_f(1-f_f)\right], \end{aligned} \tag{40}$$

where $f_f = 1/(e^{k/T}+1)$ is the distribution function for massless fermion and $\omega_{\alpha\beta}$ is the susceptibility matrix. For the Higgs phase, when sphalerons are out of thermal equilibrium, it is given by

$$\omega_{\alpha\beta} = \frac{1}{207}\begin{pmatrix} 79 & 10 & 10 \\ 10 & 79 & 10 \\ 10 & 10 & 79 \end{pmatrix}. \tag{41}$$

[6] The kinetic equations derived in the seminal work [9] unfortunately are not correct as they do not contain the transfer (last) terms in Eqs. (17), (18) and the equations for lepton asymmetries (15), (16).

Note that Δ_α are the momentum-independent quantities (neutrinos and charged leptons of a given generation are well in thermal equilibrium and thus their number densities are described well by the Fermi distribution with the chemical potentials for lepton numbers), but the density matrix for HNLs does depend on k.

Yet another effect coming from charged leptons is the flavour dependence of the neutrino dispersion relations and neutrino damping rates. It can be neglected for temperatures exceeding the τ-lepton mass.

These completes the discussion of kinetic equations for HNLs and active flavours deeply in the Higgs phase, where the main contribution to active-sterile transition comes from the mixing terms contained in $\mathcal{L}_{int}$.

At the temperatures in the region of the electroweak crossover the structure of evolution equations remains the same, but a number of kinetic coefficients has to be modified. In particular, one has to add the "direct processes" (in terminology of Ref. [32]) involving HNLs and active neutrinos. These processes occur with fermion number conservation and their contribution does not vanish when $\langle\Phi\rangle \to 0$. In fact, they dominate in baryogenesis around the sphaleron freeze-out in a part of the νMSM parameter space. The account for them results in the following modifications:

$$h_+ = \mathcal{K}(m_h)\frac{T^2}{8k} + \frac{2\langle\Phi\rangle^2 E_\nu(E_N+k)(E_N+E_\nu)}{kE_N(4(E_N+E_\nu)^2+\gamma_\nu^2)}, \tag{42}$$

$$\gamma_+ = \gamma_+^{\text{direct}} + \frac{2\langle\Phi\rangle^2 E_\nu(E_N+k)\gamma_\nu}{kE_N(4(E_N+E_\nu)^2+\gamma_\nu^2)}, \tag{43}$$

where

$$\gamma_+^{\text{direct}} = \mathcal{K}(m_h)\frac{1}{E_N}\text{Im}\,\Pi_R + \gamma_{ph}, \tag{44}$$

$$\mathcal{K}(m_h) = \frac{3}{\pi^2 T^3}\int_0^\infty dp\, p^2 f_b(E_h)(1+f_b(E_h)), \tag{45}$$

$$\gamma_{ph} = \frac{1}{E_N}\frac{m_h^2 T}{32\pi k}\ln\left\{\frac{1+e^{-\frac{m_h^2}{4kT}}}{1-e^{-\frac{1}{T}\left(k+\frac{m_h^2}{4k}\right)}}\right\}, \tag{46}$$

and $\text{Im}\,\Pi_R$ is the rate of the direct production of HNLs in the symmetric phase, which comes mainly from $2 \leftrightarrow 2$ interactions. This contribution to γ_+ dies out in the Higgs phase; this is accounted for by a function $\mathcal{K}(m_h)$ [32]. The contribution to γ_+ from the Higgs decay to $N\nu$ is given by γ_{ph} [32], $f_b(\epsilon) = 1/(e^{\epsilon/T}-1)$ is the bosonic distribution function.

In the symmetric phase these expressions simplify a lot:

$$h_+ = \frac{T^2}{8k}, \tag{47}$$

$$h_- = 0, \tag{48}$$

$$\gamma_+ = \frac{1}{E_N}\text{Im}\,\Pi_R, \tag{49}$$

$$\gamma_- = 0, \tag{50}$$

where h_+ is nothing but the Weldon high-temperature correction [49].

The last point is the modification of the susceptibility matrix in the symmetric phase, where sphalerons are in thermal equilibrium[7]:

[7] The kinetic equations (15)–(18) should be modified and supplemented by an extra equation for baryon number when the rate of sphaleron transitions becomes smaller than the HNL equilibration rates [50].

$$\omega_{\alpha\beta} = \frac{1}{333}\begin{pmatrix} 107 & -4 & -4 \\ -4 & 107 & -4 \\ -4 & -4 & 107 \end{pmatrix}. \tag{51}$$

This formula is extracted from [41]. For determination of precise temperature dependence of $\omega_{\alpha\beta}$ over the electroweak crossover see Ref. [32]. The relation between Ξ defined in Ref. [32] and our $\omega_{\alpha\beta}$ reads $\omega_{\alpha\beta} = \frac{1}{6}\Xi$.

The set of equations derived in this section allows to follow the system from very high to sufficiently small temperatures $T \simeq 1$ GeV, and address both baryon asymmetry generation around 130 GeV and late time lepton asymmetry production. At even smaller temperature one should take into account the flavour dependence of neutrino degrees of freedom in the medium, and include the decays and inverse decays of HNL which were omitted in our equations.

3. Thermal equilibrium and approximately conserved numbers

The kinetic equations (15)–(18) allow to address the question of existence of approximately conserved quantum numbers. Let us consider two combinations of the HNL and active flavour asymmetries,

$$L_\pm = \Delta_N \mp \sum_\alpha \Delta_\alpha\,, \tag{52}$$

where

$$\Delta_N = \left[\int \frac{d^3k}{(2\pi)^3}\text{Tr}(\rho_N - \rho_{\bar{N}})\right]. \tag{53}$$

Making the corresponding linear combinations of Eqs. (15)–(18), we find:

$$\begin{aligned}\frac{d}{dt}L_- = -2\int\frac{d^3k}{(2\pi)^3}\gamma_-\sum_\alpha\Big[&h_{\alpha2}h_{\alpha2}^*(\rho_{N,11}-\rho_{\bar{N},11})\\ &+h_{\alpha3}h_{\alpha3}^*(\rho_{N,22}-\rho_{\bar{N},22})\\ &-2\text{Re}(h_{\alpha2}h_{\alpha3}^*)(\text{Re}\rho_{N,12}-\text{Re}\rho_{\bar{N},12})\\ &+2\text{Im}(h_{\alpha2}h_{\alpha3}^*)(\text{Im}\rho_{N,12}+\text{Im}\rho_{\bar{N},12})\\ &+(h_{\alpha2}h_{\alpha2}^*+h_{\alpha3}h_{\alpha3}^*)\,\omega_{\alpha\beta}\left[\frac{12\Delta_\beta}{T^3}f_f(1-f_f)\right]\Big],\end{aligned} \tag{54}$$

$$\begin{aligned}\frac{d}{dt}L_+ = -2\int\frac{d^3k}{(2\pi)^3}\gamma_+\sum_\alpha\Big[&h_{\alpha2}h_{\alpha2}^*(\rho_{N,11}-\rho_{\bar{N},11})\\ &+h_{\alpha3}h_{\alpha3}^*(\rho_{N,22}-\rho_{\bar{N},22})\\ &-2\text{Re}(h_{\alpha2}h_{\alpha3}^*)(\text{Re}\rho_{N,12}-\text{Re}\rho_{\bar{N},12})\\ &+2\text{Im}(h_{\alpha2}h_{\alpha3}^*)(\text{Im}\rho_{N,12}+\text{Im}\rho_{\bar{N},12})\\ &-(h_{\alpha2}h_{\alpha2}^*+h_{\alpha3}h_{\alpha3}^*)\,\omega_{\alpha\beta}\left[\frac{12\Delta_\beta}{T^3}f_f(1-f_f)\right]\Big].\end{aligned} \tag{55}$$

The remarkable property of these relations is that the rates of the $L_\pm$ change is proportional to the corresponding $\gamma_\pm$ functions, which have very different behaviours as a function of temperature. In particular, the rate associated with γ_+ may never come into thermal equilibrium.

We plot the ratio of the different HNL production rates to the Hubble rate H in Figs. 2 and 3 for the normal hierarchy (NH) case and in Figs. 4 and 5 for the inverted hierarchy (IH). The $\Gamma_\pm$ rates defined in Eqs. (23), (24) are 2×2 matrices, so we diagonalise each of them at any given temperature and denote by $\Gamma_{\pm,i}$ with $i = 1, 2$

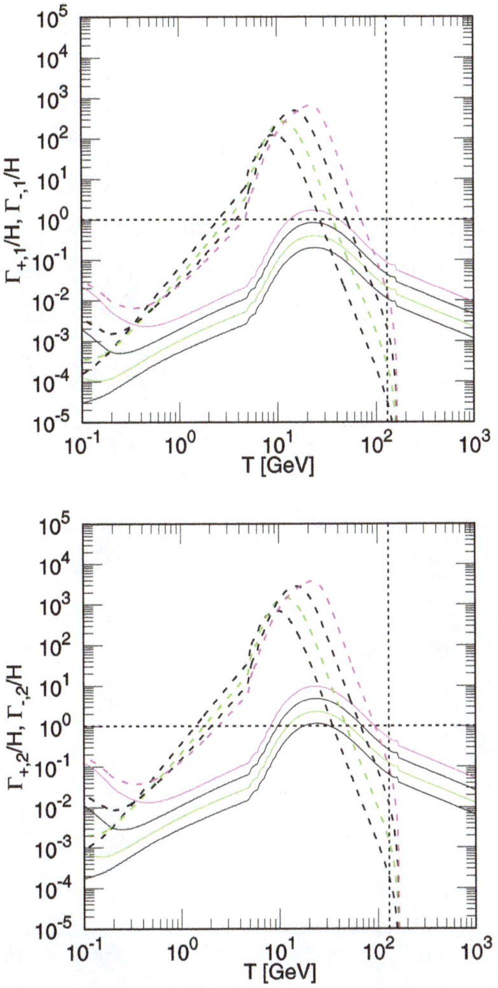

Fig. 2. Two eigen-values of momentum-averaged rates, Γ_+ (solid) and Γ_- (dashed), for $X_\omega = 1$ in NH case. Red, green, blue and magenta lines correspond to $M = 0.5, 1, 2$ and 4 GeV, respectively. In this and all subsequent figures the vertical black dotted line shows the sphaleron freeze-out temperature 130 GeV, and on the horizontal black dotted line $\Gamma = H$. (For interpretation of the references to colour in this figure legend, the reader is referred to the web version of this article.)

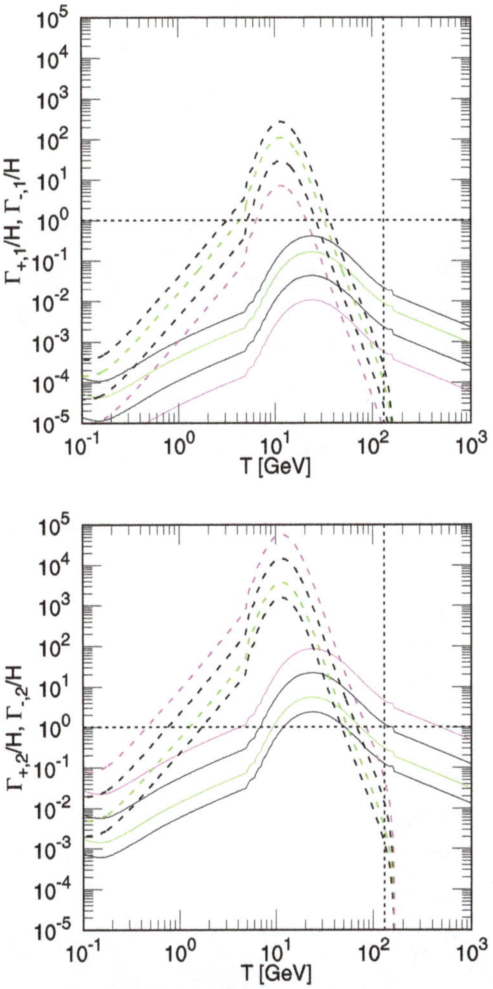

Fig. 3. Two eigen-values of momentum-averaged rates, Γ_+ (solid) and Γ_- (dashed), for $M = 1$ GeV in NH case. Red, green, blue and magenta lines correspond to $X_\omega = 1, 2, 4$ and 8, respectively. (For interpretation of the references to colour in this figure legend, the reader is referred to the web version of this article.)

marking the corresponding eigen-values. In addition, we average the rates over momentum with the use of the Fermi distribution function, f_F,

$$\langle \Gamma \rangle = \frac{1}{n_F} \int \frac{d^3k}{(2\pi)^3} f_F(k) \Gamma(k) \,, \tag{56}$$

where n_F is the fermion number density.

The rates depend on quite a number of parameters, the most important being the HNL mass M and the imaginary part of a complex mixing angle ω appearing in the Casas–Ibarra parametrisation of the HNL-neutrino mixing matrix [51]. The quantity $X_\omega \equiv \exp(\mathrm{Im}\,\omega)$ shown in the figures is a free parameter which does not change the active neutrino masses (we fix them with the available neutrino data). The amplitude of Yukawa couplings scales as $F_{\alpha I} \propto X_\omega$ for large $\mathrm{Im}\omega > 0$ or $F_{\alpha I} \propto X_\omega^{-1}$ for $\mathrm{Im}\omega < 0$. The parameter X_ω is related to the introduced previously ϵ in [10,11] as $\epsilon = 1/X_\omega^2$. More specifically, the relation between Yukawa couplings and neutrino parameters, when ΔM is negligibly small, is given by

$$\sum_\alpha h_{\alpha 2} h^*_{\alpha 2} = \frac{(m_3 + m_2)M}{2v^2} X_\omega^{-2}, \tag{57}$$

$$\sum_\alpha h_{\alpha 3} h^*_{\alpha 3} = \frac{(m_3 + m_2)M}{2v^2} X_\omega^{2}, \tag{58}$$

$$\sum_\alpha \mathrm{Re}\,(h_{\alpha 2} h^*_{\alpha 3}) = \frac{(m_3 - m_2)M}{2v^2} \cos(2\mathrm{Re}\,\omega), \tag{59}$$

$$\sum_\alpha \mathrm{Im}\,(h_{\alpha 2} h^*_{\alpha 3}) = \frac{(m_3 - m_2)M}{2v^2} \sin(2\mathrm{Re}\,\omega), \tag{60}$$

where m_3 and m_2 are the heaviest and second-heaviest active neutrino masses and $v = 174.1$ GeV is the Higgs expectation value at zero temperature. It is the largest eigenvalue of $\Gamma_\pm$ which determines the approach to thermal equilibrium for quantum numbers $L_\pm$. In general, the equilibration rates are larger for larger HNL masses, and smaller if X_ω is close to one.

In Figs. 2 and 4 we show the dependence of the rates on temperature for different HNL masses.

In Figs. 3 and 5 the behaviour of the rates as a function of temperature for different values of X_ω is shown. The large (or small) X_ω increases the magnitude of Yukawa couplings, leading to faster equilibration.

Let us comment on the splitting between the eigen-values of the rates. From Eqs. (57)–(60) for large (or small) X_ω the hierarchy between the diagonal components of the rates (23), (24), (32), (33) gets large, which is reflected in the splitting of eigen-values. Note

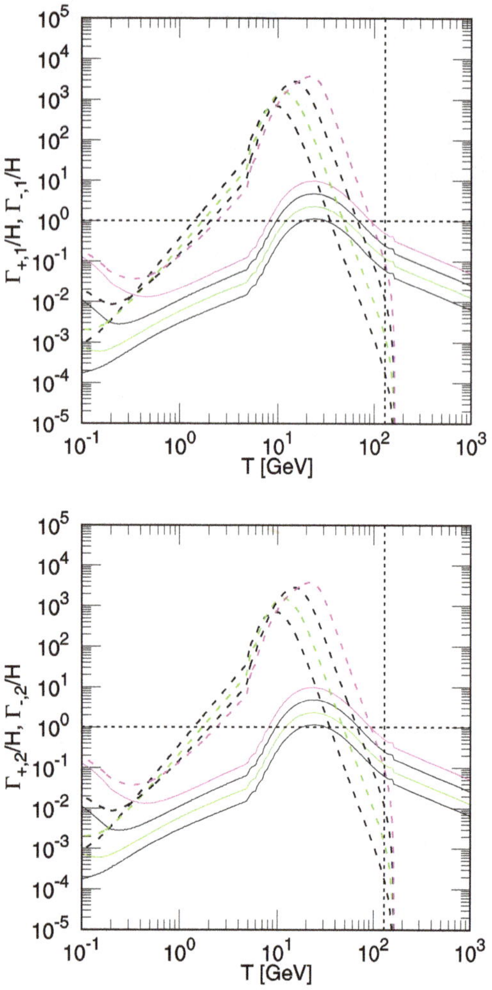

Fig. 4. Two eigen-values of momentum-averaged rates, Γ_+ (solid) and Γ_- (dashed), for $X_\omega = 1$ in IH case. Red, green, blue and magenta lines correspond to $M = 0.5, 1, 2$ and 4 GeV, respectively. (For interpretation of the references to colour in this figure legend, the reader is referred to the web version of this article.)

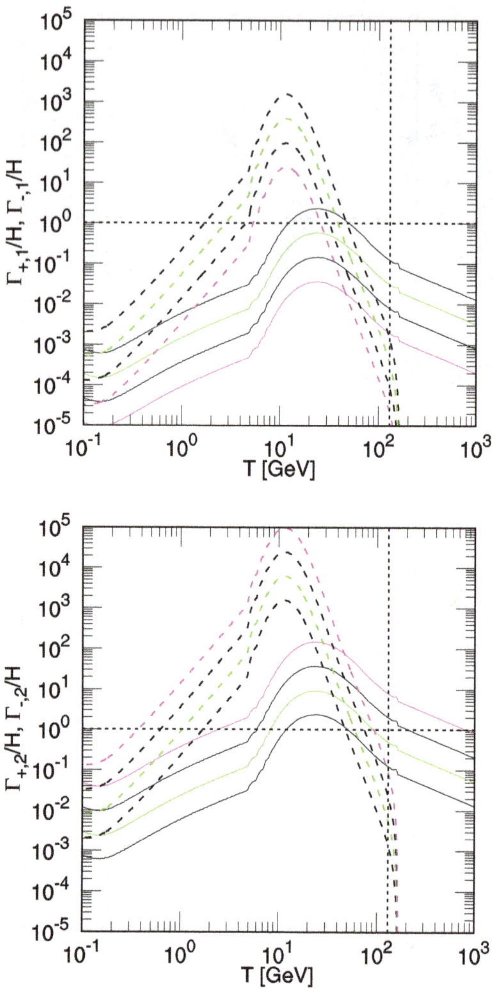

Fig. 5. Two eigen-values of momentum-averaged rates, Γ_+ (solid) and Γ_- (dashed), for $M = 1$ GeV in IH case. Red, green, blue and magenta lines correspond to $X_\omega = 1, 2, 4$ and 8, respectively. (For interpretation of the references to colour in this figure legend, the reader is referred to the web version of this article.)

that the off-diagonal components of them are independent on X_ω. The splitting is clearly visible in the NH case even if $X_\omega \simeq 1$ due to the presence of non-diagonal terms (59), (60) which are of the order of atmospheric neutrino mass scale $m_{\text{atm}} \simeq 5 \times 10^{-2}$ eV. For the IH case the eigen-values of the rates at $X_\omega \simeq 1$ are almost degenerate due to the smallness of the off-diagonal components suppressed by $(m_{\text{sol}}/m_{\text{atm}})^2$ where the solar neutrino mass scale is $m_{\text{sol}} \simeq 9 \times 10^{-3}$ eV.

In the symmetric phase at $T > 160$ GeV, $\Gamma_- = 0$ and the lepton number L_- is conserved. The generated asymmetry in the active neutrino sector is the same as that in the HNL sector with an opposite sign. In the temperature range relevant for baryogenesis, i.e. above the sphaleron freeze-out $T \simeq 130$ GeV but below $T = 160$ GeV both Γ_- and Γ_+ rates are present. However, the inspection of the figures shows that the rate Γ_+ dominates over Γ_-, meaning that the discussion above is approximately valid, with small corrections from "-" contributions. As a result the previous computations of the baryon asymmetry performed in [5,10–20] neglecting Γ_- are legitimate.

When the temperature goes down below $T \simeq 130$ GeV the generation of the baryon asymmetry stops but of the lepton asymmetry continues [11]. Eventually, the rate Γ_- starts to dominate over Γ_+. The HNLs enter in thermal equilibrium at temperature T_{in} corresponding to the intersection of the largest rate with the horizontal line $\Gamma/H = 1$, and go out of thermal equilibrium at $T = T_{\text{out}} < T_{\text{in}}$ found in a similar way. Typically, for the set of masses considered, the highest rate is achieved at $T \simeq 10$–20 GeV. The maximum of the rate Γ_- always exceeds the rate of the Universe expansion, as has been already demonstrated in [11,16,31, 32].

The most interesting range of parameters that can potentially lead to generation of large leptonic asymmetries at $T \simeq T_{\text{in}}$ corresponds to relatively small HNL masses and $X_\omega \simeq 1$. Indeed, the Figs. 2–5 show that the maximum of the ratio $\Gamma_{+,2}/H$ does not exceed $\mathcal{O}(1)$ for $M \lesssim 2$ GeV and $X_\omega \simeq 1$. At the same time, this ratio is close to 1 for these parameters, meaning that the large asymmetry in L_+ can potentially be generated but protected from washout until small temperature. The character of equilibrium which is led by the "$-$" reactions at smaller temperatures but has an (effective) L_+ conservation will ensure the relation between the asymmetries in active flavours and HNLs given by[8]

$$\frac{\Delta_N}{\sum_\alpha \Delta_\alpha} \approx -\frac{22}{69}. \tag{61}$$

[8] This is calculated following the similar analysis in [41].

Note that only the asymmetries in active flavours contribute to the resonant DM production. We leave the analysis of produced lepton asymmetries for a future publication.

4. Conclusions

In this paper we have derived kinetic equations for low scale leptogenesis, accounting for helicity effects in HNL interactions with active flavours. They are valid both for high temperatures around the sphaleron freeze-out (to describe the baryogenesis), and for smaller temperatures to account for lepton number generation.

We showed that the total lepton number can be considered as approximately conserved in the certain domain of νMSM parameters, and thus can potentially be generated at $T \simeq T_{\rm in}$ and survive until small temperatures making the resonant production of DM possible. It is still remains to be seen whether sufficiently large lepton asymmetry is indeed generated, as these depends on CP-violating effects. The work in this direction is in progress.

Quite remarkably, the requirement of existence of approximately conserved leptonic number makes the region of small HNL masses preferable, allowing to search for HNLs at SHiP or SHiP-like experiments. At the same time, it lies close to the lower bound coming from the requirement to explain neutrino masses via see-saw, urging to design the experiment in this mass region with highest possible sensitivity.

Acknowledgements

This work was supported by the ERC-AdG-2015 grant 694896. We are grateful to Mikko Laine for many illuminating discussions. We thank A. Boyarsky, O. Ruchayskiy and I. Timiryasov for helpful comments. The work of MS was supported partially by the Swiss National Science Foundation.

Once the current manuscript had been finalised, we thank M. Laine for sharing with us a draft by himself and J. Ghiglieri, in which helicity-flipping and conserving rates similar to ours are discussed at $T > 130$ GeV. Their remarks on our manuscript are appreciated.

References

[1] F. Bezrukov, M.Yu. Kalmykov, B.A. Kniehl, M. Shaposhnikov, Higgs boson mass and new physics, J. High Energy Phys. 10 (2012) 140, http://dx.doi.org/10.1007/JHEP10(2012)140, arXiv:1205.2893.

[2] D. Buttazzo, G. Degrassi, P.P. Giardino, G.F. Giudice, F. Sala, A. Salvio, A. Strumia, Investigating the near-criticality of the Higgs boson, J. High Energy Phys. 12 (2013) 089, http://dx.doi.org/10.1007/JHEP12(2013)089, arXiv:1307.3536.

[3] A.V. Bednyakov, B.A. Kniehl, A.F. Pikelner, O.L. Veretin, Stability of the electroweak vacuum: gauge independence and advanced precision, Phys. Rev. Lett. 115 (20) (2015) 201802, http://dx.doi.org/10.1103/PhysRevLett.115.201802, arXiv:1507.08833.

[4] T. Asaka, S. Blanchet, M. Shaposhnikov, The nuMSM, dark matter and neutrino masses, Phys. Lett. B 631 (2005) 151–156, http://dx.doi.org/10.1016/j.physletb.2005.09.070, arXiv:hep-ph/0503065.

[5] T. Asaka, M. Shaposhnikov, The nuMSM, dark matter and baryon asymmetry of the universe, Phys. Lett. B 620 (2005) 17–26, http://dx.doi.org/10.1016/j.physletb.2005.06.020, arXiv:hep-ph/0505013.

[6] K.A. Olive, et al., Review of particle physics, Chin. Phys. C 38 (2014) 090001, http://dx.doi.org/10.1088/1674-1137/38/9/090001.

[7] S. Dodelson, L.M. Widrow, Sterile-neutrinos as dark matter, Phys. Rev. Lett. 72 (1994) 17–20, http://dx.doi.org/10.1103/PhysRevLett.72.17, arXiv:hep-ph/9303287.

[8] X.-D. Shi, G.M. Fuller, A New dark matter candidate: nonthermal sterile neutrinos, Phys. Rev. Lett. 82 (1999) 2832–2835, http://dx.doi.org/10.1103/PhysRevLett.82.2832, arXiv:astro-ph/9810076.

[9] E.K. Akhmedov, V.A. Rubakov, A.Yu. Smirnov, Baryogenesis via neutrino oscillations, Phys. Rev. Lett. 81 (1998) 1359–1362, http://dx.doi.org/10.1103/PhysRevLett.81.1359, arXiv:hep-ph/9803255.

[10] M. Shaposhnikov, A Possible symmetry of the nuMSM, Nucl. Phys. B 763 (2007) 49–59, http://dx.doi.org/10.1016/j.nuclphysb.2006.11.003, arXiv:hep-ph/0605047.

[11] M. Shaposhnikov, The nuMSM, leptonic asymmetries, and properties of singlet fermions, J. High Energy Phys. 08 (2008) 008, http://dx.doi.org/10.1088/1126-6708/2008/08/008, arXiv:0804.4542.

[12] L. Canetti, M. Shaposhnikov, Baryon asymmetry of the universe in the NuMSM, J. Cosmol. Astropart. Phys. 1009 (2010) 001, http://dx.doi.org/10.1088/1475-7516/2010/09/001, arXiv:1006.0133.

[13] T. Asaka, H. Ishida, Flavour mixing of neutrinos and baryon asymmetry of the universe, Phys. Lett. B 692 (2010) 105–113, http://dx.doi.org/10.1016/j.physletb.2010.07.016, arXiv:1004.5491.

[14] T. Asaka, S. Eijima, H. Ishida, Kinetic equations for baryogenesis via sterile neutrino oscillation, J. Cosmol. Astropart. Phys. 1202 (2012) 021, http://dx.doi.org/10.1088/1475-7516/2012/02/021, arXiv:1112.5565.

[15] L. Canetti, M. Drewes, M. Shaposhnikov, Sterile neutrinos as the origin of dark and baryonic matter, Phys. Rev. Lett. 110 (6) (2013) 061801, http://dx.doi.org/10.1103/PhysRevLett.110.061801, arXiv:1204.3902.

[16] L. Canetti, M. Drewes, T. Frossard, M. Shaposhnikov, Dark matter, baryogenesis and neutrino oscillations from right handed neutrinos, Phys. Rev. D 87 (2013) 093006, http://dx.doi.org/10.1103/PhysRevD.87.093006, arXiv:1208.4607.

[17] B. Shuve, I. Yavin, Baryogenesis through neutrino oscillations: a unified perspective, Phys. Rev. D 89 (7) (2014) 075014, http://dx.doi.org/10.1103/PhysRevD.89.075014, arXiv:1401.2459.

[18] A. Abada, G. Arcadi, V. Domcke, M. Lucente, Lepton number violation as a key to low-scale leptogenesis, J. Cosmol. Astropart. Phys. 1511 (11) (2015) 041, http://dx.doi.org/10.1088/1475-7516/2015/11/041, arXiv:1507.06215.

[19] P. Hernández, M. Kekic, J. López-Pavón, J. Racker, J. Salvado, Testable baryogenesis in seesaw models, J. High Energy Phys. 08 (2016) 157, http://dx.doi.org/10.1007/JHEP08(2016)157, arXiv:1606.06719.

[20] M. Drewes, B. Garbrecht, D. Gueter, J. Klaric, Leptogenesis from oscillations of heavy neutrinos with large mixing angles, J. High Energy Phys. 12 (2016) 150, http://dx.doi.org/10.1007/JHEP12(2016)150, arXiv:1606.06690.

[21] F.L. Bezrukov, M. Shaposhnikov, The Standard Model Higgs boson as the inflaton, Phys. Lett. B 659 (2008) 703–706, http://dx.doi.org/10.1016/j.physletb.2007.11.072, arXiv:0710.3755.

[22] F. Bezrukov, D. Gorbunov, M. Shaposhnikov, On initial conditions for the Hot Big Bang, J. Cosmol. Astropart. Phys. 0906 (2009) 029, http://dx.doi.org/10.1088/1475-7516/2009/06/029, arXiv:0812.3622.

[23] J. Garcia-Bellido, D.G. Figueroa, J. Rubio, Preheating in the Standard Model with the Higgs-inflaton coupled to gravity, Phys. Rev. D 79 (2009) 063531, http://dx.doi.org/10.1103/PhysRevD.79.063531, arXiv:0812.4624.

[24] F. Bezrukov, J. Rubio, M. Shaposhnikov, Living beyond the edge: Higgs inflation and vacuum metastability, Phys. Rev. D 92 (8) (2015) 083512, http://dx.doi.org/10.1103/PhysRevD.92.083512, arXiv:1412.3811.

[25] V.A. Kuzmin, V.A. Rubakov, M.E. Shaposhnikov, On the anomalous electroweak baryon number nonconservation in the early universe, Phys. Lett. B 155 (1985) 36, http://dx.doi.org/10.1016/0370-2693(85)91028-7.

[26] K. Abazajian, G.M. Fuller, M. Patel, Sterile neutrino hot, warm, and cold dark matter, Phys. Rev. D 64 (2001) 023501, http://dx.doi.org/10.1103/PhysRevD.64.023501, arXiv:astro-ph/0101524.

[27] T. Asaka, M. Laine, M. Shaposhnikov, On the hadronic contribution to sterile neutrino production, J. High Energy Phys. 06 (2006) 053, http://dx.doi.org/10.1088/1126-6708/2006/06/053, arXiv:hep-ph/0605209.

[28] T. Asaka, M. Laine, M. Shaposhnikov, Lightest sterile neutrino abundance within the nuMSM, J. High Energy Phys. 01 (2007), http://dx.doi.org/10.1007/JHEP02(2015)028, Erratum; J. High Energy Phys. 02 (2015) 028, arXiv:hep-ph/0612182.

[29] M. Laine, M. Shaposhnikov, Sterile neutrino dark matter as a consequence of nuMSM-induced lepton asymmetry, J. Cosmol. Astropart. Phys. 0806 (2008) 031, http://dx.doi.org/10.1088/1475-7516/2008/06/031, arXiv:0804.4543.

[30] J. Ghiglieri, M. Laine, Improved determination of sterile neutrino dark matter spectrum, J. High Energy Phys. 11 (2015) 171, http://dx.doi.org/10.1007/JHEP11(2015)171, arXiv:1506.06752.

[31] I. Ghisoiu, M. Laine, Right-handed neutrino production rate at $T > 160$ GeV, J. Cosmol. Astropart. Phys. 1412 (12) (2014) 032, http://dx.doi.org/10.1088/1475-7516/2014/12/032, arXiv:1411.1765.

[32] J. Ghiglieri, M. Laine, Neutrino dynamics below the electroweak crossover, J. Cosmol. Astropart. Phys. 1607 (07) (2016) 015, http://dx.doi.org/10.1088/1475-7516/2016/07/015, arXiv:1605.07720.

[33] A. Boyarsky, O. Ruchayskiy, D. Iakubovskyi, A lower bound on the mass of dark matter particles, J. Cosmol. Astropart. Phys. 0903 (2009) 005, http://dx.doi.org/10.1088/1475-7516/2009/03/005, arXiv:0808.3902.

[34] D. Gorbunov, A. Khmelnitsky, V. Rubakov, Constraining sterile neutrino dark matter by phase-space density observations, J. Cosmol. Astropart. Phys. 0810 (2008) 041, http://dx.doi.org/10.1088/1475-7516/2008/10/041, arXiv:0808.3910.

[35] A. Roy, M. Shaposhnikov, Resonant production of the sterile neutrino dark matter and fine-tunings in the νMSM, Phys. Rev. D 82 (2010) 056014, http://dx.doi.org/10.1103/PhysRevD.82.056014, arXiv:1006.4008.

[36] A. Blondel, E. Graverini, N. Serra, M. Shaposhnikov, Search for Heavy Right Handed Neutrinos at the FCC-ee, in: 37th International Conference on High Energy Physics, ICHEP 2014, Valencia, Spain, July 2–9, 2014, arXiv:1411.5230.

[37] M. Shaposhnikov, I. Tkachev, The nuMSM, inflation, and dark matter, Phys. Lett. B 639 (2006) 414–417, http://dx.doi.org/10.1016/j.physletb.2006.06.063, arXiv:hep-ph/0604236.

[38] A. Kusenko, Sterile neutrinos, dark matter, and the pulsar velocities in models with a Higgs singlet, Phys. Rev. Lett. 97 (2006) 241301, http://dx.doi.org/10.1103/PhysRevLett.97.241301, arXiv:hep-ph/0609081.

[39] F. Bezrukov, D. Gorbunov, M. Shaposhnikov, Late and early time phenomenology of Higgs-dependent cutoff, J. Cosmol. Astropart. Phys. 1110 (2011) 001, http://dx.doi.org/10.1088/1475-7516/2011/10/001, arXiv:1106.5019.

[40] T. Hambye, D. Teresi, Higgs doublet decay as the origin of the baryon asymmetry, Phys. Rev. Lett. 117 (9) (2016) 091801, http://dx.doi.org/10.1103/PhysRevLett.117.091801, arXiv:1606.00017.

[41] S.Yu. Khlebnikov, M.E. Shaposhnikov, Melting of the Higgs vacuum: conserved numbers at high temperature, Phys. Lett. B 387 (1996) 817–822, http://dx.doi.org/10.1016/0370-2693(96)01116-1, arXiv:hep-ph/9607386.

[42] M. Laine, M.E. Shaposhnikov, A remark on sphaleron erasure of baryon asymmetry, Phys. Rev. D 61 (2000) 117302, http://dx.doi.org/10.1103/PhysRevD.61.117302, arXiv:hep-ph/9911473.

[43] E. Nardi, Y. Nir, E. Roulet, J. Racker, The importance of flavor in leptogenesis, J. High Energy Phys. 01 (2006) 164, http://dx.doi.org/10.1088/1126-6708/2006/01/164, arXiv:hep-ph/0601084.

[44] D. Bodeker, M. Laine, Kubo relations and radiative corrections for lepton number washout, J. Cosmol. Astropart. Phys. 1405 (2014) 041, http://dx.doi.org/10.1088/1475-7516/2014/05/041, arXiv:1403.2755.

[45] G. Sigl, G. Raffelt, General kinetic description of relativistic mixed neutrinos, Nucl. Phys. B 406 (1993) 423–451, http://dx.doi.org/10.1016/0550-3213(93)90175-O.

[46] J.G. Wright, B.S. Shastry, DiracQ: a quantum many-body physics package, arXiv:1301.4494.

[47] D. Notzold, G. Raffelt, Neutrino dispersion at finite temperature and density, Nucl. Phys. B 307 (1988) 924, http://dx.doi.org/10.1016/0550-3213(88)90113-7.

[48] J. Morales, C. Quimbay, F. Fonseca, Fermionic dispersion relations at finite temperature and nonvanishing chemical potentials in the minimal standard model, Nucl. Phys. B 560 (1999) 601–616, http://dx.doi.org/10.1016/S0550-3213(99)00459-9, arXiv:hep-ph/9906207.

[49] H.A. Weldon, Effective fermion masses of order gT in high temperature gauge theories with exact chiral invariance, Phys. Rev. D 26 (1982) 2789, http://dx.doi.org/10.1103/PhysRevD.26.2789.

[50] Y. Burnier, M. Laine, M. Shaposhnikov, Baryon and lepton number violation rates across the electroweak crossover, J. Cosmol. Astropart. Phys. 0602 (2006) 007, http://dx.doi.org/10.1088/1475-7516/2006/02/007, arXiv:hep-ph/0511246.

[51] J.A. Casas, A. Ibarra, Oscillating neutrinos and $\mu \to e\gamma$, Nucl. Phys. B 618 (2001) 171–204, http://dx.doi.org/10.1016/S0550-3213(01)00475-8, arXiv:hep-ph/0103065.

6

Generalized TMDs and the exclusive double Drell–Yan process

Shohini Bhattacharya [a], Andreas Metz [a,*], Jian Zhou [b]

[a] Department of Physics, SERC, Temple University, Philadelphia, PA 19122, USA
[b] School of Physics and Key Laboratory of Particle Physics and Particle Irradiation (MOE), Shandong University, Jinan, Shandong 250100, China

ARTICLE INFO

Editor: J.-P. Blaizot

ABSTRACT

Generalized transverse momentum dependent parton distributions (GTMDs) are the most general parton correlation functions of hadrons. By considering the exclusive double Drell–Yan process it is shown for the first time how quark GTMDs can be measured. Specific GTMDs can be addressed by means of polarization observables.

1. Introduction

Multi-dimensional imaging of strongly interacting systems is currently a very active research area. The key quantities of this field are new types of parton distribution functions (PDFs) which are extensions of the one-dimensional PDFs that became textbook material – generalized parton distributions (GPDs) and transverse momentum dependent parton distributions (TMDs), which provide multi-dimensional images of hadrons in position space and momentum space, respectively. Studying GPDs and TMDs is a core mission at several particle accelerator facilities worldwide and, in particular, at a potential future electron-ion collider (EIC) [1,2].

In this context, GTMDs [3–5] have recently attracted enormous interest. Since several GTMDs reduce to GPDs and TMDs in certain kinematical limits, they are often denoted as "mother distributions." The Fourier transform of GTMDs are Wigner functions [6,7], the quantum-mechanical counterpart of classical phase-space distributions. Partonic Wigner functions contain information on the five-dimensional parton structure – the (average) longitudinal and transverse momentum as well as transverse position of partons inside a hadron [8]. Two of the GTMDs – $F_{1,4}$ and $G_{1,1}$ in the notation of [4] – play a crucial role for the nucleon spin structure. Both functions describe the strength of spin-orbit interactions that are similar to spin-orbit interactions in atomic systems like hydrogen [9,10]. In particular, there is a direct relation between $F_{1,4}$ and the orbital angular momentum (OAM) of partons inside a longitudinally polarized nucleon [9,11,12]. It is remarkable that the same relation between $F_{1,4}$ and the quark OAM holds for both commonly used OAM definitions – the (canonical) one by Jaffe and Manohar (L_{JM}) [13], and the one by Ji (L_{Ji}) [14]. This representation of OAM also allows for an intuitive interpretation of the difference between L_{JM} and L_{Ji} [15]. Moreover, it gives access to the so far elusive L_{JM} in quantum chromodynamics (QCD) on the lattice [11,16,17].

While a number of model calculations of GTMDs is available by now [3,4,8,9,18–29], for many years it was unknown how GTMDs can be measured. Only recently it was shown that GTMDs of gluons can, in principle, be accessed via diffractive di-jet production in deep-inelastic lepton-nucleon and lepton-nucleus scattering [30–32], as well as virtual photon-nucleus quasi-elastic scattering [27]. Some numerical studies of gluon GTMDs at small x, based on a saturation model, were performed in Refs. [25,27]. Not long ago, it was also pointed out that gluon GTMDs can be studied in proton-nucleus collisions [33]. With the exception of [32], the papers on observables for GTMDs deal with the small-x region of parton saturation.

In this work we identify, for the first time, a physical process which gives access to quark GTMDs. Specifically, we show how GTMDs enter the exclusive pion-nucleon double Drell–Yan process, $\pi N \to (\ell_1^- \ell_1^+)(\ell_2^- \ell_2^+)N'$, where one detects two di-lepton pairs plus a nucleon. To this end, we perform a leading-order (LO) analysis in perturbative QCD. Our main focus is on the GTMDs $F_{1,4}$ and $G_{1,1}$, which can be measured through suitable polarization observables. We also argue that other quark GTMDs could be systematically studied in the same process.

2. Generalized TMDs

Let us briefly recall the definition of quark GTMDs for a nucleon [3,4], to the extent it is necessary for the present work.

* Corresponding author.
E-mail address: metza@temple.edu (A. Metz).

GTMDs parameterize the off-forward transverse momentum dependent quark–quark correlator

$$W^{q\,[\Gamma]}_{\lambda,\lambda'}(P,\Delta,x,\vec{k}_\perp) = \int \frac{dz^- \, d^2\vec{z}_\perp}{2(2\pi)^3} \times e^{ik\cdot z} \langle p',\lambda' | \bar{q}(-\tfrac{z}{2})\,\Gamma\,\mathcal{W}(-\tfrac{z}{2},\tfrac{z}{2})\,q(\tfrac{z}{2})\,|p,\lambda\rangle \Big|_{z^+=0}, \quad (1)$$

where q indicates the quark flavor and Γ a generic gamma matrix. The 4-momenta and the helicities of the incoming (outgoing) nucleon are denoted by $p(p')$ and $\lambda(\lambda')$, respectively. We also use the definitions $P=(p+p')/2$ and $\Delta = p'-p$. The two quark fields of the operator in (1) are separated along the light-cone minus direction z^- and the transverse direction $\vec{z}_\perp$. (We define the light-cone components of a generic 4-vector $a=(a^0,a^1,a^2,a^3)$ through $a^\pm = (a^0 \pm a^3)/\sqrt{2}$ and $\vec{a}_\perp = (a^1,a^2)$.) The Wilson line $\mathcal{W}$ makes the bi-local operator color gauge invariant. The average longitudinal and transverse quark momenta are given by x and $\vec{k}_\perp$, respectively. We also point out that, strictly speaking, some modification of the definition in (1) is needed in order to avoid the infamous light-cone singularities. One way of regulating such singularities is to invoke the scheme proposed in Ref. [34], which is widely used in the TMD case. More information on this point, which is irrelevant for the main purpose of the present work, can be found in [24] and references therein.

Here we need the parametrization of (1) in terms of GTMDs for $\Gamma = \gamma^+, \gamma^+\gamma_5$. In the notation of [4] they read

$$\begin{aligned}
W^{q\,[\gamma^+]}_{\lambda,\lambda'} &= \frac{1}{2M}\,\bar{u}(p',\lambda')\Big[F^q_{1,1} + \frac{i\sigma^{i+}k^i_\perp}{P^+}F^q_{1,2} + \frac{i\sigma^{i+}\Delta^i_\perp}{P^+}F^q_{1,3} + \frac{i\sigma^{ij}k^i_\perp\Delta^j_\perp}{M^2}F^q_{1,4}\Big]u(p,\lambda) \\
&= \frac{1}{M\sqrt{1-\xi^2}}\bigg\{\Big[M\delta_{\lambda,\lambda'} - \frac{1}{2}\Big(\lambda\Delta^1_\perp + i\Delta^2_\perp\Big)\delta_{\lambda,-\lambda'}\Big]F^q_{1,1} \\
&\quad + (1-\xi^2)\Big(\lambda k^1_\perp + ik^2_\perp\Big)\delta_{\lambda,-\lambda'}F^q_{1,2} \\
&\quad + (1-\xi^2)\Big(\lambda\Delta^1_\perp + i\Delta^2_\perp\Big)\delta_{\lambda,-\lambda'}F^q_{1,3} \\
&\quad + \frac{i\varepsilon^{ij}_\perp k^i_\perp\Delta^j_\perp}{M^2}\Big[\lambda M\delta_{\lambda,\lambda'} - \frac{\xi}{2}\Big(\Delta^1_\perp + i\lambda\Delta^2_\perp\Big)\delta_{\lambda,-\lambda'}\Big]F^q_{1,4}\bigg\},
\end{aligned} \quad (2)$$

$$\begin{aligned}
W^{q\,[\gamma^+\gamma_5]}_{\lambda,\lambda'} &= \frac{1}{2M}\,\bar{u}(p',\lambda')\Big[-\frac{i\varepsilon^{ij}_\perp k^i_\perp\Delta^j_\perp}{M^2}G^q_{1,1} + \frac{i\sigma^{i+}\gamma_5 k^i_\perp}{P^+}G^q_{1,2} + \frac{i\sigma^{i+}\gamma_5\Delta^i_\perp}{P^+}G^q_{1,3} + i\sigma^{+-}\gamma_5 G^q_{1,4}\Big]u(p,\lambda) \\
&= \frac{1}{M\sqrt{1-\xi^2}}\bigg\{-\frac{i\varepsilon^{ij}_\perp k^i_\perp\Delta^j_\perp}{M^2}\Big[M\delta_{\lambda,\lambda'} - \frac{1}{2}\Big(\lambda\Delta^1_\perp + i\Delta^2_\perp\Big)\delta_{\lambda,-\lambda'}\Big]G^q_{1,1} \\
&\quad + (1-\xi^2)\Big(k^1_\perp + i\lambda k^2_\perp\Big)\delta_{\lambda,-\lambda'}G^q_{1,2} \\
&\quad + (1-\xi^2)\Big(\Delta^1_\perp + i\lambda\Delta^2_\perp\Big)\delta_{\lambda,-\lambda'}G^q_{1,3} \\
&\quad + \Big[\lambda M\delta_{\lambda,\lambda'} - \frac{\xi}{2}\Big(\Delta^1_\perp + i\lambda\Delta^2_\perp\Big)\delta_{\lambda,-\lambda'}\Big]G^q_{1,4}\bigg\}.
\end{aligned} \quad (3)$$

In order to evaluate the first lines of Eqs. (2), (3) we considered $u(p,\lambda)$ and $u(p',\lambda')$ as light-cone helicity spinors [35,36]. Note that M is the nucleon mass, and $\xi = (p^+ - p'^+)/(p^+ + p'^+) = -\Delta^+/(2P^+)$ characterizes the longitudinal momentum transfer to the nucleon. We also use $\sigma^{\mu\nu} = i[\gamma^\mu,\gamma^\nu]/2$, and $\varepsilon^{ij}_\perp = \varepsilon^{-+ij}$ with $\varepsilon^{0123}=1$. The kinematical arguments on the l.h.s. of (2), (3) are suppressed. For a generic GTMD one has $X = X(x,\xi,\vec{k}_\perp,\vec{\Delta}_\perp)$, where the dependence on $\vec{k}_\perp$ and $\vec{\Delta}_\perp$ is through the scalar products which can be formed by these vectors. We also recall that, in general, GTMDs are complex-valued functions [3,4]. For the reasons given above our main focus will be on the GTMDs $F_{1,4}$ and $G_{1,1}$. The real part of the GTMDs $F_{1,1}$ and $G_{1,4}$ has a close connection to the distribution of unpolarized quarks in an unpolarized nucleon and the distribution of longitudinally polarized quarks in a longitudinally polarized nucleon, respectively [4,9]. Since these distributions are large we will also consider observables which are sensitive to their interference with $F_{1,4}$ and $G_{1,1}$. Below we will concentrate on the helicity-conserving terms in (2), (3) that are proportional to $\delta_{\lambda,\lambda'}$.

The cross section for the double Drell–Yan process is also sensitive to the matrix element

$$\Phi^q(x,\vec{k}^2_\perp) = \int \frac{dz^+\,d^2\vec{z}_\perp}{2(2\pi)^3} e^{i(k-p/2)\cdot z} \times \langle 0|\,\bar{q}(-\tfrac{z}{2})\,\gamma^-\gamma_5\,\mathcal{W}(-\tfrac{z}{2},\tfrac{z}{2})\,q(\tfrac{z}{2})\,|\pi(p)\rangle\Big|_{z^-=0}. \quad (4)$$

Modulo pre-factors, $\Phi^q(x,\vec{k}^2_\perp)$ is the light-cone wave function of the pion [36,37]. The double Drell–Yan process implies in both Eq. (1) and Eq. (4) a staple-like past-pointing Wilson line [38], identical to the one that appears in TMD factorization of the ordinary Drell–Yan process [39–41].

3. Double Drell–Yan process and polarization observables

To calculate observables we consider the production of two virtual photons rather than two di-lepton pairs. Specifically, we study the process

$$\pi(p_b) + N(p_a,\lambda_a) \to \gamma_1^*(q_1,\lambda_1) + \gamma_2^*(q_2,\lambda_2) + N'(p'_a,\lambda'_a). \quad (5)$$

From here on the variables of the incoming and outgoing nucleon carry an index a compared to above. We concentrate on large $s=(p_a+p_b)^2 \approx 2p_a^+p_b^-$, large photon virtualities q_1^2, q_2^2, and small transverse photon momenta, $|\vec{q}^{\,2}_{i\perp}| \ll q_i^2$. In this region one can use TMD-type factorization. The longitudinal momentum transfer to the nucleon can be written as $\xi_a = (q_1^+ + q_2^+)/(2P_a^+)$. The LO diagrams for this process are shown in Fig. 1. The scattering amplitude depends on the helicities of the nucleons and photons,

$$\mathcal{T}^{\lambda_1,\lambda_2}_{\lambda_a,\lambda'_a} = \mathcal{T}^{\mu\nu}_{\lambda_a,\lambda'_a}\,\varepsilon^*_\mu(\lambda_1)\,\varepsilon^*_\nu(\lambda_2), \quad (6)$$

where $\varepsilon^\mu(\lambda_1)$ and $\varepsilon^\mu(\lambda_2)$ are the photon polarization vectors. One finds

$$\begin{aligned}
\mathcal{T}^{\mu\nu}_{\lambda_a,\lambda'_a} &= i\sum_{q,q'} e_q e'_q e^2 \frac{1}{N_c}\int d^2\vec{k}_{a\perp}\int d^2\vec{k}_{b\perp} \\
&\quad \times \delta^{(2)}\Big(\frac{\Delta\vec{q}_\perp}{2} - \vec{k}_{a\perp} - \vec{k}_{b\perp}\Big)\Phi^{q'q}_\pi(x_b,\vec{k}^2_{b\perp}) \\
&\quad \times \Big[-i\varepsilon^{\mu\nu}_\perp\Big(W^{qq'\,[\gamma^+]}_{\lambda_a,\lambda'_a}(x_a,\vec{k}_{a\perp}) - W^{qq'\,[\gamma^+]}_{\lambda_a,\lambda'_a}(-x_a,-\vec{k}_{a\perp})\Big) \\
&\quad - g^{\mu\nu}_\perp\Big(W^{qq'\,[\gamma^+\gamma_5]}_{\lambda_a,\lambda'_a}(x_a,\vec{k}_{a\perp}) + W^{qq'\,[\gamma^+\gamma_5]}_{\lambda_a,\lambda'_a}(-x_a,-\vec{k}_{a\perp})\Big)\Big],
\end{aligned} \quad (7)$$

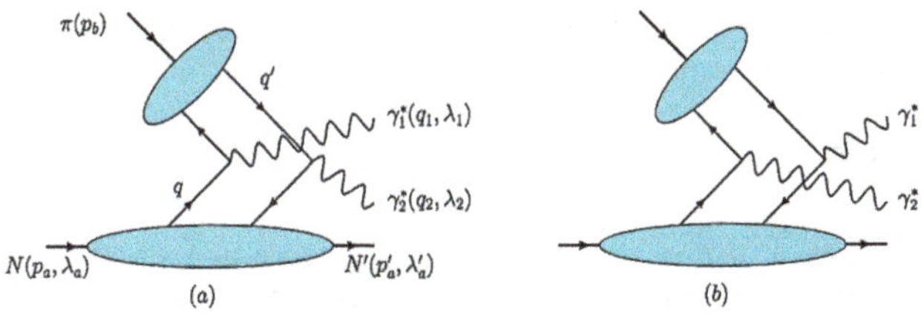

Fig. 1. LO diagrams for the exclusive double Drell-Yan process $\pi N \to \gamma_1^* \gamma_2^* N'$.

where e_q and e'_q are the quark charges in units of the elementary charge e, and N_c is the number of quark colors. The expression in (7) describes the double Drell-Yan process for all possible pion and nucleon charge states. Note that $\Phi_\pi^{q'q}$ is defined as in (4), but with the operator $\bar{q}'\gamma^-\gamma_5 q$. Isospin symmetry provides $\Phi_{\pi^+}^{du} = \Phi_{\pi^-}^{ud} = \sqrt{2}\,\Phi_{\pi^0}^{uu} = -\sqrt{2}\,\Phi_{\pi^0}^{dd}$. Likewise, $W^{qq'[\Gamma]}$ is given by (1) with the operator $\bar{q}\,\Gamma\, q'$. With this notation one can also describe transitions between different nucleons. Like in the case of transition GPDs, for the GTMDs one has $X_{p\to n}^{du} = X_{n\to p}^{ud} = X_p^u - X_p^d$ [42]. In Eq. (7) we use the vector $\Delta\vec{q}_\perp = \vec{q}_{1\perp} - \vec{q}_{2\perp}$. The transverse momenta of the photons can be expressed by $\Delta\vec{q}_\perp$ and the transverse momentum transfer to the nucleon $\vec{\Delta}_{a\perp} = -(\vec{q}_{1\perp} + \vec{q}_{2\perp})$. While the amplitude contains an integration upon the transverse momenta of the quarks, their longitudinal momenta are fixed according to $x_a = (q_1^+ - q_2^+)/(2P_a^+)$, $x_b = 1 - q_1^-/p_b^- = q_2^-/p_b^-$. The value for x_a implies the so-called ERBL region [43,44], characterized by $-\xi_a \le x_a \le \xi_a$, in which the GTMD matrix element describes the emission of a quark-antiquark pair from the nucleon. The amplitude, *a priori*, depends on both the $F_{1,i}$ and the $G_{1,i}$ $(i = 1, \ldots, 4)$. From (7) one readily sees that the dominant contribution to the amplitude is for transversely polarized photons. In this context note that $g_\perp^{\mu\nu} = g^{\mu\nu} - n_a^\mu n_b^\nu - n_a^\nu n_b^\mu$, with the light-like vectors $n_a = (1, 0, 0, -1)/\sqrt{2}$, $n_b = (1, 0, 0, 1)/\sqrt{2}$.

The relation between the scattering amplitude in (6) and the cross section in the center-of-mass frame reads

$$d\sigma_{\lambda_a,\lambda'_a}^{\lambda_1,\lambda_2} = \frac{\pi}{2s^{3/2}}\frac{1+\xi_a}{1-\xi_a}|\mathcal{T}_{\lambda_a,\lambda'_a}^{\lambda_1,\lambda_2}|^2 \delta(p_a'^0 + q_1^0 + q_2^0 - \sqrt{s}) \times \frac{d^4 q_1}{(2\pi)^4}\frac{d^4 q_2}{(2\pi)^4}, \tag{8}$$

where we have already integrated over the phase space of the outgoing nucleon. Below we consider the unpolarized cross section, single-spin asymmetries (SSAs), and double-spin asymmetries (DSAs). It is convenient to introduce

$$\tau_{UU} = \frac{1}{2}\sum_{\lambda,\lambda'}|\mathcal{T}_{\lambda,\lambda'}|^2, \tag{9}$$

$$\tau_{LU} = \frac{1}{2}\sum_{\lambda'}\left(|\mathcal{T}_{+,\lambda'}|^2 - |\mathcal{T}_{-,\lambda'}|^2\right), \tag{10}$$

$$\tau_{LL} = \frac{1}{2}\Big(\big(|\mathcal{T}_{+,+}|^2 - |\mathcal{T}_{+,-}|^2\big) - \big(|\mathcal{T}_{-,+}|^2 - |\mathcal{T}_{-,-}|^2\big)\Big), \tag{11}$$

where summation over the photon polarizations is implied. Obviously, τ_{LU} determines the numerator of the longitudinal target SSA, whereas τ_{LL} describes the longitudinal DSA with polarization of both the target and the recoil nucleon. Spin asymmetries for transverse polarization in the x-direction or y-direction are defined accordingly.

In order to get direct access to $F_{1,4}$, that is, without interference with other GTMDs, one has to consider a linear combination of (polarization) observables,

$$\frac{1}{4}(\tau_{UU} + \tau_{LL} - \tau_{XX} - \tau_{YY}) = \frac{2}{M^4}(\varepsilon_\perp^{ij}\Delta q_\perp^i \Delta_{a\perp}^j)^2\, C^{(+)}\Big[\vec{\beta}_\perp\cdot\vec{k}_{a\perp}\, F_{1,4}\,\Phi_\pi\Big] \times C^{(+)}\Big[\vec{\beta}_\perp\cdot\vec{k}_{a\perp}\, F_{1,4}^*\,\Phi_\pi^*\Big] + 2\,C^{(+)}\Big[G_{1,4}\,\Phi_\pi\Big]C^{(+)}\Big[G_{1,4}^*\,\Phi_\pi^*\Big]. \tag{12}$$

In Eq. (12) we use the shorthand notation

$$C^{(\pm)}\Big[w(\vec{k}_{a\perp},\vec{k}_{b\perp})\, X\, \Phi_\pi\Big] = \frac{e^2}{\sqrt{1-\xi_a^2}\,N_c}\sum_{q,q'} e_q e'_q \int d^2\vec{k}_{a\perp}\int d^2\vec{k}_{b\perp} \times \delta^{(2)}\Big(\frac{\Delta\vec{q}_\perp}{2} - \vec{k}_{a\perp} - \vec{k}_{b\perp}\Big)w(\vec{k}_{a\perp},\vec{k}_{b\perp}) \times \Big[X^{qq'}(x_a,\vec{k}_{a\perp}) \pm X^{qq'}(-x_a,-\vec{k}_{a\perp})\Big]\Phi_\pi^{q'q}(x_b,\vec{k}_{b\perp}^2), \tag{13}$$

with $w(\vec{k}_{a\perp},\vec{k}_{b\perp})$ a generic weight function. The vector $\vec{\beta}_\perp$ in (12) reads

$$\vec{\beta}_\perp = \frac{\vec{\Delta}_{a\perp}^2\,\Delta\vec{q}_\perp - (\vec{\Delta}_{a\perp}\cdot\Delta\vec{q}_\perp)\,\vec{\Delta}_{a\perp}}{\vec{\Delta}_{a\perp}^2\,\Delta\vec{q}_\perp^{\,2} - (\vec{\Delta}_{a\perp}\cdot\Delta\vec{q}_\perp)^2}. \tag{14}$$

We repeat that in order to obtain Eq. (12) the photon polarizations have been summed over. While in that case there is no interference between $F_{1,4}$ and other GTMDs, one still has a second term which is given by $G_{1,4}$. As already mentioned, $G_{1,4}$ is presumably large, and therefore it may actually be difficult to address $F_{1,4}$ through this observable, unless one has a reliable estimate of $G_{1,4}$. However, one can separate the two contributions in (12) by not summing over the photon polarizations. For instance, if one projects on appropriate linear polarizations of the photons, the contributions of either $F_{1,4}$ or $G_{1,4}$ can be switched off [38]. This result, which holds irrespective of the polarization states of the nucleons, follows from the expression in (7). To address $G_{1,1}$ one can study $\frac{1}{4}(\tau_{UU} + \tau_{LL} + \tau_{XX} + \tau_{YY})$. The result for this linear combination is identical to (12), but with the replacements $F_{1,4} \to G_{1,1}$ and $G_{1,4} \to F_{1,1}$. Again, the contributions from $G_{1,1}$ and $F_{1,1}$ can be separated by measuring suitable photon polarizations.

Apart from the fact that a considerable number of different polarization measurements is required, the observable in (12), as well as the corresponding observable for $G_{1,1}$, may have a drawback: In these linear combinations one has cancellations of potentially large terms [38]. Specifically, for instance the individual polarization observables entering (12) have terms proportional to $F_{1,1}\,F_{1,1}^*$, which can be expected to be large (see the discussion in the paragraph after Eq. (3)). It may therefore be beneficial to also explore interference between $F_{1,4}$ (or $G_{1,1}$) and other GTMDs. Such an interference shows up in the following linear combination of longitudinal SSAs:

$$\frac{1}{2}(\tau_{LU} + \tau_{UL}) = \frac{1}{2}\left(|\mathcal{T}_{+,+}|^2 - |\mathcal{T}_{-,-}|^2\right) = \frac{4}{M^2}\,\varepsilon_\perp^{ij}\Delta q_\perp^i \Delta_{a\perp}^j\,\mathrm{Im}\Big\{C^{(-)}\Big[F_{1,1}\,\Phi_\pi\Big]C^{(+)}\Big[\vec{\beta}_\perp\cdot\vec{k}_{a\perp}\,F_{1,4}^*\,\Phi_\pi^*\Big] - C^{(+)}\Big[G_{1,4}\,\Phi_\pi\Big]C^{(-)}\Big[\vec{\beta}_\perp\cdot\vec{k}_{a\perp}\,G_{1,1}^*\,\Phi_\pi^*\Big]\Big\}. \tag{15}$$

We point out that the expressions for τ_{LU} or τ_{UL} alone are more complicated as they contain additional GTMDs [38]. More polarization observables exist which involve interference between $F_{1,4}$ (or $G_{1,1}$) and other GTMDs, but the observable in (15) gives the simplest expression [38]. Note that on the r.h.s. of (15) the imaginary part of products of GTMDs appears. According to current knowledge the GTMDs most relevant for the spin structure of the nucleon are $\mathrm{Re}\,F_{1,4}$ and $\mathrm{Re}\,G_{1,1}$. Though these functions contribute to (15) they interfere with $\mathrm{Im}\,F_{1,4}$ and $\mathrm{Im}\,G_{1,1}$, respectively. At present, there exists no information on the latter functions, and they may in fact be small. This issue can be overcome by considering the observable $\frac{1}{2}(\tau_{XY} - \tau_{YX})$, whose result agrees with the r.h.s. of (15) but with Re{...} instead of Im{...}. Finally, we repeat that $F_{1,1}$ (and $G_{1,4}$) are presumably large. Therefore, a term of the form $F_{1,1}\,F^*_{1,4}$ can be expected to be larger than the term $F_{1,4}\,F^*_{1,4}$ which appears in Eq. (12). Presently available model calculations for GTMDs do not allow one to quantify this statement in the kinematical (ERBL) region of interest.

4. Conclusions

We have shown that GTMDs for quarks can be studied through the exclusive double Drell–Yan process. Specifically, to leading order in perturbative QCD, this process in sensitive to GTMDs in the ERBL region. The main focus was on the GTMDs $F_{1,4}$ and $G_{1,1}$ which recently attracted much attention because of their relation to the spin structure of the nucleon. The double Drell–Yan process leads to a staple-like Wilson line for the operator definition of the GTMDs [38], providing the connection to the canonical OAM (L_{JM}) [11]. We have proposed several polarization observables which can give access to these GTMDs, either directly or through interference with other GTMDs. In a similar manner, all leading-twist GTMDs could be explored via suitable polarization observables [38].

Several extensions of our work can be envisioned. An attempt should be made to numerically estimate both the unpolarized cross section and the various spin asymmetries to find out if the reaction $\pi N \to (\ell_1^- \ell_1^+)(\ell_2^- \ell_2^+)N'$ is measurable at existing facilities. We repeat that the exclusive double Drell–Yan reaction is the first known exclusive process involving the nucleon which is directly sensitive to transverse quark momenta. It therefore holds promise to give experimental access to the so far elusive GTMDs for quarks. We also note that one can perform a similar analysis for nucleon-nucleon collisions [38]. Production of heavy gauge bosons instead of photons may be considered as well. Moreover, hadronic final states typically give rise to higher count rates. One such example is the process $pp \to \eta_c \eta_c pp$, which can basically be treated along the lines discussed above, though gluon GTMDs enter the leading-order analysis [38]. One might expect that arguments similar to the ones used to justify factorization for inclusive double-parton scattering (see, e.g., Refs. [45–47]) apply to these proposed processes. We finally point out that the type of reactions discussed here could also provide constraints on GPDs in the ERBL region, where experimental information is still sparse – see Refs. [48–50] for related work on GPDs.

Acknowledgements

This work has been supported by the National Science Foundation under Contract No. PHY-1516088 (A.M.), the National Science Foundation of China under Grant No. 11675093, and by the Thousand Talents Plan for Young Professionals (J.Z.). The work of A.M. was supported by the U.S. Department of Energy, Office of Science, Office of Nuclear Physics, within the framework of the TMD Topical Collaboration.

References

[1] D. Boer, et al., arXiv:1108.1713 [nucl-th].
[2] A. Accardi, et al., Eur. Phys. J. A 52 (2016) 268, arXiv:1212.1701 [nucl-ex].
[3] S. Meissner, A. Metz, M. Schlegel, K. Goeke, J. High Energy Phys. 0808 (2008) 038, arXiv:0805.3165 [hep-ph].
[4] S. Meissner, A. Metz, M. Schlegel, J. High Energy Phys. 0908 (2009) 056, arXiv:0906.5323 [hep-ph].
[5] C. Lorcé, B. Pasquini, J. High Energy Phys. 1309 (2013) 138, arXiv:1307.4497 [hep-ph].
[6] X.-D. Ji, Phys. Rev. Lett. 91 (2003) 062001, arXiv:hep-ph/0304037.
[7] A.V. Belitsky, X.-D. Ji, F. Yuan, Phys. Rev. D 69 (2004) 074014, arXiv:hep-ph/0307383.
[8] C. Lorcé, B. Pasquini, M. Vanderhaeghen, J. High Energy Phys. 1105 (2011) 041, arXiv:1102.4704 [hep-ph].
[9] C. Lorcé, B. Pasquini, Phys. Rev. D 84 (2011) 014015, arXiv:1106.0139 [hep-ph].
[10] C. Lorcé, Phys. Lett. B 735 (2014) 344, arXiv:1401.7784 [hep-ph].
[11] Y. Hatta, Phys. Lett. B 708 (2012) 186, arXiv:1111.3547 [hep-ph].
[12] X.-D. Ji, X. Xiong, F. Yuan, Phys. Rev. Lett. 109 (2012) 152005, arXiv:1202.2843 [hep-ph].
[13] R.L. Jaffe, A. Manohar, Nucl. Phys. B 337 (1990) 509.
[14] X.-D. Ji, Phys. Rev. Lett. 78 (1997) 610, arXiv:hep-ph/9603249.
[15] M. Burkardt, Phys. Rev. D 88 (2013) 014014, arXiv:1205.2916 [hep-ph].
[16] M. Engelhardt, arXiv:1701.01536 [hep-lat].
[17] A. Rajan, A. Courtoy, M. Engelhardt, S. Liuti, Phys. Rev. D 94 (2016) 034041, arXiv:1601.06117 [hep-ph].
[18] C. Lorcé, B. Pasquini, X. Xiong, F. Yuan, Phys. Rev. D 85 (2012) 114006, arXiv:1111.4827 [hep-ph].
[19] K. Kanazawa, C. Lorcé, A. Metz, B. Pasquini, M. Schlegel, Phys. Rev. D 90 (2014) 014028, arXiv:1403.5226 [hep-ph].
[20] A. Mukherjee, S. Nair, V.K. Ojha, Phys. Rev. D 90 (2014) 014024, arXiv:1403.6233 [hep-ph];
A. Mukherjee, S. Nair, V.K. Ojha, Phys. Rev. D 91 (2015) 054018, arXiv:1501.03728 [hep-ph].
[21] Y. Hagiwara, Y. Hatta, Nucl. Phys. A 940 (2015) 158, arXiv:1412.4591 [hep-ph].
[22] C. Lorcé, B. Pasquini, Phys. Rev. D 93 (2016) 034040, arXiv:1512.06744 [hep-ph].
[23] D. Chakrabarti, T. Maji, C. Mondal, A. Mukherjee, Eur. Phys. J. C 76 (2016) 409, arXiv:1601.03217 [hep-ph];
D. Chakrabarti, T. Maji, C. Mondal, A. Mukherjee, Phys. Rev. D 95 (2017) 074028, arXiv:1701.08551 [hep-ph].
[24] M.G. Echevarria, A. Idilbi, K. Kanazawa, C. Lorcé, A. Metz, B. Pasquini, M. Schlegel, Phys. Lett. B 759 (2016) 336, arXiv:1602.06953 [hep-ph].
[25] Y. Hagiwara, Y. Hatta, T. Ueda, Phys. Rev. D 94 (2016) 094036, arXiv:1609.05773 [hep-ph].
[26] T. Gutsche, V.E. Lyubovitskij, I. Schmidt, Eur. Phys. J. C 77 (2017) 86, arXiv:1610.03526 [hep-ph].
[27] J. Zhou, Phys. Rev. D 94 (2016) 114017, arXiv:1611.02397 [hep-ph].
[28] A. Courtoy, A.S. Miramontes, Phys. Rev. D 95 (2017) 014027, arXiv:1611.03375 [hep-ph].
[29] J. More, A. Mukherjee, S. Nair, Phys. Rev. D 95 (2017) 074039, arXiv:1701.00339 [hep-ph].
[30] Y. Hatta, B.W. Xiao, F. Yuan, Phys. Rev. Lett. 116 (2016) 202301, arXiv:1601.01585 [hep-ph].
[31] Y. Hatta, Y. Nakagawa, B.W. Xiao, F. Yuan, Y. Zhao, arXiv:1612.02445 [hep-ph].
[32] X.-D. Ji, F. Yuan, Y. Zhao, arXiv:1612.02438 [hep-ph].
[33] Y. Hagiwara, Y. Hatta, B.W. Xiao, F. Yuan, arXiv:1701.04254 [hep-ph].
[34] J. Collins, Foundations of Perturbative QCD, Cambridge Monographs on Particle Physics, Nuclear Physics and Cosmology, vol. 32, Cambridge University Press, Cambridge, 2013.
[35] D.E. Soper, Phys. Rev. D 5 (1972) 1956.
[36] M. Diehl, Phys. Rep. 388 (2003) 41, arXiv:hep-ph/0307382.
[37] S.J. Brodsky, H.C. Pauli, S.S. Pinsky, Phys. Rep. 301 (1998) 299, arXiv:hep-ph/9705477.
[38] S. Bhattacharya, A. Metz, J. Zhou, in preparation.
[39] J.C. Collins, Phys. Lett. B 536 (2002) 43, arXiv:hep-ph/0204004.
[40] X.-D. Ji, F. Yuan, Phys. Lett. B 543 (2002) 66, arXiv:hep-ph/0206057.
[41] A.V. Belitsky, X.-D. Ji, F. Yuan, Nucl. Phys. B 656 (2003) 165, arXiv:hep-ph/0208038.
[42] L. Mankiewicz, G. Piller, T. Weigl, Phys. Rev. D 59 (1999) 017501, arXiv:hep-ph/9712508.
[43] A.V. Efremov, A.V. Radyushkin, Phys. Lett. B 94 (1980) 245.
[44] G.P. Lepage, S.J. Brodsky, Phys. Lett. B 87 (1979) 359.
[45] M. Diehl, A. Schäfer, Phys. Lett. B 698 (2011) 389, arXiv:1102.3081 [hep-ph].
[46] M. Diehl, D. Ostermeier, A. Schäfer, J. High Energy Phys. 1203 (2012) 089, arXiv:1111.0910 [hep-ph]; Erratum: J. High Energy Phys. 1603 (2016) 001.

[47] M. Diehl, J.R. Gaunt, D. Ostermeier, P. Plößl, A. Schäfer, J. High Energy Phys. 1601 (2016) 076, arXiv:1510.08696 [hep-ph].
[48] E.R. Berger, M. Diehl, B. Pire, Phys. Lett. B 523 (2001) 265, arXiv:hep-ph/0110080.
[49] S.V. Goloskokov, P. Kroll, Phys. Lett. B 748 (2015) 323, arXiv:1506.04619 [hep-ph].
[50] T. Sawada, W.C. Chang, S. Kumano, J.C. Peng, S. Sawada, K. Tanaka, Phys. Rev. D 93 (2016) 114034, arXiv:1605.00364 [nucl-ex].

General No-Scale Supergravity: An $\mathcal{F}$-$SU(5)$ tale

Dingli Hu [a,*], Tianjun Li [b,c], Adam Lux [d], James A. Maxin [d,e], Dimitri V. Nanopoulos [a,f,g]

[a] George P. and Cynthia W. Mitchell Institute for Fundamental Physics and Astronomy, Texas A&M University, College Station, TX 77843, USA
[b] Key Laboratory of Theoretical Physics, Institute of Theoretical Physics, Chinese Academy of Sciences, Beijing 100190, China
[c] School of Physical Sciences, University of Chinese Academy of Sciences, No. 19A Yuquan Road, Beijing 100049, China
[d] Department of Physics and Engineering Physics, The University of Tulsa, Tulsa, OK 74104, USA
[e] Department of Chemistry and Physics, Louisiana State University, Shreveport, LA 71115, USA
[f] Astroparticle Physics Group, Houston Advanced Research Center (HARC), Mitchell Campus, Woodlands, TX 77381, USA
[g] Academy of Athens, Division of Natural Sciences, 28 Panepistimiou Avenue, Athens 10679, Greece

ARTICLE INFO

Editor: M. Cvetič

ABSTRACT

We study the grand unification model flipped $SU(5)$ with additional vector-like particle multiplets, or $\mathcal{F}$-$SU(5)$ for short, in the framework of General No-Scale Supergravity. In our analysis we allow the supersymmetry (SUSY) breaking soft terms to be generically non-zero, thereby extending the phenomenologically viable parameter space beyond the highly constrained one-parameter version of $\mathcal{F}$-$SU(5)$. In this initial inquiry, the mSUGRA/CMSSM SUSY breaking terms are implemented. We find this easing away from the vanishing SUSY breaking terms enables a more broad mass range of vector-like particles, dubbed flippons, including flippons less than 1 TeV that could presently be observed at the LHC2, as well as a lighter gluino mass and SUSY spectrum overall. This presents heightened odds that the General No-Scale $\mathcal{F}$-$SU(5)$ viable parameter space can be probed at the LHC2. The phenomenology comprises both bino and higgsino dark matter, including a Higgs funnel region. Particle states emerging from the SUSY cascade decays are presented to experimentally distinguish amongst the diverse phenomenological regions.

1. Introduction

The second phase of the Large Hadron Collider (LHC) commenced in 2015, seeking to append a discovery of supersymmetry (SUSY) to the 2012 observation of the light CP-even Higgs boson. The ATLAS experiment recorded 36.0 fb^{-1} of data in 2016 at a 13 TeV center-of-mass energy, while the CMS experiment recorded 37.82 fb^{-1}. Given this rapid accumulation of luminosity in 2016 and soon to reenergize in 2017, the supersymmetric model space is expected to be probed beyond a 2 TeV gluino ($\tilde{g}$) mass. The most recently published data statistics from the 2015 LHC1 run collision data of 3.9 fb^{-1} recorded by ATLAS and 3.81 fb^{-1} recorded by CMS provide a lower search bound of about 1.9 TeV on the gluino mass [1], serving as a rather strong constraint on the SUSY model space.

The beauty of supersymmetry lies in its capacity to naturally resolve several fundamental dilemmas, such as stabilization of the electroweak scale (EW), a lightest supersymmetric particle (LSP) that is stable under R-parity serving as a natural dark matter candidate, a radiative EW scale symmetry breaking mechanism, and gauge coupling unification. SUSY thus represents a promising candidate for new physics beyond the Standard Model. The SUSY search at the LHC though has returned null results thus far, with no conclusive signals yet observed. Consequently, given an experimentally measured Higgs boson mass of $m_h = 125.1$ GeV [2,3], a rather heavy light stop ($\tilde{t}_1$) mass, and hence SUSY spectrum overall, is necessary in minimalistic models such as minimal Supergravity (mSUGRA) and the Constrained Minimal Supersymmetric Standard Model (CMSSM) in order to generate the required 1-loop and 2-loop contributions to the Higgs boson mass due to the large top Yukawa coupling. Accordingly, the experimentally viable SUSY spectra of mSUGRA/CMSSM, which are quite heavy, may be beyond the reach of the LHC2.

The GUT model flipped $SU(5)$ with additional vector-like multiplets, or $\mathcal{F}$-$SU(5)$ for short, has been thoroughly examined in the framework of No-Scale Supergravity (SUGRA) [4–8]. In these prior analyses, the strict No-Scale SUGRA boundary conditions $M_{1/2}$ and $M_0 = A_0 = B_\mu = 0$ were applied. Given these rigorous constraints at the unification scale, the vector-like particle (flippon) mass scale M_V, top quark mass m_t, and low energy ratio of Higgs vacuum expectation values (VEVs) $\tan\beta$ can be expressed as a function of the

* Corresponding author.
E-mail address: hudingli@tamu.edu (D. Hu).

sole parameter $M_{1/2}$, thus serving as a true one-parameter model. While these conditions severely constrain the model space, the resulting phenomenology uncovered is quite rich. For instance, the gluino mass scale of $M_{\tilde{g}} \geq 1.9$ TeV currently under probe at the LHC is the precise point in the No-Scale $\mathcal{F}$-$SU(5)$ model space where the light Higgs boson mass enters into its experimentally viable range of $M_h = 125.1 \pm 0.24$ GeV, offering a plausible explanation as to why no discovery of SUSY has yet surfaced [8]. Furthermore, the region of the model space presently being probed by the LHC generates a relic density Ωh^2 within the very narrowly constrained 9-year WMAP and Planck measurements, as well consistency with the latest experimental results of several rare decay processes and proton decay lifetimes [8]. Additionally, adjustments to the one-loop gauge β-function coefficients b_i induced by incorporating vector-like flippon multiplets flattens the SU(3) renormalization group equation (RGE) running ($b_3 = 0$). The effective result of a vanishing b_3 is a lighter gluino mass and lighter spectrum overall, accelerating the LHC reach into the viable parameter space. The net consequence of such strict No-Scale conditions though is a rather massive vector-like flippon mass of $M_V \sim 23$–50 TeV, well beyond the reach of the current LHC and planned future upgrades.

In an effort to search the No-Scale $\mathcal{F}$-$SU(5)$ model space beyond the highly constrained strict No-Scale condition $M_0 = A_0 = B_\mu = 0$, we now implement in this work the *General* No-Scale SUSY breaking terms, allowing the universal scalar mass M_0, trilinear A-term coupling A_0, and bilinear parameter B_μ, which is the supersymmetry breaking soft term for the $\mu H_d H_u$ term in the superpotential, to be generically non-zero. The M_0 and A_0 terms are allowed to freely float, the values of which are solely determined by the viability of the subsequent phenomenology. On the contrary, no constraint or analysis whatsoever is placed on the parameter B_μ. Therefore, our applied SUSY breaking terms are $M_{1/2}$, M_0, and A_0, where we implement the mSUGRA/CMSSM SUSY breaking parameters in this initial General No-Scale study. In contrast to $SU(5)$ with mSUGRA/CMSSM SUSY breaking soft terms, we expect here that the mSUGRA/CMSSM boundary conditions implemented conjointly with $\mathcal{F}$-$SU(5)$ RGE running will allow a lighter gluino mass and SUSY spectrum, while the vector-like flippon Yukawa coupling to the Higgs boson can lift the Higgs mass into its experimentally preferred range, likewise permitting a light and testable SUSY spectrum at the LHC2. Moreover, less constrained SUSY breaking parameters can generate more flexibility on the vector-like mass scale M_V, possibly supporting production of lighter flippon masses at the LHC2.

2. The No-Scale $\mathcal{F}$-$SU(5)$ model

The gauge group for minimal flipped $SU(5)$ model [9–11] is $SU(5) \times U(1)_X$, which may be embedded into the $SO(10)$ model. There are only two other flipped $SU(5)$ models, which are from orbifold compactification [12,13]. We refer the reader to Refs. [4–8] and references therein for a detailed description of the minimal flipped $SU(5)$ model. We introduce here the XF, $\overline{XF}$, Xl, and $\overline{Xl}$ vector-like particles (flippons) of Ref. [8] at the TeV scale to achieve string-scale gauge coupling unification [14–16]. In string models, the masses for vector-like particles cannot be generated at stringy tree level in general since the generic superpotential is a trilinear term. Interestingly though, vector-like particle masses can be generated via instanton effects and then are exponentially suppressed. Therefore, we can obtain vector-like particle masses around the TeV scale naturally by considering proper instanton effects. The alternative method to realize TeV-scale masses for vector-like particles is the Giudice–Masiero mechanism via high-dimensional operators in the Kähler potential [17].

Supersymmetry breaking must occur near the TeV scale given that mass degeneracy of the superpartners has not been observed. Supergravity models, which are GUTs with gravity mediated supersymmetry breaking, can completely characterize the supersymmetry breaking soft terms by four universal parameters (gaugino mass $M_{1/2}$, scalar mass M_0, trilinear soft term A_0, and the low energy ratio of Higgs vacuum expectation values (VEVs) $\tan\beta$), in addition to the sign of the Higgs bilinear mass term μ of the superpotential $\mu H_d H_u$ term.

3. Numerical methodology

The mSUGRA/CMSSM high-energy boundary conditions $M_{1/2}$, M_0, and A_0 are applied at the $M_{\mathcal{F}}$ scale near $M_{\mathcal{F}} \simeq 5 \times 10^{17}$ GeV (as opposed to an application at the traditional GUT scale of about 10^{16} GeV in the MSSM), along with $\tan\beta$, coupled with the vector-like flippon mass decoupling scale M_V. The General No-Scale $\mathcal{F}$-$SU(5)$ parameter space is sampled within the limits $100 \leq M_{1/2} \leq 5000$ GeV, $100 \leq M_0 \leq 5000$ GeV, $-5000 \leq A_0 \leq 5000$ GeV, $2 \leq \tan\beta \leq 65$, and $855 \leq M_V \leq 100{,}000$ GeV. The most recent LHC constraints on vector-like T and B quarks [18] establish lower limits of about 855 GeV for $(XQ,\ XQ^c)$ vector-like flippons and 735 GeV for $(XD,\ XD^c)$ vector-like flippons. Therefore, we set our lower M_V limit at $M_V \geq 855$ GeV given that we employ a universal vector-like flippon decoupling scale. A sufficient range of the top quark mass is allowed around the world average [19], implementing liberal upper and lower limits in our analysis of $171 \leq m_t \leq 175$ GeV. The WMAP 9-year [20] and 2015 Planck [21] relic density measurements are applied, where we constrain the model to be consistent with both data sets and permit the inclusion of multi-component dark matter beyond the neutralino, imposing limits of $\Omega h^2 \leq 0.1300$. Consistency with the most recent LHC gluino search is strictly implemented, imposing a hard lower limit on the gluino mass in the model space of $M_{\tilde{g}} \geq 1.9$ TeV [1]. A lower limit on the light stop mass of $M_{\tilde{t}_1} \geq 900$ GeV [1] is also imposed, though the gluino constraint just noted persists as a much stronger constraint in the $\mathcal{F}$-$SU(5)$ model space.

Our theoretical calculation of the light Higgs boson mass is allowed to float around the experimental central value of $m_h = 125.1$ GeV [2,3], where we employ the larger boundaries of $123 \leq m_h \leq 128$ GeV to account for at least a 2σ experimental uncertainty in addition to a theoretical uncertainty of 1.5 GeV in our calculations. The precise value of the flippon Yukawa coupling is unknown, thus we allow the coupling to span from minimal to maximal in our light Higgs boson mass calculations. At a minimal coupling, our theoretically computed light Higgs boson mass consists of only the 1-loop and 2-loop SUSY contributions, primarily from the coupling to the light stop. This computation must return a value of $m_h \leq 128$ GeV, where a SUSY only contribution to the Higgs mass at this maximum of 128 GeV implies a minimal vector-like flippon contribution. At the maximal coupling, the $(XD,\ XD^c)$ flippon Yukawa coupling is fixed at $Y_{XD} = 0$ and the $(XU,\ XU^c)$ flippon Yukawa coupling is set at $Y_{XU} = 1$, with the $(XD,\ XD^c)$ flippon trilinear coupling A term set at $A_{XD} = 0$ and the $(XU,\ XU^c)$ A term fixed at $A_{XU} = A_U = A_0$ [22,5]. The result of the calculation assuming a maximal coupling must give $m_h \geq 123$ GeV, as this is the maximum Higgs boson mass for any particular point in the model space. Given the intersection of these dual constraints within $123 \leq m_h \leq 128$ GeV on our theoretical computations of the light Higgs boson mass, for each discrete point in the parameter space we simultaneously uncover both the minimally and maximally allowed Higgs boson mass when coupled to the vector-like flippons.

The viable region of the model space is constrained beyond the top quark mass, light Higgs boson mass, and relic density measurements by further application of rare decay and direct dark matter detection experimental results. The rare decay experimental constraints consist of the branching ratio of the rare b-quark decay of $Br(b \to s\gamma) = (3.43 \pm 0.21^{stat} \pm 0.24^{th} \pm 0.07^{sys}) \times 10^{-4}$ [23], the branching ratio of the rare B-meson decay to a dimuon of $Br(B_s^0 \to \mu^+\mu^-) = (2.9 \pm 0.7 \pm 0.29^{th}) \times 10^{-9}$ [24], and the 3σ intervals around the SM value and experimental measurement of the SUSY contribution to the anomalous magnetic moment of the muon of $-17.7 \times 10^{-10} \leq \Delta a_\mu \leq 43.8 \times 10^{-10}$ [25]. Regarding direct dark matter detection, the constraints applied are limits on spin-independent cross-sections for neutralino-nucleus interactions derived by the Large Underground Xenon (LUX) experiment [26] and the PandaX-II Experiment[27], and limits on the proton spin-dependent cross-sections by the COUPP Collaboration [28] and XENON100 Collaboration [29].

Twenty million points in the General No-Scale $\mathcal{F}$-$SU(5)$ parameter space are sampled in a random scan applying the mSUGRA/CMSSM boundary conditions at the $M_{\mathcal{F}}$ scale. The SUSY mass spectra, relic density, rare decay processes, and direct dark matter detection cross-sections are calculated with `MicrOMEGAs 2.1` [30] utilizing a proprietary mpi modification of the `SuSpect 2.34` [31] codebase to run flippon and General No-Scale $\mathcal{F}$-$SU(5)$ enhanced RGEs, utilizing non-universal soft supersymmetry breaking parameters at the scale $M_{\mathcal{F}}$. The Particle Data Group [32] world average for the strong coupling constant is $\alpha_S(M_Z) = 0.1181 \pm 0.0011$ at 1σ, and we adopt a value in this work of $\alpha_S = 0.1172$ nearer to the lower limit.

Results of these calculations are listed in Table 1 for a set of 23 viable sample benchmark points for a given set of parameters ($M_{1/2}$, M_0, A_0, M_V, $\tan\beta$, m_t). The numerical relic density figures provided in Table 1 consist solely of a calculation of the SUSY lightest neutralino $\widetilde{\chi}_1^0$ abundance, thus those regions with values less than the combined WMAP9 and 2015 Planck 1σ measurement lower bound of about $\Omega h^2 \leq 0.1093$ are expected to admit alternate contributions to the total observed relic density by WMAP9 and Planck. To account for possible multi-component dark matter in these regions of low neutralino density, the spin-dependent and spin-independent cross-section calculations on the $\mathcal{F}$-$SU(5)$ model space shown in Table 1 have been rescaled as follows:

$$\sigma_{SI(SD)}^{\text{re-scaled}} = \sigma_{SI(SD)} \frac{\Omega h^2}{0.1138} \tag{1}$$

Each of the benchmarks models in Table 1 is categorized into five distinguishing regions of the viable model space identified by LSP composition. The vector-like flippon masses for the benchmark spectra in Table 1 are chosen to be representative of the entire viable model space, hence we showcase both light and heavy flippon masses, between the scan limits of $855 \leq M_V \leq 100{,}000$ GeV. Vector-like flippons lighter than 1 TeV could presently be produced at the LHC2, thus we bold those M_V values in Table 1.

4. Phenomenological results

The No-Scale $\mathcal{F}$-$SU(5)$ model space with the mSUGRA/CMSSM SUSY breaking terms implemented is constrained via the experimental results outlined in the prior section, with the exception of the LUX and PandaX-II spin-independent cross-sections, which we shall apply after rescaling to account for multi-component dark matter. The surviving viable parameter space consists of five distinctive regions, which we segregate based upon LSP composition. We shall show that each of these five dark matter scenarios have characteristic phenomenology and can be distinguished by means of the particle states emanating from the SUSY cascade decays. As discussed, the one-parameter version of $\mathcal{F}$-$SU(5)$ generates a unique SUSY mass spectrum of $M_{\widetilde{t}_1} < M_{\widetilde{g}} < M_{\widetilde{q}}$, which does manifest again in one of the five current scenarios, though the typical mSUGRA/CMSSM SUSY spectrum of $M_{\widetilde{g}} < M_{\widetilde{t}_1} < M_{\widetilde{q}}$ is revealed also. The five regions are: (i) bino LSP with stau coannihilation and $M_{\widetilde{t}_1} < M_{\widetilde{g}} < M_{\widetilde{q}}$; (ii) bino LSP with stau coannihilation and $M_{\widetilde{g}} < M_{\widetilde{t}_1} < M_{\widetilde{q}}$; (iii) Higgs Funnel, defined as $M_{H^0} \simeq 2M_{\widetilde{\chi}_1^0}$; (iv) Higgsino LSP; and (v) Mixed scenario, with both a Higgs Funnel and Higgsino LSP. The latter three scenarios of Higgs Funnel, Higgsino, and Mixed all possess the common SUSY spectrum mass ordering of $M_{\widetilde{g}} < M_{\widetilde{t}_1} < M_{\widetilde{q}}$, and all include regions with neutralino relic densities less than the observed value and thus would support multi-component dark matter. The LSP composition of each dark matter region is annotated in Table 2. Each LSP is nearly all bino (Stau, Higgs Funnel) or all higgsino (Higgsino, Mixed).

The five disparate regions of the model space are depicted in Figs. 1–4, highlighting the stau coannihilation, Higgs Funnel, and Higgsino LSP. All of these regions are mostly segregated from each other in Figs. 1–4, where the null space in between the regions is primarily the result of the application of experimental constraints on the gluino mass, light Higgs boson mass, and relic density. The latest constraints on the WIMP-nucleon spin-independent cross-sections published the LUX [26] and PandaX-II [27] experiments are applied as a function of the LSP mass to the General No-Scale $\mathcal{F}$-$SU(5)$ model in Fig. 1. In Fig. 1 the cross-sections have been rescaled in accordance with Eq. (1) for those points with relic densities less than the WMAP9 and Planck observations. The LUX and PandaX-II constraints are in fact strong enough to exclude a rather large swath of the model space with spin-independent cross-sections greater than about 10^{-9} pb for heavier LSP masses, that would otherwise satisfy all the alternate experimental constraints applied (gluino mass, light Higgs boson mass, relic density). Further shown is the upper boundary on coherent neutrino scattering from atmospheric neutrinos and the diffuse supernova neutrino background (DSNB), which may serve as a lower limit on direct detection probes of WIMP-nucleon scattering events. Note though that the entire viable stau coannihilation region analyzed in this work safely resides just above this neutrino scattering boundary. The LSP mass as a function of the heavy neutral Higgs pseudoscalar mass (M_{H^0/A^0}), the light chargino mass ($M_{\widetilde{\chi}_1^{\pm}}$), and light stau mass ($M_{\widetilde{\tau}_1^{\pm}}$) are shown in Figs. 2–4, respectively. In the Higgsino LSP scenario, it is clear from Fig. 3 that the chargino is essentially degenerate with the LSP.

Of particular note in Table 1 are the vector-like flippon mass scales M_V, which are allowed to be rather light, and in fact less than 1 TeV. This is in sharp contrast to the one-parameter version of No-Scale $\mathcal{F}$-$SU(5)$ where the flippons bounds must be $M_V \sim 23$–50 TeV, as this lighter 1 TeV mass scale affords possible production of flippons at the LHC2 (those with the M_V value in boldface type in Table 1). It is also significant that several of the SUSY spectra in Table 1 remain testable by the LHC2, permitting possible probing of the Stau Coannihilation, Higgs Funnel and Higgsino regions within the Run 2 schedule, a circumstance not necessarily achievable by minimal models such as $SU(5)$ with mSUGRA/CMSSM SUSY breaking soft terms that also support stau coannihilation along with a Higgs Funnel and higgsino LSP.

Testing the General No-Scale $\mathcal{F}$-$SU(5)$ model requires identifying observable signatures associated with each of the five dark matter regions. The leading cascade decay channels are highlighted in Table 3 for all regions. Note that there is only a negligible difference between the Higgsino and Mixed models with respect to the gluino branching ratios, hence we group them together in Table 3. Clearly the decay options proliferate for the gluino when it is lighter than the light stop, and therefore do not provide a dominant signature with which to identify that region of the parameter space. As such, the four models with $M_{\widetilde{g}} < M_{\widetilde{t}_1}$ show no channel with a branching ratio greater than about 30%, and thus do

Table 1
Sample benchmark spectra for General No-Scale $\mathcal{F}$-$SU(5)$. The spectra are segregated into five characteristic models based upon LSP composition: light stau coannihilation (bino LSP) with both $M_{\tilde{\tau}_1} < M_{\tilde{g}}$ and $M_{\tilde{g}} < M_{\tilde{t}_1}$, Higgs Funnel ($M_{H^0} \simeq 2M_{\tilde{\chi}_1^0}$), Higgsino LSP, and Mixed (Higgs Funnel + Higgsino). All masses are given in GeV. Those flippon masses less than 1 TeV that could presently be produced at the LHC2 are given in boldface type. The numerical values given for Δa_μ are $\times 10^{-10}$, $Br(b \to s\gamma)$ are $\times 10^{-4}$, $Br(B_s^0 \to \mu^+\mu^-)$ are $\times 10^{-9}$, spin-independent cross-sections σ_{SI} are $\times 10^{-11}$ pb, and spin-dependent cross-sections σ_{SD} are $\times 10^{-9}$ pb. The correct light Higgs boson mass around 125 GeV can be achieved by choosing the proper Yukawa coupling between the vector-like flippons and light Higgs boson, which is smaller than 1, therefore, we do not present the lightest Higgs boson mass here.

Model	$M_{1/2}$	M_0	A_0	M_V	$\tan\beta$	m_t	$M_{\tilde{\chi}_1^0}$	$M_{\tilde{\chi}_2^0/\tilde{\chi}_1^\pm}$	$M_{\tilde{\tau}_1^\pm}$	$M_{\tilde{t}_1}$	$M_{\tilde{u}_R}$	$M_{\tilde{g}}$	M_{H^0}	Ωh^2	Δa_μ	$b \to s\gamma$	$B_s^0 \to \mu^\pm$	σ_{SI}	σ_{SD}
Stau	1467	100	-1060	**855**	18.3	172.9	293	628	297	1404	2872	1882	2850	0.1130	1.48	3.49	3.11	0.4	1
Stau	1527	160	-15	**915**	24.3	173.8	308	658	311	1797	2975	1974	2580	0.1110	1.74	3.51	3.22	0.5	2
Stau	1577	210	-950	**965**	20.4	174.0	319	680	322	1576	3063	2018	2940	0.1280	1.39	3.51	3.18	0.3	1
Stau	1537	698	-990	10825	34.2	172.6	344	714	349	1570	2822	2014	2220	0.1190	2.14	3.40	3.62	0.7	3
Stau	1487	648	-1040	50373	33.7	172.8	353	725	357	1483	2624	2015	2060	0.1150	2.43	3.38	3.65	1.0	4
Stau	1617	250	75	100000	28.9	174.1	396	806	397	1802	2720	2215	2130	0.1230	2.25	3.50	3.34	1.4	6
Stau	1527	160	970	**915**	28.8	173.5	308	659	311	2009	2974	1984	2290	0.1140	2.02	3.51	3.27	0.8	4
Stau	1557	1246	3955	**945**	45.1	173.2	317	676	320	2500	3262	2052	1480	0.1180	2.53	3.57	3.73	5	25
Stau	1607	1296	4005	**995**	45.7	173.0	328	700	331	2572	3361	2114	1510	0.1170	2.39	3.57	3.74	5	21
Stau	1587	748	3000	10875	39.2	174.4	357	739	359	2197	2915	2095	1740	0.1160	2.59	3.55	3.52	3	17
Higgs Funnel	2483	4372	4900	**905**	50.0	172.4	526	1021	2795	4733	6416	3332	1120	0.1130	0.81	3.75	3.48	64	176
Higgs Funnel	2483	4900	2930	**905**	50.0	171.8	527	1049	3328	4880	6786	3349	1030	0.0962	0.70	3.76	3.89	54	123
Higgs Funnel	2493	4382	4910	**915**	51.0	174.1	529	1109	2690	4702	6433	3340	1050	0.0990	0.79	3.78	4.09	27	52
Higgs Funnel	2543	4432	3975	10865	51.6	173.4	595	1206	2741	4428	6241	3348	1190	0.1120	0.76	3.71	4.08	16	29
Higgs Funnel	1767	3667	3889	53333	51.7	174.1	431	859	2186	3243	4726	2484	928	0.1107	1.31	3.76	3.99	86	203
Higgs Funnel	1772	3505	4116	93383	51.6	173.3	441	879	2044	3162	4582	2515	945	0.1111	1.37	3.74	3.88	77	177
Higgsino	2473	4890	4890	**895**	22.3	171.7	250	256	4723	5010	6790	3388	5050	0.0090	0.24	3.60	2.98	48	1750
Higgsino	2493	4910	4910	**915**	46.6	172.5	260	265	3575	4987	6813	3383	2210	0.0096	0.52	3.64	2.88	52	1710
Higgsino	2523	4940	4940	**945**	45.1	171.9	233	239	3717	5043	6864	3417	2580	0.0097	0.49	3.62	2.88	47	1980
Higgsino	2233	5000	4556	80000	45.0	171.0	270	276	3769	4322	6225	3158	2586	0.0101	0.55	3.60	2.84	55	1670
Mixed	2243	4677	5010	40100	49.9	171.7	439	449	3005	4217	6006	3096	892	0.0019	0.78	3.84	2.69	81	382
Mixed	1972	4672	5005	66717	50.4	173.3	434	453	2937	3891	5718	2792	884	0.0008	0.86	3.86	2.74	62	294
Mixed	2105	5005	3894	86717	50.4	173.3	477	496	3296	4143	6100	2985	980	0.0013	0.75	3.81	2.88	86	406

Table 2
General No-Scale $\mathcal{F}$-$SU(5)$ lightest supersymmetric particle (LSP) composition for the five dark matter regions studied in this work, including a comparison to the previously studied one-parameter (OPM) version of the No-Scale $\mathcal{F}$-$SU(5)$ model.

OPM	100% bino
Stau ($M_{\tilde{\tau}_1} < M_{\tilde{g}}$)	100% bino
Stau ($M_{\tilde{g}} < M_{\tilde{\tau}_1}$)	100% bino
Higgs Funnel	99% bino
Higgsino	100% higgsino
Mixed	98% higgsino

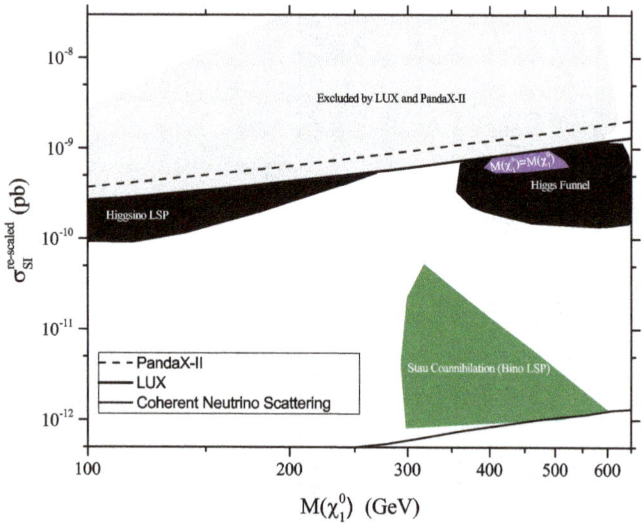

Fig. 1. Illustration of the LUX and PandaX-II WIMP-nucleon spin-independent cross-section constraints applied to the General No-Scale $\mathcal{F}$-$SU(5)$ viable parameter space. The null space in between the discrete Stau, Higgs Funnel, and Higgsino regions is primarily the result of application of the constraints on the gluino mass, light Higgs boson mass, and relic density.

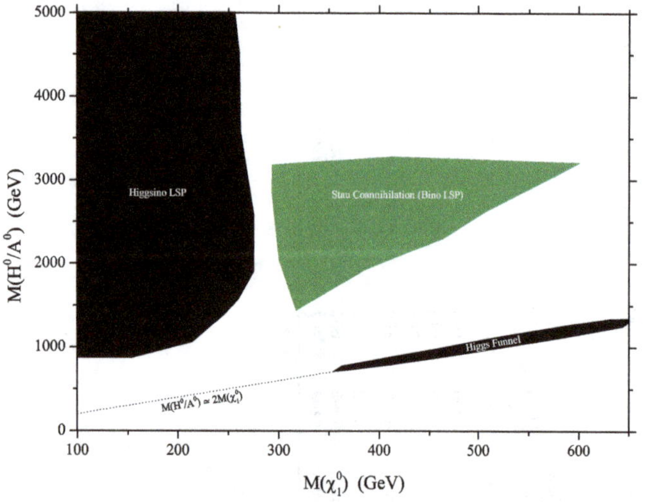

Fig. 2. Depiction of M_{H^0/A^0} as a function of the lightest neutralino mass $M_{\tilde{\chi}_1^0}$. The Stau Coannihilation (bino LSP), Higgs Funnel, and Higgsino LSP regions are annotated on the plot. The linear region that defines the Higgs Funnel, namely $M_{H^0/A^0} \simeq 2M_{\tilde{\chi}_1^0}$, is also displayed as a dashed line.

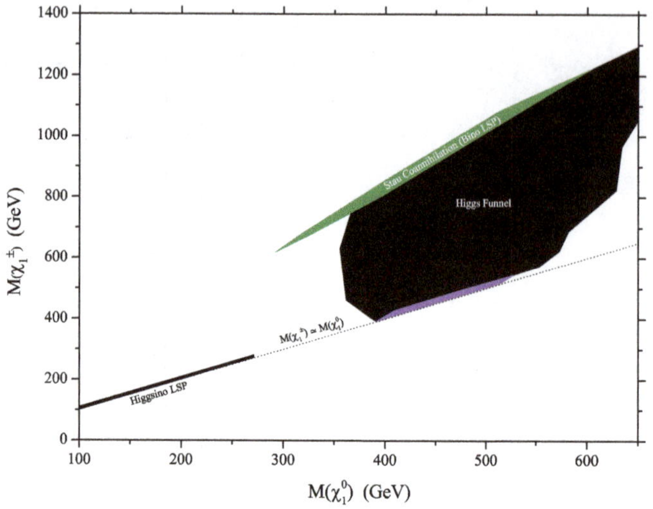

Fig. 3. Depiction of $M_{\tilde{\chi}_1^{\pm}}$ as a function of the lightest neutralino mass $M_{\tilde{\chi}_1^0}$. The Stau Coannihilation (bino LSP), Higgs Funnel, and Higgsino LSP regions are annotated on the plot. Further highlighted here is that subspace at the lower extreme of the Higgs Funnel with $M_{\tilde{\chi}_1^{\pm}} \simeq M_{\tilde{\chi}_1^0}$. The linear region that defines $M_{\tilde{\chi}_1^{\pm}} \simeq M_{\tilde{\chi}_1^0}$ is also displayed as a dashed line.

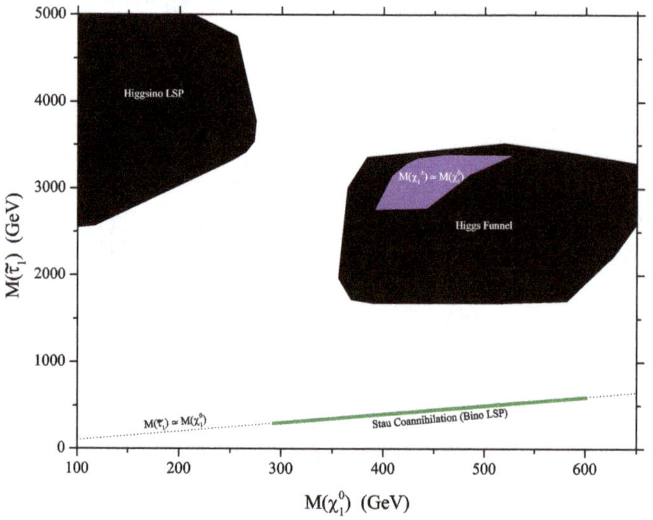

Fig. 4. Depiction of $M_{\tilde{\tau}_1^{\pm}}$ as a function of the lightest neutralino mass $M_{\tilde{\chi}_1^0}$. The Stau Coannihilation (bino LSP), Higgs Funnel, and Higgsino LSP regions are annotated on the plot. Further highlighted here is that subspace of the Higgs Funnel with $M_{\tilde{\chi}_1^{\pm}} \simeq M_{\tilde{\chi}_1^0}$. The linear region that defines $M_{\tilde{\tau}_1^{\pm}} \simeq M_{\tilde{\chi}_1^0}$ is also displayed as a dashed line.

not possess a dominant decay mode. On the contrary, the gluino in the one-parameter version of No-Scale $\mathcal{F}$-$SU(5)$ will decay to a light stop and hence $t\bar{t}$ 100% of the time [6] and the stau coannihilation region of General No-Scale SUGRA with $M_{\tilde{\tau}_1} < M_{\tilde{g}}$ also shows a reasonably large branching ratio of about 62% to a $t\bar{t}$, thus these provide a much stronger singular decay channel. The fact that the $t\bar{t}$ production does vary considerably between the five different regions does though present an opportunity to utilize the $t\bar{t}$ channel as a tool with which to discriminate between all the regions, and furthermore, differentiate the one-parameter version of No-Scale $\mathcal{F}$-$SU(5)$ [8] from General No-Scale $\mathcal{F}$-$SU(5)$. The $\tilde{g} \to t\bar{t}$ branching ratios are itemized in Table 4, displaying very evidently the divergence in this channel. Each gluino that results in a $t\bar{t}$ can produce up to six hadronic jets, with two b-jets among them, therefore manifesting as a large multijet event at the LHC, particularly with pair-produced gluinos. Large multijet events are the characteristic signature of the one-parameter version of No-Scale $\mathcal{F}$-$SU(5)$ [4], and Table 4 shows that the number of multijet

Table 3
General No-Scale $\mathcal{F}$-$SU(5)$ leading cascade decay channels for the five different dark matter regions studied in this work. The BR column represents the branching ratio.

Model	BR	Decay Mode
Stau ($M_{\widetilde{t}_1} < M_{\widetilde{g}}$)	0.62	$\widetilde{g} \to t\bar{t} + \widetilde{\chi}_1^0$
Stau ($M_{\widetilde{t}_1} < M_{\widetilde{g}}$)	0.11	$\widetilde{g} \to tb + \tau + \nu_\tau + \widetilde{\chi}_1^0$
Stau ($M_{\widetilde{g}} < M_{\widetilde{t}_1}$)	0.31	$\widetilde{g} \to t\bar{t} + \widetilde{\chi}_1^0$
Stau ($M_{\widetilde{g}} < M_{\widetilde{t}_1}$)	0.19	$\widetilde{g} \to tb + \tau + \nu_\tau + \widetilde{\chi}_1^0$
Stau ($M_{\widetilde{g}} < M_{\widetilde{t}_1}$)	0.19	$\widetilde{g} \to q\bar{q} + \tau + \nu_\tau + \widetilde{\chi}_1^0$
Stau ($M_{\widetilde{g}} < M_{\widetilde{t}_1}$)	0.15	$\widetilde{g} \to q\bar{q} + \tau^+\tau^- + \widetilde{\chi}_1^0$
Higgs Funnel	0.30	$\widetilde{g} \to tb + W + \widetilde{\chi}_1^0$
Higgs Funnel	0.11	$\widetilde{g} \to t\bar{t} + Z + \widetilde{\chi}_1^0$
Higgs Funnel	0.09	$\widetilde{g} \to t\bar{t} + h + \widetilde{\chi}_1^0$
Higgsino/Mixed	0.28	$\widetilde{g} \to tb + q\bar{q} + \widetilde{\chi}_1^0$
Higgsino/Mixed	0.12	$\widetilde{g} \to t\bar{t} + \widetilde{\chi}_1^0$

Table 4
Branching ratios of $\widetilde{g} \to t\bar{t} + \chi_1^0$ in General No-Scale $\mathcal{F}$-$SU(5)$. The models represent the five different dark matter regions studied in this work, including a comparison to the previously studied one-parameter (OPM) version of No-Scale $\mathcal{F}$-$SU(5)$. Note that the contrasting level of $t\bar{t}$ production in each of the five regions can be utilized to discriminate amongst the models.

Model	$\mathrm{Br}(\widetilde{g} \to t\bar{t} + \widetilde{\chi}_1^0)$
OPM	1.00
Stau ($M_{\widetilde{t}_1} < M_{\widetilde{g}}$)	0.62
Stau ($M_{\widetilde{g}} < M_{\widetilde{t}_1}$)	0.31
Higgsino/Mixed	0.12
Higgs Funnel	0.03

events could potentially be used to identify all the models studied here.

While the gluino decay modes can be utilized to discriminate amongst the regions of varied dark matter scenarios, in contrast, the squark channels are reasonably consistent throughout the model space. Identifying $\widetilde{q} = (\widetilde{u}, \widetilde{d}, \widetilde{c}, \widetilde{s})$, and simply computing an approximate average between branching ratios of right-handed squarks $\widetilde{q}_R$ and left-handed squarks $\widetilde{q}_L$, we find a mean branching ratio for $\widetilde{q} \to \widetilde{g} + q$ of about 75%. On the other hand, the light stop decay modes are more diverse given that it can be lighter or heavier than the gluino. Similar to the one-parameter version of No-Scale $\mathcal{F}$-$SU(5)$, the light stop in the General No-Scale $\mathcal{F}$-$SU(5)$ Stau regions will produce a top quark via $\widetilde{t}_1 \to t + \widetilde{\chi}_1^0$ 100% of the time. This can be attributed to the light stop being lighter than the gluino, or in the case of the Stau region with $M_{\widetilde{g}} < M_{\widetilde{t}_1}$, the mass delta is rather small with the two sparticles nearly degenerate. The situation is not as clean in the remaining model space where the gluino is much lighter than the light stop, as the primary channel for the light stop in each region will be $\widetilde{t}_1 \to \widetilde{g} + t$ at 40% (Higgsino), 36% (Higgs Funnel), and 31% (Mixed).

An intriguing aspect to recognize in Table 3 regards the $t\bar{t}h$ state in the Higgs Funnel, which is the production of a light Higgs boson in tandem with a $t\bar{t}$. The $t\bar{t}h$ production cross-section via off-shell top quarks in the Standard Model is well known and is used as a direct measurement of the tree-level top Yukawa coupling. This places strong limits on supersymmetric contributions to gluon fusion processes, and interestingly, strong $t\bar{t}h$ production has been observed at the LHC [33]. While the branching ratio $t\bar{t}h$ in the General No-Scale $\mathcal{F}$-$SU(5)$ Higgs Funnel is a mere 9%, this does suggest possible non-negligible production of these events in the current LHC Run 2.

Acknowledgements

The computing for this project was performed at the Tandy Supercomputing Center, using dedicated resources provided by The University of Tulsa. This research was supported in part by the Projects 11475238 and 11647601 supported by the National Natural Science Foundation of China, and by the DOE grant DE-FG02-13ER42020 (DVN).

References

[1] W. Adam, Searches for SUSY, talk at the 38th International Conference on High Energy Physics, 2016.
[2] G. Aad, et al., ATLAS Collaboration, Observation of a new particle in the search for the Standard Model Higgs boson with the ATLAS detector at the LHC, Phys. Lett. B 716 (2012) 1, arXiv:1207.7214.
[3] S. Chatrchyan, et al., CMS Collaboration, Observation of a new boson at a mass of 125 GeV with the CMS experiment at the LHC, Phys. Lett. B 716 (2012) 30, arXiv:1207.7235.
[4] T. Li, J.A. Maxin, D.V. Nanopoulos, J.W. Walker, The ultrahigh jet multiplicity signal of stringy no-scale $\mathcal{F}$-$SU(5)$ at the $\sqrt{s} = 7$ TeV LHC, Phys. Rev. D 84 (2011) 076003, arXiv:1103.4160.
[5] T. Li, J.A. Maxin, D.V. Nanopoulos, J.W. Walker, A Higgs mass shift to 125 GeV and a multi-jet supersymmetry signal: miracle of the flippons at the $\sqrt{s} =$ 7 TeV LHC, Phys. Lett. B 710 (2012) 207, arXiv:1112.3024.
[6] T. Li, J.A. Maxin, D.V. Nanopoulos, J.W. Walker, No-Scale $\mathcal{F}$-$SU(5)$ in the Light of LHC, Planck and XENON, J. Phys. G 40 (2013) 115002, arXiv:1305.1846.
[7] T. Leggett, T. Li, J.A. Maxin, D.V. Nanopoulos, J.W. Walker, Confronting electroweak fine-tuning with No-Scale Supergravity, Phys. Lett. B 740 (2015) 66, arXiv:1408.4459.
[8] T. Li, J.A. Maxin, D.V. Nanopoulos, The return of the King: No-Scale $\mathcal{F}$-$SU(5)$, Phys. Lett. B 764 (2017) 167, arXiv:1609.06294.
[9] S.M. Barr, A new symmetry breaking pattern for $SO(10)$ and proton decay, Phys. Lett. B 112 (1982) 219.
[10] J.P. Derendinger, J.E. Kim, D.V. Nanopoulos, Anti-$SU(5)$, Phys. Lett. B 139 (1984) 170.
[11] I. Antoniadis, J.R. Ellis, J.S. Hagelin, D.V. Nanopoulos, Supersymmetric flipped $SU(5)$ revitalized, Phys. Lett. B 194 (1987) 231.
[12] J.E. Kim, B. Kyae, Flipped SU(5) from Z(12-I) orbifold with Wilson line, Nucl. Phys. B 770 (2007) 47, arXiv:hep-th/0608086.
[13] J.-H. Huh, J.E. Kim, B. Kyae, SU(5) (flip) × SU(5)-prime from Z(12-I), Phys. Rev. D 80 (2009) 115012, arXiv:0904.1108.
[14] J. Jiang, T. Li, D.V. Nanopoulos, Testable flipped $SU(5) \times U(1)_X$ models, Nucl. Phys. B 772 (2007) 49, arXiv:hep-ph/0610054.
[15] J. Jiang, T. Li, D.V. Nanopoulos, D. Xie, F-SU(5), Phys. Lett. B 677 (2009) 322, arXiv:0811.2807.
[16] J. Jiang, T. Li, D.V. Nanopoulos, D. Xie, Flipped $SU(5) \times U(1)_X$ models from F-Theory, Nucl. Phys. B 830 (2010) 195, arXiv:0905.3394.
[17] G. Giudice, A. Masiero, A natural solution to the mu problem in supergravity theories, Phys. Lett. B 206 (1988) 480.
[18] ATLAS, Exotics combined summary plots, atlas.web.cern.ch/Atlas/GROUPS/PHYSICS/CombinedSummaryPlots/EXOTICS/index.html, 2016.
[19] T.A. Aaltonen, Tevatron Electroweak Working Group, CDF, D0, Combination of CDF and DO results on the mass of the top quark using up to 8.7 fb^{-1} at the Tevatron, arXiv:1305.3929, 2013.
[20] G. Hinshaw, et al., WMAP Collaboration, Nine-year Wilkinson Microwave Anisotropy Probe (WMAP) observations: cosmological parameter results, arXiv:1212.5226, 2012.
[21] P. Ade, et al., Planck, Planck 2015 results. XIII. Cosmological parameters, arXiv:1502.01589, 2015.
[22] Y. Huo, T. Li, D.V. Nanopoulos, C. Tong, The lightest CP-even Higgs boson mass in the testable flipped $SU(5) \times U(1)_X$ models from F-Theory, Phys. Rev. D 85 (2012) 116002, arXiv:1109.2329.
[23] HFAG, www.slac.stanford.edu/xorg/hfag/rare/2013/radll/OUTPUT/TABLES/radll.pdf, 2013.
[24] V. Khachatryan, et al., LHCb, CMS, Observation of the rare $B_s^0 \to \mu^+\mu^-$ decay from the combined analysis of CMS and LHCb data, Nature 522 (2015) 68, arXiv:1411.4413.
[25] T. Aoyama, M. Hayakawa, T. Kinoshita, M. Nio, Complete tenth-order QED contribution to the muon g-2, Phys. Rev. Lett. 109 (2012) 111808, arXiv:1205.5370.
[26] D.S. Akerib, et al., Results from a search for dark matter in LUX with 332 live days of exposure, arXiv:1608.07648, 2016.
[27] A. Tan, et al., PandaX-II, Dark matter results from first 98.7-day data of PandaX-II experiment, Phys. Rev. Lett. 117 (2016) 121303, arXiv:1607.07400.

[28] E. Behnke, et al., COUPP, First dark matter search results from a 4-kg CF_3I bubble chamber operated in a deep underground site, Phys. Rev. D 86 (2012) 052001, Erratum: Phys. Rev. D 90 (7) (2014) 079902, arXiv:1204.3094.
[29] E. Aprile, et al., XENON100, Limits on spin-dependent WIMP-nucleon cross sections from 225 live days of XENON100 data, Phys. Rev. Lett. 111 (2013) 021301, arXiv:1301.6620.
[30] G. Belanger, F. Boudjema, A. Pukhov, A. Semenov, Dark matter direct detection rate in a generic model with micrOMEGAs2.1, Comput. Phys. Commun. 180 (2009) 747, arXiv:0803.2360.
[31] A. Djouadi, J.-L. Kneur, G. Moultaka, SuSpect: a Fortran code for the supersymmetric and Higgs particle spectrum in the MSSM, Comput. Phys. Commun. 176 (2007) 426, arXiv:hep-ph/0211331.
[32] C. Patrignani, et al., Particle Data Group, Review of particle physics, Chin. Phys. C 40 (2016) 100001.
[33] G. Aad, et al., ATLAS, CMS, Measurements of the Higgs boson production and decay rates and constraints on its couplings from a combined ATLAS and CMS analysis of the LHC pp collision data at $\sqrt{s} = 7$ and 8 TeV, J. High Energy Phys. 08 (2016) 045, arXiv:1606.02266.

8

GPDs at non-zero skewness in ADS/QCD model

Matteo Rinaldi

Instituto de Fisica Corpuscular, CSIC-Universitat de Valencia, Parc Cientific UV, C/ Catedratico Jose Beltran 2, E-46980 Paterna, Valencia, Spain

ARTICLE INFO

Editor: J.-P. Blaizot

Keywords:
Phenomenological models
Deep inelastic scattering (phenomenology)

ABSTRACT

We study Generalized Parton Distribution functions (GPDs) usually measured in hard exclusive processes and encoding information on the three dimensional partonic structure of hadrons and their spin decomposition, for non-zero skewness within the AdS/QCD formalism. To this aim the canonical scheme to calculate GPDs at zero skewness has been properly generalized. Furthermore, we show that the latter quantities, in this non-forward regime, are sensitive to non-trivial details of the hadronic light front wave function, such as a kind of parton correlations usually not accessible in studies of form factors and GPDs at zero skewness.

1. Introduction

In the last years, much theoretical and experimental attention has been focused on the study of non-perturbative quantities in QCD encoding fundamental information on the partonic proton structure, e.g., parton distribution functions (PDFs), transverse momentum dependent PDFs (TMDs), double parton distribution functions (dPDFs) and generalized parton distribution functions (GPDs) [1–4]. From a theoretical point of view, GPDs allow to grasp information on the three dimensional partonic structure of hadrons [5], and thanks to the so called Ji's sum rule, to access the orbital angular momentum of partons inside hadrons. This knowledge is crucial to shed some light onto the still open problem of the proton spin crisis. From an experimental point of view, GPDs represent a challenge due to the difficulties in the measurements of high energy exclusive events (see [6,7] and references therein). In this paper, we study the GPDs using the AdS/QCD scheme proposed by Brodsky and de Téramond [8–14] based on the original AdS/CFT Maldacena conjecture [15]. The idea comes from the observation that for small momentum transfer the QCD coupling constant can be approximated by a constant and quark masses can be neglected [16]. Confinement can be simulated in different ways [8,10,11,17, 18]. A crucial ingredient of this proposal is the mapping between the Light Front (LF) Hamiltonian formulation of QCD and the AdS description of hadrons. This correspondence is implemented by relating the fifth dimensional variable z with the LF transverse position and longitudinal momentum fractions carried by partons [13]. This approach has been successfully applied to the calculations of the hadronic spectrum, pion TMDs, baryonic form factors and the GPDs for $\xi = 0$ (ξ represents the skewness) [8,10,11,13,19–38]. Moreover, a first analysis of hard deep inelastic scattering at small x within AdS/CFT has been proposed in Ref. [39]. The reasonable capability in describing meson static properties and nucleonic form factors motivated the calculation of other fundamental observables such as the GPDs within AdS/QCD correspondence. In the present paper we focus our attention in the calculation of the GPDs for $\xi \neq 0$, within the soft-wall model [8,10,11], proposing a suitable extension of the approaches already discussed in the literature [11, 28,40]. This step is necessary to provide a complete description of the GPDs in the AdS/QCD scheme which allows comparison of theoretical results with data usually obtained for $\xi \neq 0$. Furthermore, our results open new ways to calculate observables within the AdS/QCD framework which may lead to useful predictions on the structure of hadrons.

2. Form factor in ADS/QCD

Before we describe GPDs let us review the AdS/QCD formalism for the calculation of form factors within the soft-model. In particular, in order to construct nucleonic form factors from AdS/QCD, there are two models. One discussed by Brodsky and de Teramond in Ref. [27] (BT) and the model described in Ref. [11] (AC). The main difference in these approaches is the presence, in the second one, of non-minimal coupling term in the dressed electromagnetic operator which makes the Dirac form factor different in these two schemes. In the present paper both models are considered. In the AdS/QCD scheme within the soft-wall model, the breaking of conformal invariance is induced by introducing a quadratic dilaton term $\Phi = e^{\pm\kappa^2 z^2}$ in the AdS action. The effect of the breaking can

E-mail address: mrinaldi@ific.uv.es.

be directly incorporated into the baryonic field [16,28] and the confining potential, depending on the dilaton term, is put in by hand in the soft-wall model. The strategy to connect AdS quantities to the correspondent Light Front (LF) ones is to write the AdS Dirac equation, associated to the AdS action, as a LF equation and to obtain the corresponding baryon wave function by solving an equation in the 2×2 spinor representation [28]. Let us just show the analytic expression of such baryonic functions:

$$\phi_+(z) = \frac{\sqrt{2}\kappa^2}{R^2} z^{7/2} e^{-\kappa^2 z^2/2},$$
$$\phi_-(z) = \frac{\kappa^3}{R^2} z^{9/2} e^{-\kappa^2 z^2/2}. \tag{1}$$

In the case of the BT model $\kappa = \kappa_{BT} = 0.4066$ GeV, see Refs. [25, 28] for the details. In the case of the AC model, $\kappa = \kappa_{AC} = 0.350$ GeV, see Refs. [35,37]. The other fundamental ingredient of the calculation of form factors within the AdS/QCD scheme is the bulk-to-boundary propagator whose expression, in the soft-wall model, reads [16,41]:

$$V(Q^2, z) = \int_0^1 \tilde{V}(x, t, z)\, dx$$
$$= \kappa^2 z^2 \int_0^1 \frac{x^{Q^2/(4\kappa^2)}}{(1-x)^2} e^{-\kappa^2 z^2 x/(1-x)}\, dx. \tag{2}$$

In the following, the formal expressions of Pauli and Dirac form factors will be discussed. As shown in Refs. [11,16,28,35,37], one can realize that the expression of the Pauli form factor for the BT and AC models is formally the same. However, in the AC model [11], where a non-minimal electromagnetic coupling contribution has been introduced in the action, the proton Dirac form factor contains an extra-term depending on $\phi_-^2(z)$, at variance of that of the BT model, see Refs. [27,28]. In the present analysis, use has been made of the expressions of Pauli and Dirac form factors described in, e.g., Refs. [28,35,37]. Furthermore, as discussed in Refs. [11,35], non-minimal coupling term also produces the correct scaling behavior of nucleonic form factors at high Q^2. As will be discussed in the next section, to the aim of the present paper, it is convenient to directly write the expressions of the contribution of the parton of flavor q to the Dirac and Pauli form factors for both BT and AC models in a unified fashion way:

$$F_1^q(t) = R^4 \int \frac{dz}{z^4} \int_0^1 dx\, \tilde{V}(x, t, z) \tag{3}$$
$$\times [m_{++}^q \phi_-^2(z) + n_{++}^q \phi_+^2(z)],$$
$$F_2^q(t) = R^4 \int \frac{dz}{z^4} \int_0^1 dx\, \tilde{V}(x, t, z) \tag{4}$$
$$\times [m_{+-}^q \phi_-^2(z) + n_{+-}^q \phi_+^2(z)],$$

where in the model BT:

$$n_{++}^u = 5/3,\ m_{++}^u = 1/3,\ n_{++}^d = 1/3,\ m_{++}^d = 2/3,$$
$$n_{+-}^u = 0,\ m_{+-}^u = f_u,\ n_{+-}^d = 0,\ m_{+-}^d = f_d, \tag{5}$$

and in the model AC:

$$n_{++}^u = 1 - \eta_u T(x, z),\ m_{++}^u = 1 + \eta_u T(x, z), \tag{6}$$
$$n_{++}^d = 1/2 - \eta_d T(x, z),\ m_{++}^d = 1/2 + \eta_d T(x, z),$$
$$n_{+-}^u = 0,\ m_{+-}^u = f_u,\ n_{+-}^d = 0,\ m_{+-}^d = f_d. \tag{7}$$

Moreover, the flavor coefficient f_q is defined as follows:

$$f_u = 2\chi_p + \chi_n,\quad f_d = 2\chi_n + \chi_p, \tag{8}$$

where $\chi_{p(n)}$ is the anomalous magnetic moment of proton (neutron) and $f_u = 0.209$, $f_d = -0.254$ and $T(x, z) = 1 - \kappa^2 z^2 x/(1-x)$, see e.g. Ref. [37].

3. GPDs in AdS/QCD

As already discussed, e.g., in Ref. [28], the proton GPDs at $\xi = 0$, within the soft-wall model, can be calculated from form factors. However, for the purpose of the present analysis, it is convenient to relate the spin dependent light-cone correlator [4,42] $C^q_{\lambda,\lambda'}(x, \xi, t)$ to $\phi_\pm(z)$, being λ (λ') the third component of the proton spin in the initial (final) state. To this aim, using then the parametrization of Dirac and Pauli form factors in terms of the flavor dependent GPDs, one obtains [16,28]:

$$C_{ab}^q(x, 0, t) = R^4 \int dx_1 \frac{dz}{z^4} \tilde{V}(x_1, t, z) \tag{9}$$
$$\times [n_{ab}^q \phi_+^2(z_1) + m_{ab}^q \phi_-^2(z_1)] \delta(x - x_1)$$

Since the light cone correlator directly depends on the LF proton wave function [42], in order to match results found within the AdS/QCD framework with the same quantities evaluated within the LF approach, the baryonic functions $\phi_\pm(z)$ must be related to the LF proton wave functions [13]. To this aim the variable z turns out a function of $\vec{b}_{i\perp}$ and x_i, the transverse position and longitudinal momentum fraction carried by the i parton respectively [13]. We proceed to calculate the GPDs for $\xi \neq 0$. These quantities have been also studied in a recent paper in Ref. [40] within a different approach and using the IR improved soft-wall model [43]. To this aim since the LF proton wave function is a frame independent quantity [44], it is useful to work in the intrinsic frame where $z = z_1 = \sqrt{x_1/(1-x_1)}|\vec{b}_{1\perp}|$ can be considered as a one body variable (for details on the chosen frame see, e.g., Ref. [45]). In this scenario, Eqs. (9) can be written in terms of z_1, associated to the interacting parton, and z_2 associated to a second spectator particle. Within this choice, the light cone correlator can be written:

$$C_{ab}^q(x, 0, t) = R^8 \int dx_1 dx_2 \frac{dz_1}{z_1^4} \frac{dz_2}{z_2^4} \tilde{V}(x_1, t, z_1)$$
$$\times [n_{ab}^q \phi_+^2(z_1, z_2) + m_{ab}^q \phi_-^2(z_1, z_2)] \delta(x - x_1), \tag{10}$$

being $a, b = \pm$. As one can see, the expression (9) is recovered by introducing the following normalization conditions:

$$R^4 \int \frac{dz_2}{z_2^4} \phi_\pm^2(z_1, z_2) = \phi_\pm^2(z_1);$$
$$R^8 \int \frac{dz_1}{z_1^4} \frac{dz_2}{z_2^4} \phi_\pm^2(z_1, z_2) = 1. \tag{11}$$

Let us call $\phi_\pm(z_1, z_2)$ the "two body intrinsic" proton function, where the spin-flavor part is already included. As will be discussed later on, the expression Eq. (10) is quite suitable for the generalization of the correlator at $\xi \neq 0$. Following the line of Ref. [42], in order to include the ξ dependence, it is sufficient to properly change the argument of the two body intrinsic proton function, appearing in Eq. (10). To this aim, let us introduce the following variable $\bar{z}_i$, in the initial state and $\tilde{z}_i$ in the final one:

$$\bar{z}_i = \sqrt{\frac{1-\bar{x}_i}{\bar{x}_i} \frac{x_i}{1-x_i}} z_i,\quad \tilde{z}_i = \sqrt{\frac{1-\tilde{x}_i}{\tilde{x}_i} \frac{x_i}{1-x_i}} z_i; \tag{12}$$

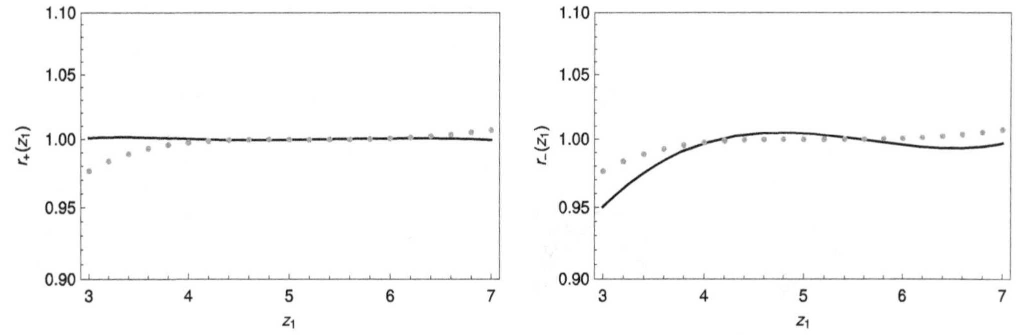

Fig. 1. Left panel: the ratio $r_+(z_1)$, Eq. (22). Right panel: the ratio $r_-(z_1)$, Eq. (23). Full line for the BT model and dotted line for the AC model.

where here

$$\bar{x}_1 = \frac{x_1+\xi}{1+\xi}, \quad \bar{x}_2 = \frac{x_2}{1+\xi}, \quad \tilde{x}_1 = \frac{x_1-\xi}{1-\xi}, \quad \tilde{x}_2 = \frac{x_2}{1-\xi}. \tag{13}$$

Starting from the generalization of the correlator, Eq. (10), in the $\xi \neq 0$ case, using as argument the variables described in Eqs. (12), (13), one finds:

$$E_v^q(x,\xi,t) = \sqrt{\frac{t_0-t}{-t}}\frac{R^8}{\sqrt{1-\xi^2}}\int dx_1 dx_2 dz_1 dz_2\, z_1 z_2$$

$$\times\, \tilde{V}(x_1,t,z_1)\frac{f_q}{(2\pi)^2}\phi_-(\bar{z}_1,\bar{z}_2)\phi_-(\tilde{z}_1,\tilde{z}_2)\frac{\delta(x-x_1)}{[\bar{z}_1\bar{z}_2\tilde{z}_1\tilde{z}_2]^{5/2}}$$

$$\times \sqrt{\frac{\bar{x}_1\bar{x}_2\tilde{x}_1\tilde{x}_2}{(1-\bar{x}_1)(1-\bar{x}_2)(1-\tilde{x}_1)(1-\tilde{x}_2)}\frac{(1-x_1)(1-x_2)}{x_1x_2}}, \tag{14}$$

where, in the last line, functions of $\bar{x}_i$ and $\tilde{x}_i$ do not cancel the jacobian due to the transformation between $b_{i\perp}$ and z_i, in this not diagonal case. For the GPD H one finds:

$$H_v^q(x,\xi,t) = \frac{R^8}{1-\xi^2}\int dx_1 dx_2 dz_1 dz_2\, z_1 z_2 \tag{15}$$

$$\times\, \tilde{V}(x_1,t,z_1)\left[\frac{\xi^2}{\sqrt{1-\xi^2}}\sqrt{\frac{t_0-t}{-t}}\phi_-(\bar{z}_1,\bar{z}_2)\phi_-(\tilde{z}_1,\tilde{z}_2)f_q + \right.$$

$$\left. +\, F_q(\bar{z}_1,\bar{z}_2,\tilde{z}_1,\tilde{z}_2)\right]\frac{\delta(x-x_1)}{(2\pi)^2}\frac{1}{[\bar{z}_1\bar{z}_2\tilde{z}_1\tilde{z}_2]^{5/2}}$$

$$\times\, \frac{(1-x_1)(1-x_2)}{x_1x_2}\sqrt{\frac{\bar{x}_1\bar{x}_2\tilde{x}_1\tilde{x}_2}{(1-\bar{x}_1)(1-\bar{x}_2)(1-\tilde{x}_1)(1-\tilde{x}_2)}}\,.$$

where in this case the following flavor function has been introduced:

$$F_q(z_1,z_2,z_2,z_4) = \left[n^q_{++}\phi^2_+(z_1,z_2) + m^q_{++}\phi^2_-(z_1,z_2)\right]^{1/2}$$

$$\times \left[n^q_{++}\phi^2_+(z_3,z_4) + m^q_{++}\phi^2_-(z_3,z_4)\right]^{1/2}. \tag{16}$$

Let us remark that the above expressions are general and can be used with all the approaches in which Dirac and Pauli form factor are evaluated within the AdS/QCD soft wall models. For example, results, obtained by means of the model where non-minimal coupling term is neglected, are recovered by properly setting $T(x,z)=0$ and the coefficients in Eq. (16) equal to the ones in Eqs. (5).

4. Numerical solutions for the two body intrinsic function

In order to evaluate GPDs at $\xi \neq 0$ the calculation of the two body intrinsic function is necessary, however, the only constraint

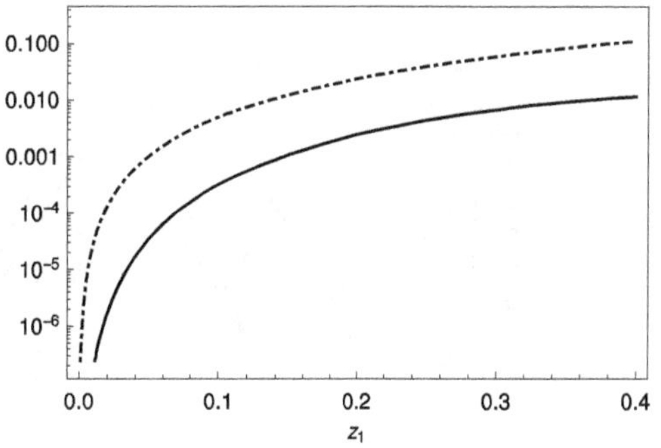

Fig. 2. The difference between the calculations of the two body intrinsic functions within the correlated and uncorrelated ansatz for, e.g., the BT. Full line: ${}_1\phi_+(z_1, z_2 = 5\ \mathrm{GeV}^{-1}) - {}_2\phi_+(z_1, z_2 = 5\ \mathrm{GeV}^{-1})$; dot-dashed line: ${}_1\phi_-(z_1, z_2 = 5\ \mathrm{GeV}^{-1}) - {}_2\phi_-(z_1, z_2 = 5\ \mathrm{GeV}^{-1})$.

for the evaluation of $\phi_\pm(z_1,z_2)$ is the integral Eq. (11). Due to this condition different two body intrinsic distributions, corresponding to the same 1 body one ($\phi_\pm(z)$), can be found. In particular, in the present analysis, two physically different scenarios have been considered, i.e. a fully uncorrelated ansatz, ${}_1\phi_\pm(z_1,z_2)$ and a correlated one, ${}_2\phi_\pm(z_1,z_2)$. For the uncorrelated case, one can straightforwardly consider a function where the z_1 and z_2 dependence is fully factorized:

$${}_1\phi_\pm(z_1,z_2) = \phi_\pm(z_1)\phi_\pm(z_2)\,. \tag{17}$$

For the evaluation of the correlated plus distribution ${}_2\phi_+(z_1,z_2)$, a numerical solution of Eq. (11) has been used:

$${}_2\phi_+(z_1,z_2) = \frac{C_5}{R^4}e^{-\kappa^2\beta(z_1,z_2)/2}\kappa^5 z_1^4 z_2^4 \tag{18}$$

$$\times \left[C_1U^2(-1,1,C_4\,\beta(z_1,z_2)) + C_2U^2(-2,1,C_4\,\beta(z_1,z_2))\right.$$

$$\left. +\, C_3U(-1,1,C_4\,\beta(z_1,z_2))U(-3,1,C_4\,\beta(z_1,z_2))\,\right]^{1/2}$$

being $\beta(z_1,z_2) = z_1^2 + z_2^2 + z_1z_2$ and $U(a,b,z)$ the Tricomi confluent hypergeometric function and the coefficients read for κ_{BT} (κ_{AC}):

$$C_1 = 1.68944\,(15.309);\quad C_2 = 17.977\,(-2.0252);$$

$$C_3 = -18.5158\,(6.04302);\quad C_4 = 0.082599\,(-0.10063);$$

$$C_5 = 7.9937\,(0.263131189). \tag{19}$$

For the minus component, one finds

$${}_2\phi_-(z_1,z_2) = \frac{D_4}{R^4}e^{-\kappa^2\beta(z_1,z_2)/2}\kappa^7 z_1^5 z_2^5 \tag{20}$$

$$\times \left[D_2U^2(-1,1,D_1\,\beta(z_1,z_2)) + D_3U^2(-2,1,D_1\,\beta(z_1,z_2))\right.$$

$$\left. +\, D_5U(-1,1,D_1\,\beta(z_1,z_2))U(-3,1,D_1\,\beta(z_1,z_2))\,\right]^{1/2}$$

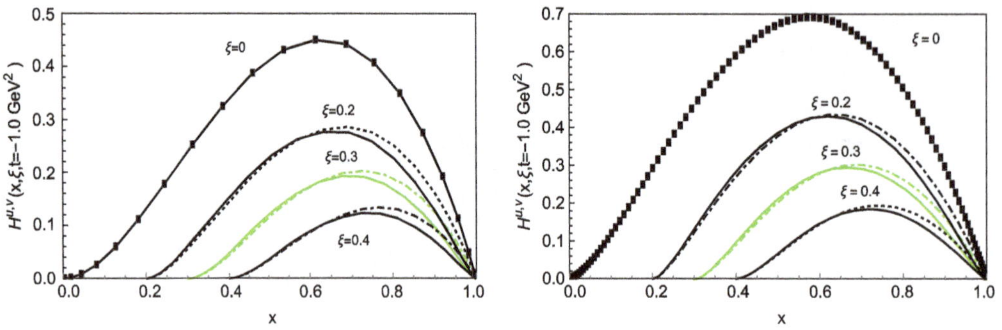

Fig. 3. The GPDs $H_v^u(x, \xi, t)$ evaluated at $t = -0.1$ GeV^2 for four values of $\xi = 0, 0.2, 0.3, 0.4$. The full line is obtained by means of the correlated two body intrinsic function, the dot-dashed lines is obtained by means of the uncorrelated one and dotted is obtained for $\xi = 0$. Left panel: the GPD is evaluated in AC model. Right panel: same quantity of left panel for the BT model.

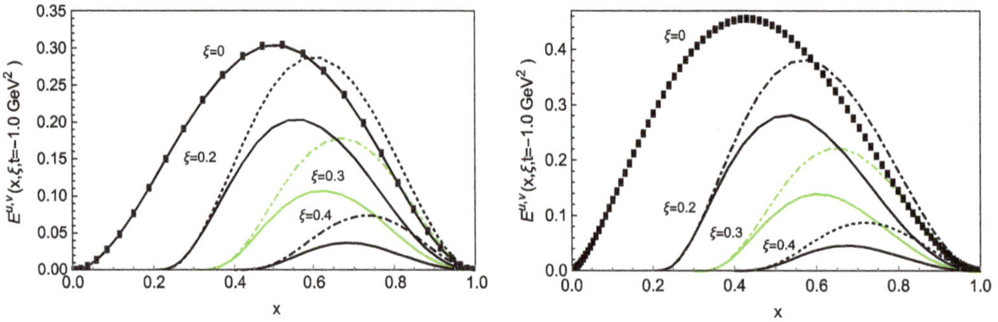

Fig. 4. Same of Fig. 3 but for the GPDs $E_v^u(x, \xi, t)$.

where now the coefficients read:

$$D_1 = 1.48814\,(-0.33029);\quad D_2 = 38.6679\,(1.339837);$$
$$D_3 = 0.208628\,(0.103095);\quad D_4 = 0.01847\,(0.3037);$$
$$D_5 = 3.73552\,(-0.26531). \tag{21}$$

In order to qualitatively show the accuracy of the procedure, the ratios:

$$r_+(z_1) = \frac{\int dz_2\ {}_2\phi_+^2(z_1, z_2)/z_2^4}{\int dz_2\ {}_1\phi_+^2(z_1, z_2)/z_2^4}; \tag{22}$$

$$r_-(z_1) = \frac{\int dz_2\ {}_2\phi_-^2(z_1, z_2)/z_2^4}{\int dz_2\ {}_1\phi_-^2(z_1, z_2)/z_2^4}, \tag{23}$$

have been plotted in Fig. 1.

As one can see the accuracy is quite good for the relevant region, where distributions $\phi_\pm(z)$ are picked, i.e. $3 < z_1 < 8$ GeV^{-1}. With the comfort of these successful checks, in the next section, the result of the calculation of the GPDs at $\xi \neq 0$, within the correlated and uncorrelated ansatz, will be discussed.

5. Numerical analysis of the GPDs at $\xi \neq 0$

In this section, the main results of the calculations of the valence GPDs at $\xi \neq 0$ are presented. The difference between the calculations of the GPDs, performed by means of the uncorrelated and correlated scenarios can be considered as the theoretical error of the present approach. In Fig. 3, the GPD $H_v^u(x, \xi, t)$ has been shown for $t = -1.0$ GeV^2 (left panel for the GPD evaluated within the AC model and right panel within the BT model) for four values of ξ (see caption of Fig. 3). As one can see differences between the GPDs evaluated within the two models can be appreciated also at $t = -1.0$ GeV^2. As one can notice, the shape of GPDs, in the ξ dependence, is basically decreasing, qualitatively in agreement with results discussed in Refs. [42,45], where GPDs have been calculated within constituent quark models. In particular, thanks to the correspondence with the Light-Front approach, the support of the calculated GPDs is correct, i.e., they vanish for $x \geq 1$, at variance of the case where non-relativistic models are adopted. Moreover, by comparing GPDs calculated with correlated and uncorrelated distributions, it is clear that, in this scenario, the fully decreasing shape of the GPDs, w.r.t. the ξ dependence, is related to the presence of correlations. This trend is also confirmed for the GPD $E_v^u(x, \xi, t)$, see Fig. 4. Same results are found for the flavor d. Moreover, for $\xi = 0$, as expected, results discussed in Ref. [28] for the BT model and in Ref. [35] for the AC model, are fully recovered using both the correlated and uncorrelated form of $\phi_\pm(z_1, z_2)$. Furthermore, for the GPDs H, one can notice that the difference between the calculations with the correlated and the uncorrelated two body intrinsic proton functions is quite small. This feature is related to the fact that in the relevant region of the integrals Eqs. (14), (15), the plus component is dominant w.r.t. the minus, for both the correlated and uncorrelated ansatz. Let us remark that since the bulk-to-boundary propagator, Eq. (2), is peaked around $z_1 \sim 0$, the integrals Eqs. (14), (15) are dominated by the small z_1 region. Furthermore, in this range of z_1, the difference between the correlated and uncorrelated calculations of the plus component of the two body intrinsic function (see the full line in Fig. 2) is dramatically small, a feature that explains the small difference in the evaluation of the GPD H in Fig. 3. As one can see in Fig. 2, in fact, in the small z_1 region, the difference between the distributions calculated within the correlated and uncorrelated is less then 10^{-3}, making the calculation of the GPD H with analytic and numerical solutions to Eq. (11) basically the same. Regarding the GPD E, since in this case, as one can see in Eq. (14), this quantity depends only on the minus component of the two body intrinsic proton function, the error between the correlated and uncorrelated ansatz is bigger then in the plus component case. One can realize such feature by looking at the dot-dashed line in Fig. 2. This aspect explains why, in the same kinematic conditions, the error in the calculation of the GPD E is bigger then that in the GPD H case. Let us stress that for both GPDs for very small values of ξ and t, the kinematic condition useful to study Ji's sum rule and the three dimensional partonic structure of the proton, the error in the present approach is quite small. In particular, in this framework, the decreasing trend of the GPDs in the ξ dependence is due to partonic correlations. Furthermore, our analysis shows that if high values of ξ and t are reached in experiments, the GPDs at $\xi \neq 0$ are more sensitive to details of the full proton wave function than form factors or GPDs at $\xi = 0$, allowing to access parton correlations usually integrated out in the diagonal case. This analysis has taught us how distributions sensitive to correlations, like GPDs in non-forward regions, evaluated thanks to the experience gained in the LF and AdS/QCD approaches, can be used to find the importance of correlations in the structure of hadrons. In further studies, this analysis will be completed by calculating, within the present scheme, other distributions, like dPDFs already investigated within the LF approach [46–48], and which have shown to be sensitive to the kind of correlations here addressed. Work is in progress in that direction. A first evaluation of an approximated expression of dPDFs, through the AdS/QCD correspondence, together with the calculation of an experimental observable is discussed in Ref. [49].

6. Conclusion

In the present analysis, GPDs have been calculated in a fully non-forward region, namely ξ and t different from zero. To this aim, the usual strategy, developed to evaluate GPDs at $\xi = 0$, within the AdS/QCD correspondence together with the soft-wall model, has been extended in order to evaluate the full proton Light-Front wave function from AdS/QCD, including, in principle, two body correlations. As shown, within this approach, results previously discussed in other analyses have been successfully recovered and the ξ dependence found for the leading twist GPDs H and E, evaluated for different flavors, is compatible with the one discussed in calculations with constituent quark models. Furthermore, since in the present study the full proton wave function is obtained by solving an integral equation, different solutions have been scrutinized for the numerical evaluations of GPDs and a discussion on the theoretical error of the approach has been provided. In particular, for small values ξ, using different proton wave functions, leading to same form factors and GPDs at $\xi = 0$, similar results have been found. However, since at high values of ξ, in particular for the GPD E, these quantities start to be sensitive to details of the proton structure, e.g., to two body correlations, differences in the calculations of the latter with correlated and uncorrelated distributions become sizable. Results presently discussed demonstrate that in principle new information on partonic structure of hadrons can be obtained from GPDs at $\xi \neq 0$, indirectly accessing two parton correlations, usually studied, e.g. with double parton distribution function in double parton scattering. Moreover, in further studies, thanks to our approach, AdS/QCD predictions will be compared with data on GPDs, e.g., calculating Compton form factors. In closing, our analysis shows that AdS/QCD can be used in the future to estimate other fundamental observables and parton distributions.

Acknowledgements

This work was supported by MINECO under contract FPA2013-47443-C2-1-P and SEV-2014-0398. We warmly thank Sergio Scopetta, Vicente Vento and Marco Traini for many useful discussions.

References

[1] D. Müller, D. Robaschik, B. Geyer, F.-M. Dittes, J. Hořejši, Fortschr. Phys. 42 (1994) 101.
[2] A.V. Radyushkin, Phys. Rev. D 56 (1997) 5524.
[3] X.D. Ji, Phys. Rev. Lett. 78 (1997) 610.
[4] M. Diehl, Phys. Rep. 388 (2003) 41.
[5] M. Burkardt, Int. J. Mod. Phys. A 18 (2003) 173.
[6] A. Airapetian, et al., HERMES Collaboration, Phys. Rev. Lett. 87 (2001) 182001.
[7] S. Stepanyan, et al., CLAS Collaboration, Phys. Rev. Lett. 87 (2001) 182002.
[8] S.J. Brodsky, G.F. de Teramond, Subnucl. Ser. 45 (2009) 139.
[9] S.J. Brodsky, G.F. de Téramond, Phys. Lett. B 582 (2004) 211.
[10] G.F. de Teramond, S.J. Brodsky, AIP Conf. Proc. 1257 (2010) 59.
[11] Z. Abidin, C.E. Carlson, Phys. Rev. D 79 (2009) 115003.
[12] G.F. de Teramond, S.J. Brodsky, Phys. Rev. Lett. 94 (2005) 201601.
[13] S.J. Brodsky, G.F. de Teramond, Phys. Rev. D 77 (2008) 056007.
[14] S.J. Brodsky, G.F. de Teramond, Phys. Rev. D 78 (2008) 025032.
[15] J.M. Maldacena, Int. J. Theor. Phys. 38 (1999) 1113;
J.M. Maldacena, Adv. Theor. Math. Phys. 2 (1998) 231;
S.S. Gubser, I.R. Klebanov, A.M. Polyakov, Phys. Lett. B 428 (1998) 105;
E. Witten, Adv. Theor. Math. Phys. 2 (1998) 253.
[16] G.F. de Teramond, S.J. Brodsky, Phys. Rev. Lett. 102 (2009) 081601.
[17] J. Polchinski, M.J. Strassler, Phys. Rev. Lett. 88 (2002) 031601.
[18] A. Karch, E. Katz, D.T. Son, M.A. Stephanov, Phys. Rev. D 74 (2006) 015005.
[19] T. Branz, T. Gutsche, V.E. Lyubovitskij, I. Schmidt, A. Vega, Phys. Rev. D 82 (2010) 074022.
[20] A. Bacchetta, S. Cotogno, B. Pasquini, arXiv:1703.07669 [hep-ph].
[21] A. Karch, E. Katz, D.T. Son, M.A. Stephanov, Phys. Rev. D 74 (2006) 015005.
[22] A. Vega, I. Schmidt, Phys. Rev. D 78 (2008) 017703.
[23] A. Vega, I. Schmidt, AIP Conf. Proc. 1265 (2010) 226.
[24] A. Vega, I. Schmidt, Phys. Rev. D 79 (2009) 055003.
[25] D. Chakrabarti, C. Mondal, Eur. Phys. J. C 73 (2013) 2671.
[26] A. Ballon-Bayona, G. Krein, C. Miller, arXiv:1702.08417 [hep-ph].
[27] G.F. de Teramond, S.J. Brodsky, arXiv:1203.4025 [hep-ph].
[28] D. Chakrabarti, C. Mondal, Phys. Rev. D 88 (7) (2013) 073006.
[29] S.J. Brodsky, G.F. de Teramond, Phys. Rev. Lett. 96 (2006) 201601.
[30] H. Forkel, M. Beyer, T. Frederico, Int. J. Mod. Phys. E 16 (2007) 2794.
[31] W. de Paula, T. Frederico, H. Forkel, M. Beyer, Phys. Rev. D 79 (2009) 075019.
[32] A. Vega, I. Schmidt, T. Branz, T. Gutsche, V.E. Lyubovitskij, Phys. Rev. D 80 (2009) 055014.
[33] S.J. Brodsky, F.G. Cao, G.F. de Teramond, Phys. Rev. D 84 (2011) 033001.
[34] M. Ahmady, R. Sandapen, Phys. Rev. D 88 (2013) 014042.
[35] A. Vega, I. Schmidt, T. Gutsche, V.E. Lyubovitskij, Phys. Rev. D 83 (2011) 036001.
[36] S.J. Brodsky, G.F. de Téramond, H.G. Dosch, Nuovo Cimento C 036 (2013) 265.
[37] C. Mondal, Eur. Phys. J. C 76 (2) (2016) 74.
[38] J.R. Forshaw, R. Sandapen, Phys. Rev. Lett. 109 (2012) 081601.
[39] J. Polchinski, M.J. Strassler, J. High Energy Phys. 0305 (2003) 012.
[40] M.C. Traini, Eur. Phys. J. C 77 (4) (2017) 246.
[41] M. Ahmady, R. Sandapen, Phys. Rev. D 87 (5) (2013) 054013.
[42] S. Boffi, B. Pasquini, M. Traini, Nucl. Phys. B 649 (2003) 243.
[43] Z. Fang, D. Li, Y.L. Wu, Phys. Lett. B 754 (2016) 343.
[44] S.J. Brodsky, H.C. Pauli, S.S. Pinsky, Phys. Rep. 301 (1998) 299.
[45] S. Scopetta, V. Vento, Eur. Phys. J. A 16 (2003) 527.
[46] M. Rinaldi, S. Scopetta, M. Traini, V. Vento, J. High Energy Phys. 12 (2014) 028.
[47] M. Rinaldi, S. Scopetta, M.C. Traini, V. Vento, J. High Energy Phys. 1610 (2016) 063.
[48] M. Rinaldi, S. Scopetta, M. Traini, V. Vento, Phys. Lett. B 752 (2016) 40.
[49] M. Traini, M. Rinaldi, S. Scopetta, V. Vento, Phys. Lett. B 768 (2017) 270.

9

GUT models at current and future hadron colliders and implications to dark matter searches

Giorgio Arcadi [a,*], Manfred Lindner [a], Yann Mambrini [b], Mathias Pierre [b], Farinaldo S. Queiroz [a]

[a] Max-Planck-Institut für Kernphysik, Saupfercheckweg 1, 69117 Heidelberg, Germany
[b] Laboratoire de Physique Théorique, CNRS, Univ. Paris-Sud, Université Paris-Saclay, 91405 Orsay, France

ARTICLE INFO

Editor: A. Ringwald

Dedicated to the memory of Pierre Binétruy (1955–2017)

ABSTRACT

Grand Unified Theories (GUT) offer an elegant and unified description of electromagnetic, weak and strong interactions at high energy scales. A phenomenological and exciting possibility to grasp GUT is to search for TeV scale observables arising from Abelian groups embedded in GUT constructions. That said, we use dilepton data (ee and $\mu\mu$) that has been proven to be a golden channel for a wide variety of new phenomena expected in theories beyond the Standard Model to probe GUT-inspired models. Since heavy dilepton resonances feature high signal selection efficiencies and relatively well-understood backgrounds, stringent and reliable bounds can be placed on the mass of the Z' gauge boson arising in such theories. In this work, we obtain 95% C.L. limits on the Z' mass for several GUT-models using current and future proton–proton colliders with $\sqrt{s} = 13$ TeV, 33 TeV, and 100 TeV, and put them into perspective with dark matter searches in light of the next generation of direct detection experiments.

1. Introduction

Many popular extensions of the Standard Model (SM) rely on the existence of new spin-1 states, possibly with sizable couplings with the SM fermions. This new fields can be promptly interpreted as the gauge bosons of a new $U(1)$ symmetry, spontaneously broken above the Electroweak (EW) scale.[1] An appealing case would be a broken scale not above few TeV, since this would make the new particle, typically dubbed Z', accessible by collider searches.[2]

The simplest way to couple the Z' to SM fermions consists in assuming that the latter are charged under the new symmetry group. Even if this is not the case a coupling with the SM can be provided by a kinetic mixing term [3–5] between the Hypercharge field strength and the one of the new boson. It might be already argued that a natural embedding of this setup is represented by Grand Unified Theories (GUT). Interestingly the minimal viable GUT groups, like e.g. $SO(10)$, have higher rank as the SM group $SU(3)_c \times SU(2)_L \times U(1)_Y$. Under suitable conditions the GUT group can be spontaneously broken, at a first stage, into the SM and an additional U(1) component with the latter, spontaneously broken at some scale above the EW one. As already pointed, a phenomenologically interesting scenario is represented by the case in which the intermediate group is broken to the SM at scales not exceeding few TeVs.

In case the Z' boson can be produced at current colliders, and if the Z' boson possessed sizable couplings to the SM fermions, it would provide a clear signal represented by resonances in the dilepton, dijet final states peaked at the Z' mass. These searches have been intensively conducted at the LHC and many of the results have been made available [6–8].

Indeed, at the LHC, the high selection efficiencies and relatively small and well understood backgrounds make the dilepton channel a great laboratory probe of new physics at the TeV scale. For these reasons ATLAS and CMS collaborations have reported the most stringent bounds on some Z' models that have sizable couplings to charged leptons [9,10]. Recently ATLAS collaboration has collected at $\sqrt{s} = 13$ TeV, 13.3 fb^{-1} of integrated luminosity to exclude Z' masses below 3–4 TeV at 95% C.L. [11]. For the Sequential Standard Model, which stands for a Z' boson that couples to SM fermions precisely like the Z boson, the lower mass limit of 4.05 TeV was derived. Additional models were investigated such as the GUT-inspired model $U(1)_\eta$, Z'_η for short, which yielded a lower mass limit of 3.43 TeV.

* Corresponding author.
E-mail address: arcadi@mpi-hd.mpg.de (G. Arcadi).

[1] Alternatively, the Stueckelberg mechanism [1,2], could be considered for the generation of the mass of the new bosons.

[2] This work is focussed on Abelian extensions of the SM. Non-Abelian extensions of the SM are similarly interesting, featuring the presence of both Z' and W' bosons; however they will not be explicitly discussed here.

Motivated by the theoretical importance of these models and the upcoming data collection at the LHC, and future generation of proton–proton colliders, in this letter we cast 95% C.L. bounds on mass of the Z' gauge boson arising in many GUT-inspired models, complementary to previous collider studies. In particular, we will extend the present LHC limits to GUT-inspired models, and also provide projected limits for $\sqrt{s} = 13, 14, 33, 100$ TeV, for a variety of integrated luminosities, reaching up to 5 ab^{-1} in the case of a 100 TeV collider.[3]

Moreover, we put our finding into perspective with dark matter endeavors in light of ongoing and next generation of direct detection experiments, namely XENONnT, LZ, and Darwin [13–19].[4]

Since both collider and direct dark matter detection observables are dictated by the Z' interactions an interesting degree of complementarity between these searches is expected [26–35] as we discuss further on.

The paper is organized as follows. In the next section we will describe in more detail the benchmark scenarios adopted in our study. In section 3 we will discuss our analysis procedure and present the limits we obtained. In section 4, we exploit the complementarity between direct dark matter detection and collider searches for Z' bosons before concluding.

2. GUT models

A Grand Unified Theory (GUT) is a model where the three gauge interactions of the SM which govern the electromagnetic, weak, and strong interactions degenerate into one value, i.e. an unified interaction. This unified description of these forces is characterized by a larger gauge group, such as $SO(10)$ and E_6, spontaneously broken at a scale M_{GUT}, typically above 10^{16} GeV to respect the proton lifetime constraints.

New particles predicted by GUT models are expected to have masses around the GUT scale, thus beyond the reach of any foreseen collider experiments. Nevertheless, signs of grand unification at high energy scales take place via (for instance) fast proton decay or electric dipole moments of elementary particles [36]. It is however possible that the breaking of large groups like $SO(10)$ or E_6 to the SM gauge group occurs through different phases, opening the possibility of the existence of states at an intermediate lower scales, possibly accessible to collider experiments. TeV scale manifestations of Grand Unification can be searched via the signal of a Z' gauge boson that possesses coupling strength with SM fermions as predicted by GUT constructions. In our work we will consider generic Z' models which correspond to Grand Unification through $SO(10)$ and E_6 symmetry groups as proposed in [37,38].

$SO(10)$ is a rank-5 group, thus allowing for an extra $U(1)$ component with respect to the SM gauge group. A very natural one is represented by $B-L$, with B and L being respectively the baryon and lepton numbers, as new (spontaneously broken) symmetry. We will consider, in alternative, the case in which the Z' originates from the Left-Right symmetry, which can be described by the following breaking pattern for $SO(10)$ [36,39–49]: $SO(10) \to SU(3)_C \times SU(2)_L \times U(1)_R \times U(1)_{B-L}$.[5] The $U(1)_R \times U(1)_{B-L}$ is then broken to $U(1)_Y$ at a scale $M_{Z'} > M_Z$. The Z' particle relevant for DM phenomenology is a mixture of the gauge bosons of the two $U(1)$ components (as a consequence its coupling with SM fermions rely on a linear combination of their R and $B-L$ charges).

[3] Our analysis is focused on hadron colliders. Linear e^+e^- would be also an interesting option [12].

[4] There are other important direct detection experiments planned for the future but not particularly sensitive to ours models [20–25].

[5] One could also consider $SO(10) \to SU(3)_C \times SU(2)_L \times SU(2)_R \times U(1)_{B-L}$.

Table 1
Table of couplings of the SSM and GUT-inspired models under investigation.

D	Z'_χ	Z'_ψ	Z'_η	Z'_{LR}	Z'_{B-L}	Z'_{SSM}
	$2\sqrt{10}$	$2\sqrt{6}$	$2\sqrt{15}$	$\sqrt{5/3}$	1	1
$\hat{\epsilon}^u_L$	−1	1	−2	−0.109	1/6	$\frac{1}{2} - \frac{2}{3}\sin^2\theta_W$
$\hat{\epsilon}^d_L$	−1	1	−2	−0.109	1/6	$-\frac{1}{2} + \frac{1}{3}\sin^2\theta_W$
$\hat{\epsilon}^u_R$	1	−1	2	0.656	1/6	$-\frac{2}{3}\sin^2\theta_W$
$\hat{\epsilon}^d_R$	−3	−1	−1	−0.874	1/6	$\frac{1}{3}\sin^2\theta_W$
$\hat{\epsilon}^\nu_L$	3	1	1	0.327	−1/2	$\frac{1}{2}$
$\hat{\epsilon}^l_L$	3	1	1	0.327	−1/2	$-\frac{1}{2} + \sin^2\theta_W$
$\hat{\epsilon}^e_R$	1	−1	2	−0.438	−1/2	$\sin^2\theta_W$

A larger variety of Z' models is based on the E_6 gauge group. Indeed, given its higher rank, two extra $U(1)$'s, with respect to the SM gauge group, can be embedded in it. Among the many possible decompositions of E_6, two anomaly free gauge groups arise by the following breaking pattern [37]: $E_6 \to SO(10) \times U(1)_\psi$, $SO(10) \to SU(5) \times U(1)_\chi$. The Z' associated to the collider phenomenology is, in general, a linear combination of the two components associated to the two $U(1)$'s and schematically expressed as: $Z' = \cos\theta_{E_6} Z'_\chi + \sin\theta_{E_6} Z'_\psi$. In this work we will consider three specific assignations for the angle θ_{E_6}: pure Z'_χ and Z'_ψ, thus corresponding, respectively, to $\theta_{E_6} = \pi/2$ and $\theta_{E_6} = 0$, and a string theory inspired scenario, $Z_\eta = \sqrt{\frac{3}{8}} Z_\chi - \sqrt{\frac{5}{8}} Z_\psi$. Interestingly, the Z'_χ model features very similar interactions to models based on the $SU(3)_L$ gauge group [50–58].

As comparison we will also include in our analysis the so called Sequential Standard Model (SSM) consisting in the same assignation as the SM Z-boson of the couplings of the Z' with SM fermions.

For our phenomenological study we can encode all the considered scenarios in a Lagrangian of the form:

$$\mathcal{L} = \sum_f g_f \bar{f} \gamma^\mu \left(\epsilon^f_L P_L + \epsilon^f_R P_R\right) f Z'_\mu \tag{1}$$

where $g_f = g \approx 0.65$ in the case of the SSM and $g_f = g_{GUT} = \sqrt{\frac{5}{3}} g \tan\theta_W \approx 0.46$ for the GUT inspired constructions. ϵ^f_L (ϵ^f_R) are the couplings associated to the left (right)-handed and their values, for the considered models, are reported in Table 1 (notice that we have used the parametrization $\epsilon^f_{L,R} = \frac{\hat{\epsilon}^f_{L,R}}{D}$ [38]).

Notice that we are considering the case where the couplings of the Z' with the SM fermions are determined only by the quantum numbers of the latter with respect to the new symmetry groups. Additional couplings might arise from Z/Z' mixing induced either by direct mass mixing or by kinetic mixing. We remark, in particular, that kinetic mixing between the field strengths of different $U(1)$ gauge bosons is not forbidden neither by Lorentz not by gauge invariance. Furthermore, even once set to zero at the three level, it could be radiatively generated in presence of fermions charged under both the $U(1)$ components. A kinetic mixing term between the Z' and the Z would induce, after EW symmetry breaking mass mixing between the two states and change the couplings of the Z' with respect to the values reported in Table 1. For simplicity we will assume that kinetic mixing between the Z' and the Z boson is negligible. If this is not the case interesting phenomenological possibilities would be accessible. Mixing between the Z and the Z' would induce for the latter trilinear couplings with W^+W^- and with Zh. This last coupling could be used to explain the excess in $qqbb$, recently reported by the ATLAS collaboration [59,60], through the resonant production of a Z' with mass of 3–3.2 TeV decaying into Zh. Interestingly one of model which we are going to present, the $E_{6\psi}$ (see Table 2) allows values of the Z' mass, still compatible with limits from dileptons, in this ballpark. An interpretation of the excess reported in [59] would be also feasible in LR models based on the $SU(2)_R \times SU(2)_L \times U(1)_{B-L}$ group [61]. Have thus far presented the benchmark models under study we will now move to the collider analysis.

3. Dilepton limits

Searches for isolated lepton pairs in the final state are considered to be a clean environment to probe new physics at the TeV scale, we will briefly review the reason for such. The most relevant background contributions arise from the Drell–Yann processes. In the dielectron data, Top Quarks, Diboson, Multi-jet and W+jets also subdominantly contribute to the background. The misidentification of jets as electrons also known as jet-fake rate give rise to the multi-jet and W+jets channels as background events. To suppress background from misidentified jets as well as from hadron decays inside jets, electrons are required to satisfy some isolation criteria. It is required that the transverse energy (E_T) deposited by the electron to be contained in a cone of size $\Delta R = 0.2$. Moreover, events with electrons with pseudo-rapidity ($1.37 < |\eta| < 1.52$) are removed from the analysis because this transition region between the central and forward regions of the calorimeters feature degraded energy resolution. Lastly, with no charge identification requirement, the electrons have to be isolated, within a cone of size $\Delta R = 10$ GeV$/p_T$, where p_T is the transverse momentum of the electron track. These criteria lead to a selection efficiency of about 87%, and product of acceptance-efficiency of nearly 70% for TeV scale dielectron resonances [11].

In the dimuon channel, the background from multi-jet and W+jets are irrelevant, since muon misidentification rate is relatively small. In this case only Drell–Yann, Top Quarks, Diboson are important. As for muons opposite-charge assignments are applied. The efficiency is about 94%, but the product acceptance-efficiency is degraded compared to the dielectron, being about 44% for TeV scale dimuon resonances [11].

Systematic uncertainties in the dielectron channel are at the level of 7% (25%) for the signal (background), whereas for the dimuon channel, systematic uncertainties read about 17% (25%) for the signal (background). The much larger systematic uncertainties in the dimuon channel are due to uncertainties in the reconstruction efficiency which are in the ballpark of 16% [11].

That said, to clearly see the power that heavy dilepton resonance searches have at probing new physics we need to also compute the number of signal events. To evaluate the impact of the 13 TeV LHC search for dilepton resonances with 13.3 fb^{-1} of integrated luminosity [11], we simulate the process $pp \to Z' \to e^+e^-$ allowing for the presence of jets but requiring the charged leptons to be isolated using MadGraph5 [62], clustering and hadronizing jets were held within Pythia [63], and simulating detector effects accounted for with Delphes3 [64]. We have adopted the CTEQ6L parton distribution functions throughout.

The signal events were selected with the same criteria used by [11] which in summary read:

- $E_T(e_1) > 30$ GeV, $E_T(e_2) > 30$ GeV, $|\eta_e| < 2.5$,
- $p_T(\mu_1) > 30$ GeV, $p_T(\mu_2) > 30$ GeV, $|\eta_\mu| < 2.5$,
- 80 GeV $< M_{ll} < 6000$ GeV,

where M_{ll} is the invariant mass of the lepton pair used to enhance signal-to-noise ratio. In the analysis, the presence of a narrow resonance in the dilepton invariant mass has been assumed. In Fig. 1 we show the differential cross section as function of the dilepton invariant mass for the B-L model at 13 TeV. A narrow resonance is rather visible. From Fig. 1, one can clearly see the pronounced peak

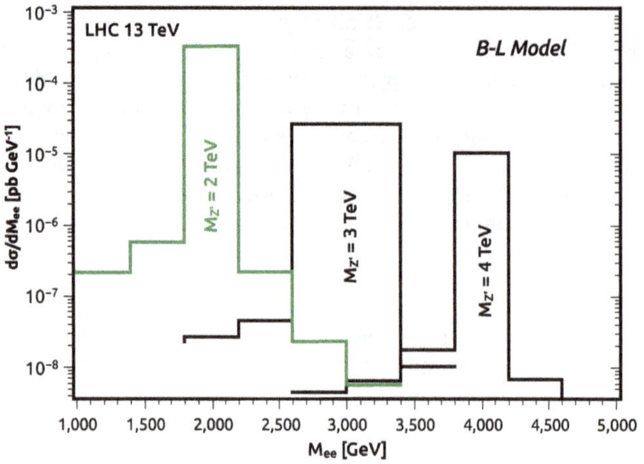

Fig. 1. Differential cross section for the dielectron channel at 13 TeV. The p_T and E_T cuts given in Sec. 3 were applied to the figure. (For interpretation of the references to color in this figure legend, the reader is referred to the web version of this article.)

in the dilepton invariant mass coinciding with the mass of the Z' gauge boson. A similarly behavior appears also in the remaining models except for the SSM, which has a mildly large decay width. Anyways, we will see further that our results fully agree with ones presented by ATLAS collaboration regarding the SSM.

Having described in detail all important ingredients for study, we can provide a more quantitative taste of dilepton searches. By looking back at Fig. 1 and taking the bin with invariant dielectron invariant mass of 3000–6000 GeV, one can infer that with 13.3 fb^{-1} of integrated luminosity at 13 TeV around 100 events are expected from the B-L model. However, ATLAS collaboration has measured only 0.1 ± 0.026 events [11]. Since the new physics contribution far exceeds any statistical or systematic error, a reliable bound can be derived on the B-L model. A similar reasoning can be applied to all models under study. By repeatedly doing this comparison in a bin-by-bin model independent basis ATLAS collaboration placed 95% C.L. limits on the underlying particle physics input quantity namely production cross section times branching ration into charged leptons (dielectron + dimuon), represented by a black solid line in Fig. 2. That said, we computed the production cross section for the six models under study following the receipt aforementioned and compared with the 95% upper limit from ATLAS collaboration as shown in Fig. 2. Our results for the SSM, Z_χ and Z_η models agree well with the ones reported by ATLAS collaboration in [11], as well as with the analysis of [65,66] concerning the B-L model. See [67–69] for other complementary studies of GUT models at the LHC.

Moreover, we obtain projected sensitivities for $\sqrt{s} = 33$ and 100 TeV energies having in mind the proposed proton–proton colliders, namely the high-energy LHC [70–72] and 100 TeV collider [73–77]. The former is proposed to reach a center-of-energy of 33 TeV and up to 300 fb^{-1}, whereas the latter is projected to reach from $\mathcal{L} \sim 1$–30 ab^{-1} [73]. We conservatively adopt $\mathcal{L} = 5$ ab^{-1}. In order to derive projected bounds we follow the recommendation of the CERN code that yields reasonable predictions [78], and we solve for $M_{\rm new}$ the equation,

$$\frac{N_{\rm signal\ events}(M_{\rm new}^2, E_{\rm new}, \mathcal{L}_{\rm new})}{N_{\rm signal\ events}(M^2, 13\ {\rm TeV}, 13.3\ {\rm fb}^{-1})} = 1, \tag{2}$$

where M is the current bound on the Z' mass, and the number of events is estimated by computing the production cross section at a given center-of-energy with certain luminosity. This procedure has

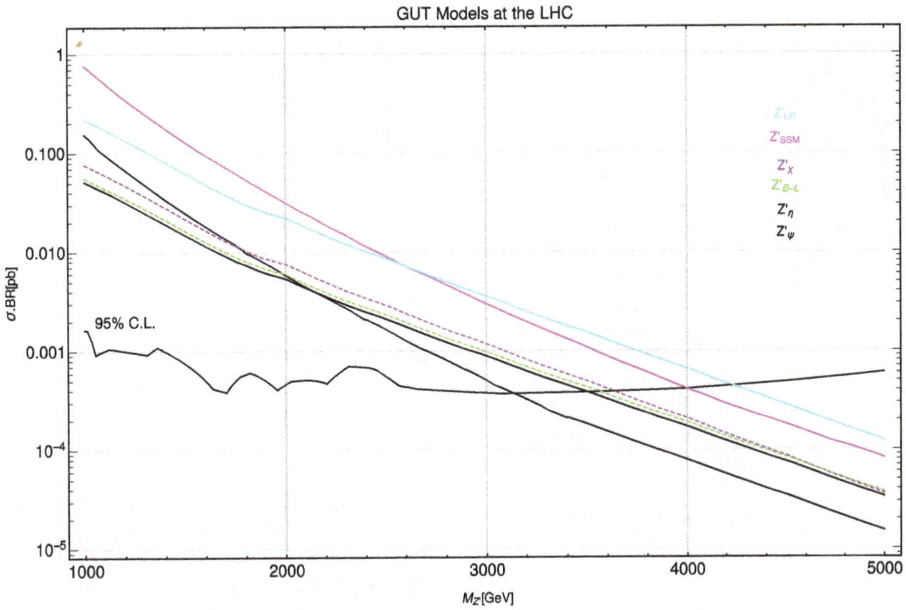

Fig. 2. Production cross section times branching ratio for the combined dilepton channel ($ee + \mu\mu$) at 13 TeV. The black curve is the 95% C.L. limits from ATLAS using 13.3 fb^{-1} of integrated luminosity. From *bottom to top* the curves delimit the results for Z'_ψ, Z'_η, Z'_{B-L}, Z'_χ, Z'_{SSM} and Z'_{LR} models. A table with lower mass bounds for current and planned future hadron colliders can be found below. (For interpretation of the references to color in this figure legend, the reader is referred to the web version of this article.)

Table 2
Summary of current and projected bounds on the Z' mass in the SSM and various GUT models, in light of current and future proton–proton colliders.

Model	13 TeV, 13.3 fb^{-1}	13 TeV, 37 fb^{-1}	14 TeV, 100 fb^{-1}	14 TeV, 300 fb^{-1}	33 TeV, 100 fb^{-1}	33 TeV, 300 fb^{-1}	100 TeV, 5 ab^{-1}
Z'_ψ	3.13 TeV	3.68 TeV	4.46 TeV	5.13 TeV	7.98 TeV	9.47 TeV	30.54 TeV
Z'_η	3.47 TeV	4.04 TeV	4.85 TeV	5.51 TeV	8.85 TeV	10.38 TeV	33.25 TeV
Z'_{B-L}	3.55 TeV	4.11 TeV	5.55 TeV	5.59 TeV	9.03 TeV	10.56 TeV	33.8 TeV
Z'_χ	3.63 TeV	4.19 TeV	5.55 TeV	5.68 TeV	9.23 TeV	10.76 TeV	34.41 TeV
Z'_{SSM}	4.02 TeV	4.59 TeV	6.05 TeV	6.09 TeV	10.21 TeV	11.75 TeV	37.36 TeV
Z'_{LR}	4.23 TeV	4.8 TeV	6.27 TeV	6.31 TeV	10.73 TeV	12.28 TeV	38.92 TeV

been validated in [79], where the predictions for 13 TeV results agree well with experimental limits.

Our results are summarized in Table 2. It is clear that a major sensitivity boost is expected when ramping up the center-of-energy from 14 TeV to 33 TeV, where Z' masses near 10 TeV become available. Furthermore, a 100 TeV collider with the modest luminosity of 5 ab^{-1} is sensitivity to Z' masses between 30–39 TeV. Hopefully these discovery machines will be built and spot a signal at the multi-TeV scale [73–77,80]. See [81–90] for other interesting sensitivity reach of a 100 TeV collider.

In the next section, our results based on collider physics will be put into perspective with dark matter searches at direct detection experiments.[6]

4. Connection to dark matter

The nature of dark matter is one of the most fascinating puzzle in science [103,104]. In order to unveil its nature it is desirable to collect data across different but complementary search strategies, such as collider and direct detection. Vector mediators are a special example in this direction since both collider and direct detection observables are strongly dictated by the Z' properties. In particular, a Z' boson represents an attractive portal for interaction within the WIMP paradigm.[7] In a GUT inspired framework the DM would be represented by a new particle state belonging to a suitable representation such that it has not trivial quantum numbers with respect to the symmetry group associated to the Z' while being singlet with respect to the SM group.[8] Interestingly the stability of the DM does not require the imposition of ad hoc discrete (or global) symmetries for its stability, as customary done in simplified realizations [26,28,65,107,108], since they might naturally arise as remnants in the different breaking steps of the GUT group [106].[9] The eventual presence of a DM candidate might have a sizable impact on collider phenomenology. Indeed, in case the $Z' \to$ DM DM decay process is kinematically allowed, a sizable invisible branching ratio would weaken the limit from searches of dilepton resonances [111] since the corresponding cross section is reduced by a factor $1 - \mathrm{BR}(Z' \to \mathrm{DM\ DM})$. This creates an interesting complementarity with Dark Matter searches as well as the DM relic density, since they constraint the possible values of BR($Z' \to$ DM DM). LHC limits have, in turn, impact on the DM phenomenology. For example, too strong limits on the

[6] We will ignore indirect dark matter detection limits [91–93], as well as limits from flavor physics since for the models under study they are rather subdominant [31,45,94–102].

[7] Viable DM can be accommodated in GUT frameworks also without relying on the WIMP paradigm [105,106].

[8] Notice that, in general, in this kind of construction the DM is actually part of a multiples so other states might be relevant for its phenomenology.

[9] See e.g. here [57,99,109,110] for alternative examples of natural emergence of DM stability.

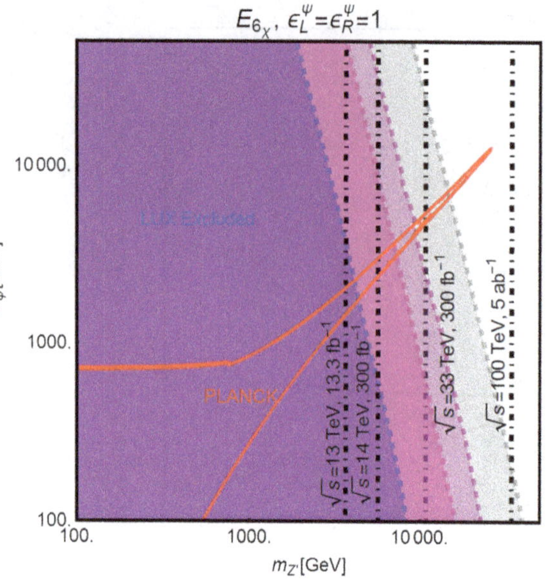

Fig. 3. Comparison between DM current/projected constraints and current/projected constraints on the mass of the Z' from collider searches. The DM has been chosen to be a Dirac fermion and the couplings of the Z' with the SM fermions are dictated by the $E_{6,\chi}$ model. In the plot the red line represents the isocontour of the correct DM relic density. The region at the left of the blue dashed line is ruled out by DD constraints by LUX [115]. Regions at the left of the magenta, purple and gray dashed lines correspond, respectively, to the projected sensitivities of XENON1T [15], LZ [18] and Darwin [19]. The black lines represent current (first line on the left) and projected exclusions by LHC of dilepton resonances (the corresponding values of center of mass energy and luminosity are reported in vicinity of the lines). The region at the left of each line should be regarded as experimentally ruled out in case no signal is detected at the values of center of mass energy and luminosity reported in proximity of the line itself. (For interpretation of the references to color in this figure legend, the reader is referred to the web version of this article.)

mass of the Z' would correspond in general to a suppressed pair annihilation cross section, hence implying an overabundant DM[10] (see next section for more details).

We have shown, in Fig. 3, an example of this kind of complementarity (this topic has been more extensively reviewed e.g. in [35]).

We have focused here on the case of a Dirac DM candidate ψ coupled to the Z'_χ. The relevant Lagrangian for DM interactions can be written in analogous way as the one of the SM fermions:

$$\mathcal{L} = g_f \bar{\psi}\gamma^\mu \left(\epsilon_L^\psi P_L + \epsilon_R^\psi P_R\right)\psi Z'_\mu \tag{3}$$

We have set for simplicity $\epsilon_L^\psi = \epsilon_R^\psi = 1$ (this choice is actually rather special since would imply only vectorial coupling of the DM with the Z' but it is nevertheless not problematic given the illustrative purposes of the Fig. 3[11])

For what regards the DM phenomenology we have required the correct DM relic density according the WIMP paradigm, i.e. $\Omega h^2 \propto 1/\langle\sigma v\rangle$ where $\langle\sigma v\rangle$ is the thermally averaged DM pair annihilation cross-section. The experimentally favored value $\Omega h^2 \approx$ 0.12 [118] corresponds to an annihilation cross-section of the order of $10^{-26}\text{cm}^3\text{s}^{-1}$ (for details on the quantitative determination of the relic density we refer to [35]). Regarding DM searches we have focused on Direct Detection, which provides the strongest constraints in the considered scenarios, which relies on Spin Independent (SI) interactions of the DM with nucleons which are described by a cross-section of the form (for definiteness we consider the case of scattering on protons):

$$\sigma_{\psi,p}^{\text{SI}} = \frac{g_f^4 \mu_{\psi p}^2}{\pi m_{Z'}^4} V_\psi^2 \left[f_p \frac{Z}{A} + f_n \left(1 - \frac{Z}{A}\right)\right]^2 \tag{4}$$

where $f_p = 2V_u + V_d$, $f_n = V_u + 2V_d$, $\mu_{\psi p}$ is the DM-proton reduced mass while Z and A represent the number of protons and the total number of nucleons of the detector material (we will consider Xenon-type detectors). The parameters $V_{\psi,u,d}$ in the equation above represent the vectorial coupling of the DM, up and down quarks to the Z', i.e. $V_{f=u,d,\psi} = \frac{1}{2}(\epsilon_L^f + \epsilon_R^f)$.

Fig. 3 reports in a bidimensional plane of the DM and Z' masses the curve of the correct DM relic density, and the current limits, provided at present times by the LUX experiment [115], and projected limits next future experiments. For these we have considered the Xenon1T [15], LZ [18] and Darwin [19] experiments. Notice that the sensitivity of the Darwin experiment is comparable to the expected value of the cross-section associated to coherent neutrino scattering on nuclei, which represents somehow the ultimate reach of experiment probing elastic scattering of WIMPs on nuclei. The DM constraints have been compared with the current LHC exclusion limit[12] from dilepton resonance searches as well as the maximal reaches for the three values of the center of mass energy considered in this work (see Table 2). As evident future collider limits can overcome current and future limits by Direct Detection. For the chosen assignation of the parameters, an absence of signals at a 100 TeV collider would completely rule out the WIMP hypothesis in this framework.

5. Conclusions

Grand Unified Theories (GUT) provide an unified description of electromagnetic, weak and strong interactions at high energy scales around 10^{16} GeV. These kind of theories can be nevertheless probed through collider studies in the case the GUT gauge symmetry group is broken, before EW symmetry breaking, in some subgroup larger than the SM gauge group. A phenomenological and gripping method to probe this kind of scenario consist in to study Abelian groups, which can be embedded in GUT frameworks, which thus predict the existence of a new neutral gauge boson, a Z', whose couplings with the SM fermions are dictated by the breaking patter of the GUT group itself. The observation of a signal of a Z' at TeV scale, with interactions strength as predicted by GUT-inspired models, would then constitute a hint of GUT at high energy scales. In light of the current null results, we used up-to-date dilepton data from LHC to derive the lower mass bounds for several GUT models. Moreover, we casted projected limits having in mind possible future colliders namely, the high-energy LHC and 100 TeV collider, with the latter being able to probe Z' masses around 38 TeV. Lastly, we put our findings into the perspective of a connection with the DM problem. Interpreting the Z' at the mediator (portal) of the interactions between the DM and the SM fermions, we have exploited, in a simple example with

---

10 This issue could be overcome in the case that additional particle states influence DM relic density, e.g. through coannihilations, or by invoking non-thermal DM production [112,113] or more generally, modified cosmological histories [114].

11 Notice also that an axial coupling of the DM with the Z' [116] could potentially lead to violation of unitarity from the annihilation process $\psi\psi \to Z'Z'$. [117]. This problem is automatically cured in UV complete frameworks, like a GUT theory, by the presence of the Higgs fields responsible of the breaking of the $U(1)$ gauge symmetry associated to the Z'. A proper treatment would then require an explicit construction of the GUT model, which is not in the purpose of this letter.

12 For parameters adopted in the analysis the invisible branching fraction of the Z' is typically rather small and the impacts in a negligible way limits from resonance searches.

Dirac fermion DM, the complementarity between collider searches and DM direct detection experiments.

Acknowledgements

The authors warmly thank Alexandre Alves, Carlos Yaguna, Werner Rodejohann for fruitful discussions. This work is also supported by the Spanish MICINN's Consolider-Ingenio 2010 Programme under grant Multi-Dark **CSD2009-00064**, the contract **FPA2010-17747**, the France-US PICS no. 06482 and the LIA-TCAP of CNRS. Y. M. acknowledges partial support the ERC advanced grants Higgs@LHC and MassTeV. This research was also supported in part by the Research Executive Agency (REA) of the European Union under the Grant Agreement **PITN-GA2012-316704** ("HiggsTools").

References

[1] E.C.G. Stueckelberg, Helv. Phys. Acta 11 (1938) 299.
[2] H. Ruegg, M. Ruiz-Altaba, Int. J. Mod. Phys. A 19 (2004) 3265, arXiv:hep-th/0304245.
[3] B. Holdom, Phys. Lett. B 166 (1986) 196.
[4] B. Holdom, Phys. Lett. B 259 (1991) 329.
[5] K.S. Babu, C.F. Kolda, J. March-Russell, Phys. Rev. D 57 (1998) 6788, arXiv:hep-ph/9710441.
[6] G. Aad, et al., ATLAS, J. High Energy Phys. 11 (2012) 138, arXiv:1209.2535.
[7] ATLAS Collaboration, http://inspirehep.net/search?ln=it&ln=it&p=ATLAS%3A2012ipa&of=hb&action_search=Cerca&sf=earliestdate&so=d&rm=&rg=25&sc=0, 2013.
[8] G. Aad, et al., ATLAS, Phys. Rev. D 90 (2014) 052005, arXiv:1405.4123.
[9] M. Aaboud, et al., ATLAS, Eur. Phys. J. C 76 (2016) 541, arXiv:1607.08079.
[10] M. Aaboud, et al., ATLAS, Phys. Lett. B 761 (2016) 372, arXiv:1607.03669.
[11] ATLAS Collaboration, http://inspirehep.net/search?ln=it&ln=it&p=ATLAS%3A2016cyf&of=hb&action_search=Cerca&sf=earliestdate&so=d&rm=&rg=25&sc=0, 2016.
[12] F. Richard, G. Arcadi, Y. Mambrini, Eur. Phys. J. C 75 (2015) 171, arXiv:1411.0088.
[13] E. Aprile, XENON1T collaboration, arXiv:1206.6288, 2012.
[14] L. Baudis, DARWIN Consortium, J. Phys. Conf. Ser. 375 (2012) 012028, arXiv:1201.2402.
[15] E. Aprile, et al., XENON, J. Cosmol. Astropart. Phys. 1604 (2016) 027, arXiv:1512.07501.
[16] A. Rizzo, XENON, EPJ Web Conf. 121 (2016) 06009.
[17] D.N. McKinsey, LZ, J. Phys. Conf. Ser. 718 (2016) 042039.
[18] M. Szydagis, LUX, LZ, in: 38th International Conference on High Energy Physics, ICHEP 2016, Chicago, IL, USA, August 03–10, 2016, 2016, arXiv:1611.05525.
[19] J. Aalbers, et al., DARWIN, J. Cosmol. Astropart. Phys. 1611 (2016) 017, arXiv:1606.07001.
[20] P. Agnes, et al., DarkSide, Phys. Lett. B 743 (2015) 456, arXiv:1410.0653.
[21] P. Agnes, et al., DarkSide, Phys. Rev. D 93 (2016) 081101, Addendum: Phys. Rev. D 95 (6) (2017) 069901, arXiv:1510.00702.
[22] G. Angloher, et al., CRESST, Eur. Phys. J. C 76 (2016) 25, arXiv:1509.01515.
[23] L. Hehn, et al., EDELWEISS, Eur. Phys. J. C 76 (2016) 548, arXiv:1607.03367.
[24] C. Amole, et al., PICO, Phys. Rev. D 93 (2016) 061101, arXiv:1601.03729.
[25] C. Amole, et al., PICO, arXiv:1702.07666, 2017.
[26] M.T. Frandsen, F. Kahlhoefer, A. Preston, S. Sarkar, K. Schmidt-Hoberg, J. High Energy Phys. 07 (2012) 123, arXiv:1204.3839.
[27] G. Arcadi, Y. Mambrini, M.H.G. Tytgat, B. Zaldivar, J. High Energy Phys. 03 (2014) 134, arXiv:1401.0221.
[28] A. Alves, S. Profumo, F.S. Queiroz, J. High Energy Phys. 04 (2014) 063, arXiv:1312.5281.
[29] O. Buchmueller, M.J. Dolan, S.A. Malik, C. McCabe, J. High Energy Phys. 01 (2015) 037, arXiv:1407.8257.
[30] A. De Simone, G.F. Giudice, A. Strumia, J. High Energy Phys. 06 (2014) 081, arXiv:1402.6287.
[31] A. Alves, A. Berlin, S. Profumo, F.S. Queiroz, Phys. Rev. D 92 (2015) 083004, arXiv:1501.03490.
[32] M. Chala, F. Kahlhoefer, M. McCullough, G. Nardini, K. Schmidt-Hoberg, J. High Energy Phys. 07 (2015) 089, arXiv:1503.05916.
[33] T. Jacques, A. Katz, E. Morgante, D. Racco, M. Rameez, A. Riotto, J. High Energy Phys. 10 (2016) 071, arXiv:1605.06513.
[34] M. Fairbairn, J. Heal, F. Kahlhoefer, P. Tunney, J. High Energy Phys. 09 (2016) 018, arXiv:1605.07940.
[35] G. Arcadi, M. Dutra, P. Ghosh, M. Lindner, Y. Mambrini, M. Pierre, S. Profumo, F.S. Queiroz, arXiv:1703.07364, 2017.
[36] P. Langacker, Phys. Rep. 72 (1981) 185.
[37] P. Langacker, Rev. Mod. Phys. 81 (2009) 1199, arXiv:0801.1345.
[38] T. Han, P. Langacker, Z. Liu, L.-T. Wang, arXiv:1308.2738, 2013.
[39] J.L. Hewett, T.G. Rizzo, Phys. Rep. 183 (1989) 193.
[40] A. Abada, P. Hosteins, F.-X. Josse-Michaux, S. Lavignac, Nucl. Phys. B 809 (2009) 183, arXiv:0808.2058.
[41] M. Lindner, M. Weiser, Phys. Lett. B 383 (1996) 405, arXiv:hep-ph/9605353.
[42] A. Dueck, W. Rodejohann, J. High Energy Phys. 09 (2013) 024, arXiv:1306.4468.
[43] C. Arbeláez, M. Hirsch, M. Malinský, J.C. Romão, Phys. Rev. D 89 (2014) 035002, arXiv:1311.3228.
[44] R.L. Awasthi, M.K. Parida, S. Patra, J. High Energy Phys. 08 (2013) 122, arXiv:1302.0672.
[45] S. Patra, F.S. Queiroz, W. Rodejohann, Phys. Lett. B 752 (2016) 186, arXiv:1506.03456.
[46] F.F. Deppisch, L. Graf, S. Kulkarni, S. Patra, W. Rodejohann, N. Sahu, U. Sarkar, Phys. Rev. D 93 (2016) 013011, arXiv:1508.05940.
[47] F.F. Deppisch, C. Hati, S. Patra, P. Pritimita, U. Sarkar, Phys. Lett. B 757 (2016) 223, arXiv:1601.00952.
[48] K.S. Babu, B. Bajc, S. Saad, J. High Energy Phys. 02 (2017) 136, arXiv:1612.04329.
[49] C. Hati, S. Patra, M. Reig, J.W.F. Valle, C.A. Vaquera-Araujo, arXiv:1703.09647, 2017.
[50] R. Foot, H.N. Long, T.A. Tran, Phys. Rev. D 50 (1994) R34, arXiv:hep-ph/9402243.
[51] A.G. Dias, R. Martinez, V. Pleitez, Eur. Phys. J. C 39 (2005) 101, arXiv:hep-ph/0407141.
[52] A.G. Dias, V. Pleitez, Phys. Rev. D 80 (2009) 056007, arXiv:0908.2472.
[53] S. Profumo, F.S. Queiroz, Eur. Phys. J. C 74 (2014) 2960, arXiv:1307.7802.
[54] C. Kelso, H.N. Long, R. Martinez, F.S. Queiroz, Phys. Rev. D 90 (2014) 113011, arXiv:1408.6203.
[55] D. Cogollo, A.X. Gonzalez-Morales, F.S. Queiroz, P.R. Teles, J. Cosmol. Astropart. Phys. 1411 (2014) 002, arXiv:1402.3271.
[56] P.V. Dong, D.T. Huong, F.S. Queiroz, N.T. Thuy, Phys. Rev. D 90 (2014) 075021, arXiv:1405.2591.
[57] A. Alves, G. Arcadi, P.V. Dong, L. Duarte, F.S. Queiroz, J.W.F. Valle, arXiv:1612.04383, 2016.
[58] M. Lindner, M. Platscher, F.S. Queiroz, arXiv:1610.06587, 2016.
[59] Tech. Rep. ATLAS-CONF-2017-018, CERN, Geneva, 2017, http://cds.cern.ch/record/2258132.
[60] K. Cheung, W.-Y. Keung, C.-T. Lu, P.-Y. Tseng, arXiv:1704.02087, 2017.
[61] J. Brehmer, J. Hewett, J. Kopp, T. Rizzo, J. Tattersall, J. High Energy Phys. 10 (2015) 182, arXiv:1507.00013.
[62] J. Alwall, M. Herquet, F. Maltoni, O. Mattelaer, T. Stelzer, J. High Energy Phys. 06 (2011) 128, arXiv:1106.0522.
[63] T. Sjostrand, S. Mrenna, P.Z. Skands, J. High Energy Phys. 05 (2006) 026, arXiv:hep-ph/0603175.
[64] J. de Favereau, C. Delaere, P. Demin, A. Giammanco, V. Lemaître, A. Mertens, M. Selvaggi, DELPHES 3, J. High Energy Phys. 02 (2014) 057, arXiv:1307.6346.
[65] A. Alves, A. Berlin, S. Profumo, F.S. Queiroz, arXiv:1506.06767, 2015.
[66] M. Klasen, F. Lyonnet, F.S. Queiroz, arXiv:1607.06468, 2016.
[67] E. Accomando, C. Coriano, L. Delle Rose, J. Fiaschi, C. Marzo, S. Moretti, J. High Energy Phys. 07 (2016) 086, arXiv:1605.02910.
[68] L. Cerrito, D. Millar, S. Moretti, F. Spanò, arXiv:1609.05540, 2016.
[69] E. Accomando, J. Fiaschi, F. Hautmann, S. Moretti, C. Shepherd-Themistocleous, arXiv:1609.07788.
[70] A. Avetisyan, et al., in: Proceedings, Community Summer Study 2013: Snowmass on the Mississippi, CSS2013, Minneapolis, MN, USA, July 29–August 6, 2013, 2013, arXiv:1308.1636, http://lss.fnal.gov/archive/test-fn/0000/fermilab-fn-0965-t.pdf.
[71] T. Cohen, T. Golling, M. Hance, A. Henrichs, K. Howe, J. Loyal, S. Padhi, J.G. Wacker, J. High Energy Phys. 04 (2014) 117, arXiv:1311.6480.
[72] G. Apollinari, I. Béjar Alonso, O. Brüning, M. Lamont, L. Rossi, http://dx.doi.org/10.5170/CERN-2015-005, 2015.
[73] I. Hinchliffe, A. Kotwal, M.L. Mangano, C. Quigg, L.-T. Wang, Int. J. Mod. Phys. A 30 (2015) 1544002, arXiv:1504.06108.
[74] N. Arkani-Hamed, T. Han, M. Mangano, L.-T. Wang, Phys. Rep. 652 (2016) 1, arXiv:1511.06495.
[75] R. Contino, et al., arXiv:1606.09408, 2016.
[76] M.L. Mangano, et al., arXiv:1607.01831, 2016.
[77] D. Goncalves, T. Plehn, J.M. Thompson, arXiv:1702.05098, 2017.
[78] Collider reach code, http://collider-reach.web.cern.ch/.
[79] T.G. Rizzo, Eur. Phys. J. C 75 (2015) 161, arXiv:1501.05583.
[80] J. Baglio, A. Djouadi, J. Quevillon, Rep. Prog. Phys. 79 (2016) 116201, arXiv:1511.07853.
[81] F.F. Deppisch, P.S. Bhupal Dev, A. Pilaftsis, New J. Phys. 17 (2015) 075019, arXiv:1502.06541.
[82] P.S.B. Dev, R.N. Mohapatra, Y. Zhang, J. High Energy Phys. 05 (2016) 174, arXiv:1602.05947.
[83] S.V. Chekanov, M. Beydler, A.V. Kotwal, L. Gray, S. Sen, N.V. Tran, S.S. Yu, J. Zuzelski, arXiv:1612.07291, 2016.

[84] A. Kobakhidze, M. Talia, L. Wu, Phys. Rev. D 95 (2017) 055023, arXiv:1608.03641.
[85] G. Grilli di Cortona, E. Hardy, A.J. Powell, J. High Energy Phys. 08 (2016) 014, arXiv:1606.07090.
[86] N. Craig, J. Hajer, Y.-Y. Li, T. Liu, H. Zhang, J. High Energy Phys. 01 (2017) 018, arXiv:1605.08744.
[87] X. Zhao, Q. Li, Z. Li, Q.-S. Yan, Chin. Phys. C 41 (2017) 023105, arXiv:1604.04329.
[88] B. Lillard, T.M.P. Tait, P. Tanedo, Phys. Rev. D 94 (2016) 054012, arXiv:1602.08622.
[89] S. Antusch, C. Sluka, Int. J. Mod. Phys. A 31 (2016) 1644011, arXiv:1604.00212.
[90] C. Englert, Q. Li, M. Spannowsky, M. Wang, L. Wang, arXiv:1702.01930, 2017.
[91] D. Hooper, C. Kelso, F.S. Queiroz, Astropart. Phys. 46 (2013) 55, arXiv:1209.3015.
[92] F.S. Queiroz, W. Shepherd, Phys. Rev. D 89 (2014) 095024, arXiv:1403.2309.
[93] F.S. Queiroz, C. Siqueira, J.W.F. Valle, Phys. Lett. B 763 (2016) 269, arXiv:1608.07295.
[94] F.S. Queiroz, K. Sinha, Phys. Lett. B 735 (2014) 69, arXiv:1404.1400.
[95] A.X. Gonzalez-Morales, S. Profumo, F.S. Queiroz, Phys. Rev. D 90 (2014) 103508, arXiv:1406.2424.
[96] B. Allanach, F.S. Queiroz, A. Strumia, S. Sun, arXiv:1511.07447, 2015.
[97] M.G. Baring, T. Ghosh, F.S. Queiroz, K. Sinha, arXiv:1510.00389, 2015.
[98] F.S. Queiroz, C.E. Yaguna, J. Cosmol. Astropart. Phys. 1602 (2016) 038, arXiv:1511.05967.
[99] Y. Mambrini, S. Profumo, F.S. Queiroz, Phys. Lett. B 760 (2016) 807, arXiv:1508.06635.
[100] F.S. Queiroz, C.E. Yaguna, C. Weniger, J. Cosmol. Astropart. Phys. 1605 (2016) 050, arXiv:1602.05966.
[101] S. Profumo, F.S. Queiroz, C.E. Yaguna, Mon. Not. R. Astron. Soc. 461 (2016) 3976, arXiv:1602.08501.
[102] M.D. Campos, F.S. Queiroz, C.E. Yaguna, C. Weniger, arXiv:1702.06145, 2017.
[103] F.S. Queiroz, W. Rodejohann, C.E. Yaguna, arXiv:1610.06581, 2016.
[104] G. Bertone, D. Hooper, Rev. Mod. Phys. (2016), http://inspirehep.net/search?ln=it&ln=it&p=Bertone%3A2016nfn&of=hb&action_search=Cerca&sf=earliestdate&so=d&rm=&rg=25&sc=0, submitted for publication, arXiv:1605.04909.
[105] Y. Mambrini, K.A. Olive, J. Quevillon, B. Zaldivar, Phys. Rev. Lett. 110 (2013) 241306, arXiv:1302.4438.
[106] Y. Mambrini, N. Nagata, K.A. Olive, J. Quevillon, J. Zheng, Phys. Rev. D 91 (2015) 095010, arXiv:1502.06929.
[107] J. Abdallah, et al., Phys. Dark Universe 9–10 (2015) 8, arXiv:1506.03116.
[108] A. Alves, G. Arcadi, Y. Mambrini, S. Profumo, F.S. Queiroz, arXiv:1612.07282, 2016.
[109] C. Gross, O. Lebedev, Y. Mambrini, J. High Energy Phys. 08 (2015) 158, arXiv:1505.07480.
[110] G. Arcadi, C. Gross, O. Lebedev, Y. Mambrini, S. Pokorski, T. Toma, J. High Energy Phys. 12 (2016) 081, arXiv:1611.00365.
[111] G. Arcadi, Y. Mambrini, F. Richard, J. Cosmol. Astropart. Phys. 1503 (2015) 018, arXiv:1411.2985.
[112] G. Arcadi, P. Ullio, Phys. Rev. D 84 (2011) 043520, arXiv:1104.3591.
[113] X. Chu, Y. Mambrini, J. Quevillon, B. Zaldivar, J. Cosmol. Astropart. Phys. 1401 (2014) 034, arXiv:1306.4677.
[114] F. D'Eramo, N. Fernandez, S. Profumo, arXiv:1703.04793, 2017.
[115] D.S. Akerib, et al., LUX, Phys. Rev. Lett. 118 (2017) 021303, arXiv:1608.07648.
[116] O. Lebedev, Y. Mambrini, Phys. Lett. B 734 (2014) 350, arXiv:1403.4837.
[117] F. Kahlhoefer, K. Schmidt-Hoberg, T. Schwetz, S. Vogl, J. High Energy Phys. 02 (2016) 016, arXiv:1510.02110.
[118] P.A.R. Ade, et al., Planck, Astron. Astrophys. 594 (2016) A13, arXiv:1502.01589.

10

Left–right model with TeV fermionic dark matter and unification

Triparno Bandyopadhyay *, Amitava Raychaudhuri

Department of Physics, University of Calcutta, 92 Acharya Prafulla Chandra Road, Kolkata 700009, India

ARTICLE INFO

Editor: A. Ringwald

ABSTRACT

The ingredients for a model with a TeV right-handed scale, gauge coupling unification, and suitable dark matter candidates lie at the heart of left–right symmetry with broken D-parity. After detailing the contents of such a model, with $SU(2)_R$ self-conjugate fermions at the right-handed scale aiding in unification of couplings, we explore its dark matter implications and collider signatures.

1. Introduction

These are indeed exciting times for particle physics as the Large Hadron Collider (LHC) at CERN is all set to run at its machine configuration of $\sqrt{s} = 14$ TeV. With experiments at this highest energy facility in a hunt for new physics at TeV scales, it is no surprise that the community is particularly focussed on models with phenomenological signatures in the $\mathcal{O}$(TeV) range. Of the various models that try to explain natural phenomena beyond the scope of the standard model (SM), those based on left–right (LR) symmetry [1–4] have withstood the tests of time as they extend the SM electroweak sector in well motivated ways. These models explain the origin of parity violation and at the same time gauge the global $U(1)_{(B-L)}$ symmetry inherent in SM and in the process explain the smallness of the neutrino mass.

Hypothesised primarily in the context of visible sector physics, LR models do not have any *de facto* dark matter (DM) candidate built into their bare bones structure. However, the group theoretic configuration of LR symmetry has the provision of a naturally arising discrete symmetry, remnant after the spontaneous breaking of $U(1)_{(B-L)}$ [5–10], which facilitates the building of a plethora of DM models [11–16].

The LR gauge symmetry and particle content, along with gauge coupling unification (GCU), can be embedded in $SO(10)$ "grand unified theories" (GUTs) [17,18] having numerous desired features such as quark-lepton unification, unification of the SM interactions, and explanation of the arbitrary $U(1)_Y$ assignment of the SM, among others. However, in models with the left–right symmetry breaking scale $M_R \sim \mathcal{O}$(TeV), and a minimal scalar sector, GCU is impossible [19–24]. To achieve unification one either needs to add scalar multiplets redundant to their primary function of symmetry breaking and mass generation, or larger symmetries intermediate between the Left–Right symmetry (LRS) and GUT scales. These modifications end up introducing additional scalar fine tunings and a degree of arbitrariness.

In this letter, we show that the three requirements of $\mathcal{O}$(TeV) right-handed breaking scale, unification of LRS couplings, and the presence of a suitable dark matter candidate can be achieved with a single stroke by the careful appraisal of fermion masses in a class of left–right models where the exact $L \leftrightarrow R$ symmetry is spontaneously broken at a scale different from the one where the right-handed gauge symmetry is broken [25,26]. While focussing on model mechanics, we discuss dark matter phenomenology and show that though its direct detection prospects are not bright, the collider signatures of the model are testable.

2. Model

The left–right symmetry is defined by the gauge group, $SU(3)_C \times SU(2)_L \times SU(2)_R \times U(1)_{(B-L)}$, and a discrete $SU(2)_L \leftrightarrow SU(2)_R$ symmetry, $\mathcal{P}$. Under this, the SM quarks, leptons, and a right-handed (RH) neutrino of one family transform as:

$$l_L \equiv (1_C, 2_L, 1_R, -1_{(B-L)});\ l_R \equiv (1_C, 1_L, 2_R, -1_{(B-L)});$$
$$q_L \equiv (3_C, 2_L, 1_R, 1/3_{(B-L)});\ q_R \equiv (3_C, 1_L, 2_R, 1/3_{(B-L)})\,; \quad (1)$$

with $(B-L)$ being normalised by the relation:

$$Q_{em} = T_{3R} + T_{3L} + \frac{B-L}{2}\ . \quad (2)$$

The scalar sector is given by:

$$\Phi \equiv (1_C, 2_L, 2_R, 0_{(B-L)});\ \eta \equiv (1_C, 1_L, 1_R, 0_{(B-L)});$$
$$\Delta_R \equiv (1_C, 1_L, 3_R, 2_{(B-L)});\ \Delta_L \equiv (1_C, 3_L, 1_R, 2_{(B-L)})\ . \quad (3)$$

* Corresponding author.
E-mail addresses: gondogolegogol@gmail.com (T. Bandyopadhyay), palitprof@gmail.com (A. Raychaudhuri).

Under $\mathcal{P}$ the multiplets transform as:

$$l_L \leftrightarrow l_R; \qquad q_L \leftrightarrow q_R; \qquad \Delta_L \leftrightarrow \Delta_R;$$
$$\Phi \leftrightarrow \Phi^\dagger; \qquad \eta \leftrightarrow -\eta \ . \tag{4}$$

The scalar sector is modified to accommodate spontaneous breaking of $\mathcal{P}$ at a scale $M_\mathcal{P}$, where the $\mathcal{P}$ odd gauge singlet, η, acquires a vacuum expectation value (vev), v_η. Thus, symmetry breaking takes place in three steps. The first being the breaking of $\mathcal{P}$, followed by the breaking of $SU(2)_R \otimes U(1)_{(B-L)}$ to SM by the vev, v_R, of Δ_R, and finally electroweak symmetry breaking is achieved through the $vevs$ k_1 and k_2 of the bi-doublet Φ, with $\sqrt{k_1^2+k_2^2} = 246$ GeV. We show that the other mass scales of the model are $M_R \sim \mathcal{O}$(TeV) and $M_\mathcal{P}$ at the GUT scale.

The gauge bosons related to $SU(2)_R \otimes U(1)_{(B-L)}$ breaking, $W_R^\pm$ and Z', acquire mass at M_R, and for $\left(M_W/M_{W_R}\right)^2 \ll 1$ are given by:

$$M_{W_R} = \frac{g_R}{\sqrt{2}} v_R \ ; \qquad M_{Z'} = \frac{\sqrt{2}}{\cos\phi} M_{W_R} \ , \tag{5}$$

with

$$\sin\phi = \frac{g_L}{g_R} \tan\theta_W \ ,$$

where θ_W is the weak mixing angle. For $\left(M_W/M_{W_R}\right)^2 \ll 1$, W_L-W_R mixing is negligible. The physical states of the bi-doublet other than the SM Higgs are constrained to be $\geq \mathcal{O}(10$ TeV) from lepton flavour violation limits [27], and the scalars from Δ_L are all heavy at $M_\mathcal{P}$ [25,26]. There are no stringent constraints on the masses of the Δ_R scalars and they can be lighter than M_{W_R} and even $\mathcal{O}(100$ GeV). Here, for simplicity we take them to be heavier than M_{W_R}.

$U(1)_{(B-L)}$ being broken by a scalar with $(B-L)=2$, leaves behind a remnant $\mathbb{Z}_2$ symmetry, defined by: $\mathcal{Z} \equiv (-1)^{3(B-L)}$ [9]. LRS fermions (scalars) have odd (even) $3(B-L)$ and hence are odd (even) under $\mathcal{Z}$. As a result, fermions with even $3(B-L)$ are forbidden to decay only to SM fermions and/or bosons, and hence the lightest one of them is stable. If this state is neutral, then subject to relic density and direct/indirect detection constraints, it can be taken to be a dark matter candidate.

Fermions in self-conjugate representations of $SU(2)_L \otimes SU(2)_R$, $X_L \oplus X_R \equiv (1_C, (2m+1)_L, 1_R, 0_{(B-L)}) \oplus (1_C, 1_L, (2m+1)_R, 0_{(B-L)})$, typify this scenario with $m \in \mathbb{N}$. Each multiplet consists of a Majorana fermion and m pairs of Dirac fermions and their antiparticles with electric charges 1 to m. Thus, these multiplets must be assigned $B = L = 0$.

The left–right symmetric bare mass and Yukawa terms of these multiplets for a general case of n_g such 'generations' is given by:

$$\mathcal{L}_{X_M} = \frac{\mathcal{M}_i}{2}\left(\overline{X_L^{i}}^{c} X_L^i + R \leftrightarrow L\right) + \frac{h_i}{2}(v_\eta + \eta)\left(\overline{X_L^{i}}^{c} X_L^i - R \leftrightarrow L\right) + h.c. , \tag{6}$$

where summation over $i = 1, \cdots, n_g$ is implicit. The negative signs pertaining to interactions of X_R^i with η are according to eq. (4). Because the multiplets can always be rotated into a diagonal basis, we can do away with the cross terms without any loss of generality. From eq. (6), we see that the breaking of $\mathcal{P}$ enforces a separation of the masses of the multiplets transforming under $SU(2)_L$ and $SU(2)_R$ with the corresponding masses given by:

$$M_i^L = \mathcal{M}_i + h_i v_\eta; \ M_i^R (=: M_i) = \mathcal{M}_i - h_i v_\eta. \tag{7}$$

With $\mathcal{M}_i \sim h_i v_\eta$, X_L^i multiplets remain heavy at $M_\mathcal{P}$ and the X_R^i become light with the exact mass scale dependent on the couplings. We want to underscore that in general for $(1_C, (2m+1)_L, 1_R, 0_{(B-L)}) \oplus (1_C, 1_L, (2m+1)_R, 0_{(B-L)})$ fermion multiplets, the mass scale of either one will be at the larger of $\mathcal{M}_i$ and $h_i v_\eta$ while the other can be tuned to be at lower values. During the evolution of the Universe the superheavy $SU(2)_L$ multiplets are Boltzmann suppressed and annihilate and co-annihilate rapidly to lighter states through their couplings to W_L and Z.

The framework being discussed can lead to a variety of DM models, which we label as (m, n_g), with the DM particle(s) completely separated from the SM and interacting only with the RH sector. For the rest of the letter we focus on the $(1,2)$ case as this model simultaneously provides a suitable DM sector, gauge coupling unification, and $\mathcal{O}$(TeV) M_R.

3. Gauge coupling unification

With the self-conjugate $SU(2)_R$ generations of an $(m, n_g) \equiv (1,2)$ model, i.e., the model with a pair of $(1_C, 3_L, 1_R, 0_{(B-L)}) + (1_C, 1_L, 3_R, 0_{(B-L)})$, in the TeV range, the gauge couplings unify with $M_\mathcal{P} = M_U$. The LR gauge group is a subgroup of $SO(10)$ and with GCU we can embed the model in an $SO(10)$ unified theory. The LRS multiplets of the model belong in the following $SO(10)$ representations:

$$(3_C, 2_L, 1_R, 1/3_{(B-L)}) + (\bar{3}_C, 1_L, 2_R, -1/3_{(B-L)})$$
$$+ (1_C, 2_L, 1_R, -1_{(B-L)}) + (1_C, 1_L, 2_R, 1_{(B-L)}) \subseteq 16_F;$$
$$(1_C, 2_L, 2_R, 0_{(B-L)}) \subset 10_H;$$
$$(1_C, 3_L, 1_R, 2_{(B-L)}) + (1_C, 1_L, 3_R, 2_{(B-L)}) \subset 126_H;$$
$$(1_C, 1_L, 1_R, 0_{(B-L)}) \subset 210_H;$$
$$(1_C, 3_L, 1_R, 0_{(B-L)}) + (1_C, 1_L, 3_R, 0_{(B-L)}) \subset 45_F; \tag{8}$$

where the subscripts F and H denote whether the multiplets contain fermions or scalars, respectively. There is an element of the $SO(10)$ algebra, '$\mathcal{D}$' [28], which in the case that all the couplings of the lagrangian are real, plays the role of the parity symmetry $\mathcal{P}$. $\eta \subset 210_H$ is odd under '$\mathcal{D}$'.

The fermion triplets reside in 45-plets. The $SO(10)$ symmetric mass term for which is:

$$\mathcal{L}_{\text{Mass}} = -\frac{\mathcal{M}_{1,2}}{2} \overline{45_F^{1,2}}^{C} 45_F^{1,2} + h.c. \ , \tag{9}$$

with $\mathcal{M}_{1,2} \sim M_U = M_\mathcal{P}$. Under the Pati–Salam [1,2] symmetry, $SU(4)_C \otimes SU(2)_L \otimes SU(2)_R$, 45_F is decomposed as: $45 \supset (15_4, 1_L, 1_R) + (6_4, 2_L, 2_R) + (1_4, 3_L, 1_R) + (1_4, 1_L, 3_R)$. Since $(15_4, 1_L, 1_R)$ and $(6_4, 2_L, 2_R)$ transform identically under $SU(2)_L$ and $SU(2)_R$, they have masses at $\mathcal{M}_{1,2}$, while $(1_4, 3_L, 1_R)$ and $(1_4, 1_L, 3_R)$ are split according to the previous discussion. As for the scalars, all submultiplets not required to be either at the right-handed or the electroweak scale are at the unification scale according to the minimal fine-tuning principle of the extended-survival hypothesis [29,30].

In Fig. 1, we show the running of the inverses of the fine structure constants ($\alpha = g^2/(4\pi)$), as obtained from 2-loop perturbation theory. As inputs at the Z-pole, $M_Z = 91.1876(21)$, we take $\alpha_s = 0.1181(11)$, $\sin^2\theta_W = 0.23129(5)$, and $\alpha_{EM} = 1/128$ [31]. We find that when $\alpha_s(M_Z)$ and $\sin^2\theta_W$ are varied over their 1σ allowed ranges the unification scale varies between $(0.81\text{–}1.05) \times 10^{16}$ GeV and the unification coupling comes out to be, $g_U = 0.53$. The $SU(2)_R$ breaking scale lies between 3.78–9.40 TeV, with the

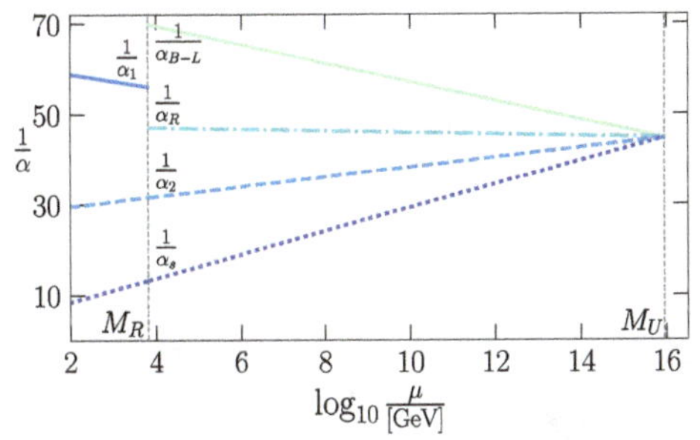

Fig. 1. 2-loop running of the inverse of the gauge couplings α_i with mass scale, μ.

$SU(2)_R$ coupling, $g_R = 0.52$. Fig. 1 has been drawn using the central value. The $U(1)$ couplings of the theory are normalised according to GUT (canonical) normalisation, resulting in the matching condition at M_R:

$$\frac{1}{g_Y^2} = \frac{3}{5}\frac{1}{g_R^2} + \frac{2}{5}\frac{1}{g_{(\text{B-L})}^2} + \frac{1}{20\pi}. \tag{10}$$

In between M_R and M_U the particles flowing in the loops and hence contributing to the β-coefficients are Φ, Δ_R, l_L, l_R, q_L, q_R as in traditional D-parity broken models, and the pair of dark sector $SU(2)_R$ triplets $X_R^{1,2}$. The system of running equations are given by [32,33]:

$$\frac{\partial g_i}{\partial \log\mu} = \frac{a_i}{16\pi^2}g_i^3 + \sum_j \frac{b_{ij}}{(16\pi^2)^2}g_i^3 g_j^2. \tag{11}$$

The 1-loop β-coefficients a_i and the 2-loop β-coefficients, b_{ij}, for the couplings of the SM are readily available [33], the same for the LRS stage are given in eq. (12). Since the only additions on top of the usual LRS particle content are the self-conjugate $SU(2)_R$ triplets which transform trivially under the other symmetries, the only change in the β-coefficients are for the $SU(2)_R$ coupling for the 1-loop case and the diagonal coefficient corresponding to $SU(2)_R$ for the two loop case.

$$\begin{array}{lcccccc} & B-L & 2R & 2L & 3C & \\ a_i \equiv & [\ \frac{11}{2} & \frac{1}{3} & -3 & -7\ ] & \\ b_{ij} \equiv & \left[\begin{array}{c} \frac{61}{2} \\ \frac{27}{2} \\ \frac{3}{2} \\ \frac{1}{2}\end{array}\right. & \begin{array}{c}\frac{81}{2} \\ \frac{208}{3} \\ 3 \\ \frac{9}{2}\end{array} & \begin{array}{c}\frac{9}{2} \\ 3 \\ 8 \\ \frac{9}{2}\end{array} & \left.\begin{array}{c}4 \\ 12 \\ 12 \\ -26\end{array}\right] & \begin{array}{c}B-L \\ 2R \\ 2L \\ 3C\end{array} \end{array} \tag{12}$$

In principle, a complete treatment of 2-loop RGE running should take into account threshold effects [34,35] at all the symmetry breaking scales. However, in this work we do not include threshold effects, as in demanding exact unification of the couplings, we are establishing GCU for a more restricted case. Threshold corrections introduce more parameters to the model and hence such situations are bound to follow suit.

We next estimate the lifetime of the proton in our model. In non-supersymmetric GUTs, scalar induced $d = 6$ and the $d > 6$ operators contributing to proton decay are generally highly suppressed in comparison to the gauge induced $d = 6$ operators [36–38], and here we concentrate only on the latter. The decay rate of the proton in the $p \to e^+\pi^0$ channel is expressed as [36, 39]:

$$\Gamma(p \to e^+\pi^0) = \frac{m_p g_U^4}{16\pi f_\pi^2 M_U^4} R_L^2 (A_{SL}^2 + A_{SR}^2)|\alpha_H|^2 (1 + D + F)^2, \tag{13}$$

where $m_p = 938.3$ MeV [31] is the mass of the proton, $f_\pi = 130.41(23)$ MeV [40] is the pion decay constant, $\alpha_H = -0.0118(0.0021)$ GeV^3 denotes the relevant hadronic matrix element, $D = 0.8(2)$ and $F = 0.47(1)$ are chiral lagrangian parameters calculated from lattice gauge theory [41–43]. g_U is the unified coupling constant, M_U the unification scale. $R_L = 1.46$ is the two-loop long range running effect on the effective proton decay operator, corresponding to running from M_Z to m_p, while $A_{SL(R)}$ is the short range left-(right-) handed short range renormalisation factor of the proton decay operator corresponding to running from M_U to M_Z [44]. $A_{SL(R)}$ is a function of the anomalous dimensions and β-coefficients of the running couplings, and also the values of the couplings at the symmetry breaking scales and are taken to be $A_{SL} \simeq A_{SR} = 2.0$ [45–49]. We set the masses of the leptoquark gauge bosons to be degenerate and at M_U. Further, the flavour matrices associated with baryon and lepton flavour changing currents have been set to unity [50,51].

With $M_U = 10^{15.97}$ GeV we get from eq. (13) a proton decay lifetime in this channel, $\tau_{p\to e^+\pi^0} \sim 1.5 \times 10^{35}$ years, which is larger than the present bound of $\tau_{p\to e^+\pi^0} = 1.6 \times 10^{34}$ years [52], but testable at the Hyper-Kamiokande experiment [53], which is expected to probe lifetimes $\sim 2 \times 10^{35}$ yrs. As indicated by eq. (13), this value is extremely sensitive to the unification scale M_U. Still, we have checked that for the above chosen values of the parameters, $\tau_{p\to e^+\pi^0}$ remains below the Hyper-Kamiokande projection with M_U varying between its allowed range, i.e., $(0.81–1.05) \times 10^{16}$ GeV. However, extreme choices of the different parameters may make the model not falsifiable even by this experiment.

The mass scales, predicted by unification, are particularly ingratiating for neutrino seesaw masses. In minimal LR models, the left-handed neutrino has both type-I and type-II seesaw [54–57] contributions. L $\leftrightarrow$ R symmetry breaking induces a nonzero $SU(2)_L$ triplet vev [58]:

$$v_L \simeq \frac{v_R}{v_\eta}\frac{\mathcal{O}(k_i^2)}{2M_{\eta\Delta}}. \tag{14}$$

Here $M_{\eta\Delta}$ is the dimensionful coefficient of the $\eta\Delta\Delta$ type term in the potential. The left- and right-handed neutrino masses are given by [59–61]:

$$M_{\nu_R} = f v_R \quad , \quad M_{\nu_L} = f v_L + \frac{v^2}{v_R} y f^{-1} y^T. \tag{15}$$

f is the Yukawa coupling matrix of the leptons with the triplet scalars $\Delta_{L,R}$ while y is the Yukawa matrix of the neutrinos with the bidoublet Φ. From the values of the symmetry breaking scales as given above, we see that the left-handed neutrino gets a mass of the order of 0.1 eV, with $f \sim \mathcal{O}(1)$, when the Yukawa matrix y is set at the order of that of the up quark, in the spirit of quark–lepton unification.[1] The seesaw is predominantly type I.

[1] Grand unification implies the same Yukawa couplings for up-type quarks and the neutrinos. However, the contributions to the masses in the two sectors can be the same, for the $(1_4, 2_L, 2_R) \subset 10_H$, or unequal and of opposite sign, for $(15_4, 2_L, 2_R) \subset 126_H$. For the second and the third generation a fine-tuned cancellation between the two contributions (at the level of 1 in 10^5 for the third generation) is needed to keep the Type I seesaw neutrino masses in the desired range.

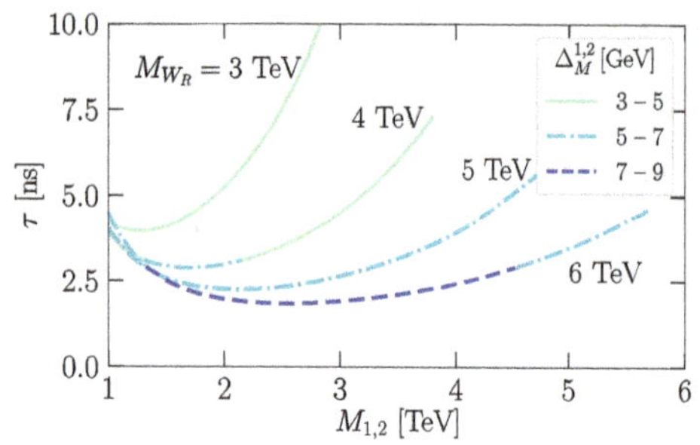

Fig. 2. The decay time of $\chi_{1,2}^{\pm}$ as a function of its mass for different choices of M_{W_R}. Linestyles distinguish different mass splittings between the charged and neutral states of the multiplet.

4. Dark matter phenomenology

The triplets, $\chi_R^{1,2}$, each contribute a singly-charged Dirac fermion–anti-fermion pair ($\chi_{1,2}^{\pm}$), and a Majorana fermion ($\chi_{1,2}^{0}$). The charged and neutral states are mass degenerate at tree level, with mass $M_{1,2}$. At one-loop order, gauge interactions induce the mass splitting, $\Delta_M^{1,2} = M_{\chi^{\pm}}^{1,2} - M_{\chi^0}^{1,2}$ [62–64].

The interaction lagrangian for the constituents of the triplets, $\chi_R^{1,2}$, for the LR stage is given by:

$$\mathcal{L}_{\rm int} = -g_R \left(\overline{\chi_i^+} \not{W}_R^+ \chi_i^0 + h.c.\right) - e\overline{\chi_i^+} \not{A} \chi_i^+ - g_R \cos\phi_0 \, \overline{\chi_i^+} \not{Z}' \chi_i^+ + e \tan\theta_W \, \overline{\chi_i^+} \not{Z} \chi_i^+ \quad (i = 1, 2), \tag{16}$$

where e is the electromagnetic coupling. Presence of charged heavy fermions during big bang nucleosynthesis (BBN) would imply the existence of atom-like bound states, in the present epoch, containing such particles [65,66]. The non-observance of such entities in deep sea water searches [67–71] rules out their existence. This implies positive $\Delta_M^{1,2}$, large enough to produce a lifetime for the charged states smaller than the time at BBN ~ 1 s.

In Fig. 2 we plot the decay time of $\chi_{1,2}^{\pm}$ as a function of its mass for different M_{W_R} near M_R. The intra-multiplet mass splitting, $\Delta_M^{1,2}$, calculated using expressions in [13,14], is indicated by the line-styles of the curves, i.e., short-dashed, long-dashed, or solid. Notice that for each curve $\Delta_M^{1,2}$ changes with $M_{1,2}$. We find that the lifetime of the charged states for all masses near M_{W_R} is $\mathcal{O}$(ns), and the mass splitting is $\mathcal{O}$(GeV). Hence the heavy charged states of our model decay well before BBN. As the mass difference is tiny with respect to the masses themselves, $\chi_2^{\pm} \to \chi_2^0 \chi_1^{\pm} \chi_1^0$ decay is forbidden from kinematics. Of course the same argument also applies the other way round. Hence, although we have a single stabilising $\mathbb{Z}_2$, we end up with two component ($\chi_{1,2}^0$) dark matter.

The behaviour of dark matter relic density for this model is illustrated in Fig. 3. The allowed regions in the M_1–M_2 plane are those points which fall on the ellipse-like or semi-circle-like plots. We show only the region for which $M_1 < M_2$. The allowed values with $M_1 > M_2$ can be readily obtained by a reflection. In the inset of Fig. 3 we exhibit the relic density as a function of the dark matter mass for $M_{W_R} = 4$ TeV. The observed value of the relic density, $\Omega h^2 = 0.1198 \pm 0.0015$ [72], is indicated by the dashed horizontal straight line. As noted, in the model under discussion, there are two dark matter candidates, χ_1^0 and χ_2^0. In the inset, for simplicity, they have been taken to be degenerate. The dips in the curve reflect resonant $\chi_i^{\pm}\chi_i^0 \to W_R^{\pm}$ or $\chi_i^+\chi_i^- \to Z'$ production. Without these dips, the relic density in this model would have been about an order of magnitude larger than the observed limit. The points where the curves agree with the observation are near the two resonant dips. In Fig. 3 the closed ellipse-type curves with an asterisk in the middle correspond to regions where the dark matter candidate χ_2^0 is near the Z' resonance (i.e., $M_2 \simeq M_{Z'}/2$) while

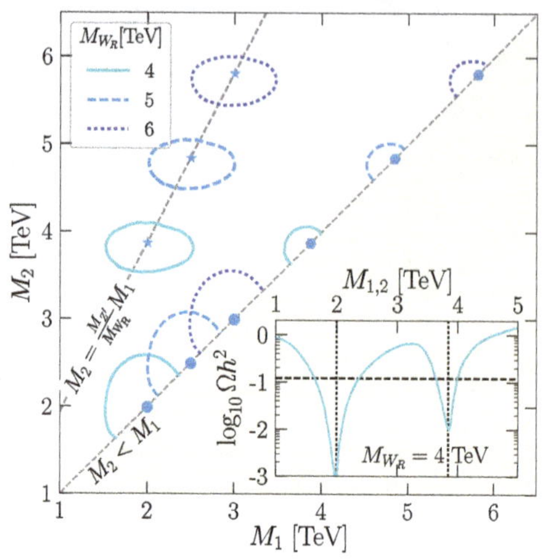

Fig. 3. The points in the M_1–M_2 plane consistent with the measured dark matter relic density lie on the curves (see text). Only the solutions with $M_2 > M_1$ are displayed. Plots are shown for different M_{W_R}. Inset: The dark matter relic density as a function of its mass for $M_{W_R} = 4$ TeV. The curve is for the case when the two dark matter candidates are degenerate.

χ_1^0 is close to the W_R resonance point (i.e., $M_1 \simeq M_{W_R}/2$). The semicircle-like curves with a dot (hexagon) within correspond to the situation where the dark matter particles χ_1^0 and χ_2^0 are near degenerate and also close in mass to $M_{W_R}/2$ ($M_{Z'}/2$). We have kept the lower bound of $M_{1,2} > 547$ GeV, as set by recent searches for heavy singly charged particles [73,74]. For these relic density computations we have utilized the MicrOMEGAS 4.3 [75] package. The model file was written using FeynRules 2.0 [76], modifying the version in [77] to our needs.

At freeze-out temperature, the charged states, $\chi_{1,2}^{\pm}$, did not have enough time to decay to the neutral ones, and hence annihilation and co-annihilation of all the triplet states contribute to the net annihilation cross section $\langle \sigma v \rangle$. Near the W_R resonance, $\langle \sigma v \rangle$ is saturated by co-annihilation of $\chi_{1,2}^{\pm}$ with $\chi_{1,2}^0$ and around the Z' resonance by both co-annihilation and annihilation of $\chi_{1,2}^{\pm}$. As the neutral $\chi_{1,2}^0$ have no interaction with the Z' or Z, it can only annihilate to a W_R pair through the t-channel exchange of $\chi_{1,2}^{\pm}$. This channel, however, opens up only when $M_{1,2} \gtrsim M_{W_R}$, and even then it accounts for a minute fraction of the total $\langle \sigma v \rangle$. g_R is essentially fixed from the running of gauge couplings and is no more a free parameter while calculating cross sections. Furthermore, the relic density constraint fixes a narrow range for $M_{1,2}$ given M_{W_R}. The scale of M_{W_R} itself is fixed by gauge coupling unification. This makes the model remarkably predictive and free of parameters which can be altered at will. Thus, falsifying the model is quite straightforward.

Present and proposed DM direct detection experiments such as LUX, LZ, XENON1T [78–80], are all based on detecting elastic scattering of WIMP DM candidates with nucleons. The dark matter candidates of this model, $\chi_{1,2}^0$, do not have any neutral current interactions, neither do they couple to the Higgs boson. Their only possible interaction with nucleons ($\mathcal{N}$) are through charged current processes, $\chi_i^0 \mathcal{N}^0 \to \chi_i^- \mathcal{N}^+$ or $\chi_i^0 \mathcal{N}^+ \to \chi_i^+ \mathcal{N}^0$. At the direct detection experiments, the $\mathcal{N}$ is initially at rest and the DM kinetic energy alone is not large enough to surmount the $\mathcal{O}$(GeV) mass difference between the $\chi_{1,2}^{\pm}$ and $\chi_{1,2}^0$. Therefore, an on-shell $\chi_{1,2}^{\pm}$ in the final state is disallowed from kinematic considerations. An off-shell $\chi_{1,2}^{\pm}$ decaying to $\chi_{1,2}^0$ and (l ν_l) or pions through W_R^* is in principle possible but highly suppressed due to lack of available phase space and $\mathcal{O}$(TeV) masses in the propagators. Neutral current NLO cross sections for $\chi_i^0 \mathcal{N} \to \chi_i^0 \mathcal{N}$, involving W_R and

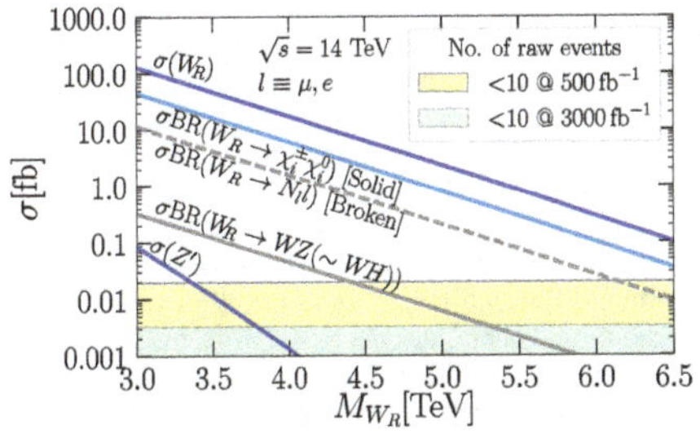

Fig. 4. Cross section times branching fraction for production of W_R and Z' and their subsequent decays into different channels, as labelled, at $\sqrt{s} = 14$ TeV. The thick lines represent total cross sections. The deeply (lightly) shaded regions delimit the cross sections for which the total number of raw events drops below 10 at 3000 (500) fb^{-1}.

$\chi^{\pm}_{1,2}$ in the loop are naturally negligible. A detailed discussion in case of $SU(2)_L$ triplets and a possible way of circumventing the difficulty in detection can be found in [64]. In the absence of any annihilation channels at tree level, the DM parameter region is not constricted by indirect detection [81,82] constraints.

5. Collider studies

As noted previously, the dark matter relic density constraint restricts the masses of $\chi^{\pm}_{1,2}$ and $\chi^{0}_{1,2}$ to near $M_{W_R}/2$ or $M_{Z'}/2$. The $\chi^{\pm}_{1,2}$ particles, if produced, for example, through W_R or Z' decay, will be observed as tracks in the CMS and ATLAS pixel detectors and silicon trackers. These particles will typically be at sub-relativistic velocities and can be distinguished from SM charged particles from the higher rate of ionization energy loss (dE/dx). For most of the allowed mass region, the final state particles have $0.3 < \beta\gamma (= p/M) < 1.5$ and hence the average energy loss with distance travelled can be modelled by the Bethe-Bloch distribution. Given a lifetime of $\mathcal{O}$(ns) for $\chi^{\pm}_{1,2}$, as can be seen from Fig. 2, we find their decay lengths to be of the order ~ 0.1–1 m. The charged particles will hence decay almost exclusively in the trackers of CMS and ATLAS. The only decay mode of $\chi^{\pm}_{1,2}$ is to $\chi^{0}_{1,2}$, and the mass difference being $\sim \mathcal{O}$(GeV), the associated jets will be too soft to be reconstructed for a displaced vertex analysis. Hence, the signal of the charged particles will be the observation of disappearing tracks.[2] An energetic initial state radiation jet can be effectively used as a trigger for the event. The neutral states will obviously be missed completely. The charged particle decay length and $\beta\gamma$ are also favourable for detection at the MoEDAL detector [84] at LHC. If observed, the masses of the particles can be calculated from information about average energy loss and reconstructed transverse momentum as measured from the curvature of the charged tracks in the magnetic fields [85,86].

The vindication of TeV scale $SU(2)_R$ breaking will be the discovery of the W_R and the heavy neutrino, in the $lljj$ channel, the event topology being given by: $pp \to W_R \to N_l l \to lljj$ [87]. If the neutrino is a Majorana particle as predicted by the LRS model, one should observe equal same-sign and opposite-sign final states. We ask to what extent this signal is affected by the presence of the self-conjugate triplets $\chi_{1,2}$? In Fig. 4 we show the cross section times branching fractions of W_R production and its subsequent decays in different channels.[3] For this purpose, leading order cross sections were calculated in CalcHep 3.4 [89] using the CTEQ6L1 parton distribution functions [90], and multiplied by the corresponding K-factors, as obtained from [91]. For the sake of comparison, we have chosen the DM multiplets to be mass degenerate and having the smallest mass as allowed by relic density constraints and taken $M_{N_l} = M_{\chi^{\pm}_{1,2}}$.

The dominant decay mode of W_R is obviously to two jets. As can be seen from Fig. 4 the decay $W^{\pm}_R \to \chi^{\pm}_i \chi^0_i$ is a few times larger than the subdominant but often searched for leptonic decays ($l \equiv e, \mu$). Nonetheless, the leptonic branching remains substantial and we find that a W_R with mass ~ 6.5 TeV can still be discovered by the ATLAS and CMS collaborations in this channel with $\sqrt{s} = 14$ TeV and an integrated luminosity of 3000 fb^{-1}. Indirect detection of a heavier W_R with masses up to ~ 8 TeV is possible in the studies of K and B meson decays at LHCb [27] where this model has no distinction from the canonical LRS model. Another promising mode for the detection of W_R is the di-boson channel ($W_R \to WZ$ or $W_R \to WH$). The branching ratios are almost the same for these two channels. However, due to the suppressed W_L-W_R mixing, they are small, see Fig. 4, and as M_{W_R} approaches 6 TeV this channel becomes unfeasible. Note that as the masses of the triplet fermions are related to the mass of the W_R boson from relic density constraints, and since the $\chi^{\pm}_{1,2}$ do not interact with the SM particles, the detection of W_R or Z' without detection of these will essentially falsify the model.

With $M_{Z_R} \sim 1.94 \times M_{W_R}$, the discovery potential of Z' is bleak at the LHC. For W_R masses above 3.5 TeV, the Z' becomes too heavy to be detected at LHC-II as can be seen from Fig. 4. For HL-LHC luminosities of 3000 fb^{-1}, the sensitivity increases slightly.

6. Conclusion

In this work we have presented a model which rests on left-right symmetry, is amenable to gauge coupling unification, and provides suitable dark matter candidates. Aided by two distinct discrete symmetries inherent to the left-right symmetric theory the stability of dark matter and the scales of symmetry breaking are ensured. The model is falsifiable at both the GUT scale and the LRS breaking scale at the Hyper-Kamiokande experiment and the LHC respectively. The model predicts a 'desert' between the LRS and GUT scales. In the absence of multiple symmetry breaking thresholds, the variable parameters of the model viz. the $SU(2)_R$ coupling, g_R, and scale of the W_R mass are essentially fixed from unification. The Dark Matter candidates satisfy the relic density constraint aided by resonant enhancements of the cross section and hence allowed masses are intertwined with $M_{W_R}/2$ and $M_{Z'}/2$. Their direct detection in ongoing and planned experiments is unlikely. Nonetheless, with a very small leeway for the parameters to vary, the model is remarkably predictive, making falsification or vindication more or less straightforward at colliders.

Acknowledgements

TB acknowledges a Senior Research Fellowship from UGC, India. AR is partially funded by the Science and Engineering Research Board Grant No. SR/S2/JCB-14/2009.

References

[1] J.C. Pati, A. Salam, Is baryon number conserved?, Phys. Rev. Lett. 31 (1973) 661.
[2] J.C. Pati, A. Salam, Lepton number as the fourth color, Phys. Rev. D 10 (1974) 275; Phys. Rev. D 11 (1975) 703 (Erratum).
[3] R.N. Mohapatra, J.C. Pati, "Natural" left-right symmetry, Phys. Rev. D 11 (1975) 2558.
[4] G. Senjanovic, R.N. Mohapatra, Exact left-right symmetry and spontaneous violation of parity, Phys. Rev. D 12 (1975) 1502.
[5] L.M. Krauss, F. Wilczek, Discrete gauge symmetry in continuum theories, Phys. Rev. Lett. 62 (1989) 1221.

[2] For a recent discussion of the sensitivity of the LHC detectors to such disappearing charged tracks, see for example, [83].

[3] The possibility of detecting a virtual heavy W_R signal through much lighter RH 'neutrino jets' has recently been examined in [88].

[6] L.E. Ibanez, G.G. Ross, Discrete gauge symmetry anomalies, Phys. Lett. B 260 (1991) 291.
[7] L.E. Ibanez, G.G. Ross, Discrete gauge symmetries and the origin of baryon and lepton number conservation in supersymmetric versions of the standard model, Nucl. Phys. B 368 (1992) 3.
[8] S.P. Martin, Some simple criteria for gauged R-parity, Phys. Rev. D 46 (1992) R2769, arXiv:hep-ph/9207218.
[9] M. Kadastik, K. Kannike, M. Raidal, Dark matter as the signal of grand unification, Phys. Rev. D 80 (2009) 085020, arXiv:0907.1894 [hep-ph]; Phys. Rev. D 81 (2010) 029903 (Erratum).
[10] M. Frigerio, T. Hambye, Dark matter stability and unification without supersymmetry, Phys. Rev. D 81 (2010) 075002, arXiv:0912.1545 [hep-ph].
[11] Y. Mambrini, K.A. Olive, J. Quevillon, B. Zaldivar, Gauge coupling unification and nonequilibrium thermal dark matter, Phys. Rev. Lett. 110 (2013) 241306, arXiv:1302.4438 [hep-ph].
[12] Y. Mambrini, N. Nagata, K.A. Olive, J. Quevillon, J. Zheng, Dark matter and gauge coupling unification in nonsupersymmetric SO(10) grand unified models, Phys. Rev. D 91 (2015) 095010, arXiv:1502.06929 [hep-ph].
[13] J. Heeck, S. Patra, Minimal left–right symmetric dark matter, Phys. Rev. Lett. 115 (2015) 121804, arXiv:1507.01584 [hep-ph].
[14] C. Garcia-Cely, J. Heeck, Phenomenology of left–right symmetric dark matter, J. Cosmol. Astropart. Phys. 1603 (2016) 021, arXiv:1512.03332 [hep-ph].
[15] A. Berlin, P.J. Fox, D. Hooper, G. Mohlabeng, Mixed dark matter in left–right symmetric models, J. Cosmol. Astropart. Phys. 1606 (2016) 016, arXiv:1604.06100 [hep-ph].
[16] P.S.B. Dev, R.N. Mohapatra, Y. Zhang, Naturally stable right-handed neutrino dark matter, arXiv:1608.06266 [hep-ph], 2016.
[17] H. Georgi, The state of the art gauge theories, AIP Conf. Proc. 23 (1975) 575.
[18] H. Fritzsch, P. Minkowski, Unified interactions of leptons and hadrons, Ann. Phys. 93 (1975) 193.
[19] D. Chang, R.N. Mohapatra, J. Gipson, R.E. Marshak, M.K. Parida, Experimental tests of new SO(10) grand unification, Phys. Rev. D 31 (1985) 1718.
[20] B. Brahmachari, U. Sarkar, K. Sridhar, Ruling out low-energy left–right symmetry in unified theories, Phys. Lett. B 297 (1992) 105.
[21] N.G. Deshpande, E. Keith, P.B. Pal, Implications of LEP results for SO(10) grand unification, Phys. Rev. D 46 (1993) 2261.
[22] S. Bertolini, L. Di Luzio, M. Malinsky, Intermediate mass scales in the non-supersymmetric SO(10) grand unification: a reappraisal, Phys. Rev. D 80 (2009) 015013, arXiv:0903.4049 [hep-ph].
[23] J. Chakrabortty, A. Raychaudhuri, GUTs with dim-5 interactions: gauge unification and intermediate scales, Phys. Rev. D 81 (2010) 055004, arXiv:0909.3905 [hep-ph].
[24] T. Bandyopadhyay, B. Brahmachari, A. Raychaudhuri, Implications of the CMS search for W_R on grand unification, J. High Energy Phys. 02 (2016) 023, arXiv:1509.03232 [hep-ph].
[25] D. Chang, R.N. Mohapatra, M.K. Parida, Decoupling parity and SU(2)-R breaking scales: a new approach to left–right symmetric models, Phys. Rev. Lett. 52 (1984) 1072.
[26] D. Chang, R.N. Mohapatra, M.K. Parida, A new approach to left–right symmetry breaking in unified gauge theories, Phys. Rev. D 30 (1984) 1052.
[27] S. Bertolini, A. Maiezza, F. Nesti, Present and future K and B meson mixing constraints on TeV scale left–right symmetry, Phys. Rev. D 89 (2014) 095028, arXiv:1403.7112 [hep-ph].
[28] R. Slansky, Group theory for unified model building, Phys. Rep. 79 (1981) 1.
[29] F. del Aguila, L.E. Ibanez, Higgs bosons in SO(10) and partial unification, Nucl. Phys. B 177 (1981) 60.
[30] R.N. Mohapatra, G. Senjanovic, Higgs boson effects in grand unified theories, Phys. Rev. D 27 (1983) 1601.
[31] C. Patrignani, et al., Review of particle physics, Chin. Phys. C 40 (2016) 100001.
[32] M.E. Machacek, M.T. Vaughn, Two loop renormalization group equations in a general quantum field theory. 1. Wave function renormalization, Nucl. Phys. B 222 (1983) 83.
[33] D.R.T. Jones, The two loop beta function for a G(1) × G(2) gauge theory, Phys. Rev. D 25 (1982) 581.
[34] S. Weinberg, Effective gauge theories, Phys. Lett. B 91 (1980) 51.
[35] L.J. Hall, Grand unification of effective gauge theories, Nucl. Phys. B 178 (1981) 75.
[36] S. Bertolini, L. Di Luzio, M. Malinsky, Light color octet scalars in the minimal SO(10) grand unification, Phys. Rev. D 87 (2013) 085020, arXiv:1302.3401 [hep-ph].
[37] S. Bertolini, L. Di Luzio, M. Malinsky, Seesaw scale in the minimal renormalizable SO(10) grand unification, Phys. Rev. D 85 (2012) 095014, arXiv:1202.0807 [hep-ph].
[38] K.S. Babu, S. Khan, Minimal nonsupersymmetric SO(10) model: gauge coupling unification, proton decay, and fermion masses, Phys. Rev. D 92 (2015) 075018, arXiv:1507.06712 [hep-ph].
[39] P. Nath, P. Fileviez Perez, Proton stability in grand unified theories, in strings and in branes, Phys. Rep. 441 (2007) 191, arXiv:hep-ph/0601023.
[40] T. Blum, et al., Domain wall QCD with physical quark masses, Phys. Rev. D 93 (2016) 074505, arXiv:1411.7017 [hep-lat].
[41] Y. Aoki, C. Dawson, J. Noaki, A. Soni, Proton decay matrix elements with domain-wall fermions, Phys. Rev. D 75 (2007) 014507, arXiv:hep-lat/0607002.
[42] Y. Aoki, P. Boyle, P. Cooney, L. Del Debbio, R. Kenway, C.M. Maynard, A. Soni, R. Tweedie, Proton lifetime bounds from chirally symmetric lattice QCD, Phys. Rev. D 78 (2008) 054505, arXiv:0806.1031 [hep-lat].
[43] Y. Aoki, E. Shintani, A. Soni, Proton decay matrix elements on the lattice, Phys. Rev. D 89 (2014) 014505, arXiv:1304.7424 [hep-lat].
[44] T. Nihei, J. Arafune, The two loop long range effect on the proton decay effective Lagrangian, Prog. Theor. Phys. 93 (1995) 665, arXiv:hep-ph/9412325.
[45] A.J. Buras, J.R. Ellis, M.K. Gaillard, D.V. Nanopoulos, Aspects of the grand unification of strong, weak and electromagnetic interactions, Nucl. Phys. B 135 (1978) 66.
[46] J.T. Goldman, D.A. Ross, A new estimate of the proton lifetime, Phys. Lett. B 84 (1979) 208.
[47] J.R. Ellis, D.V. Nanopoulos, S. Rudaz, GUTs 3: SUSY GUTs 2, Nucl. Phys. B 202 (1982) 43.
[48] L.E. Ibanez, C. Munoz, Enhancement factors for supersymmetric proton decay in the Wess–Zumino gauge, Nucl. Phys. B 245 (1984) 425.
[49] C. Munoz, Enhancement factors for supersymmetric proton decay in SU(5) and SO(10) with superfield techniques, Phys. Lett. B 177 (1986) 55.
[50] P. Fileviez Perez, Fermion mixings versus $d = 6$ proton decay, Phys. Lett. B 595 (2004) 476, arXiv:hep-ph/0403286.
[51] I. Dorsner, P. Fileviez Perez, How long could we live?, Phys. Lett. B 625 (2005) 88, arXiv:hep-ph/0410198.
[52] K. Abe, et al., Search for proton decay via $p \to e^+\pi^0$ and $p \to \mu^+\pi^0$ in 0.31 megaton·years exposure of the Super-Kamiokande water Cherenkov detector, Phys. Rev. D 95 (2017) 012004, arXiv:1610.03597 [hep-ex].
[53] K. Abe, et al., Physics potentials with the second hyper-Kamiokande detector in Korea, arXiv:1611.06118 [hep-ex], 2016.
[54] P. Minkowski, $\mu \to e\gamma$ at a rate of one out of 10^9 muon decays?, Phys. Lett. B 67 (1977) 421.
[55] T. Yanagida, Horizontal symmetry and masses of neutrinos, Conf. Proc. C 7902131 (1979) 95.
[56] R.N. Mohapatra, G. Senjanovic, Neutrino mass and spontaneous parity violation, Phys. Rev. Lett. 44 (1980) 912.
[57] J. Schechter, J.W.F. Valle, Neutrino masses in SU(2) × U(1) theories, Phys. Rev. D 22 (1980) 2227.
[58] D. Borah, S. Patra, U. Sarkar, TeV scale left right symmetry with spontaneous D-parity breaking, Phys. Rev. D 83 (2011) 035007, arXiv:1006.2245 [hep-ph].
[59] M. Magg, C. Wetterich, Neutrino mass problem and gauge hierarchy, Phys. Lett. B 94 (1980) 61.
[60] G. Lazarides, Q. Shafi, C. Wetterich, Proton lifetime and fermion masses in an SO(10) model, Nucl. Phys. B 181 (1981) 287.
[61] R.N. Mohapatra, G. Senjanovic, Neutrino masses and mixings in gauge models with spontaneous parity violation, Phys. Rev. D 23 (1981) 165.
[62] H.-C. Cheng, B.A. Dobrescu, K.T. Matchev, Generic and chiral extensions of the supersymmetric standard model, Nucl. Phys. B 543 (1999) 47, arXiv:hep-ph/9811316.
[63] J.L. Feng, T. Moroi, L. Randall, M. Strassler, S.-f. Su, Discovering supersymmetry at the Tevatron in wino LSP scenarios, Phys. Rev. Lett. 83 (1999) 1731, arXiv:hep-ph/9904250.
[64] M. Cirelli, N. Fornengo, A. Strumia, Minimal dark matter, Nucl. Phys. B 753 (2006) 178, arXiv:hep-ph/0512090.
[65] H. Goldberg, Constraint on the photino mass from cosmology, Phys. Rev. Lett. 50 (1983) 1419; Phys. Rev. Lett. 103 (2009) 099905 (Erratum).
[66] J.R. Ellis, J.S. Hagelin, D.V. Nanopoulos, K.A. Olive, M. Srednicki, Supersymmetric relics from the Big Bang, Nucl. Phys. B 238 (1984) 453.
[67] P.F. Smith, J.R.J. Bennett, A search for heavy stable particles, Nucl. Phys. B 149 (1979) 525.
[68] P.F. Smith, J.R.J. Bennett, G.J. Homer, J.D. Lewin, H.E. Walford, W.A. Smith, A search for anomalous hydrogen in enriched D-20, using a time-of-flight spectrometer, Nucl. Phys. B 206 (1982) 333.
[69] T.K. Hemmick, et al., A search for anomalously heavy isotopes of low Z nuclei, Phys. Rev. D 41 (1990) 2074.
[70] P. Verkerk, G. Grynberg, B. Pichard, M. Spiro, S. Zylberajch, M.E. Goldberg, P. Fayet, Search for superheavy hydrogen in sea water, Phys. Rev. Lett. 68 (1992) 1116.
[71] T. Yamagata, Y. Takamori, H. Utsunomiya, Search for anomalously heavy hydrogen in deep sea water at 4000-m, Phys. Rev. D 47 (1993) 1231.
[72] P.A.R. Ade, et al., Planck 2015 results. XIII. Cosmological parameters, Astron. Astrophys. 594 (2016) A13, arXiv:1502.01589 [astro-ph.CO].
[73] S. Chatrchyan, et al., Searches for long-lived charged particles in pp collisions at $\sqrt{s} = 7$ and 8 TeV, J. High Energy Phys. 07 (2013) 122, arXiv:1305.0491 [hep-ex].
[74] V. Khachatryan, et al., Search for long-lived charged particles in proton–proton collisions at $\sqrt{s} = 13$ TeV, Phys. Rev. D 94 (2016) 112004, arXiv:1609.08382 [hep-ex].
[75] G. Bélanger, F. Boudjema, A. Pukhov, A. Semenov, micrOMEGAs4.1: two dark matter candidates, Comput. Phys. Commun. 192 (2015) 322, arXiv:1407.6129 [hep-ph].

[76] A. Alloul, N.D. Christensen, C. Degrande, C. Duhr, B. Fuks, FeynRules 2.0 — a complete toolbox for tree-level phenomenology, Comput. Phys. Commun. 185 (2014) 2250, arXiv:1310.1921 [hep-ph].
[77] A. Roitgrund, G. Eilam, S. Bar-Shalom, Implementation of the left-right symmetric model in FeynRules, Comput. Phys. Commun. 203 (2016) 18, arXiv:1401.3345 [hep-ph].
[78] D.S. Akerib, et al., The Large Underground Xenon (LUX) experiment, Nucl. Instrum. Methods, Sect. A 704 (2013) 111, arXiv:1211.3788 [physics.ins-det].
[79] D.C. Malling, et al., After LUX: the LZ program, arXiv:1110.0103 [astro-ph.IM], 2011.
[80] E. Aprile, et al., Physics reach of the XENON1T dark matter experiment, J. Cosmol. Astropart. Phys. 1604 (2016) 027, arXiv:1512.07501 [physics.ins-det].
[81] A. Abramowski, et al., Search for a dark matter annihilation signal from the Galactic Center halo with H.E.S.S., Phys. Rev. Lett. 106 (2011) 161301, arXiv:1103.3266 [astro-ph.HE].
[82] A. Abramowski, et al., Search for photon-linelike signatures from dark matter annihilations with H.E.S.S., Phys. Rev. Lett. 110 (2013) 041301, arXiv:1301.1173 [astro-ph.HE].
[83] R. Mahbubani, P. Schwaller, J. Zurita, Closing the window for compressed dark sectors with disappearing charged tracks, arXiv:1703.05327 [hep-ph], 2017.
[84] N.E. Mavromatos, V.A. Mitsou, Physics reach of MoEDAL at LHC: magnetic monopoles, supersymmetry and beyond, in: 5th International Conference on New Frontiers in Physics Kolymbari, Crete, Greece, July 6–14, 2016, 2016, arXiv:1612.07012 [hep-ph].
[85] M. Drees, X. Tata, Signals for heavy exotics at hadron colliders and supercolliders, Phys. Lett. B 252 (1990) 695.
[86] M. Fairbairn, A.C. Kraan, D.A. Milstead, T. Sjostrand, P.Z. Skands, T. Sloan, Stable massive particles at colliders, Phys. Rep. 438 (2007) 1, arXiv:hep-ph/0611040.
[87] W.-Y. Keung, G. Senjanovic, Majorana neutrinos and the production of the right-handed charged gauge boson, Phys. Rev. Lett. 50 (1983) 1427.
[88] M. Mitra, R. Ruiz, D.J. Scott, M. Spannowsky, Neutrino jets from high-mass W_R gauge bosons in TeV-scale left-right symmetric models, Phys. Rev. D 94 (2016) 095016, arXiv:1607.03504 [hep-ph].
[89] A. Belyaev, N.D. Christensen, A. Pukhov, CalcHEP 3.4 for collider physics within and beyond the standard model, Comput. Phys. Commun. 184 (2013) 1729, arXiv:1207.6082 [hep-ph].
[90] J. Pumplin, D.R. Stump, J. Huston, H.L. Lai, P.M. Nadolsky, W.K. Tung, New generation of parton distributions with uncertainties from global QCD analysis, J. High Energy Phys. 07 (2002) 012, arXiv:hep-ph/0201195.
[91] Qing-Hong Cao, Zhao Li, Jiang-Hao Yu, C.-P. Yuan, Discovery and identification of W' and Z' in $SU(2)_1 \times SU(2)_2 \times U(1)_X$ models at the LHC, Phys. Rev. D 86 (2012) 095010.

11

Lepton portal limit of inert Higgs doublet dark matter with radiative neutrino mass

Debasish Borah [a], Soumya Sadhukhan [b,*], Shibananda Sahoo [a]

[a] Department of Physics, Indian Institute of Technology Guwahati, Assam 781039, India
[b] Physical Research Laboratory, Ahmedabad 380009, India

ARTICLE INFO

Editor: G.F. Giudice

ABSTRACT

We study an extension of the Inert Higgs Doublet Model (IHDM) by three copies of right handed neutrinos and heavy charged leptons such that both the inert Higgs doublet and the heavy fermions are odd under the Z_2 symmetry of the model. The neutrino masses are generated at one loop in the scotogenic fashion. Assuming the neutral scalar of the inert Higgs to be the dark matter candidate, we particularly look into the region of parameter space where dark matter relic abundance is primarily governed by the inert Higgs coupling with the leptons. This corresponds to tiny Higgs portal coupling of dark matter as well as large mass splitting within different components of the inert Higgs doublet suppressing the coannihilations. Such lepton portal couplings can still produce the correct relic abundance even if the Higgs portal couplings are arbitrarily small. Such tiny Higgs portal couplings may be responsible for suppressed dark matter nucleon cross section as well as tiny invisible branching ratio of the standard model Higgs, to be probed at ongoing and future experiments. We also briefly discuss the collider implications of such a scenario.

1. Introduction

The observational evidence suggesting the presence of dark matter (DM) in the Universe are irrefutable, with the latest data from the Planck experiment [1] indicating that approximately 27% of the present Universe is composed of dark matter. The observed abundance of DM is usually represented in terms of density parameter Ω as

$$\Omega_{\rm DM}h^2 = 0.1187 \pm 0.0017 \quad (1)$$

where $h =$ (Hubble Parameter)/100 is a parameter of order unity. In spite of astrophysical and cosmological evidences confirming the presence of DM, the fundamental nature of DM is not yet known. Since none of the particles in the Standard Model (SM) can fulfil the criteria of a DM candidate, several beyond Standard Model (BSM) proposals have been put forward in the last few decades. Among them, the weakly interacting massive particle (WIMP) paradigm is the most popular one. Such WIMP dark matter candidates can interact with the SM particles through weak interactions and hence can be produced at the Large Hadron Collider (LHC) or can scatter off nuclei at dark matter direct detection experiments like the ongoing LUX [2] and PandaX-II experiment [3].

Among different BSM proposals to incorporate dark matter, the inert Higgs doublet model (IHDM) [4–6] is one of the simplest extensions of the SM with an additional scalar field transforming as doublet under $SU(2)$ and having hypercharge $Y = 1$, odd under an imposed Z_2 discrete symmetry. As shown by the earlier works on IHDM, there are typically two mass ranges of DM mass satisfying the correct relic abundance criteria: one below the W boson mass and the other around 550 GeV or above. Among these, the low mass regime is particularly interesting due to stronger direct detection bounds. For example, the latest data from the LUX experiment rules out DM-nucleon spin independent cross section above around 2.2×10^{-46} cm^2 for DM mass of around 50 GeV [2]. In this mass range, as we discuss in details below, the tree level DM-SM interaction through the SM Higgs (h) portal is interesting as it can simultaneously control the relic abundance as well as the DM-nucleon scattering cross section. In this mass range, only a narrow region near the resonance $m_{\rm DM} \approx m_h/2$ is currently allowed by the LUX data. Though future DM direct detection experiments will be able to probe this region further, it could also be true that the DM-Higgs interaction is indeed too tiny to be observed at experiments. Such a tiny Higgs portal interaction will also be insufficient

* Corresponding author.
E-mail addresses: dborah@iitg.ernet.in (D. Borah), soumyas@prl.res.in (S. Sadhukhan), shibananda@iitg.ernet.in (S. Sahoo).

to produce the correct relic abundance of DM in this low mass regime. This almost rules out the low mass regime of DM in IHDM $m_{DM} \lesssim 70$ GeV.

Here we consider a simple extension of IHDM by singlet leptons (both neutral and charged) odd under the Z_2 symmetry such that the inert scalar dark matter can interact with the SM particles through these singlet leptons. This new interaction through lepton portal can revive the low mass regime of inert scalar DM even if future direct detection experiment rules out the Higgs portal interaction completely. The lepton portal interactions can also remain unconstrained from the limits on DM-nucleon interactions. Such a scenario is particularly interesting if LHC finds some signatures corresponding to the low mass regime of inert scalar DM while the direct detection continues to give null results. The dominant lepton portal interactions can explain correct relic abundance, null results at direct detection experiments and also give rise to interesting signatures at colliders. The neutral leptons added to IHDM can also give rise to tiny neutrino masses at one-loop level through scotogenic fashion [12]. We discuss the constraints on the model parameters from neutrino mass, DM constraints and also make some estimates of some interesting collider signatures while comparing them with the pure IHDM.

This article is organised as follows. In section 2, we discuss the IHDM and then consider the lepton portal extension of it in section 3. In section 4, we discuss the dark matter related studies followed by our collider estimates in section 5. We finally conclude in section 6.

2. Inert Higgs Doublet Model

The inert Higgs Doublet Model (IHDM) [4–6] is an extension of the Standard Model (SM) by an additional Higgs doublet Φ_2 and a discrete Z_2 symmetry under which all SM fields are even while $\Phi_2 \to -\Phi_2$. This Z_2 symmetry not only prevents the coupling of SM fermions to Φ_2 at renormalisable level but also forbids those terms in the scalar potential which are linear or trilinear in Φ_2. Therefore, the second Higgs doublet Φ_2 can interact with the SM particles only through its couplings to the SM Higgs doublet and the electroweak gauge bosons. The Z_2 symmetry also prevents the lightest component of Φ_2 from decaying, making it stable on cosmological scale. If one of the neutral components of Φ_2 happen to be the lightest Z_2 odd particle, then it can be a potential dark matter candidate. The scalar potential of the model involving the SM Higgs doublet Φ_1 and the inert doublet Φ_2 can be written as

$$\begin{aligned} V(\Phi_1, \Phi_2) &= \mu_1^2|\Phi_1|^2 + \mu_2^2|\Phi_2|^2 \\ &+\frac{\lambda_1}{2}|\Phi_1|^4 + \frac{\lambda_2}{2}|\Phi_2|^4 + \lambda_3|\Phi_1|^2|\Phi_2|^2 \\ &+\lambda_4|\Phi_1^\dagger\Phi_2|^2 + \{\frac{\lambda_5}{2}(\Phi_1^\dagger\Phi_2)^2 + \text{h.c.}\}, \end{aligned} \tag{2}$$

To ensure that none of the neutral components of the inert Higgs doublet acquire a non-zero vacuum expectation value (vev), $\mu_2^2 > 0$ is assumed. This also prevents the Z_2 symmetry from being spontaneously broken. The electroweak symmetry breaking (EWSB) occurs due to the non-zero vev acquired by the neutral component of Φ_1. After the EWSB these two scalar doublets can be written in the following form in the unitary gauge.

$$\Phi_1 = \begin{pmatrix} 0 \\ \frac{v+h}{\sqrt{2}} \end{pmatrix}, \ \Phi_2 = \begin{pmatrix} H^+ \\ \frac{H+iA}{\sqrt{2}} \end{pmatrix} \tag{3}$$

The masses of the physical scalars at tree level can be written as

$$\begin{aligned} m_h^2 &= \lambda_1 v^2, \\ m_{H^+}^2 &= \mu_2^2 + \frac{1}{2}\lambda_3 v^2, \\ m_H^2 &= \mu_2^2 + \frac{1}{2}(\lambda_3+\lambda_4+\lambda_5)v^2 = m_{H^\pm}^2 + \frac{1}{2}(\lambda_4+\lambda_5)v^2, \\ m_A^2 &= \mu_2^2 + \frac{1}{2}(\lambda_3+\lambda_4-\lambda_5)v^2 = m_{H^\pm}^2 + \frac{1}{2}(\lambda_4-\lambda_5)v^2. \end{aligned} \tag{4}$$

Here m_h is the SM like Higgs boson mass, m_H, m_A are the masses of the CP even and CP odd scalars from the inert doublet. Without loss of generality, we consider $\lambda_5 < 0, \lambda_4 + \lambda_5 < 0$ so that the CP even scalar is the lightest Z_2 odd particle and hence a stable dark matter candidate.

The new scalar fields discussed above can be constrained from the LEP I precision measurement of the Z boson decay width. In order to forbid the decay channel $Z \to HA$, one arrives at the constraint $m_H + m_A > m_Z$. In addition to this, the LEP II constraints roughly rule out the triangular region [7]

$$m_H < 80 \text{ GeV}, \quad m_A < 100 \text{ GeV}, \quad m_A - m_H > 8 \text{ GeV}$$

The LEP collider experiment data restrict the charged scalar mass to $m_{H^+} > 70$–90 GeV [8]. The Run 1 ATLAS dilepton limit is discussed in the context of IHDM in Ref. [9] taking into consideration of specific masses of charged Higgs. Another important restriction on m_{H^+} comes from the electroweak precision data (EWPD). Since the contribution of the additional doublet Φ_2 to electroweak S parameter is always small [4], we only consider the contribution to the electroweak T parameter here. The relevant contribution is given by [4]

$$\Delta T = \frac{1}{16\pi^2\alpha v^2}[F(m_{H^+}, m_A) + F(m_{H^+}, m_H) - F(m_A, m_H)] \tag{5}$$

where

$$F(m_1, m_2) = \frac{m_1^2 + m_2^2}{2} - \frac{m_1^2 m_2^2}{m_1^2 - m_2^2}\ln\frac{m_1^2}{m_2^2} \tag{6}$$

The EWPD constraint on ΔT is given as [10]

$$-0.1 < \Delta T + T_h < 0.2 \tag{7}$$

where $T_h \approx -\frac{3}{8\pi\cos^2\theta_W}\ln\frac{m_h}{m_Z}$ is the SM Higgs contribution to the T parameter [11].

3. Lepton portal extensions of IHDM

As discussed in the introduction, considering lepton portal extensions of IHDM is very well motivated, specially from the origin of neutrino mass, dark matter direct detections and other flavour physics observables in the lepton sector. The inert Higgs doublet of the IHDM can couple to the SM leptons, if the model is suitably extended either by Z_2 odd neutral Majorana fermions or by charged vector like leptons, none of which introduce any chiral anomalies. The addition of three copies of neutral heavy singlet fermions N_i, odd under the Z_2 symmetry leads to the upgradation of the IHDM to the scotogenic model [12]. Apart from providing another dark matter candidate in terms of the lightest N_i, the model also can explain tiny neutrino masses at one loop level. In the set up we study here, all of these singlet neutral fermions are assumed to be heavier than the neutral component of the inert Higgs doublet and hence our dark matter analysis is confined to the scalar dark matter only. The relevant interaction terms of these singlet fermions can be written as

$$\mathcal{L} \supset M_N NN + \left((Y_N)_{ij}\bar{L}_i\tilde{\Phi}_2 N_j + \text{h.c.}\right). \tag{8}$$

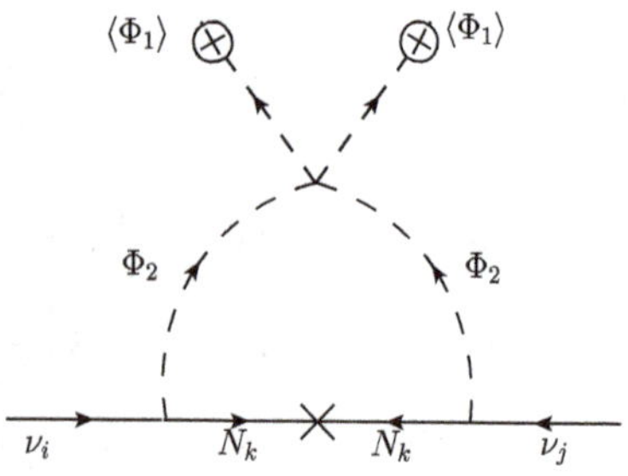

Fig. 1. One-loop contribution to neutrino mass.

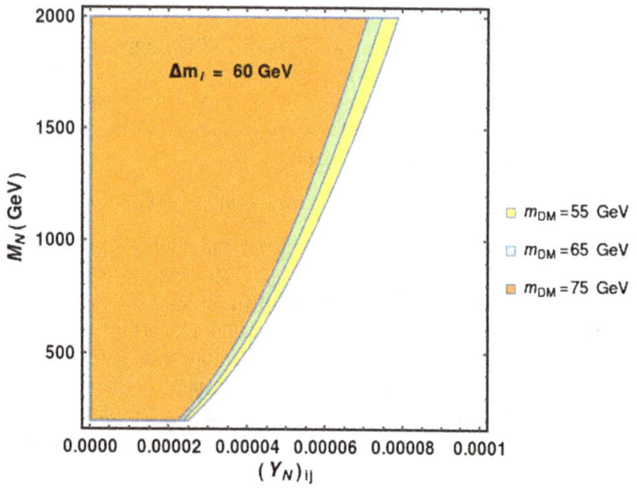

Fig. 2. Allowed model parameters for neutrino mass generation.

The Feynman diagram for such one loop neutrino mass is shown in Fig. 1. Using the expression from [12] of one-loop neutrino mass

$$(m_\nu)_{ij} = \frac{(Y_N)_{ik}(Y_N)_{jk}M_k}{16\pi^2}\left(\frac{m_R^2}{m_R^2-M_k^2}\ln\frac{m_R^2}{M_k^2} - \frac{m_I^2}{m_I^2-M_k^2}\ln\frac{m_I^2}{M_k^2}\right) \tag{9}$$

Here $m_{R,I}^2 = m_{H,A}^2$ are the masses of scalar and pseudo-scalar part of Φ_2^0 and M_k the mass of singlet fermion N in the internal line. The index $i, j = 1, 2, 3$ runs over the three fermion generations as well as three copies of N. For $m_H^2 + m_A^2 \approx M_k^2$, the above expression can be simply written as

$$(m_\nu)_{ij} \approx \frac{\lambda_5 v^2}{32\pi^2}\frac{(Y_N)_{ik}(Y_N)_{jk}}{M_k} = \frac{m_A^2 - m_H^2}{32\pi^2}\frac{(Y_N)_{ik}(Y_N)_{jk}}{M_k} \tag{10}$$

In this model for the neutrino mass to match with experimentally observed limits (~ 0.1 eV), very tiny Yukawa couplings are required for the right handed neutrino mass of order of 1 TeV. Taking the mass difference $m_A - m_H = m_{H^\pm} - m_H = 60$ GeV, we show the constraints on neutral singlet fermion mass and corresponding Yukawa coupling from correct neutrino mass requirement in Fig. 2. It can be seen that for low mass regime of DM, the neutrino mass constraints force the Yukawa couplings to be smaller than 10^{-4}, too small to have any impact on dark matter relic abundance calculation, to be discussed below. These neutral fermions can also contribute to charged lepton flavour violation (LFV) at one loop involving $N, \Phi_2^\pm$. The LFV processes like $\mu \to e\gamma$ remain suppressed in the SM due to the smallness of neutrino masses. Such LFV decays like $\mu \to e\gamma$ are being searched for at experiments like MEG [13]. The latest bound from the MEG collaboration is $\mathrm{BR}(\mu \to e\gamma) < 4.2 \times 10^{-13}$ at 90% confidence level [13]. However, due to small Yukawa couplings, as required by tiny neutrino mass constraints discussed above, keeps this new contribution to $\mu \to e\gamma$ way below this latest experimental bound, as discussed in the recent works [14,15].

Similar to neutral singlet fermions, one can also incorporate charged singlet leptons $\chi_{L,R}$ with hypercharge $Y = 2$ and odd under the Z_2 symmetry. The relevant Lagrangian is

$$\mathcal{L} \supset M_\chi \bar{\chi}_L \chi_R + (Y_\chi)_{ij}\bar{L}_i \Phi_2 \chi_R + \text{h.c.} \tag{11}$$

These leptons can contribute both to dark matter relic abundance as well as LFV decays mentioned above. Since the corresponding Yukawa couplings are not restricted to be small from neutrino mass constraints, they can be sizeable and hence play a non-trivial role in generating DM relic abundance as we discuss below. Such large Yukawa couplings can however give a large contribution to LFV decays like $\mu \to e\gamma$, with χ, Φ_2^0 in loop. As shown in a recent work [16], the above MEG bound can constrain the product of two relevant Yukawa couplings to be below 10^{-9} for χ mass around 100 GeV–1 TeV, too small to have any impact on DM relic abundance. These strict bounds from MEG can however be evaded by choosing diagonal structure of singlet lepton mass matrix M_χ and relevant Yukawa coupling Y. Such a structure can still have non-trivial impact on DM relic abundance, to be discussed below.

Although the addition of singlet fermions in this fashion may appear ad-hoc, they have very interesting phenomenological consequences as we discuss below. From UV completeness point of view, such low energy set up can in principle, be realised within well motivated BSM frameworks. For example, such exotic fermions can be realised within E_6 grand unified theories [17]. In the recent work [16], within the framework of left-right symmetric model it was shown that gauge singlet vector like fermions are necessary in order to generate all charged fermion masses through a common universal seesaw. The vector like charged leptons in the model discussed by [16] can generate light charged lepton masses at tree level and light Dirac neutrino masses at one loop. Thus, the low energy effective theory we study in this work from phenomenological point of view can have very well motivated UV completions. In another recent work [18], within the framework of an $SU(6)/Sp(6)$ little Higgs model, it is discussed how the vector like fermions appear along with an extended scalar sector, stabilizing the Higgs mass at electroweak scale. Even if this model (similar to all little Higgs models) has its own cut off scale which is incompatible with a strict idea of UV completion, it can naturally accommodate the vector like fermions instead of including it in an ad-hoc manner.

4. Dark matter

The relic abundance of a dark matter particle ψ which was in thermal equilibrium at some earlier epoch can be calculated by solving the Boltzmann equation

$$\frac{dn_\psi}{dt} + 3Hn_\psi = -\langle\sigma v\rangle(n_\psi^2 - (n_\psi^{\mathrm{eqb}})^2) \tag{12}$$

where n_ψ is the number density of the dark matter particle ψ and n_ψ^{eqb} is the number density when ψ was in thermal equilibrium. H is the Hubble expansion rate of the Universe and $\langle\sigma v\rangle$ is the thermally averaged annihilation cross section of the dark matter particle ψ. In terms of partial wave expansion $\langle\sigma v\rangle = a + bv^2$. Clearly, in the case of thermal equilibrium $n_\psi = n_\psi^{\mathrm{eqb}}$, the number density is decreasing only by the expansion rate H of the Universe. The approximate analytical solution of the above Boltzmann equation gives [19,20]

$$\Omega_\psi h^2 \approx \frac{1.04 \times 10^9 x_F}{M_{Pl}\sqrt{g_*}(a+3b/x_F)} \quad (13)$$

where $x_F = m_\psi/T_F$, T_F is the freeze-out temperature, g_* is the number of relativistic degrees of freedom at the time of freeze-out and $M_{Pl} \approx 10^{19}$ GeV is the Planck mass. Here, x_F can be calculated from the iterative relation

$$x_F = \ln \frac{0.038 g M_{\rm Pl} m_\psi < \sigma v >}{g_*^{1/2} x_F^{1/2}} \quad (14)$$

The expression for relic density also has a more simplified form given as [21]

$$\Omega_\psi h^2 \approx \frac{3 \times 10^{-27} \text{ cm}^3 \text{ s}^{-1}}{\langle \sigma v \rangle} \quad (15)$$

The thermal averaged annihilation cross section $\langle \sigma v \rangle$ is given by [22]

$$\langle \sigma v \rangle = \frac{1}{8 m_\psi^4 T K_2^2(m_\psi/T)} \int_{4m_\psi^2}^{\infty} \sigma(s - 4m_\psi^2)\sqrt{s} K_1(\sqrt{s}/T) ds \quad (16)$$

where K_i's are modified Bessel functions of order i, m_ψ is the mass of Dark Matter particle and T is the temperature.

If we consider the neutral component of the scalar doublet Φ_2 to be the dark matter candidate, the details of relic abundance calculation is similar to the inert doublet model studied extensively in the literature [4,6,10,12,23–26]. In the low mass regime $m_H = m_{DM} \leq M_W$, dark matter annihilation into the SM fermions through s-channel Higgs mediation dominates over other channels. As pointed out by [27], the dark matter annihilations $HH \to WW^* \to W f \bar{f}'$ can also play a role in the $m_{DM} \leq M_W$ region. Also, depending on the mass differences $m_{H^+} - m_H, m_A - m_H$, the coannihilations of H, H^+ and H, A can also play a role in generating the relic abundance of dark matter. The relic abundance calculation incorporating these effects were studied by several groups in [28,29]. Beyond the W boson mass threshold, the annihilation channel of scalar doublet dark matter into W^+W^- pairs opens up suppressing the relic abundance below what is observed by Planck experiment, unless the dark matter mass is heavier than around 500 GeV, depending on the DM-Higgs coupling. Apart from the usual annihilation channels of inert doublet dark matter, in this model there is another interesting annihilation channel where dark matter annihilates into a pair of neutrinos (charged leptons) through the heavy fermion N_i (χ) in the t-channel.

Apart from the relic abundance constraints from Planck experiment, there exists strict bounds on the dark matter nucleon cross section from direct detection experiments like Xenon100 [30] and more recently LUX [2,31]. For scalar dark matter considered in this work, the relevant spin independent scattering cross section mediated by SM Higgs is given as [4]

$$\sigma_{\rm SI} = \frac{\lambda_L^2 f^2}{4\pi} \frac{\mu^2 m_n^2}{m_h^4 m_{DM}^2} \quad (17)$$

where $\mu = m_n m_{DM}/(m_n + m_{DM})$ is the DM-nucleon reduced mass and $\lambda_L = (\lambda_3 + \lambda_4 + \lambda_5)$ is the quartic coupling involved in DM-Higgs interaction. A recent estimate of the Higgs-nucleon coupling f gives $f = 0.32$ [32] although the full range of allowed values is $f = 0.26 - 0.63$ [33]. The latest LUX bound [2] on $\sigma_{\rm SI}$ constrains the η_L-Higgs coupling λ significantly, if η_L gives rise to most of the dark matter in the Universe. According to this latest bound, at a dark matter mass of 50 GeV, dark matter nucleon scattering cross sections above 1.1×10^{-46} cm^2 are excluded at 90% confidence level. Similar but slightly weaker bound has been reported by the PandaX-II experiment recently [3]. We however include only the LUX bound in our analysis. One can also constrain the DM-Higgs coupling λ from the latest LHC constraint on the invisible decay width of the SM Higgs boson. This constraint is applicable only for dark matter mass $m_{DM} < m_h/2$. The invisible decay width is given by

$$\Gamma(h \to \text{Invisible}) = \frac{\lambda_L^2 v^2}{64\pi m_h}\sqrt{1 - 4m_{DM}^2/m_h^2} \quad (18)$$

The latest ATLAS constraint on invisible Higgs decay is [34]

$$\text{BR}(h \to \text{Invisible}) = \frac{\Gamma(h \to \text{Invisible})}{\Gamma(h \to \text{Invisible}) + \Gamma(h \to \text{SM})} < 22\%$$

As we will discuss below, this bound is weaker than the LUX 2016 bound.

It should be noted that, there can be sizeable DM-nucleon scattering cross section at one loop level as well, which does not depend on the Higgs portal coupling discussed above. Even in the minimal IHDM such one loop scattering can occur with charged scalar and electroweak gauge bosons in loop [35]. The contributions of such one loop scattering can be kept even below future direct detection experiments like Xenon-1T by choosing large mass differences between the components of the inert scalar doublet [35]. Such large mass splittings also minimise the role of coannihilation between different inert scalar components on the DM relic abundance. This is in the spirit of the present work's motivation, as the DM abundance is primarily determined by the lepton portal couplings, rather than gauge and Higgs portal couplings. Another one loop scattering can occur, in principle, due to the exchange of photons or Z boson. This is possible through an effective coupling of the form $C\partial^\mu \Phi_2^0 \partial^\nu \Phi_2^{0\dagger} F_{\mu\nu}$ with C being the loop factor [36]. However, since we have broken the degeneracy of our complex DM candidate Φ_2^0 and reduced it to one scalar and pseudoscalar, we can avoid such one loop scattering by choosing a mass splitting. In fact, one requires a non-zero mass splitting, at least greater than of the order of $\mathcal{O}(100$ keV), typical kinetic energy of DM particles, in order to avoid tree level inelastic scattering of DM off nuclei mediated by Z boson [37].

We implement the model in micrOMEGA 4.3.1 [38] to calculate the relic abundance of DM. We first reproduce the known results in IHDM by considering the neutral scalar H to be the DM candidate having mass below the W boson mass threshold. In the left panel of Fig. 3, we first show the parameter space of pure IHDM in $\lambda_L - m_{DM}$ plane that satisfies the condition $\Omega_{\rm DM} h^2 \leq 0.1187$. We have taken both the mass difference $m_A - m_H = m_{H^\pm} - m_H = 60$ GeV as a typical benchmark value satisfying all other constraints. Such a large benchmark point reduces the coannihilation effects and show the dependence of relic abundance on Higgs portal coupling λ_L in a visible manner.[1] The blue region in the left panel of Fig. 3 therefore indicates the parameter space where the DM annihilation is either just enough or more than the required one to produce the correct relic abundance. Therefore, considering the additional lepton portal couplings for such values of λ_L will further suppress the relic abundance. Therefore, we choose benchmark values of $\lambda_L - m_{DM}$ for our next analysis, from that region of this plot which overproduces the DM in pure IDM, so that an efficient lepton portal annihilation can bring down the relic abundance to the observed range. In the right panel of Fig. 3, we further impose the relic abundance criteria $\Omega_{\rm DM} h^2 \in 0.1187 \pm 0.0017$

[1] We have not considered low mass differences in this work as that will make the coannihilations more efficient reducing the dependence of relic abundance on Higgs or lepton portal couplings and here our main motivation is to show the importance of lepton portal couplings.

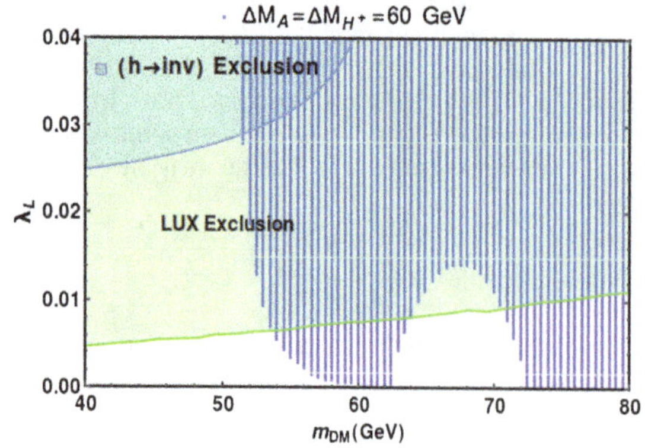

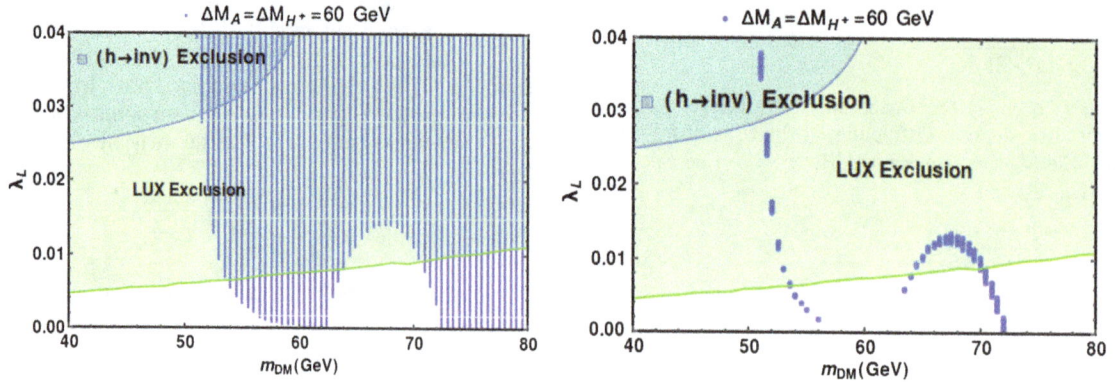

Fig. 3. Parameter space in the $\lambda_L - m_{DM}$ plane giving rise to dark matter relic abundance $\Omega_{\rm DM}h^2 \leq 0.1187$ (left panel) and $\Omega_{\rm DM}h^2 \in 0.1187 \pm 0.0017$ (right panel) in pure IHDM. (For interpretation of the references to colour in this figure, the reader is referred to the web version of this article.)

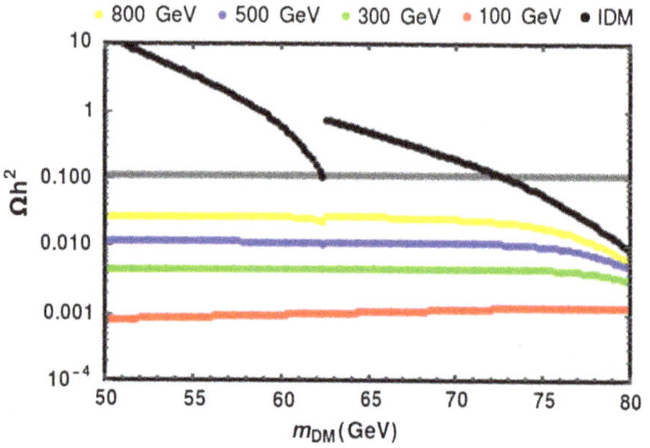

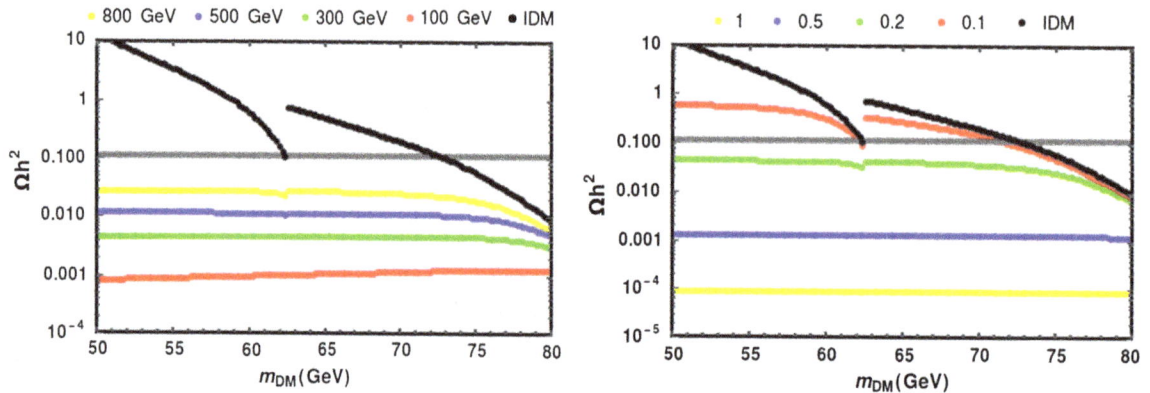

Fig. 4. Effect of lepton portal couplings on dark matter relic abundance, for specific dark matter Higgs coupling λ_L. Left: Relic density vs. m_{DM} for different M_N with fixed $Y_N = 0.2$. Right: Relic density vs. m_{DM} for different Y_N with fixed $M_N = 1000$ GeV. (For interpretation of the references to colour in this figure, the reader is referred to the web version of this article.)

which reduces the number of allowed points significantly from the one in the left panel. In both the plots we also show the LUX 2016 exclusion line based on the upper bound on DM nucleon scattering cross section. We also show the LHC limit on Higgs invisible decay width which remains weaker than the LUX 2016 bound. The tiny allowed region near $m_{\rm DM} \approx m_h/2$ corresponds to the s-channel resonance mediated by the SM Higgs while the allowed region of $m_{\rm DM}$ close to W boson mass threshold corresponds to the dominance of DM annihilation into three body SM final states mentioned above.

After reproducing the known results of IHDM in the low mass regime for a benchmark value of mass splitting, we calculate the DM relic abundance by incorporating the Z_2 odd heavy leptons. In Fig. 4, we show the effect of vector like neutral heavy leptons on relic abundance. To make DM annihilations through lepton portal more efficient, we choose the Higgs portal coupling to be very small $\lambda_L = 0.0001$ and also keep both the mass splitting within the components of the inert scalar doublet as 60 GeV like before. In the left panel of Fig. 4, the effect of heavy neutral fermion mass on the relic abundance is shown for a fixed value of Yukawa coupling $Y_N = 0.2$. In the right panel of Fig. 4, the effect of lepton portal Yukawa couplings on DM relic abundance is shown for fixed value of heavy neutral fermion mass $M_N = 1000$ GeV. From both these panels of Fig. 4, it is clear that the leptonic portal can play a non-trivial role in generating the DM relic abundance. While the benchmark values of Higgs portal coupling and mass splitting chosen above produce correct DM abundance only for two different masses, the introduction of lepton portal can result in new allowed region of DM masses. As expected, the maximum effect of lepton portal on DM relic abundance occurs for smaller values of heavy lepton mass or equivalently large values of Yukawa couplings. Since neutral heavy fermion couplings with SM leptons are required to be tiny from neutrino mass constraints as can be seen from Fig. 2, we consider only the effect of heavy charged leptons on DM relic abundance. The effect of charged lepton portal on DM relic abundance will be similar to that of neutral case discussed above.

After showing the effect of lepton portal on DM relic abundance for specific values of Yukawa and heavy neutral fermion masses, we do a general scan of these two parameters from the requirement of generating correct abundance. Since neutral heavy fermion portal is not efficient after neutrino mass constraints are incorporated, we do the general scan only for charged heavy lepton portal here. In Fig. 5, we show the allowed parameter space satisfying relic density in the $Y_\chi - m_{DM}$ plane for a benchmark point of IHDM parameters like before and taking the heavy charged fermion mass to be 100 GeV. The left panel of Fig. 5 considers the lepton portal couplings to be of general non-diagonal type while the right panel considers the couplings to be diagonal. As discussed before, such diagonal couplings will evade the constraints from LFV decay. Since a diagonal structure of Yukawa couplings reduces the total number of annihilation channels, one requires larger values of Yukawa couplings to produce the correct relic abundance, compared to the ones in the non-diagonal case. In Fig. 6, we show the allowed parameter space in $Y_\chi - M_\chi$ plane for two specific dark matter masses $m_{DM} = 55, 65$ GeV with general non-diagonal Yukawa couplings. The corresponding result for diagonal Yukawa couplings are shown in Fig. 7. It should be noted that these two benchmark values of DM masses in pure IHDM can not give rise to correct relic abundance for small values of Higgs

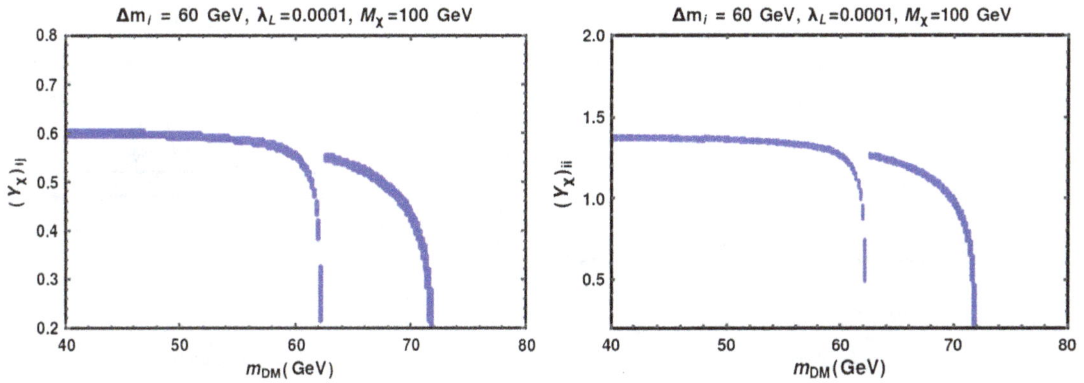

Fig. 5. Parameter space in the $Y_\chi - m_{DM}$ plane giving rise to the correct dark matter relic abundance with 3σ range for specific choice of $\lambda_L = 0.0001$ and $M_\chi = 100$ GeV. Left: nonzero off-diagonal Yukawa coupling scenario, Right: Diagonal Yukawa coupling scenario.

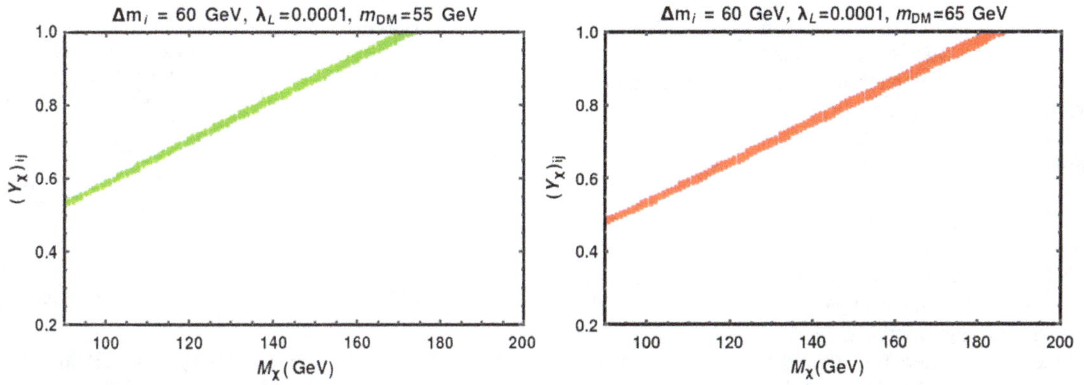

Fig. 6. Parameter space in the $Y_\chi - M_\chi$ plane giving rise to the correct dark matter relic abundance with 3σ range for specific choice of $\lambda_L = 0.0001$ and m_{DM} for nonzero off-diagonal Yukawa coupling scenario. Left: for $m_{DM} = 55$ GeV, Right: for $m_{DM} = 65$ GeV.

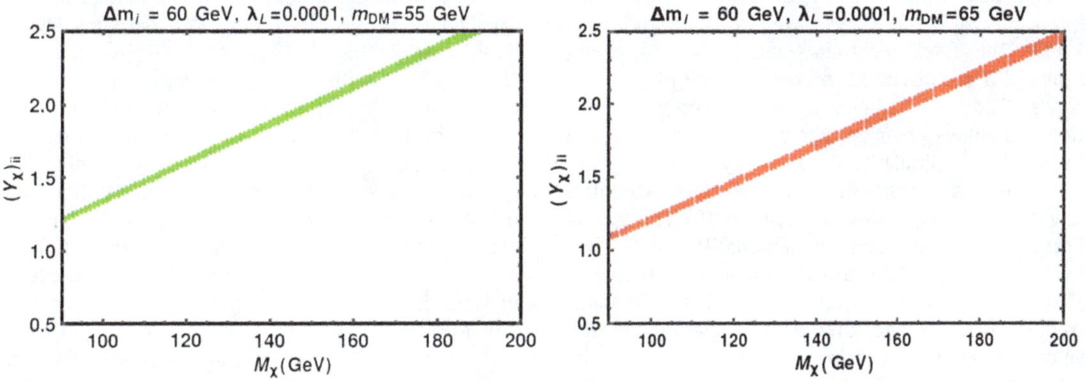

Fig. 7. Parameter space in the $Y_\chi - M_\chi$ plane giving rise to the correct dark matter relic abundance with 3σ range for specific choice of $\lambda_L = 0.0001$ and m_{DM} for Diagonal Yukawa coupling scenario. Left: for $m_{DM} = 55$ GeV, Right: for $m_{DM} = 65$ GeV.

portal couplings as seen from Fig. 3. However, after allowing the lepton portal couplings, we can generate correct relic abundance for such values of DM masses which remain disallowed in the pure IHDM.

The large lepton portal Yukawa couplings of order one required for giving correct relic abundance of dark matter can have serious implications for the renormalisation group evolution (RGE) of the quartic couplings of the scalar potential. Several works have appeared in the literature studying the implications of vector like fermions on the RGE of couplings some of which can be found in [49,50]. Using the RGE equations for the IHDM given in [51] and extending it by three copies of vector like charged leptons we study the evolution of different parameters of the model. The RGE equations used in our numerical analysis are given in Appendix A. We choose benchmark values of couplings at low energy that satisfy all relevant constraints. We consider $m_{DM} = 55$ GeV, $m_A = m_{H^+} = 115$ GeV, $\lambda_L = 0.0001, \lambda_1 = 0.258, \lambda_2 = 0.1, \lambda_3 = 0.338, \lambda_4 = \lambda_5 = -0.1685$ with the Higgs boson mass being kept at 125 GeV. The vector like charged lepton mass is 100 GeV with the corresponding Yukawa coupling being $Y_\chi = 1.4$. The neutral heavy fermion mass is also kept at 100 GeV with the corresponding Yukawa couplings being fixed to 0.001. The evolution of the quartic and the Yukawa couplings with energy μ is shown in Fig. 8. It can be seen that some of the Yukawa couplings including the one involving vector like charged leptons χ become non-perturbative around $\mu \approx 10^9$ GeV implying that some new dynamics should take over at that energy scale. This is due to the large value of Yukawa coupling between χ and Φ_2 chosen at low energy scale.

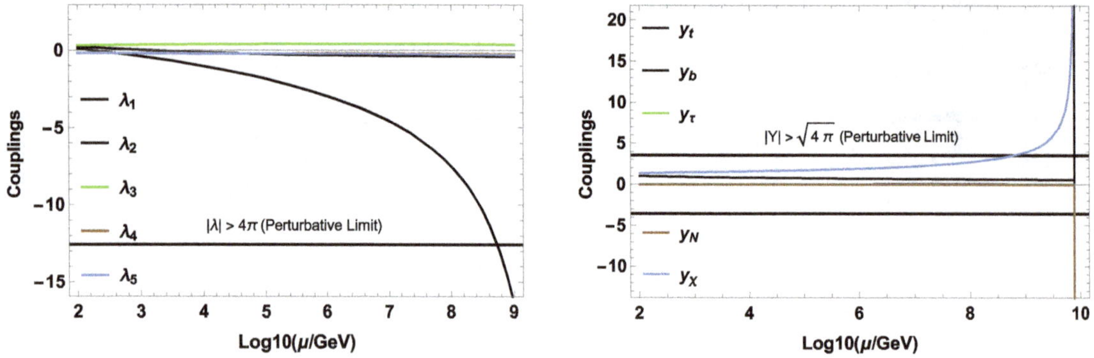

Fig. 8. Evolution of quartic and Yukawa couplings under RGE. The benchmark values chosen at low energy which satisfy all relevant constraints are $m_{DM} = 55$ GeV, $m_A = m_{H^+} = 115$ GeV, $\lambda_L = 0.0001$, $\lambda_1 = 0.258$, $\lambda_2 = 0.1$, $\lambda_3 = 0.338$, $\lambda_4 = \lambda_5 = -0.1685$. (For interpretation of the references to colour in this figure, the reader is referred to the web version of this article.)

The same non-perturbative nature is also seen in quartic coupling λ_2 around $\mu \approx 10^9$ GeV. Due to the coupled nature of the RGE equations, all other couplings including quartic and gauge also become non-perturbative beyond $\mu \approx 10^{10}$ GeV.

5. Collider implications

In pure IHDM, the pseudoscalar A can decay into Z and H whereas $H^\pm$ can decay to either $W^\pm H$ or $W^\pm A$. When $m_{H^\pm}$ is close to m_A, then the first decay mode of $H^\pm$ almost dominates. Depending upon the decay mode of $W^\pm$ and Z, we have either pure leptonic plus missing transverse energy (MET) or hadronic plus MET or mixed final states from pair production of the inert scalars. Earlier studies in the IHDM [39–41] focussed on pair production of inert scalars and their decays into leptons and MET. In another recent work [42], the authors studied dijet plus MET final states in the context of IHDM at LHC. The dilepton plus dijet plus MET and trilepton plus MET final states have also been studied in a recent work [43]. The 8 TeV constraints and 13 TeV projection from monojet plus MET are discussed in another work [44].

In the presence of both Z_2 odd neutral and charged vector like leptons, additional channels open up. For example, now $H^\pm$ can decay to $\chi^\pm\ \nu_i$ or $N_i\ l^\pm$. Similarly, A can decay into $l^\pm\ \chi^\mp$ or $N_i\ \bar{\nu}_i$. Since neutrino mass constraints push the mass of neutral leptons typically to the order of TeV range, both $H^\pm$ and A will mainly decay through charged vector like leptons (VLL) $\chi^\pm$. Then $\chi^\pm$ will further decay into $l^\pm H$. One can find earlier studies in the context of vector like leptons in references [45–49]. To highlight the difference in collider signatures with comparison to pure IHDM, we have considered a few benchmark points. We choose the following benchmark points all of which correspond to the fixed values of $m_h = 125$ GeV, $\lambda_L = 0.0001$, $\lambda_2 = 0.1$, $M_N = 1000$ GeV, $Y_N = 0.001$.

$$\text{BP1}: m_H = 55\ \text{GeV}, m_{H^+} = m_A = 115\ \text{GeV},$$
$$M_\chi = 100\ \text{GeV}, (Y_\chi)_{ii} = 1.5$$
$$\text{BP2}: m_H = 65\ \text{GeV}, m_{H^+} = m_A = 125\ \text{GeV},$$
$$M_\chi = 100\ \text{GeV}, (Y_\chi)_{ii} = 1.5$$
$$\text{BP3}: m_H = 65\ \text{GeV}, m_{H^+} = m_A = 200\ \text{GeV},$$
$$M_\chi = 150\ \text{GeV}, (Y_\chi)_{ii} = 2.0$$
$$\text{BP4}: m_H = 65\ \text{GeV}, m_{H^+} = m_A = 300\ \text{GeV},$$
$$M_\chi = 150\ \text{GeV}, (Y_\chi)_{ii} = 2.0.$$

In Table 1, we have listed the parton level cross sections for final states that contribute to dilepton+MET final states at detector level in both IHDM and IHDM+VLL models for the above benchmark points. It should be noted that for BP1 and BP2, $H^\pm$ will go through off-shell decay that is, $H^\pm \to W^{*\pm} H$ with $W^{*\pm}$ decaying leptonically in pure IHDM case due to limited phase space availability. But for BP3 and BP4, $H^\pm$ will go through on-shell decay that is, $H^\pm \to W^\pm H$ with $W^\pm$ decaying leptonically in pure IHDM case. In IHDM+VLL model, $H^\pm$ will decay to $\chi^\pm$ that is, $H^\pm \to \chi^\pm\ \nu_l$ with $\chi^\pm$ further decaying into $l^\pm H$. It is clearly evident from this table that we have enhancement of the cross section in IHDM+VLL due to opening of new decay modes of $H^\pm$. We must highlight one point that it is very difficult to probe heavier charged Higgs mass (like the ones in BP3 and BP4) in pure IHDM case due to small cross section. But in the IHDM+VLL model discussed here, we have sufficient cross section to probe these heavier masses of charged Higgs. Apart from the channels listed in Table 1, there is another process which contributes to dilepton plus MET final states that is $\chi^+\chi^-$ production with $\chi^\pm$ decays to $l^\pm H$. So as a whole, the dilepton plus MET final state will be an important collider signature to probe the modified IHDM that we discussed in this article. This inspires us to do a full signal versus background study at detector level which we will come up in a separate work [52].

Table 1
The parton level cross section for final states that contribute to dilepton+MET final states at detector level in both IHDM and IHDM+VLL models at the LHC ($\sqrt{s} = 14$ TeV) for different BPs considered.

Benchmark points	$\sigma(pp \to H^+H^- \to 2l + 2\nu + 2H)$ (in fb)	
	IHDM	IHDM+VLL
BP1	8.1	126
BP2	6.1	93.5
BP3	1.7	13.8
BP4	0.3	2.1

6. Conclusion

We have studied a very specific region of parameter space in IHDM where the Higgs portal coupling of DM is very small, as suggested by null results in dark matter direct detection experiments so far. In the low mass regime of DM that is $m_{DM} < M_W$, such small value of Higgs portal coupling λ_L may not be sufficient to produce the correct relic abundance of DM except for a few specific values of m_{DM}. We then extend this model by heavy neutral and charged leptons which are also odd under the Z_2 symmetry of the IHDM. These heavy leptons can be motivating from neutrino mass as well as LHC phenomenology point of view, apart from their role in producing the correct DM relic abundance in those region of parameter space which can not produce correct relic in pure IHDM. The neutral heavy fermions can generate tiny neutrino masses at one loop level via scotogenic mechanism, requiring the corresponding Yukawa couplings to be small ($< 10^{-4}$) for TeV scale heavy neutral fermion masses. This keeps the contribution of neutral heavy leptons to DM abundance suppressed. The heavy charged fermion couplings to DM are however, not constrained to be tiny from neutrino mass point of view and hence can be sizeable enough to play a role in DM abundance. We show that the entire low mass regime of IHDM is allowed from relic abundance criteria if the lepton portal parameters are suitably chosen. This does not affect the DM direct detection scattering rates as there are no tree level or one loop couplings of DM with nuclei through leptons. The heavy leptons can also give rise to observable LFV decay rates like $\mu \to e\gamma$ as well as interesting collider signatures like dilepton plus missing energy. Although for simplicity, we choose particular type of Yukawa structure which does not contribute to LFV decay rates, it is in principle possible to choose some structure of the Yukawa couplings which can simultaneously produce correct DM abundance as well as keep the decay rate of LFV decays like $\mu \to e\gamma$ within experimental reach. We check the evolution of different couplings of the model under RGE and find that the model remains perturbative all the way upto around 10^9 GeV. Due to the requirement of large order one Yukawa coupling to keep the lepton portal annihilation of dark matter more efficient, different couplings of the model receive large corrections from RGE leading to non-perturbative nature at high energy scale. We also show how the lepton portal extension of IHDM enhances dilepton plus missing energy signals at the LHC, for chosen benchmark points. There can also be lepton number violating signal like same sign dilepton plus dijet plus missing energy in this model, but remain suppressed for the benchmark values chosen in our analysis.

Acknowledgements

We thank P. Poulose for useful discussions while carrying out this work. SS would like to thank Nirakar Sahoo for his constant help in resolving issues in micrOMEGA. Also SS thanks Biswajit Karmakar, Abhijit Saha for technical help in using Mathematica and Ashis Kundu, Sourav Chattopadhyay for help in shell scripting.

Appendix A. RGE equations for different couplings

Here we list the RGE equations for the model IHDM plus vector like charged lepton singlets discussed in the work. For the gauge couplings, they are given by

$$16\pi^2\frac{dg_c}{dt} = -7g_c^3,\quad 16\pi^2\frac{dg_L}{dt} = -3g_L^3,\quad 16\pi^2\frac{dg_Y}{dt} = 11g_Y^3 \quad \text{(A1)}$$

where g_c, g_L, g_Y are the gauge couplings of $SU(3)_c, SU(2)_L, U(1)_Y$ gauge groups respectively and $t = \ln\mu$, μ being the energy scale. The quartic couplings of the scalar potential evolve as

$$\begin{aligned}16\pi^2\frac{1}{2}\frac{d\lambda_1}{dt} &= 3\lambda_1^2 + 4\lambda_3^2 + 4\lambda_3\lambda_4 + 2\lambda_4^2 + 2\lambda_5^2\\ &+ \frac{3}{4}(3g_L^4 + g_Y^4 + 2g_L^2g_Y^2) - \frac{\lambda_1}{2}(9g_L^2 + 3g_Y^2 - 12y_t^2\\ &- 12y_b^2 - 4y_\tau^2) - 12y_t^4 \qquad \text{(A2)}\end{aligned}$$

$$\begin{aligned}16\pi^2\frac{1}{2}\frac{d\lambda_2}{dt} &= 3\lambda_2^2 + 4\lambda_3^2 + 4\lambda_3\lambda_4 + 2\lambda_4^2 + 2\lambda_5^2\\ &+ \frac{3}{4}(3g_L^4 + g_Y^4 + 2g_L^2g_Y^2) - \frac{3}{2}\lambda_2(3g_L^2 + g_Y^2 - \frac{4}{3}y_N^2)\\ &- 4y_N^4 - 4y_\chi^4 \qquad \text{(A3)}\end{aligned}$$

$$\begin{aligned}16\pi^2\frac{d\lambda_3}{dt} &= (\lambda_1 + \lambda_2)(3\lambda_3 + \lambda_4) + 4\lambda_3^2 + 2\lambda_4^2 + 2\lambda_5^2\\ &+ \frac{3}{4}(3g_L^4 + g_Y^4 - 2g_L^2g_Y^2) - \lambda_3(9g_L^2 + 3g_Y^2 - 6y_t^2\\ &- 6y_b^2 - 2y_\tau^2 - 2y_N^2 - 2y_\chi^2) \qquad \text{(A4)}\end{aligned}$$

$$\begin{aligned}16\pi^2\frac{d\lambda_4}{dt} &= (\lambda_1 + \lambda_2)\lambda_4 + 8\lambda_3\lambda_4 + 4\lambda_4^2 + 8\lambda_5^2 + 3g_L^2g_Y^2\\ &- \lambda_4(9g_L^2 + 3g_Y^2 - 6y_t^2 - 6y_b^2 - 2y_\tau^2 - 2y_N^2 - 2y_\chi^2) \qquad \text{(A5)}\end{aligned}$$

$$\begin{aligned}16\pi^2\frac{d\lambda_5}{dt} &= (\lambda_1 + \lambda_2 + 8\lambda_3 + 12\lambda_4)\lambda_5 - \lambda_5(9g_L^2 + 3g_Y^2 - 6y_t^2\\ &- 6y_b^2 - 2y_\tau^2 - 2y_N^2 - 2y_\chi^2) \qquad \text{(A6)}\end{aligned}$$

Here y_t, y_b, y_τ are the top quark, bottom quark and tau lepton Yukawa couplings with the Higgs field Φ_1. The Yukawa couplings have the following RGE equations

$$16\pi^2\frac{dy_t}{dt} = y_t\left(-8g_c^2 - \frac{9}{4}g_L^2 - \frac{17}{12}g_Y^2 + \frac{9}{2}y_t^2 + y_\tau^2 + \frac{3}{2}y_b^2\right) \qquad \text{(A7)}$$

$$16\pi^2\frac{dy_b}{dt} = y_b\left(-8g_c^2 - \frac{9}{4}g_L^2 - \frac{5}{12}g_Y^2 + \frac{9}{2}y_b^2 + y_\tau^2 + \frac{3}{2}y_t^2\right) \qquad \text{(A8)}$$

$$\begin{aligned}16\pi^2\frac{dy_\tau}{dt} &= y_\tau\left(-\frac{9}{4}g_L^2 - \frac{15}{4}g_Y^2 + 3y_t^2 + 3y_b^2\right.\\ &\left.+ \frac{1}{2}y_N^2 + \frac{1}{2}y_\chi^2 + \frac{5}{2}y_\tau^2\right) \qquad \text{(A9)}\end{aligned}$$

$$16\pi^2\frac{dy_N}{dt} = y_N\left(-\frac{9}{4}g_L^2 - \frac{3}{4}g_Y^2 - \frac{3}{4}y_\tau^2 + \frac{5}{2}y_N^2\right) \qquad \text{(A10)}$$

$$16\pi^2\frac{dy_\chi}{dt} = y_\chi\left(-\frac{9}{4}g_L^2 - \frac{3}{4}g_Y^2 - \frac{3}{4}y_\tau^2 + \frac{5}{2}y_\chi^2\right) \qquad \text{(A11)}$$

References

[1] P.A.R. Ade, et al., Planck Collaboration, Astron. Astrophys. 594 (2016) A13.
[2] Talk on "Dark-matter results from 332 new live days of LUX data" by A. Manalaysay, LUX Collaboration, IDM, Sheffield, July 2016;
D.S. Akerib, et al., LUX Collaboration, arXiv:1608.07648.
[3] A. Tan, et al., PandaX-II Collaboration, Phys. Rev. Lett. 117 (2016) 121303.
[4] R. Barbieri, L.J. Hall, V.S. Rychkov, Phys. Rev. D 74 (2006) 015007, arXiv:hep-ph/0603188.
[5] M. Cirelli, N. Fornengo, A. Strumia, Nucl. Phys. B 753 (2006) 178, arXiv:hep-ph/0512090.
[6] L. Lopez Honorez, E. Nezri, J.F. Oliver, M.H.G. Tytgat, JCAP 0702 (2007) 028, arXiv:hep-ph/0612275.
[7] E. Lundstrom, M. Gustafsson, J. Edsjo, Phys. Rev. D 79 (2009) 035013, arXiv:0810.3924.
[8] A. Pierce, J. Thaler, JHEP 0708 (2007) 026.
[9] G. Belanger, B. Dumont, A. Goudelis, B. Herrmann, S. Kraml, D. Sengupta, Phys. Rev. D 91 (11) (2015) 115011, arXiv:1503.07367 [hep-ph].
[10] L. Lopez Honorez, C.E. Yaguna, JCAP 1101 (2011) 002, arXiv:1011.1411.
[11] M.E. Peskin, T. Takeuchi, Phys. Rev. D 46 (1992) 381.
[12] E. Ma, Phys. Rev. D 73 (2006) 077301.
[13] A.M. Baldini, et al., MEG Collaboration, Eur. Phys. J. C 76 (2016) 434.
[14] D. Borah, A. Dasgupta, JCAP 1612 (2016) 034.
[15] D. Borah, A. Dasgupta, JHEP 1701 (2017) 072.
[16] D. Borah, A. Dasgupta, arXiv:1702.02877.
[17] T.G. Rizzo, Phys. Rev. D 34 (1986) 1438.
[18] S. Gopalakrishna, T.S. Mukherjee, S. Sadhukhan, Phys. Rev. D 94 (1) (2016) 015034, arXiv:1512.05731 [hep-ph].
[19] E.W. Kolb, M.S. Turner, Front. Phys. 69 (1990) 1.
[20] R.J. Scherrer, M.S. Turner, Phys. Rev. D 33 (1986) 1585.
[21] G. Jungman, M. Kamionkowski, K. Griest, Phys. Rep. 267 (1996) 195, arXiv:hep-ph/9506380.

[22] P. Gondolo, G. Gelmini, Nucl. Phys. B 360 (1991) 145.
[23] D. Majumdar, A. Ghosal, Mod. Phys. Lett. A 23 (2008) 2011, arXiv:hep-ph/0607067.
[24] T.A. Chowdhury, M. Nemevsek, G. Senjanovic, Y. Zhang, JCAP 1202 (2012) 029.
[25] D. Borah, J.M. Cline, Phys. Rev. D 86 (2012) 055001.
[26] A. Dasgupta, D. Borah, Nucl. Phys. B 889 (2014) 637.
[27] L.L. Honorez, C.E. Yaguna, JHEP 1009 (2010) 046.
[28] K. Griest, D. Seckel, Phys. Rev. D 43 (1991) 3191.
[29] J. Edsjo, P. Gondolo, Phys. Rev. D 56 (1997) 1879;
N.F. Bell, Y. Cai, A.D. Medina, Phys. Rev. D 89 (2014) 115001.
[30] E. Aprile, et al., Phys. Rev. Lett. 109 (2012) 181301.
[31] D.S. Akerib, et al., LUX Collaboration, Phys. Rev. Lett. 112 (2014) 091303.
[32] J. Giedt, A.W. Thomas, R.D. Young, Phys. Rev. Lett. 103 (2009) 201802, arXiv:0907.4177.
[33] Y. Mambrini, Phys. Rev. D 84 (2011) 115017.
[34] G. Aad, et al., ATLAS Collaboration, JHEP 1511 (2015) 206.
[35] M. Klasen, C.E. Yaguna, J.D. Ruiz-Alvarez, Phys. Rev. D 87 (2013) 075025.
[36] S. Chang, R. Edezhath, J. Hutchinson, M. Luty, Phys. Rev. D 90 (2014) 015011;
Y. Bai, J. Berger, JHEP 1408 (2014) 153.
[37] C. Arina, F.-S. Ling, M.H.G. Tytgat, JCAP 0910 (2009) 018.
[38] G. Belanger, F. Boudjema, A. Pukhov, A. Semenov, Comput. Phys. Commun. 185 (2014) 960.
[39] X. Miao, S. Su, B. Thomas, Phys. Rev. D 82 (2010) 035009, arXiv:1005.0090 [hep-ph].
[40] M. Gustafsson, S. Rydbeck, L. Lopez-Honorez, E. Lundstrom, Phys. Rev. D 86 (2012) 075019, arXiv:1206.6316 [hep-ph].
[41] A. Datta, N. Ganguly, N. Khan, S. Rakshit, Phys. Rev. D 95 (1) (2017) 015017, arXiv:1610.00648 [hep-ph].
[42] P. Poulose, S. Sahoo, K. Sridhar, Phys. Lett. B 765 (2017) 300, arXiv:1604.03045 [hep-ph].
[43] M. Hashemi, S. Najjari, arXiv:1611.07827 [hep-ph].
[44] A. Belyaev, G. Cacciapaglia, I.P. Ivanov, F. Rojas, M. Thomas, arXiv:1612.00511 [hep-ph].
[45] S. Bhattacharya, N. Sahoo, N. Sahu, Phys. Rev. D 93 (11) (2016) 115040, arXiv:1510.02760 [hep-ph].
[46] N. Kumar, S.P. Martin, Phys. Rev. D 92 (11) (2015) 115018, arXiv:1510.03456 [hep-ph].
[47] A. Falkowski, D.M. Straub, A. Vicente, JHEP 1405 (2014) 092, arXiv:1312.5329 [hep-ph], and references therein.
[48] S. Gopalakrishna, T.S. Mukherjee, S. Sadhukhan, Phys. Rev. D 93 (5) (2016) 055004, arXiv:1504.01074 [hep-ph].
[49] A. Angelescu, G. Arcadi, arXiv:1611.06186 [hep-ph].
[50] K. Blum, R.T. D'Agnolo, J. Fan, JHEP 1503 (2015) 166, arXiv:1502.01045 [hep-ph];
E. Bertuzzo, P.A.N. Machado, M. Taoso, Phys. Rev. D 94 (2016) 115006, arXiv:1601.07508 [hep-ph].
[51] N. Chakrabarty, D.K. Ghosh, B. Mukhopadhyaya, I. Saha, Phys. Rev. D 92 (2015) 015002.
[52] Debasish Borah, P. Poulose, Soumya Sadhukhan, Shibananada Sahoo, in preparation.

12

Measurement of $\hat{q}$ in Relativistic Heavy Ion Collisions using di-hadron correlations

M.J. Tannenbaum

Brookhaven National Laboratory, Upton, NY, 11973, USA

ARTICLE INFO

Editor: V. Metag

ABSTRACT

The propagation of partons from hard scattering through the Quark Gluon Plasma produced in A+A collisions at RHIC and the LHC is represented in theoretical analyses by the transport coefficient $\hat{q}$ and predicted to cause both energy loss of the outgoing partons, observed as suppression of particles or jets with large transverse momentum p_T, and broadening of the azimuthal correlations of the outgoing di-jets or di-hadrons from the outgoing parton-pair, which has not been observed. The widths of azimuthal correlations of di-hadrons with the same trigger particle p_{Tt} and associated p_{Ta} transverse momenta in p+p and Au+Au are so-far statistically indistinguishable as shown in recent as well as older di-hadron measurements and also with jet-hadron and hadron-jet measurements. The azimuthal width of the di-hadron correlations in p+p collisions, beyond the fragmentation transverse momentum, j_T, is dominated by k_T, the so-called intrinsic transverse momentum of a parton in a nucleon, which can be measured. The broadening should produce a larger k_T in A+A than in p+p collisions. The present work introduces the observation that the k_T measured in p+p collisions for di-hadrons with p_{Tt} and p_{Ta} must be reduced to compensate for the energy loss of both the trigger and away parent partons when comparing to the k_T measured with the same di-hadron p_{Tt} and p_{Ta} in Au+Au collisions. This idea is applied to a recent STAR di-hadron measurement, with result $\langle\hat{q}L\rangle = 2.1 \pm 0.6$ GeV^2. This is more precise but in agreement with a theoretical calculation of $\langle\hat{q}L\rangle = 14^{+42}_{-14}$ GeV^2 using the same data. Assuming a length $\langle L\rangle \approx 7$ fm for central Au+Au collisions the present result gives $\hat{q} \approx 0.30 \pm 0.09$ GeV^2/fm, in fair agreement with the JET collaboration result from single hadron suppression of $\hat{q} \approx 1.2 \pm 0.3$ GeV^2/fm at an initial time $\tau_0 = 0.6$ fm/c in Au+Au collisions at $\sqrt{s_{NN}} = 200$ GeV.

1. Introduction

In the original BDMPSZ formalism [1,2], the energy loss of an outgoing parton, $-dE/dx$, per unit length (x) of a medium with total length L, is proportional to the 4-momentum transfer-squared, q^2, and takes the form:

$$\frac{-dE}{dx} \simeq \alpha_s \langle q^2(L)\rangle = \alpha_s\, \mu^2\, L/\lambda_{\rm mfp} = \alpha_s\, \hat{q}\, L \quad , \qquad (1)$$

where μ, is the mean momentum transfer per collision, and the transport coefficient $\hat{q} = \mu^2/\lambda_{\rm mfp}$ is the 4-momentum-transfer-squared to the medium per mean free path, $\lambda_{\rm mfp}$. Additionally, the accumulated momentum-squared, $\langle p^2_{\perp W}\rangle$ transverse to a parton traversing a length L in the medium is well approximated by [1] $\langle p^2_{\perp W}\rangle \approx \langle q^2(L)\rangle = \hat{q}\, L$. This results in a direct and simple relationship between the parton energy loss (Eq. (1)) and the di-jet azimuthal broadening, $\langle p^2_{\perp W}\rangle/2$, because only one of the components of the accumulated momentum transverse to the outgoing parton is in the scattering plane, the other being along the beam axis for mid-rapidity di-jets.

It has long been established [3] that even in p+p collisions, or in the initial hard-scattered parton pair in A+A collisions, the mid-rapidity di-jets from hard-scattering are not back-to-back in azimuth but are acollinear with a net transverse momentum, $\langle p^2_T\rangle_{\rm pair} = 2\langle k^2_T\rangle$, where $\langle k_T\rangle$ is the average 'intrinsic' transverse momentum of a quark or gluon in a nucleon as defined by Feynman, Field and Fox [4]. Again, only the component of $\langle p^2_T\rangle_{\rm pair}$ perpendicular to the di-jet axis leads to acoplanarity. Thus in an A+A collision the relationship in Eq. (2) should hold:

$$\langle\hat{q}L\rangle/2 = \left\langle k^2_T\right\rangle_{AA} - \left\langle k'^2_T\right\rangle_{pp} \qquad (2)$$

E-mail address: mjt@bnl.gov.

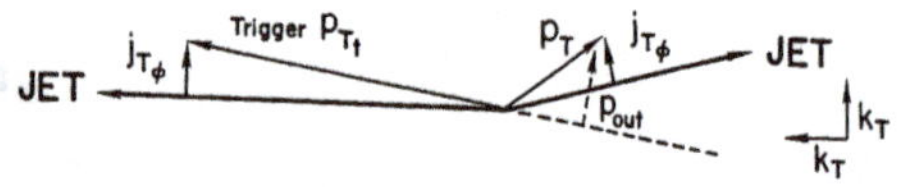

Fig. 1. Azimuthal projection of di-jet with trigger particle p_{Tt} and associated away-side particle p_{Ta}, and the azimuthal components $j_{T\phi}$ of the fragmentation transverse momentum. The initial state k_T of a parton in each nucleon is shown schematically: one vertical which gives an azimuthal decorrelation of the jets and one horizontal which changes the transverse momentum of the trigger jet.

for azimuthal correlations of a trigger particle with p_{Tt} and away-side particles with p_{Ta}. It is important to note the $'$ in $\langle k'^2_T \rangle_{pp}$, introduced here, which indicates that the k_T measured in p+p collisions for di-hadrons with p_{Tt} and p_{Ta} must be reduced to compensate for the energy loss of both the trigger and away parent partons when comparing to the k_T calculated with the same di-hadron p_{Tt} and p_{Ta} in Au+Au collisions.

Many experiments at RHIC, including recent experiments with di-hadron [5], jet-hadron [6] and di-jet [7] azimuthal correlations have searched for azimuthal broadening in Au+Au collisions compared to p+p collisions but have not found a significant difference in the azimuthal angular Gaussian width of the away-peaks. Here we shall reexamine the STAR di-hadron measurement [5] in terms of the out of plane component, p_{out} rather than the azimuthal angular width, taking account of the energy lost by the original parton-pair in Au+Au collisions when comparing to the p+p measurement.

2. How information about the initial partons can be derived from two-particle correlations

We shall calculate $\langle k_T^2 \rangle$ from p+p and Au+Au di-hadron measurements with the same trigger particle transverse momentum, p_{Tt}, away-side p_{Ta} and $x_h = p_{Ta}/p_{Tt}$. The di-hadrons are assumed to be fragments of jets with transverse momenta $\hat{p}_{Tt}$ and $\hat{p}_{Ta}$ with ratio $\hat{x}_h = \hat{p}_{Ta}/\hat{p}_{Tt}$, where $z_t \simeq p_{Tt}/\hat{p}_{Tt}$ is the fragmentation variable, the fraction of momentum of the trigger particle in the trigger jet, and j_T is the jet fragmentation transverse momentum. The standard equation at RHIC comes from PHENIX [8], which we write in a slightly different form in Eq. (3):

$$\sqrt{\langle k_T^2 \rangle} = \frac{\hat{x}_h}{\langle z_t \rangle} \sqrt{\frac{\langle p_{out}^2 \rangle - (1 + x_h^2)\langle j_T^2 \rangle /2}{x_h^2}} \quad . \tag{3}$$

Here $p_{out} \equiv p_{Ta} \sin \Delta\phi$ (see Fig. 1) and we have taken $\left\langle j^2_{Ta\phi} \right\rangle = \left\langle j^2_{Tt\phi} \right\rangle = \langle j_T^2 \rangle /2$. The variable x_h (which STAR calls z_T) is used as an approximation of the variable $x_E = x_h \cos\phi$ of the original terminology from the CERN ISR where k_T was discovered and measured 40 years ago [3,4,9,10].

2.1. Bjorken parent–child relation and 'trigger-bias' [11]

If the fragmentation function of the jet is a function only of the fragmentation variable z and not of the jet $\hat{p}$, then the single particle cross section has the same power law shape, $d^3\sigma/2\pi p_T dp_T dy \propto p_T^{-n}$, as the parent jet cross section.

Furthermore, large values of $\langle z_t \rangle = p_{Tt}/\hat{p}_{Tt}$ dominate the single-particle cross section (e.g. π^0) used as the trigger for the di-hadron (e.g. π^0-h) measurement. This is called trigger-bias but is valid also for the simple single-particle measurements. Calculations of $\langle z_t \rangle$ vs. p_{Tt} for π^0 at $\sqrt{s_{NN}} = 200$ GeV are given in Ref. [12].

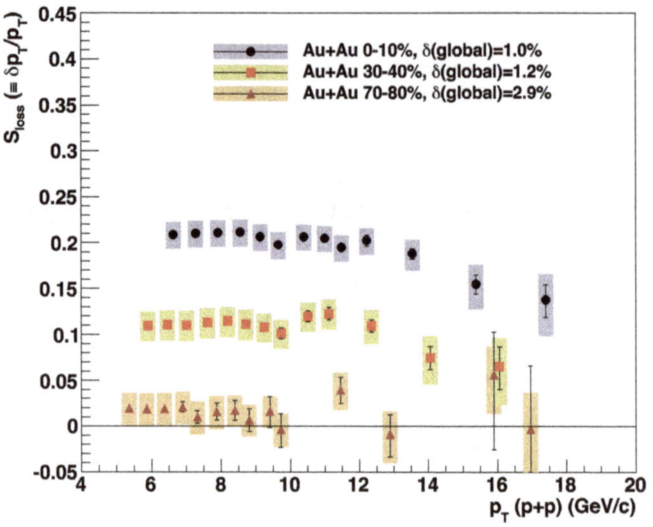

Fig. 2. p_T^{pp} dependence of $\delta p_T/p_T^{pp}$ of π^0 from p+p to Au+Au for the centralities indicated at $\sqrt{s_{NN}} = 200$ GeV [14].

2.2. The energy loss of the trigger jet from p+p to Au+Au can be measured

At RHIC, in p+p and Au+Au collisions as a function of centrality the π^0 p_T spectra with $5 < p_T \lesssim 20$ GeV/c all follow the same power law with $n \approx 8.10 \pm 0.05$ [13]. From the Bj parent–child relation, the energy loss of the trigger jet is found by measuring $\delta p_T/p_T^{pp}$, the shift in the π^0 spectra in Au+Au at a given p_T from the $\langle T_{AA} \rangle$ corrected p+p cross section (Fig. 2) [14]. The small dropoff of $\delta p_T/p_T^{pp}$ for $p_T \geq 14$ GeV/c indicates a small increase of n with increasing p_T.

It is important to note that the same value of n for the π^0 spectra in p+p and Au+Au collisions implies the same value of n for the original parton in p+p and the one that has lost energy in Au+Au. However $\langle z_t \rangle$ for p+p and Au+Au measurements may differ slightly because the maximum possible parton energy $\sqrt{s_{NN}}/2$ is reduced by the energy loss. The effect on $\langle z_t \rangle$ from p+p to Au+Au was estimated by increasing p_{Tt} in the calculation of $\langle z_t \rangle$ in p+p collisions [12] by the largest $\delta p_T/p_T^{pp} = 0.20$ for centrality 0–10% (Fig. 2) with result for $p_{Tt} = 7.8$ GeV/c, $\langle z_t \rangle = 0.63 \pm 0.07$, and for $p_{Tt} = 7.8/0.80 = 8.78$ GeV/c, $\langle z_t \rangle = 0.66 \pm 0.06$. Since the difference for the largest $\delta p_T/p_T^{pp} = 0.20$ is considerably less than the error in the calculation, we shall use the measured or calculated $\langle z_t \rangle$ in p+p also for Au+Au with the same p_{Tt}.

2.3. The away particles from a hadron-trigger do not measure the fragmentation function [8]

It was generally assumed, as implied by Feynman, Field and Fox in 1977 [4], that the p_{Ta} (or x_E, or x_h) distribution of away-side hadrons from a single hadron trigger with p_{Tt}, corrected for $\langle z_t \rangle$, would be the same as that from a jet-trigger and would measure the away-jet fragmentation function as it does for direct photon triggers [15]. However, attempts to try this at RHIC led to the discovery [8] that the x_E distribution does not measure the fragmentation function. The good news was that it measured the ratio of the away jet to the trigger jet transverse momenta, $\hat{x}_h = \hat{p}_{Ta}/\hat{p}_{Tt}$, Eq. (4)

$$\left. \frac{dP_\pi}{dx_E} \right|_{p_{Tt}} = N(n-1) \frac{1}{\hat{x}_h} \frac{1}{(1 + \frac{x_E}{\hat{x}_h})^n} \quad , \tag{4}$$

with the value of $n = 8.10$ (± 0.05) fixed as determined in Ref. [13], where n is the power-law of the inclusive π^0 spectrum and is ob-

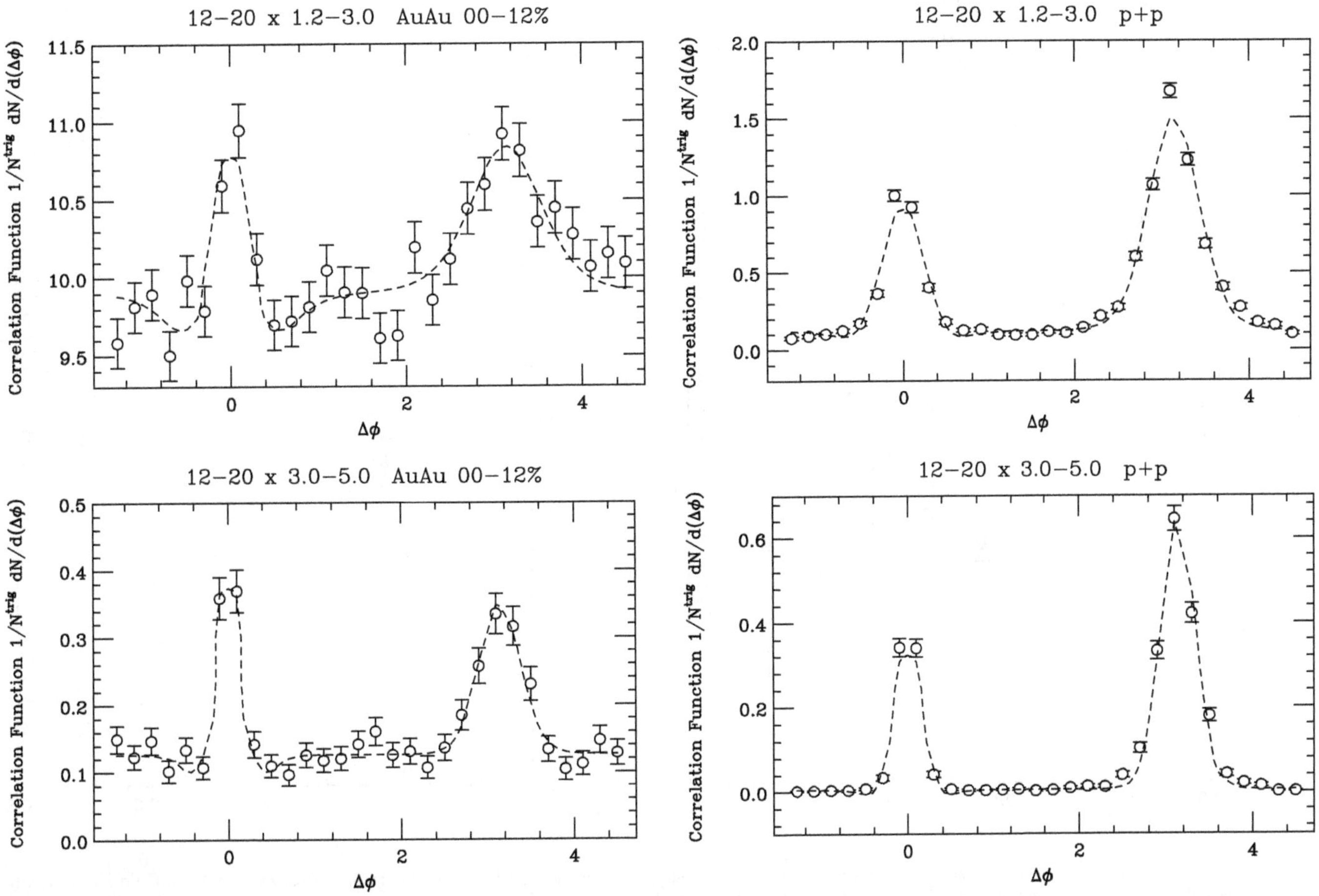

Fig. 3. Fits to STAR π^0-h correlation functions for $12 < p_{Tt} < 20$ GeV/c [5] measured in central (0–20%) Au+Au collisions (left) and p+p collisions (right): (top) $1.2 < p_{Ta} < 3$ GeV/c; (bottom) $3 < p_{Ta} < 5$ GeV/c.

served to be the same in p+p and Au+Au collisions in the p_{T_t} range of interest as noted in section 2.2 above.

3. How to apply this information to find $\hat{q}$ from p+p and Au+Au di-hadron measurements

A recent STAR π^0+h di-hadron measurement in p+p and Au+Au collision at $\sqrt{s_{NN}} = 200$ GeV [5] is used to measure $\langle \hat{q}L \rangle$ by calculating k_T in each case as in Eq. (2). For a di-jet produced in a hard scattering, the initial $\hat{p}_{Tt}$ and $\hat{p}_{Ta}$ will both be reduced by energy loss in the medium to become $\hat{p}'_{Tt}$ and $\hat{p}'_{Ta}$ that will be measured by the di-hadron correlations with p_{Tt} and p_{Ta} in Au+Au collisions. As both jets from the initial di-jet lose energy in the medium, the azimuthal angle between the di-jets from the $\langle k_T^2 \rangle$ in the original collision should not change unless the medium induces multiple scattering from $\hat{q}$. Thus, without $\hat{q}$ and assuming the same fragmentation transverse momentum $\langle j_T^2 \rangle$ in the original jets and those that have lost energy, the $p_{\rm out}$ between the away hadron with p_{Ta} and the trigger hadron with p_{Tt} will not change (Fig. 1), but the $\langle k'^2_T \rangle$ will be reduced according to Eq. (3) because the ratio of the away to the trigger jets $\hat{x}'_h = \hat{p}'_{Ta}/\hat{p}'_{Tt}$ will be reduced. Thus the calculation of k'_T from the di-hadron p+p measurement to compare with Au+Au measurement with the same di-hadron trigger p_{Tt} and p_{Ta} must use the values of $\hat{x}_h$, and $\langle z_T \rangle$ from the Au+Au measurement to compensate for the energy lost by the original dijet in p+p collisions.

4. Calculation of $\langle \hat{q}L \rangle$ from the STAR measurement [5]

4.1. Determine $\langle p^2_{\rm out} \rangle$ from the π^0-h correlation function

This is accomplished by fitting the π^0-h correlation functions for $12 < p_{Tt} < 20$ GeV/c [5] to a gaussian in $\sin \Delta\phi$ for the away-side, $\pi/2 \leq \Delta\phi \leq 3\pi/2$ [8,12,16]; and a gaussian in $\Delta\phi$ for the trigger side $-\pi/2 \leq \Delta\phi \leq \pi/2$ (Fig. 3). The results for Au+Au 0–20% centrality are $\langle p^2_{\rm out} \rangle = 0.547 \pm 0.163$ (GeV/c)^2, $\chi^2/{\rm dof} = 33/21$ for $1.2 < p_{Ta} < 3$ GeV/c and $\langle p^2_{\rm out} \rangle = 0.851 \pm 0.203$ (GeV/c)^2, $\chi^2/{\rm dof} = 23/21$ for $3 < p_{Ta} < 5$ GeV/c. The same fits to the p+p measurements with the same p_{Tt} and p_{Ta} yielded $\langle p^2_{\rm out} \rangle = 0.263 \pm 0.113$ (GeV/c)^2, $\chi^2/{\rm dof} = 186/21$ for $1.2 < p_{Ta} < 3$ GeV/c and $\langle p^2_{\rm out} \rangle = 0.576 \pm 0.167$ (GeV/c)^2, $\chi^2/{\rm dof} = 137/21$ for $3 < p_{Ta} < 5$ GeV/c where the errors have been corrected up by $\sqrt{\chi^2/{\rm dof}}$.

4.2. Determine $\hat{x}_h = \hat{p}_{Ta}/\hat{p}_{Tt}$

This is done by a fit of Eq. (4) to the STAR measurements of what they call the away-side z_T distribution [5] (called the x_h or x_E distribution here) for $12 < p_{Tt} < 20$ GeV/c in p+p and Au+Au 0–12% centrality collisions (Fig. 4). The fit [17] takes account of the statistical and correlated systematic errors, σ_i and σ_{b_i}, for each data point with $dP/dx_E = y_i$:

$$\chi^2 = \left[\sum_{i=1}^{n} \frac{(y_i + \epsilon_b \sigma_{b_i} - y_i^{\rm fit})^2}{\tilde{\sigma}_i^2} \right] + \epsilon_b^2 \quad , \tag{5}$$

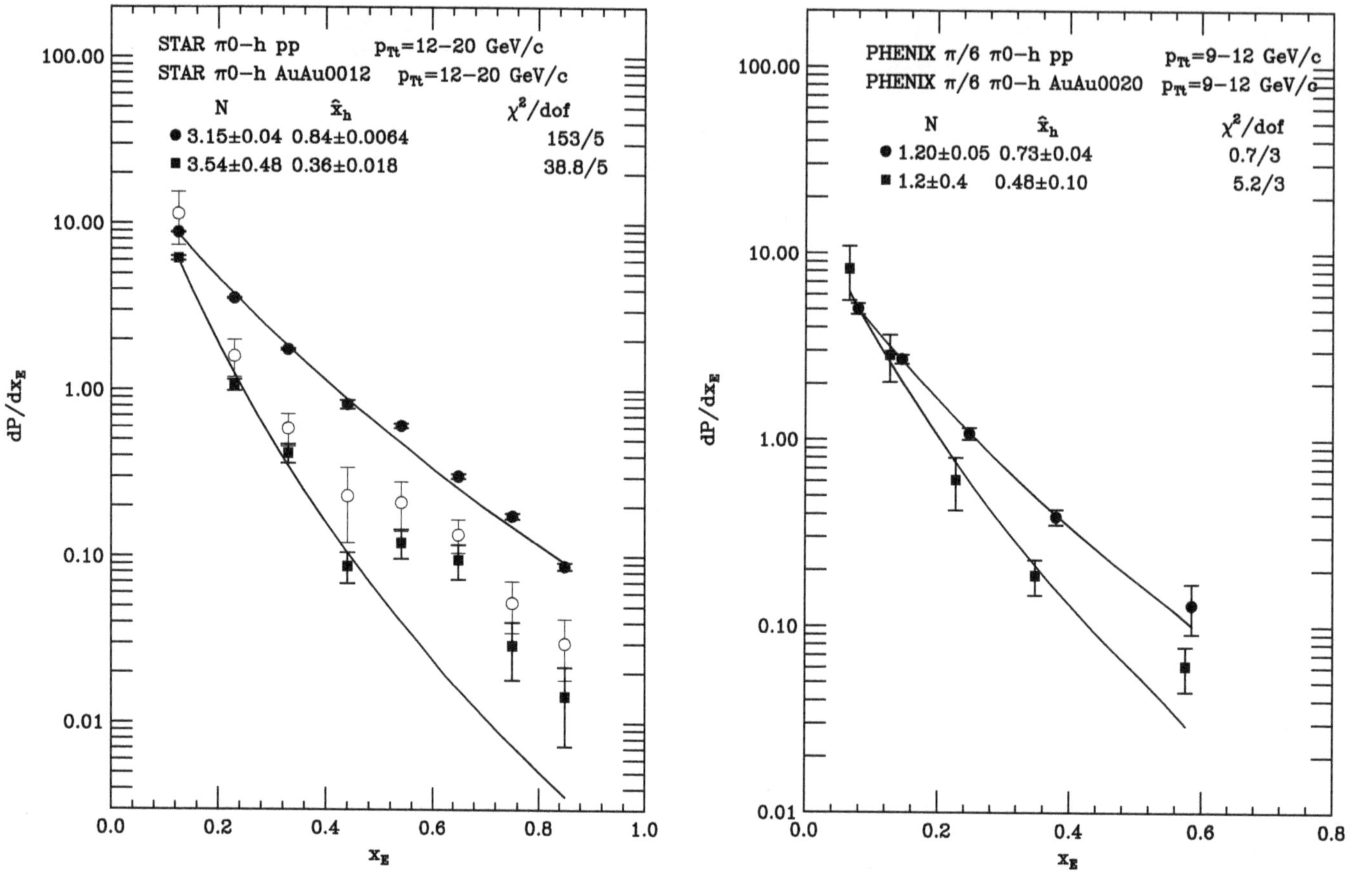

Fig. 4. a) (left) Fits of Eq. (4) to the STAR away-side z_T distributions [5] in Au+Au 0–12% centrality and p+p for $12 < p_{Tt} < 20$ GeV/c. The Au+Au curve is a fit with $\hat{x}_h^{AA} = 0.36 \pm 0.05$ with error corrected by $\sqrt{\chi^2/\text{dof}}$. The points with the open circles are the y_i and systematic errors σ_{b_i} of the data points while the filled points are $y_i + \epsilon_b \sigma_{b_i}$ with errors $\tilde{\sigma}_i$ and $\epsilon_b = -1.3 \pm 0.5$. b) (right) Fits to PHENIX x_E distributions [18,19] in p+p and Au+Au for π^0-h correlations with $9 \le p_{Tt} \le 12$ GeV/c.

where $\tilde{\sigma}_i$ is the statistical error, σ_i, scaled by the shift in y_i such that the fractional error remains unchanged: $\tilde{\sigma}_i = \sigma_i \left(1 + \epsilon_b \sigma_{b_i}/y_i\right)$, where ϵ_b is to be fit.

The fit worked very well with a result for Au+Au of $\hat{x}_h = 0.36 \pm 0.05$ with $\chi^2/\text{dof} = 38.8/5$ where the error has been corrected upward by $\sqrt{\chi^2/\text{dof}}$. This is consistent with the value $\hat{x}_h = 0.48 \pm 0.10$ for $9 \le p_{Tt} \le 12$ GeV/c from a PHENIX measurement [18,19] (see Fig. 4).

The value of $\hat{x}_h$ for the p+p measurement, although not needed for determining $\langle \hat{q}L \rangle$ in the present method, was determined for the STAR p+p data with fitted result $\hat{x}_h^{pp} = 0.84 \pm 0.04$ which is in decent agreement with the result $\hat{x}_h^{pp} = 0.73 \pm 0.04$ for $9 \le p_{Tt} \le 12$ GeV/c from the PHENIX measurement (Fig. 4).

4.3. Determine $\langle z_t \rangle$

This was the easiest part of the calculation because STAR [5] had determined that $\langle z_t \rangle = 0.80 \pm 0.05$ in their p+p collisions for π^0 with $12 < p_{Tt} < 20$ GeV/c.

4.4. Calculate $\langle k_T^2 \rangle_{AA}$, $\langle k'^2_T \rangle_{pp}$, $\langle \hat{q}L \rangle /2$

The $\langle p_{\text{out}}^2 \rangle$ values from the fits to the correlation functions in p+p and Au+Au plus the results for $\hat{x}_h^{AA} = 0.36 \pm 0.05$, $\langle z_t \rangle = 0.80 \pm 0.05$ above are used to calculate $\sqrt{\langle k_T^2 \rangle}$ using Eq. (3) with the value $\sqrt{\langle j_T \rangle^2} = 0.62 \pm 0.04$ GeV/c [8,12] for both p+p and Au+Au. Equation (6) is used for $\langle \hat{q}L \rangle /2$. The results are given in Table 1.

$$\langle \hat{q}L \rangle /2 = \left[\frac{\hat{x}_h}{\langle z_t \rangle}\right]^2 \left[\frac{\langle p_{\text{out}}^2 \rangle_{AA} - \langle p_{\text{out}}^2 \rangle_{pp}}{x_h^2}\right] \quad . \tag{6}$$

Table 1
Tabulations for $\hat{q}$–STAR π^0-h [5].

STAR PLB760			
$\sqrt{s_{NN}} = 200$	$\langle p_{Tt} \rangle$	$\langle p_{Ta} \rangle$	$\langle p_{\text{out}}^2 \rangle$
Reaction	GeV/c	GeV/c	(GeV/c)^2
p+p	14.71	1.72	0.263 ± 0.113
p+p	14.71	3.75	0.576 ± 0.167
Au+Au 0–12%	14.71	1.72	0.547 ± 0.163
Au+Au 0–12%	14.71	3.75	0.851 ± 0.203
p+p comp	14.71	1.72	0.263 ± 0.113
p+p comp	14.71	3.75	0.576 ± 0.167
	$\sqrt{\langle k_T^2 \rangle_{AA}}$	$\sqrt{\langle k'^2_T \rangle_{pp}}$	$\langle \hat{q}L \rangle$
Reaction	GeV/c	GeV/c	GeV^2
Au+Au 0–12%	2.28 ± 0.35	1.006 ± 0.18	8.41 ± 2.66
Au+Au 0–12%	1.42 ± 0.22	1.076 ± 0.18	1.71 ± 0.67

Table 2
Tabulations for $\hat{q}$-STAR π^0-h [5].

	$\langle p_{Tt}\rangle$	$\langle p_{Ta}\rangle$	$\sqrt{\langle k_T^2\rangle}_{pp}$
Reaction	GeV/c	GeV/c	GeV/c
p+p	14.71	1.72	2.34 ± 0.34
p+p	14.71	3.75	2.51 ± 0.31

For completeness, the results for $\sqrt{\langle k_T^2\rangle}_{pp}$ with the p+p values $\hat{x}_h^{pp} = 0.84 \pm 0.04$, $\langle z_t\rangle = 0.80 \pm 0.05$ are given in Table 2.

5. Discussion and conclusion

For the $12 < p_{Tt} < 20$ ($\langle p_{Tt}\rangle = 14.71$) GeV/c, $1.2 < p_{Ta} < 3$ ($\langle p_{Ta}\rangle = 1.72$) GeV/c bin, the result of $\langle \hat{q}L\rangle = 8.41 \pm 2.66$ GeV^2 agrees with the Ref. [20] result, $\langle \hat{q}L\rangle = 14^{+42}_{-14}$ GeV^2, but is not consistent with zero because of the much smaller error. The result for the $3 < p_{Ta} < 5$ ($\langle p_{Ta}\rangle = 3.75$) GeV/c bin, $\langle \hat{q}L\rangle = 1.71 \pm 0.67$ GeV^2, is at the edge of agreement, 2.4 σ below the value in the lower p_{Ta} bin, but also 2.6 σ from zero. If the different p_{Ta} ranges do not change the original di-jet configuration, then the value of $\langle \hat{q}L\rangle$ should be equal in both ranges and can be weighted averaged with a result of $\langle \hat{q}L\rangle = 2.11 \pm 0.64$ GeV^2. Taking a guess for $\langle L\rangle$ in an Au+Au central collision as 7 fm, half the diameter of an Au nucleus, the result would be $\hat{q} = 1.20 \pm 0.38$ GeV^2/fm for the lowest p_{Ta} bin, $\hat{q} = 0.24 \pm 0.096$ GeV^2/fm for the higher p_{Ta} bin, with weighted average $\hat{q} = 0.30 \pm 0.09$ GeV^2/fm. These results are close to or lower than the result of the JET collaboration [21] $\hat{q} = 1.2 \pm 0.3$ GeV^2/fm at $\tau_0 = 0.6$ fm/c.

The new method presented here gives results for $\langle \hat{q}L\rangle$ comparable with the theoretical calculations noted [20,21] but is more straightforward and transparent for experimentalists. This is possibly the first experimental evidence for the predicted di-jet azimuthal broadening [1,2]. It is noteworthy that the value of $\hat{x}_h^{AA} = \hat{p}_{Ta}/\hat{p}_{Tt} \approx 0.4$ combined with the 20% loss of $\hat{p}_{Tt}$ for the trigger jet (Fig. 2), which is surface biased [22], implies that the away jet has lost ≈ 3 times more energy than the trigger jet and thus traveled a longer distance so spent a longer time in the QGP. This may affect [23] the value of $\hat{q}$ used for comparison from the JET collaboration which used only single (trigger) hadrons for their calculation.

It is important to emphasize that the calculated values of $\langle \hat{q}L\rangle$ are proportional to the square of the value of $\hat{x}_h$ derived from the measured away-side z_T (i.e. x_E) distribution using Eq. (4). Although in the literature for more than a decade in a well-cited paper [8] and referenced in an important QCD Resource Letter [24], Eq. (4) has neither been verified nor falsified by a measurement of di-jet correlations with a di-hadron trigger. Future measurements at RHIC [25,26] will be able to do this and thus greatly improve the understanding of di-jet and di-hadron azimuthal broadening.

Acknowledgement

Research supported by U.S. Department of Energy, Contract No. DE-SC0012704.

References

[1] R. Baier, Yu.L. Dokshitzer, A.H. Mueller, S. Peigne, D. Schiff, Nucl. Phys. B 484 (1997) 265–282.
[2] R. Baier, D. Schiff, B.G. Zakharov, Annu. Rev. Nucl. Part. Sci. 50 (2000) 37–69.
[3] CCHK Collab., M. Della Negra, et al., Nucl. Phys. B 127 (1977) 1–42.
[4] R.P. Feynman, R.D. Field, G.C. Fox, Nucl. Phys. B 128 (1977) 1–65.
[5] STAR Collab., L. Adamczyk, et al., Phys. Lett. B 760 (2016) 689–696.
[6] STAR Collab., L. Adamczyk, et al., Phys. Rev. Lett. 112 (2014) 122301.
[7] STAR Collab., P.M. Jacobs, A. Schmah, et al., Nucl. Phys. A 956 (2016) 641–644.
[8] PHENIX Collab., S.S. Adler, et al., Phys. Rev. D 74 (2006) 072002.
[9] P. Darriulat, et al., Nucl. Phys. B 107 (1976) 429–456.
[10] CCOR Collab., A.L.S. Angelis, et al., Phys. Lett. B 97 (1977) 163–168.
[11] M. Jacob, P.V. Landshoff, Phys. Rep. 48 (1978) 285–350.
[12] PHENIX Collab., A. Adare, et al., Phys. Rev. D 81 (2010) 012002.
[13] PHENIX Collab., A. Adare, et al., Phys. Rev. Lett. 101 (2008) 232301.
[14] PHENIX Collab., A. Adare, et al., Phys. Rev. C 87 (2013) 034911.
[15] PHENIX Collab., A. Adare, et al., Phys. Rev. D 82 (2010) 072001.
[16] PHENIX Collab., A. Adare, et al., arXiv:1609.04769 [hep-ex].
[17] PHENIX Collab., A. Adare, et al., Phys. Rev. C 77 (2008) 064907.
[18] PHENIX Collab., A. Adare, et al., Phys. Rev. Lett. 104 (2010) 252301.
[19] M.J. Tannenbaum, J. Phys. Conf. Ser. 589 (2015) 012019.
[20] L. Chen, G.-Y. Qin, S.-Y. Wei, B.-W. Xiao, H.-Z. Zhang, arXiv:1607.01932.
[21] JET Collab, K.M. Burke, et al., Phys. Rev. C 90 (2014) 014909.
[22] PHENIX Collab., S.S. Adler, et al., Phys. Rev. C 76 (2007) 034904.
[23] Thanks to Ivan Vitev for alerting me that $\hat{q}$ calculated at a fixed τ_0 will likely be larger than if integrated over the expansion or time spent in the QGP.
[24] A.S. Kronfeld, C. Quigg, Am. J. Phys. 78 (2010) 1081.
[25] A.H. Mueller, B. Wu, B-W. Xiao, F. Yuan, Phys. Lett. B 763 (2016) 208–212.
[26] sPHENIX Collab., A. Adare, et al., arXiv:1501.06197 [nucl-ex].

13

ϕ meson mass and decay width in nuclear matter and nuclei

J.J. Cobos-Martínez [a,d,*], K. Tsushima [a], G. Krein [b], A.W. Thomas [c]

[a] Laboratório de Física Teórica e Computacional – LFTC, Universidade Cruzeiro do Sul, 01506-000, São Paulo, SP, Brazil
[b] Instituto de Física Teórica, Universidade Estadual Paulista, Rua Dr. Bento Teobaldo Ferraz, 271 – Bloco II, 01140-070, São Paulo, SP, Brazil
[c] CSSM and ARC Centre of Excellence for Particle Physics at the Terascale, Department of Physics, University of Adelaide, Adelaide, SA 5005, Australia
[d] Instituto de Física y Matemáticas, Universidad Michoacana de San Nicolás de Hidalgo, Edificio C-3, Ciudad Universitaria, Morelia, Michoacán 58040, México

ARTICLE INFO

Editor: J.-P. Blaizot

ABSTRACT

The mass and decay width of the ϕ meson in cold nuclear matter are computed in an effective Lagrangian approach. The medium dependence of these properties are obtained by evaluating kaon–antikaon loop contributions to the ϕ self-energy, employing the medium-modified kaon masses, calculated using the quark-meson coupling model. The loop integral is regularized with a dipole form factor, and the sensitivity of the results to the choice of cutoff mass in the form factor is investigated. At normal nuclear matter density we find a downward shift of the ϕ mass by a few percent, while the decay width is enhanced by an order of magnitude. For a large variation of the cutoff mass parameter, the results for the ϕ mass and the decay width turn out to vary very little. Our results support results in the literature which suggest that one should observe a small downward mass shift and a large broadening of the decay width. In order to explore the possibility of studying the binding and absorption of ϕ mesons in nuclei, we also present the single-particle binding energies and half-widths of ϕ-nucleus bound states for some selected nuclei.

1. Introduction

The study of the changes in light vector meson properties in a nuclear medium have attracted much experimental and theoretical interest—see Refs. [1–3] for recent reviews. Amongst the arguments motivating these studies we mention the interest in chiral symmetry restoration at high density and the possible role of QCD van der Waals forces. In particular, there is special interest on the ϕ meson, the main reasons being: (i) despite its nearly pure $s\bar{s}$ content, the ϕ does interact strongly with a nucleus, composed predominantly of light u and d quarks, through the excitation of below-threshold virtual kaon and anti-kaon states that might have their properties changed in medium, the latter issue in itself being also of current interest [4–8]; (ii) the ϕN interaction in vacuum [9–12] and a possible in-medium mass shift of the ϕ are related to the strangeness content of the nucleon [13], which may have implications beyond the physics of the strong interaction, affecting, for example, the experimental searches for dark matter [14–16]; (iii) medium modifications of ϕ properties have been proposed [17] as a possible source for the anomalous nuclear mass number A-dependence observed in ϕ production from nuclear targets [18]; (iv) furthermore, as the ϕ is a nearly pure $s\bar{s}$ state and gluonic interactions are flavor blind, studying it serves to test theories of the multi-gluon exchange interactions, including long range QCD van der Waals forces [19], which are believed to play a role in the binding of the J/Ψ and other exotic heavy-quarkonia to matter [20–31].

Heavy-ion collisions and photon- or proton-induced reactions on nuclear targets have been used to extract information on the in-medium properties of hadrons. Although the medium modifications of hadron properties are expected to be stronger in heavy-ion collisions, they are also expected to be large enough in photon- or proton-induced reactions to enable the study of in-medium properties by fixed-target experiments. Several experiments have focused on the light vector mesons ρ, ω, and ϕ, since their mean-free paths can be comparable with the size of a nucleus after being produced inside the nucleus. However, a unified consensus has not yet been reached among the different experiments—see Refs. [1–3] for comprehensive reviews of the current status.

For the ϕ meson, although the precise values are different, a large in-medium broadening of the width has been reported by most of the experiments performed, while only a few of them find

* Corresponding author at: Laboratório de Física Teórica e Computacional, Universidade Cruzeiro do Sul, 01506-000, São Paulo, SP, Brazil.

E-mail addresses: javiercobos@ifm.umich.mx (J.J. Cobos-Martínez), kazuo.tsushima@gmail.com (K. Tsushima), gkrein@ift.unesp.br (G. Krein), anthony.thomas@adelaide.edu.au (A.W. Thomas).

evidence for a substantial mass shift. For example, the KEK-E325 collaboration [32] reported a mass reduction of 3.4% and an in-medium decay width of ≈ 14.5 MeV at normal nuclear matter density. The latter disagrees with the SPring8 [18] result, which reported a large in-medium ϕN cross section leading to a decay width of 35 MeV. But this 35 MeV is in close agreement with the two JLab CLAS collaboration measurements reported in Refs. [33] and [34].

In an attempt to clarify the situation, the CLAS collaboration at JLab [35] performed new measurements of nuclear transparency ratios, and estimated in-medium widths in the range of 23–100 MeV. These values overlap with that of the SPring8 measurement [18]. More recently, the ANKE-COSY collaboration [36] has measured the ϕ meson production from proton-induced reactions on various nuclear targets. The comparison of data with model calculations suggests an in-medium ϕ width of ≈ 50 MeV. This result is consistent with that of SPring8 [18], as well as the one deduced from CLAS at JLab [35]. However, the value is clearly larger than that of the KEK-E325 collaboration [32].

From the discussions above, it is obvious that the search for evidence of a light vector meson mass shift is indeed complicated. It certainly requires further experimental efforts to understand better the changes of ϕ properties in a nuclear medium. For example, the J-PARC E16 collaboration [37] intends to perform a more systematic study for the mass shift of vector mesons with higher statistics. Furthermore, the E29 collaboration at J-PARC has recently put forward a proposal [38,39] to study the in-medium mass modification of ϕ via the possible formation of the ϕ-nucleus bound states [26], using the primary reaction $\overline{p}p \to \phi\phi$. Finally, there is a proposal at JLab, following the 12 GeV upgrade, to study the binding of ϕ (and η) to ^{4}He [40].

On the theoretical side, various authors predict a downward shift of the in-medium ϕ meson mass and a broadening of the decay width. The possible decrease of the light vector meson masses in a nuclear medium was first predicted by Brown and Rho [41]. Thereafter, many theoretical investigations have been conducted, some of them focused on the self-energies of the ϕ due to the kaon–antikaon loop. Ko et al. [42] used a density-dependent kaon mass determined from chiral perturbation theory and found that at normal nuclear matter density, ρ_0, the ϕ mass decreases very little, by at most 2%, and the width $\Gamma_\phi \approx 25$ MeV and broadens drastically for large densities. Hatsuda and Lee calculated the in-medium ϕ mass based on QCD sum rule approach [43,44], and predicted a decrease of 1.5%–3% at normal nuclear matter density. Other investigations also predict a large broadening of the ϕ width: Ref. [45] reports a negative mass shift of < 1% and a decay width of 45 MeV at ρ_0; Ref. [46] predicts a decay width of 22 MeV but does not report a result on the mass shift; and Ref. [47] gives a rather small negative mass shift of $\approx 0.81\%$ and a decay width of 30 MeV. More recently, Ref. [48] reported a downward mass shift of < 2% and a large broadening width of 45 MeV; and finally, in Ref. [49], extending the work of Refs. [46,47], the authors reported a negative mass shift of 3.4% and a large decay width of 70 MeV at ρ_0. The reason for these differences may lie in the different approaches used to estimate the kaon–antikaon loop contributions for the ϕ self-energy.

In the present article we report results for the ϕ mass shift and decay width in nuclear matter, taking into account the medium dependence of the K and $\overline{K}$ masses. The latter are included by an explicit calculation based upon the quark-meson coupling (QMC) model [50,51]. The QMC model is a quark-based model of finite nuclei and nuclear matter, and has been very successful in describing the nuclear matter saturation properties, hadron properties in nuclear medium, as well as the properties of finite nuclei [52] and hypernuclei [53]—for a comprehensive review of the QMC model, see Ref. [54].

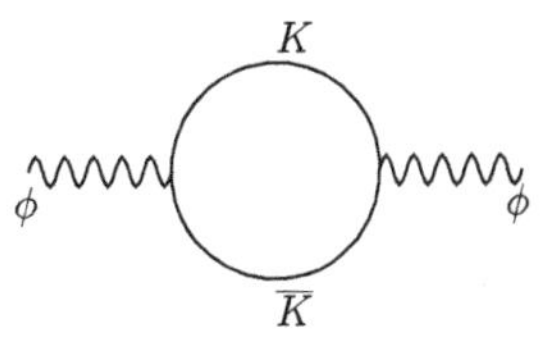

Fig. 1. $K\overline{K}$-loop contribution to the ϕ meson self-energy.

The paper is organized as follows. In Sec. 2 we present the effective Lagrangian used to calculate the ϕ-meson self-energy in vacuum, and give explicit expressions for its real and imaginary parts. Since the in-medium properties of the ϕ are dependent on the kaon and anti-kaon masses in a nuclear medium calculated within the QMC model, we briefly review this model in Sec. 3, and provide the necessary detail to understand the dressing of the kaons in nuclear medium. In Sec. 4 we calculate the ϕ-meson self-energy in nuclear matter and report the in-medium ϕ-meson mass and decay width, as well as the binding energies and widths of selected ϕ-nucleus bound states. Finally, conclusions and perspectives are given in Sec. 5.

2. ϕ meson self-energy in vacuum

We use the effective Lagrangian of Refs. [42,55] to compute the ϕ self-energy; the interaction Lagrangian $\mathcal{L}_{int}$ involves $\phi K\overline{K}$ and $\phi\phi K\overline{K}$ couplings dictated by a local gauge symmetry principle:

$$\mathcal{L}_{int} = \mathcal{L}_{\phi K\overline{K}} + \mathcal{L}_{\phi\phi K\overline{K}}, \tag{1}$$

where

$$\mathcal{L}_{\phi K\overline{K}} = \mathrm{i}g_\phi \phi^\mu \left[\overline{K}(\partial_\mu K) - (\partial_\mu \overline{K})K\right], \tag{2}$$

and

$$\mathcal{L}_{\phi\phi K\overline{K}} = g_\phi^2 \phi^\mu \phi_\mu \overline{K} K. \tag{3}$$

We use the convention:

$$K = \begin{pmatrix} K^+ \\ K^0 \end{pmatrix}, \; \overline{K} = \begin{pmatrix} K^- & \overline{K}^0 \end{pmatrix}. \tag{4}$$

We note that the use of the effective interaction Lagrangian of Eq. (1) without the term given in Eq. (3) may be considered as being motivated by the hidden gauge approach in which there are no four-point vertices, such as Eq. (3), that involve two pseudoscalar mesons and two vector mesons [56,57]. This is in contrast to the approach of using the minimal substitution to introduce vector mesons as gauge particles where such four-point vertices do appear. However, these two methods have been shown to be consistent if both the vector and axial vector mesons are included [58–61]. Therefore, we present results with and without such an interaction. We consider first the contribution from the $\phi K\overline{K}$ coupling given by Eq. (2) to the scalar part of the ϕ self-energy, $\Pi_\phi(p)$; Fig. 1 depicts this contribution. For a ϕ meson at rest, it is given by

$$\mathrm{i}\Pi_\phi(p) = -\frac{8}{3} g_\phi^2 \int \frac{\mathrm{d}^4 q}{(2\pi)^4} \vec{q}^{\,2} D_K(q) D_K(q-p), \tag{5}$$

where $D_K(q) = \left(q^2 - m_K^2 + \mathrm{i}\epsilon\right)^{-1}$ is the kaon propagator; $p = (p^0 = m_\phi, \vec{0})$ is the ϕ meson four-momentum vector, with m_ϕ the ϕ meson mass; $m_K (= m_{\overline{K}})$ is the kaon mass. When $m_\phi < 2m_K$ the self-energy $\Pi_\phi(p)$ is real. However, when $m_\phi > 2m_K$, which is the case here, $\Pi_\phi(p)$ acquires an imaginary part. The mass of the ϕ is determined from the real part of $\Pi_\phi(p)$

$$m_\phi^2 = \left(m_\phi^0\right)^2 + \Re\Pi_\phi(m_\phi^2), \tag{6}$$

with m_ϕ^0 being the bare mass of the ϕ and

$$\Re\Pi_\phi = -\frac{2}{3} g_\phi^2 \mathcal{P} \int \frac{d^3q}{(2\pi)^3} \vec{q}^{\,2} \frac{1}{E_K(E_K^2 - m_\phi^2/4)}. \tag{7}$$

Here $\mathcal{P}$ denotes the Principal Value part of the integral Eq. (5) and $E_K = (\vec{q}^{\,2} + m_K^2)^{1/2}$. The decay width of ϕ to a $K\overline{K}$ pair is given in terms of the imaginary part of $\Pi_\phi(p)$

$$\Im\Pi_\phi = -\frac{g_\phi^2}{24\pi} m_\phi^2 \left(1 - \frac{4m_K^2}{m_\phi^2}\right)^{3/2}, \tag{8}$$

as

$$\Gamma_\phi = -\frac{1}{m_\phi} \Im\Pi_\phi = \frac{g_\phi^2}{24\pi} m_\phi \left(1 - \frac{4m_K^2}{m_\phi^2}\right)^{3/2}. \tag{9}$$

The integral in Eq. (7) is divergent and needs regularization; we use a phenomenological form factor, with a cutoff parameter Λ_K, as in Ref. [62]. The coupling constant g_ϕ is determined by the experimental width of the ϕ in vacuum [63]. For the ϕ mass, m_ϕ, we use its experimental value: $m_\phi^{\text{expt}} = 1019.461$ MeV [63]. For the kaon mass m_K, there is a small ambiguity since $m_{K^+} \neq m_{K^0}$, as a result of charge symmetry breaking and electromagnetic interactions. The experimental values for the K^+ and K^0 meson masses in vacuum are $m_{K^+}^{\text{expt}} = 493.677$ MeV and $m_{K^0}^{\text{expt}} = 497.611$ MeV, respectively [63]. For definiteness we use the average of $m_{K^+}^{\text{expt}}$ and $m_{K^0}^{\text{expt}}$ as the value of m_K in vacuum. The effect of this tiny mass ambiguity on the in-medium kaon (antikaon) properties is negligible. Then, we get the coupling $g_\phi = 4.539$, and can fix the bare mass m_ϕ^0.

3. The quark-meson coupling model and the in-medium kaon mass

Essential to our results for the in-medium ϕ mass, m_ϕ^*, and decay width, Γ_ϕ^*, at finite baryon density $\rho_B = \rho_p + \rho_n$ (sum of the proton and neutron densities), is the in-medium kaon mass, m_K^*, which is driven by the interactions of the kaon with the nuclear medium—we denote with an asterisk an in-medium quantity. The in-medium kaon mass is calculated in the QMC model. This model has been successfully applied to investigate the properties of infinite nuclear matter and finite nuclei. Here we briefly present the necessary details needed to understand our results. For a more in depth discussion of the model see Refs. [4,50,54] and references therein.

We consider nuclear matter in its rest frame, where all the scalar and vector mean field potentials, which are responsible for the nuclear many-body interactions, are constants in Hartree approximation. The Dirac equations for the quarks and antiquarks ($q = u$ or d, and s) in a hadron bag in nuclear matter at the position $x = (t, \vec{r})$ (with $|\vec{r}| \leq R_h^*$, the in medium bag radius) are given by [4,54]:

$$\left[i\partial\!\!\!/_x - m_q^* \mp \gamma^0 V_+\right] \begin{pmatrix} \psi_u \\ \psi_{\bar{u}} \end{pmatrix} = 0,$$

$$\left[i\partial\!\!\!/_x - m_q^* \mp \gamma^0 V_-\right] \begin{pmatrix} \psi_d \\ \psi_{\bar{d}} \end{pmatrix} = 0,$$

$$[i\partial\!\!\!/_x - m_s] \begin{pmatrix} \psi_s \\ \psi_{\bar{s}} \end{pmatrix} = 0, \tag{10}$$

where $m_q^* = m_q - V_\sigma^q$ and $V_\pm = V_\omega^q \pm 1/2\, V_\rho^q$. Here we neglect the Coulomb force, and assume SU(2) symmetry for the light quarks ($m_q = m_u = m_d$). The constant mean-field potentials in nuclear matter are defined by $V_\sigma^q \equiv g_\sigma^q \sigma$, $V_\omega^q \equiv g_\omega^q \omega$, and $V_\rho^q \equiv g_\rho^q b$, where b is the time component of the ρ mean field, with g_σ^q, g_ω^q, and g_ρ^q the corresponding quark-meson coupling constants. Note that $V_\rho^q \propto (\rho_p - \rho_n) = 0$ in symmetric nuclear matter, although this is not true in a nucleus where the Coulomb force may induce an asymmetry between the proton and neutron distributions even in a nucleus with the same number of protons and neutrons, resulting in $V_\rho^q \propto (\rho_p - \rho_n) \neq 0$ at a given position in a nucleus.

The normalized, static solution for the ground-state quarks or antiquarks with flavor f in the hadron h may be written as $\psi_f(x) = N_f e^{-i\epsilon_f t/R_h^*} \psi_f(\vec{r})$, where N_f and $\psi_f(\vec{r})$ are the normalization factor and the corresponding spin and spatial part of the wave function, respectively. The in-medium bag radius R_h^* of hadron h is determined through the stability condition for the mass of the hadron against the variation of the bag radius [50,54]—see Eq. (15) below. The eigenenergies in units of $1/R_h^*$ are given by

$$\begin{pmatrix} \epsilon_u \\ \epsilon_{\bar{u}} \end{pmatrix} = \Omega_q^* \pm R_h^* V_+, \tag{11}$$

$$\begin{pmatrix} \epsilon_d \\ \epsilon_{\bar{d}} \end{pmatrix} = \Omega_q^* \pm R_h^* V_-, \tag{12}$$

$$\epsilon_s = \epsilon_{\bar{s}} = \Omega_s. \tag{13}$$

Recall that $V_\rho^q = 0$, as explained earlier. The in-medium hadron mass, m_h^*, is calculated by

$$m_h^* = \sum_{j=q,\bar{q},s,\bar{s}} \frac{n_j \Omega_j^* - z_h}{R_h^*} + \frac{4\pi}{3} R_h^{*3} B, \tag{14}$$

$$\frac{\partial m_h^*}{\partial R_h^*} = 0, \tag{15}$$

where $\Omega_q^* = \Omega_{\bar{q}}^* = [x_q^2 + (R_h^* m_q^*)^2]^{1/2}$ with $\Omega_s^* = \Omega_{\bar{s}}^* = [x_s^2 + (R_h^* m_s)^2]^{1/2}$, $x_{q,s}$ being the lowest bag eigenfrequencies; and $n_q(n_{\bar{q}})$, $n_s(n_{\bar{s}})$ are the quark (antiquark) numbers for the quark flavors q and s, respectively. The MIT bag quantities, z_h, B, $x_{q,s}$, and $m_{q,s}$ are the parameters for the sum of the c.m. and gluon fluctuation effects, bag constant, lowest eigenvalues for the quarks q or s, respectively, and the corresponding current quark masses. The parameters z_N (z_h) and B are fixed by fitting the nucleon (hadron) mass in free space.

For the current quark masses relevant for this study, we use $(m_{u,d}, m_s) = (5, 250)$ MeV, where these values were used in Refs. [4,54] and many studies made in the standard version of the QMC model. Since the effects of the current-quark mass values on the final results are very small, we use the same values as those used in the past, so that we can compare and discuss the results with those obtained previously. The bag radius of the nucleon in vacuum is taken to be $R_N = 0.8$ fm, and the parameter z_N, simulating the zero-point and c.m. energy, is obtained $z_N = 3.295$. For the kaon, the values in vacuum calculated here are $(R_K, z_K) = (0.574$ fm$, 3.295)$. The bag constant calculated for the present study is $B = (170 \text{ MeV})^4$. The quark-meson coupling constants, which are determined so as to reproduce the saturation properties of symmetric nuclear matter—the binding energy per nucleon of 15.7 MeV at $\rho_0 = 0.15$ fm^{-3}—are $(g_\sigma^q, g_\omega^q, g_\rho^q) = (5.69, 2.72, 9.33)$. In addition, the incompressibility obtained is $K = 279.3$ MeV. The σ coupling at the nucleon level, which is not trivial, is related by $g_\sigma \equiv g_\sigma^N \equiv 3g_\sigma^q S_N(\sigma = 0) = 3 \times 5.69 \times 0.483 = 8.23$ [4,54], where

$$S_N(\sigma) = \int d^3r\, \bar{\psi}_q(\vec{r}) \psi_q(\vec{r}), \tag{16}$$

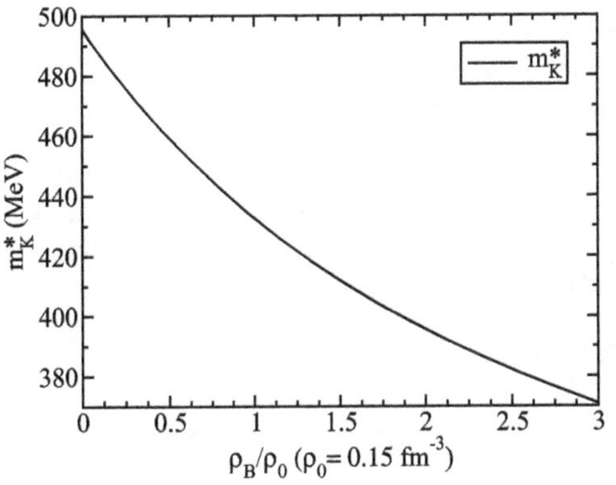

Fig. 2. In-medium kaon mass m^*_K.

Table 1
ϕ mass and width at normal nuclear matter density, ρ_0. All quantities are given in MeV.

	$\Lambda_K = 1000$	$\Lambda_K = 2000$	$\Lambda_K = 3000$
m^*_ϕ	1009.3	1000.9	994.9
Γ^*_ϕ	37.7	34.8	32.8

with the ground state light-quark wave functions evaluated self-consistently in-medium.

The resulting in-medium kaon (Lorentz-scalar) mass, calculated via Eqs. (14) and (15), is shown in Fig. 2, with the parameters fixed by the nuclear matter saturation properties. The kaon effective mass at normal nuclear matter density $\rho_0 = 0.15$ fm^{-3} decreases by about 13%. This is a little larger than the 10% decrease used in Ref. [42]. Note that, the isoscalar-vector ω-mean-field potentials arise both for the kaon and antikaon. However, they have opposite signs and cancel each other (or they can be eliminated by a variable shift) in the calculation of the ϕ self-energy, and therefore we do not show here—see Ref. [4] for details.

4. ϕ meson in matter

The in-medium ϕ mass is calculated by solving Eq. (6) by replacing m_K by m^*_K and m_ϕ by m^*_ϕ, and the width is obtained by using the solutions in Eq. (9). We regularize the associated loop integral with a dipole form factor using a cutoff mass parameter Λ_K. In principle, this parameter may be determined phenomenologically using, for example, a quark model—see Ref. [62] for more details. However, for simplicity we keep it free and vary its value over a wide interval, namely 1000–3000 MeV.

In Table 1, we present the values for m^*_ϕ and Γ^*_ϕ at normal nuclear matter density ρ_0. A negative kaon mass shift of 13% induces only $\approx 2\%$ downward mass shift of the ϕ. On the other hand, Γ^*_ϕ is very sensitive to the change in the kaon mass; at $\rho_B = \rho_0$, the broadening of the ϕ becomes an order of magnitude larger than its vacuum value and it increases rapidly with increasing nuclear density, up to a factor of ~ 20 enhancement for the largest nuclear matter density treated, $\rho_B = 3\rho_0$. This can be seen in Fig. 3, where we plot m^*_ϕ and Γ^*_ϕ as a function of the ratio ρ_B/ρ_0. The effect of the in-medium kaon mass change gives a negative shift of the ϕ meson mass. However, even for the largest value of density treated in this study, the downward mass shift is only a few percent for all values of the cutoff parameter Λ_K. For m^*_ϕ at normal nuclear matter density, the average downward mass shift is 1.8% with a 0.7% standard deviation from the averaged value, while Γ^*_ϕ broadens in average by a factor of 10 with a 0.7 standard deviation from the average.

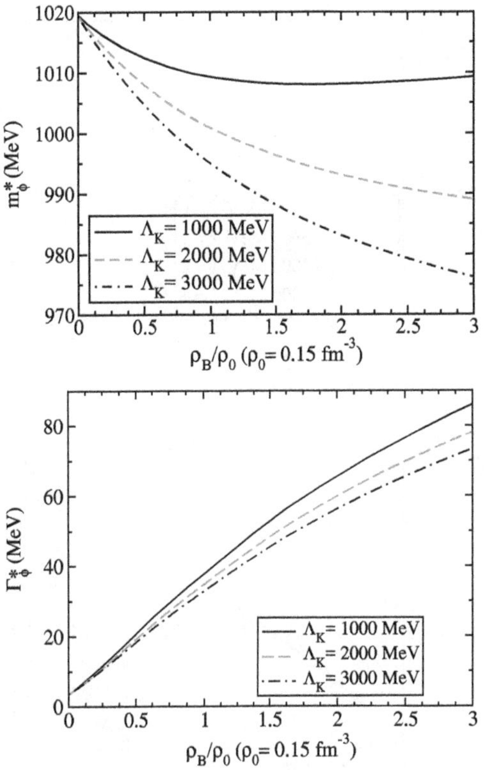

Fig. 3. In-medium ϕ mass (upper panel) and width (lower panel) for three values of the cutoff parameter Λ_K.

Table 2
ϕ-nucleus bound state single-particle energies E and half widths $\Gamma/2$, calculated by solving the Schrödinger equation with and without the imaginary part of the ϕ-nucleus potential $V_{\phi A}(r)$. The cutoff value used is $\Lambda_K = 3000$ MeV. Quantities are all in MeV. "n" in the entry ^{4}He denotes that we find no bound state.

		E	Γ/2	E	Γ/2
^{4}He	1s	−1.39	0	n	
^{12}C	1s	−7.70	0	−6.47	11.00
^{208}Pb	1s	−21.22	0	−21.06	16.25
	1p	−17.69	0	−17.35	15.76
	1d	−13.34	0	−12.78	15.06
	2s	−11.68	0	−10.97	14.67

Next, we present predictions for single-particle energies and half widths for ϕ-nucleus bound states for several selected nuclei. We solve the Schrödinger equation for a complex ϕ-nucleus scalar potential determined by a local-density approximation using the ϕ mass shift and decay width in nuclear matter. This amounts to using the following for the complex ϕ-nucleus (A) potential

$$V_{\phi A}(r) = \Delta m^*_\phi(\rho_B(r)) - (i/2)\Gamma^*_\phi(\rho_B(r)), \tag{17}$$

where $\Delta m^*_\phi(\rho_B(r)) \equiv m^*_\phi(\rho_B(r)) - m_\phi$, r is the distance from the center of the nucleus and $\rho_B(r)$ is the density profile of the given nucleus, which we calculate in the QMC model. Table 2 shows the results for the real and imaginary parts of the single-particle energies $\mathcal{E} = E - (i/2)\Gamma$ in ^{4}He, ^{12}C and ^{208}Pb. We present results with and without the imaginary (absorptive) part of the ϕ-nucleus potential $V_{\phi A}(r)$. One sees that ϕ is not bound to ^{4}He when the imaginary part of the potential is included. For larger nuclei, the ϕ does bind but while the binding is substantial the energy levels are quite broad; the half widths being roughly the same size as the central values of the real parts.

To conclude and for completeness, we show the impact of adding the $\phi\phi K\overline{K}$ interaction of Eq. (3) on the in-medium ϕ mass

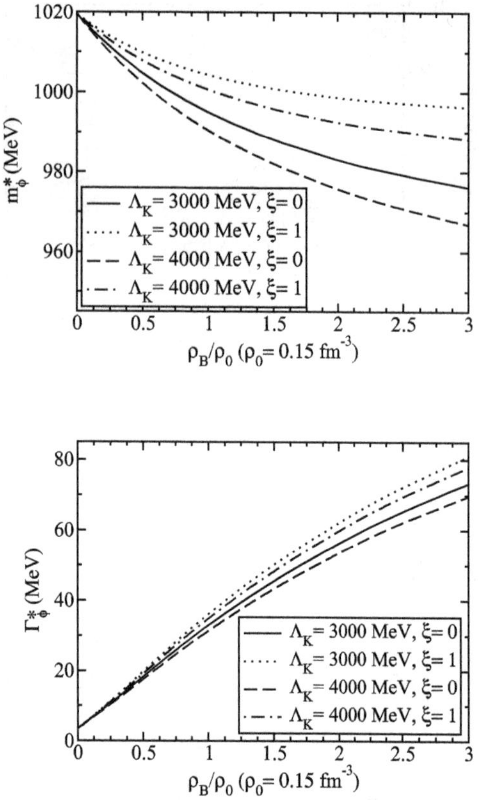

Fig. 4. Effect of adding ($\xi = 1$) the $\phi\phi K\overline{K}$ interaction of Eq. (3) on the in-medium ϕ mass (upper panel) and width (lower panel) for two values of the cutoff parameter Λ_K.

and width. Fig. 4 presents the results. We have used the notation that $\xi = 1(0)$ means that this interaction is (not) included in the calculation of the ϕ self-energy. One still gets a downward shift of the in-medium ϕ mass when $\xi = 1$, although the absolute value is slightly different from $\xi = 0$. The in-medium width is not very sensitive to this interaction.

5. Conclusions and perspectives

We have calculated the ϕ meson mass and width in nuclear matter within an effective Lagrangian approach up to three times of normal nuclear matter density. Essential to our results are the in-medium kaon masses, which are calculated in the quark-meson coupling (QMC) model, where the scalar and vector meson mean fields couple directly to the light u and d quarks (antiquarks) in the K ($\overline{K}$) mesons.

At normal nuclear matter density, allowing for a very large variation of the cutoff parameter Λ_K, although we have found a sizable negative mass shift of 13% in the kaon mass, this induces only a few percent (1.8% on average) downward shift of the ϕ meson mass. On the other hand, it induces an order-of-magnitude broadening of the decay width.

Given the nuclear matter results, we have used a local density approximation to infer the position dependent attractive complex scalar potential, $V_s(\rho_B(r)) = \Delta m^*_\phi(\rho_B(r)) - (i/2)\Gamma^*_\phi(\rho_B(r))$, in a finite nucleus. This allowed us to study the binding and absorption of a number of ϕ-nuclear systems, given the nuclear density profiles, $\rho_B(r)$, also calculated using the QMC model. While the results found in this study show that one should expect the ϕ meson to be bound in all but the lightest nuclei, the broadening of these energy levels, which is comparable to the amount of binding, may introduce challenges in observing such states experimentally.

In the present study, we have focused on the ϕ self-energy in medium due to the medium modified kaon–antikaon loop. However, more study of gluonic color forces is needed on binding of the ϕ-meson to a nucleus.

As a possible extension of this work, we note that the medium effects on the ϕ meson may lead to some enhancement of the strangeness content of the bound nucleon, with consequences, for example, for dark matter detection.

Acknowledgements

This work was partially supported by Conselho Nacional de Desenvolvimento Científico e Tecnológico—CNPq, Grant Nos. 152348/2016-6 (J.J.C-M.), 400826/2014-3 and 308088/2015-8 (K.T.), 305894/2009-9 (G.K.), and 313800/2014-6 (A.W.T.), and Fundação de Amparo à Pesquisa do Estado de São Paulo—FAPESP, Grant Nos. 2015/17234-0 (K.T.) and 2013/01907-0 (G.K.). This research was also supported by the Australian Research Council through the ARC Centre of Excellence for Particle Physics at the Terascale (CE110001104), and through Grant No. DP151103101 (A.W.T.).

References

[1] S. Leupold, V. Metag, U. Mosel, Int. J. Mod. Phys. E 19 (2010) 147, arXiv:0907.2388 [nucl-th].
[2] R.S. Hayano, T. Hatsuda, Rev. Mod. Phys. 82 (2010) 2949, arXiv:0812.1702 [nucl-ex].
[3] G. Krein, AIP Conf. Proc. 1701 (2016) 020012.
[4] K. Tsushima, K. Saito, A.W. Thomas, S.V. Wright, Phys. Lett. B 429 (1998) 239; Erratum: Phys. Lett. B 436 (1998) 453, arXiv:nucl-th/9712044.
[5] F. Laue, et al., KaoS Collaboration, Phys. Rev. Lett. 82 (1999) 1640, arXiv:nucl-ex/9901005.
[6] J. Schaffner-Bielich, V. Koch, M. Effenberger, Nucl. Phys. A 669 (2000) 153, arXiv:nucl-th/9907095.
[7] Y. Akaishi, T. Yamazaki, Phys. Rev. C 65 (2002) 044005.
[8] C. Fuchs, Prog. Part. Nucl. Phys. 56 (2006) 1, arXiv:nucl-th/0507017.
[9] A.I. Titov, Y.s. Oh, S.N. Yang, Phys. Rev. Lett. 79 (1997) 1634, arXiv:nucl-th/9702015.
[10] A.I. Titov, Y.s. Oh, S.N. Yang, T. Morii, Phys. Rev. C 58 (1998) 2429, arXiv:nucl-th/9804043.
[11] Y.s. Oh, A.I. Titov, S.N. Yang, T. Morii, Phys. Lett. B 462 (1999) 23, arXiv:nucl-th/9905044.
[12] Y.s. Oh, H.C. Bhang, Phys. Rev. C 64 (2001) 055207, arXiv:nucl-th/0104068.
[13] P. Gubler, K. Ohtani, Phys. Rev. D 90 (9) (2014) 094002, arXiv:1404.7701 [hep-ph].
[14] A. Bottino, F. Donato, N. Fornengo, S. Scopel, Astropart. Phys. 18 (2002) 205, arXiv:hep-ph/0111229.
[15] J.R. Ellis, K.A. Olive, C. Savage, Phys. Rev. D 77 (2008) 065026, arXiv:0801.3656 [hep-ph].
[16] J. Giedt, A.W. Thomas, R.D. Young, Phys. Rev. Lett. 103 (2009) 201802, http://dx.doi.org/10.1103/PhysRevLett.103.201802, arXiv:0907.4177 [hep-ph].
[17] A. Sibirtsev, H.W. Hammer, U.G. Meissner, A.W. Thomas, Eur. Phys. J. A 29 (2006) 209, arXiv:nucl-th/0606044.
[18] T. Ishikawa, et al., Phys. Lett. B 608 (2005) 215, arXiv:nucl-ex/0411016.
[19] T. Appelquist, W. Fischler, Phys. Lett. B 77 (1978) 405.
[20] S.J. Brodsky, I.A. Schmidt, G.F. de Teramond, Phys. Rev. Lett. 64 (1990) 1011.
[21] M.E. Luke, A.V. Manohar, M.J. Savage, Phys. Lett. B 288 (1992) 355, arXiv:hep-ph/9204219.
[22] A. Sibirtsev, K. Tsushima, K. Saito, A.W. Thomas, Phys. Lett. B 484 (2000) 23, http://dx.doi.org/10.1016/S0370-2693(00)00635-3, arXiv:nucl-th/9904015.
[23] D. Kawama, J-PARC E16 Collaboration, in: PoS Hadron, vol. 2013, 2013, p. 178.
[24] D. Kawama, et al., J-PARC E16 Collaboration, JPS Conf. Proc. 1 (2014) 013074.
[25] H. Ohnishi, et al., Acta Phys. Pol. B 45 (2014) 819.
[26] K. Aoki, J-PARC E16 Collaboration, arXiv:1502.00703 [nucl-ex].
[27] Y. Morino, et al., J-PARC E16 Collaboration, JPS Conf. Proc. 8 (2015) 022009.
[28] H. Gao, T.S.H. Lee, V. Marinov, Phys. Rev. C 63 (2001) 022201, arXiv:nucl-th/0010042.
[29] S.R. Beane, E. Chang, S.D. Cohen, W. Detmold, H.-W. Lin, K. Orginos, A. Parreño, M.J. Savage, Phys. Rev. D 91 (11) (2015) 114503, arXiv:1410.7069 [hep-lat].
[30] N. Brambilla, G. Krein, J. Tarrús Castellà, A. Vairo, Phys. Rev. D 93 (2016) 054002, arXiv:1510.05895 [hep-ph].
[31] H. Gao, H. Huang, T. Liu, J. Ping, F. Wang, Z. Zhao, arXiv:1701.03210 [hep-ph].
[32] R. Muto, et al., KEK-PS-E325 Collaboration, Phys. Rev. Lett. 98 (2007) 042501, arXiv:nucl-ex/0511019.

[33] T. Mibe, et al., CLAS Collaboration, Phys. Rev. C 76 (2007) 052202, arXiv:nucl-ex/0703013.
[34] X. Qian, et al., CLAS Collaboration, Phys. Lett. B 680 (2009) 417, arXiv:0907.2668 [nucl-ex].
[35] M.H. Wood, et al., CLAS Collaboration, Phys. Rev. Lett. 105 (2010) 112301, arXiv:1006.3361 [nucl-ex].
[36] A. Polyanskiy, et al., Phys. Lett. B 695 (2011) 74, arXiv:1008.0232 [nucl-ex].
[37] http://rarfaxp.riken.go.jp/~yokkaich/paper/jparc-proposal-0604.pdf.
[38] http://j-parc.jp/researcher/Hadron/en/pac_0907/pdf/Ohnishi.pdf.
[39] http://j-parc.jp/researcher/Hadron/en/pac_1007/pdf/KEK_J-PARC-PAC2010-02.pdf.
[40] https://www.jlab.org/exp_prog/PACpage/PAC42/PAC42_FINAL_Report.pdf.
[41] G.E. Brown, M. Rho, Phys. Rev. Lett. 66 (1991) 2720.
[42] C.M. Ko, P. Levai, X.J. Qiu, C.T. Li, Phys. Rev. C 45 (1992) 1400.
[43] T. Hatsuda, S.H. Lee, Phys. Rev. C 46 (1) (1992) R34.
[44] T. Hatsuda, H. Shiomi, H. Kuwabara, Prog. Theor. Phys. 95 (1996) 1009, arXiv:nucl-th/9603043.
[45] F. Klingl, T. Waas, W. Weise, Phys. Lett. B 431 (1998) 254, arXiv:hep-ph/9709210.
[46] E. Oset, A. Ramos, Nucl. Phys. A 679 (2001) 616, arXiv:nucl-th/0005046.
[47] D. Cabrera, M.J. Vicente Vacas, Phys. Rev. C 67 (2003) 045203, arXiv:nucl-th/0205075.
[48] P. Gubler, W. Weise, Phys. Lett. B 751 (2015) 396, arXiv:1507.03769 [hep-ph].
[49] D. Cabrera, A.N. Hiller Blin, M.J. Vicente Vacas, Phys. Rev. C 95 (1) (2017) 015201, arXiv:1609.03880 [nucl-th].
[50] P.A.M. Guichon, Nucl. Phys. A 497 (1989) 265C.
[51] P.A.M. Guichon, K. Saito, E.N. Rodionov, A.W. Thomas, Nucl. Phys. A 601 (1996) 349, http://dx.doi.org/10.1016/0375-9474(96)00033-4, arXiv:nucl-th/9509034.
[52] J.R. Stone, P.A.M. Guichon, P.G. Reinhard, A.W. Thomas, Phys. Rev. Lett. 116 (9) (2016) 092501, http://dx.doi.org/10.1103/PhysRevLett.116.092501, arXiv:1601.08131 [nucl-th];
K. Saito, K. Tsushima, A.W. Thomas, Nucl. Phys. A 609 (1996) 339, http://dx.doi.org/10.1016/S0375-9474(96)00263-1, arXiv:nucl-th/9606020;
K. Saito, K. Tsushima, A.W. Thomas, Phys. Rev. C 55 (1997) 2637, http://dx.doi.org/10.1103/PhysRevC.55.2637, arXiv:nucl-th/9612001.
[53] K. Tsushima, K. Saito, J. Haidenbauer, A.W. Thomas, Nucl. Phys. A 630 (1998) 691, http://dx.doi.org/10.1016/S0375-9474(98)00806-9, arXiv:nucl-th/9707022;
K. Tsushima, K. Saito, A.W. Thomas, Phys. Lett. B 411 (1997) 9, http://dx.doi.org/10.1016/S0370-2693(97)00944-1; Erratum: Phys. Lett. B 421 (1998) 413, http://dx.doi.org/10.1016/S0370-2693(98)00065-3, arXiv:nucl-th/9701047;
K. Tsushima, F.C. Khanna, Phys. Rev. C 67 (2003) 015211, http://dx.doi.org/10.1103/PhysRevC.67.015211, arXiv:nucl-th/0207077;
K. Tsushima, F.C. Khanna, J. Phys. G 30 (2004) 1765, http://dx.doi.org/10.1088/0954-3899/30/12/001, arXiv:nucl-th/0303073;
P.A.M. Guichon, A.W. Thomas, K. Tsushima, Nucl. Phys. A 814 (2008) 66, http://dx.doi.org/10.1016/j.nuclphysa.2008.10.001, arXiv:0712.1925 [nucl-th].
[54] K. Saito, K. Tsushima, A.W. Thomas, Prog. Part. Nucl. Phys. 58 (2007) 1, arXiv:hep-ph/0506314.
[55] F. Klingl, N. Kaiser, W. Weise, Z. Phys. A 356 (1996) 193, arXiv:hep-ph/9607431.
[56] Z.w. Lin, C.M. Ko, B. Zhang, Phys. Rev. C 61 (2000) 024904.
[57] S.H. Lee, C. Song, H. Yabu, Phys. Lett. B 341 (1995) 407, arXiv:hep-ph/9408266.
[58] K. Yamawaki, Phys. Rev. D 35 (1987) 412.
[59] U.G. Meissner, I. Zahed, Z. Phys. A 327 (1987) 5.
[60] U.G. Meissner, Phys. Rep. 161 (1988) 213.
[61] S. Saito, K. Yamawaki, Nagoya, Japan: Univ., Phys. Dept., 1987, 225p.
[62] G. Krein, A.W. Thomas, K. Tsushima, Phys. Lett. B 697 (2011) 136, arXiv:1007.2220 [nucl-th].
[63] K.A. Olive, et al., Particle Data Group, The review of particle physics, Chin. Phys. C 38 (2014) 090001 and 2015 update, http://pdg.lbl.gov/.

14

Model-independent extraction of $|V_{cb}|$ from $\bar{B} \to D^{*}\ell\bar{\nu}$

Benjamín Grinstein, Andrew Kobach *

Physics Department, University of California, San Diego, La Jolla, CA 92093, USA

ARTICLE INFO

Editor: J. Hisano

ABSTRACT

We fit the unfolded data of $\bar{B}^0 \to D^{*+}\ell\bar{\nu}$ from the Belle experiment, where $\ell \equiv e, \mu$, using a method independent of heavy quark symmetry to extrapolate to zero-recoil and extract the value of $|V_{cb}|$. This results in $|V_{cb}| = (41.9^{+2.0}_{-1.9}) \times 10^{-3}$, which is robust to changes in the theoretical inputs and very consistent with the value extracted from inclusive semileptonic B decays.

1. Introduction

The discrepancy between the measured values of the CKM matrix element $|V_{cb}|$ from inclusive versus exclusive semileptonic B decays has been an ongoing dilemma for a few decades; for a recent review, see Ref. [1]. Currently, the world averages for $|V_{cb}|$ are [2,3]:

$$|V_{cb}| = (39.18 \pm 0.99) \times 10^{-3} \quad (\bar{B} \to D\ell\bar{\nu}) \tag{1}$$

$$|V_{cb}| = (38.71 \pm 0.75) \times 10^{-3} \quad (\bar{B} \to D^{*}\ell\bar{\nu}) \tag{2}$$

$$|V_{cb}| = (42.19 \pm 0.78) \times 10^{-3} \quad (\bar{B} \to X_c\ell\bar{\nu},\ \text{kinetic scheme}) \tag{3}$$

$$|V_{cb}| = (41.98 \pm 0.45) \times 10^{-3} \quad (\bar{B} \to X_c\ell\bar{\nu},\ \text{1S scheme}) \tag{4}$$

The extracted values of $|V_{cb}|$ from exclusive decays, especially the value of $|V_{cb}|$ measured in $\bar{B} \to D^{*}\ell\bar{\nu}$, are systematically lower than those from inclusive decays. When measuring $|V_{cb}|$ in exclusive $\bar{B} \to D^{(*)}\ell\bar{\nu}$ decays, it has been commonplace to use the parameterization developed by Caprini, Lellouch, and Neubert (CLN) [4,5], which utilizes not only dispersion relations to bound the hadronic form factors, but also relations at $1/m_Q$ in the heavy quark expansion, and claims to describe the full differential decay to within a few percent given only 4 parameters. Interestingly, it has now become clear that the CLN parameterization is no longer a good fit to the $\bar{B} \to D\ell\bar{\nu}$ data from current experiments [6]. One would naively expect the same to be true, in general, for $\bar{B} \to D^{*}\ell\bar{\nu}$ as well.

A recent analysis by Belle [7] used the CLN parameterization to extract a value for $|V_{cb}|$ at zero recoil given a dataset of $\bar{B}^0 \to D^{*+}\ell\bar{\nu}$, resulting in $|V_{cb}| = (37.4 \pm 1.3) \times 10^{-3}$. This is one of the first times an experiment with a large dataset has published unfolded data for $\bar{B} \to D^{*}\ell\bar{\nu}$, so one can easily explore if a different parameterization of the hadronic form factors gives rise to a different value of $|V_{cb}|$. We use the method developed by Boyd, Grinstein, and Lebed (BGL) in Refs. [8–12], from which the original CLN parameterization was based, but which does not utilize any assumptions from heavy quark symmetry.[1] We find that while both the CLN and BGL parameterizations give a good fit to the Belle data, the BGL parameterization gives a larger value for $|V_{cb}|$ when extrapolating to zero recoil, consistent with the value of $|V_{cb}|$ measured in inclusive analyses [2,3]. As we prepared this manuscript, a similar analysis was made public, using the same BGL parameterization [14]. We are pleased to see the results match nearly identically. Another analysis was recently made public, which highlighted the importance of including theoretical uncertainties from Λ_{QCD}/m_Q corrections in the relations between the form factors in the heavy quark expansion [15], which suggests that the uncertainties associated with the extracted value of $|V_{cb}|$ using the CLN method may be underestimated. Moreover, in their analysis of $B \to D\ell\nu$ lattice data, the Fermilab Lattice and MILC collaborations decide not to quote results of their CLN fits because they are more confident in the errors obtained from the BGL parameterization which, they state, can be used to obtain $|V_{cb}|$ even as the uncertainties become arbitrarily more precise [16].

In all, the current evidence points to the strong possibility that the tension between inclusive and exclusive measurements of $|V_{cb}|$ in semileptonic B decays may be due to the use of the CLN parameterization of $\bar{B} \to D^{(*)}\ell\bar{\nu}$, and a different parameterization should be employed.

* Corresponding author.
E-mail address: akobach@gmail.com (A. Kobach).

[1] A brief history of the developments that led to the BGL parametrization and a more complete list of references can be found in Ref. [13].

2. Differential decay rate and BGL parameterization

Using the notation in Ref. [12], the $\bar{B} \to D^*$ matrix elements are defined as

$$\langle D^*(\varepsilon, p')|\bar{c}\gamma^\mu b|\bar{B}(p)\rangle = ig\epsilon^{\mu\nu\alpha\beta}\varepsilon^*_\nu p_\alpha p'_\beta, \tag{5}$$

$$\begin{aligned}&\langle D^*(\varepsilon, p')|\bar{c}\gamma^\mu\gamma^5 b|\bar{B}(p)\rangle\\ &= f\varepsilon^{*\mu} + (\varepsilon^*\cdot p)[a_+(p+p')^\mu + a_-(p-p')^\mu],\end{aligned} \tag{6}$$

where ε^μ is the polarization tensor of the vector D^* meson. In the limit when the final-state leptons are massless, the full differential decay rate for $\bar{B} \to D^*\ell\bar{\nu}$ is

$$\begin{aligned}&\frac{d\Gamma(\bar{B}\to D^*\ell\bar{\nu})}{dw\, d\cos\theta_\ell\, d\cos\theta_v\, d\chi}\\
&= \frac{3\eta_{ew}^2 G_F^2|V_{cb}|^2}{1024\pi^4}|\mathbf{p}_{D^*}|q^2 r\Big((1-\cos\theta_\ell)^2\sin^2\theta_v H_+^2\\
&+(1+\cos\theta_\ell)^2\sin^2\theta_v H_-^2 + 4\sin^2\theta_\ell\cos^2\theta_v H_0^2\\
&-2\sin^2\theta_\ell\sin^2\theta_v\cos 2\chi H_+H_-\\
&-4\sin\theta_\ell(1-\cos\theta_\ell)\sin\theta_v\cos\theta_v\cos\chi H_+H_0\\
&+4\sin\theta_\ell(1+\cos\theta_\ell)\sin\theta_v\cos\theta_v\cos\chi H_-H_0\Big),\end{aligned} \tag{7}$$

where q^μ is the 4-momentum of the lepton system, $r \equiv m_{D^*}/m_B$, and $|\mathbf{p}_{D^*}|$ is the magnitude of the D^* 3-momentum in the rest frame of the $\bar{B}$:

$$\begin{aligned}&w \equiv \frac{m_B^2 + m_{D^*}^2 - q^2}{2m_B m_{D^*}}, \qquad q^2 = m_B^2 + m_{D^*}^2 - 2m_B m_{D^*} w,\\ &|\mathbf{p}_{D^*}| = m_{D^*}\sqrt{w^2-1}.\end{aligned} \tag{8}$$

Here, H_+, H_-, and H_0 are form factors associated with each of the three helicity states of the D^*, all of which are functions of q^2. Also, θ_ℓ is the angle between the anti-neutrino and the direction antiparallel to the D^* in the rest frame of the leptonic system, θ_v is the angle between the D^* momentum and its daughter D meson, and χ is the angle between the planes defined by the leptonic system and the D^* system. The factor η_{ew} incorporates the leading electroweak corrections [17], $\eta_{ew} = 1 + \alpha/\pi \ln(M_Z/m_B) \simeq 1.0066$. In terms of the form factors in Eqs. (5) and (6),

$$H_+ = f - m_B|\mathbf{p}_{D^*}|g, \tag{9}$$

$$H_- = f + m_B|\mathbf{p}_{D^*}|g, \tag{10}$$

$$\begin{aligned}H_0 &= \frac{1}{m_{D^*}\sqrt{q^2}}\left[2m_B^2|\mathbf{p}_{D^*}|^2 a_+ - \frac{1}{2}(q^2 - m_B^2 + m_{D^*}^2)f\right]\\ &\equiv \frac{\mathcal{F}_1}{\sqrt{q^2}}.\end{aligned} \tag{11}$$

A detailed discussion about the BGL method for parameterizing the form factors f, g, and $\mathcal{F}_1$ can be found in Ref. [12]. The final result gives a parametrization of each form factor in terms of $N+1$ coefficients:

$$\begin{aligned}&g(z) = \frac{1}{P_g(z)\phi_g(z)}\sum_{n=0}^{N} a_n z^n, \quad f(z) = \frac{1}{P_f(z)\phi_f(z)}\sum_{n=0}^{N} b_n z^n,\\ &\mathcal{F}_1(z) = \frac{1}{P_{\mathcal{F}_1}(z)\phi_{\mathcal{F}_1}(z)}\sum_{n=0}^{N} c_n z^n,\end{aligned} \tag{12}$$

where the conformal variable z is defined as

Table 1

B_c^* masses used in this analysis from [18]. Only states that have a mass less than $m_B + m_{D^*}$ are included.

Type	Mass [GeV]
vector	6.337, 6.899, 7.012, 7.280
axial vector	6.730, 6.736, 7.135, 7.142

$$z \equiv \frac{\sqrt{w+1} - \sqrt{2a}}{\sqrt{w+1} + \sqrt{2a}}. \tag{13}$$

Here, $a = 1$ can be chosen such that $z = 0$ corresponds to zero recoil, and the coefficients a_n, b_n, and c_n are bounded by unitarity [10],

$$\sum_{n=0}^{N}|a_n|^2 \le 1, \quad \text{and} \quad \sum_{n=0}^{N}\left(|b_n|^2 + |c_n|^2\right) \le 1. \tag{14}$$

From Eq. (11), $\mathcal{F}_1(0) = (m_B - m_{D^*})f(0)$; hence b_0 and c_0 are not independent, i.e.,

$$c_0 = \left(\frac{(m_B - m_{D^*})\phi_{\mathcal{F}_1}(0)}{\phi_f(0)}\right)b_0. \tag{15}$$

The Blaschke factors $P(z)$ remove poles for $q^2 < (m_B + m_{D^*})^2$ associated with on-shell production of B_c^* bound states:

$$P_g(z) = \prod_i^4 \frac{z - z_{P_i}}{1 - z z_{P_i}}, \quad P_f(z) = P_{\mathcal{F}_1}(z) = \prod_i^4 \frac{z - z_{P_i}}{1 - z z_{P_i}}, \tag{16}$$

$$z_P \equiv \frac{\sqrt{(m_B + m_{D^*})^2 - m_P^2} - \sqrt{(m_B + m_{D^*})^2 - (m_B - m_{D^*})^2}}{\sqrt{(m_B + m_{D^*})^2 - m_P^2} + \sqrt{(m_B + m_{D^*})^2 - (m_B - m_{D^*})^2}}. \tag{17}$$

For the form factor g, the index i runs over the 4 vector B_c^* states, and for the f and $\mathcal{F}_1$, it runs over the 4 axial vector states, as listed in Table 1. The outer functions ϕ are defined as:

$$\phi_g(z) = \sqrt{\frac{256 n_I}{3\pi\chi^T(+u)}\frac{r^2(1+z)^2(1-z)^{-1/2}}{[(1+r)(1-z) + 2\sqrt{r}(1+z)]^4}}, \tag{18}$$

$$\phi_f(z) = \frac{1}{m_B^2}\sqrt{\frac{16 n_I}{3\pi\chi^T(-u)}\frac{r(1+z)(1-z)^{3/2}}{[(1+r)(1-z) + 2\sqrt{r}(1+z)]^4}}, \tag{19}$$

$$\phi_{\mathcal{F}_1}(z) = \frac{1}{m_B^3}\sqrt{\frac{8 n_I}{3\pi\chi^T(-u)}\frac{r(1+z)(1-z)^{5/2}}{[(1+r)(1-z) + 2\sqrt{r}(1+z)]^5}}, \tag{20}$$

where n_I is the effective number of light quarks, $u \equiv m_c/m_b$, and $\chi^T(\pm u)$ is related to a perturbative calculation at $q^2 = 0$. We use the result for $\chi^T(\pm u)$ in the pole mass scheme, and take numerical values to be those listed in Table 2, as done in Ref. [12], ignoring the small contributions from condensates:

$$\begin{aligned}&\chi^T(+0.33) = 5.28 \times 10^{-4}\,\text{GeV}^{-2},\\ &\chi^T(-0.33) = 3.07 \times 10^{-4}\,\text{GeV}^{-2}\end{aligned} \tag{21}$$

We do not incorporate any uncertainties associated with the numerical inputs to the BGL theory parameterization. Specifically, the uncertainties associated with m_B and m_{D^*} are negligible in this analysis. Also, varying the inputs that only affect the overall normalization of the outer function, namely n_I and those used in the perturbative calculation of χ^T, amounts to changing the right-hand side of the bounds in Eq. (14) by a corresponding fractional amount. We impose the bounds Eq. (14) as constraints, however,

Table 2
Numerical inputs for this analysis. The meson masses are taken from Ref. [19]. See the text at the end of Section 2 for a discussion regarding the choices of the other parameters.

m_B	5.279 GeV
m_{D^*}	2.010 GeV
α_s	0.22
m_b	4.9 GeV
m_c/m_b	0.33
n_I	2.6

the resultant best-fit values are orders of magnitude from saturating the unitarity bounds, and these constraints effectively play no role. Because of this, changing the values of, say, α_s, m_b, etc., will change the best-fit values for the free parameters in the theory, but will not change the extracted value $|V_{cb}|$, to a very good approximation. The remaining inputs are B_c^* pole masses. They affect both the effectiveness of the unitarity bound and the shape of form factors. For example, for $F \equiv g, f, \mathcal{F}_1$, the unitarity bound implies that the truncation error for finite N in Eq. (12) is bounded over the physical region $0 \leq z \leq z_{\max} \approx 0.056$ by

$$\Delta F(z) < \frac{1}{|P_F(z_{\max})\phi_F(z_{\max})|} \frac{z_{\max}^{N+1}}{\sqrt{1-z_{\max}^2}}.$$

Since $P_F(z)$ varies by less than 15% when the input pole masses are individually varied by 1%, the ability to control the expansion is unaffected by uncertainties in pole masses. Also, the precise location of the pole masses affects the shape of the form factors in the physical region, but the parameters of the fit accommodate small changes in pole masses easily. For example, consider the $N=0$ parametrization of the form factor f. Varying the value of the lowest pole mass by 1%, the change in f is compensated by a multiplicative factor of $1.07 - 0.17z$ to better than 1% over the whole physical region. For these reasons, we expect that while the values of the fit parameters may depend somewhat sensitively on the inputs, the extracted value of $|V_{cb}|$ does not. We illustrate this by using the same input values as Ref. [12], as listed in Tables 1 and 2, which are about 20 years old, but find that our extracted value of $|V_{cb}|$ is nearly identical to that of a recent analysis by the authors of Ref. [14], which uses the current state-of-the-art theoretical inputs. Extracting the value of $|V_{cb}|$ using the BGL parameterization is robust to changes in the numerical inputs. We discuss this more in Section 3.

3. Fitting results and extracting $|V_{cb}|$

We fit to the unfolded data provided by Belle in Ref. [7], using the numerical inputs found in Tables 1 and 2. We choose to truncate the series in Eq. (12) at $\mathcal{O}(z^2)$ for f and g, and at order $\mathcal{O}(z^3)$ for $\mathcal{F}_1$, as well as define the value c_0 in terms of b_0, as in Eq. (15). This results in 6 free parameters, defined as $\tilde{a}_i \equiv |V_{cb}|\eta_{ew}a_i$, $\tilde{b}_i \equiv |V_{cb}|\eta_{ew}b_i$, where $i=0,1$, and $\tilde{c}_i \equiv |V_{cb}|\eta_{ew}c_i$, where $i=1,2$. We choose to truncate to these orders because it results in a good fit to the data. A Markov Chain Monte Carlo (MCMC) is used to estimate the shape of the likelihood and the shape around its extremum. The result of the fit is shown in Fig. 1. We find $\chi^2_{\min}/\text{dof} \simeq 27.7/34$. Comparing to the Belle data, our best fit value for the BGL parameterization can be found in Fig. 2, compared to the fit performed by Belle using the CLN parameterization.

One requires input from the lattice to extract the value of $|V_{cb}|$. The $\bar{B} \to D^*\ell\bar{\nu}$ rate is given in terms of a function $\mathcal{F}(w)$ defined as [20]:

$$\frac{d\Gamma}{dw} = \frac{\eta_{ew}^2 G_F^2 |V_{cb}|^2}{48\pi^3}(m_B - m_{D^*})^2 m_{D^*}^3 \sqrt{w^2-1}(w+1)^2 \times \left[1 + \left(\frac{4w}{w+1}\right)\left(\frac{m_B^2 + m_{D^*}^2 - 2wm_Bm_{D^*}}{(m_B - m_{D^*})^2}\right)\right]|\mathcal{F}(w)|^2. \tag{22}$$

One can estimate on the lattice the value of $\mathcal{F}(1)$. Given the definition of the form factors in Eqs. (7), (9), and (12), the relationship between $|V_{cb}|\eta_{ew}\mathcal{F}(1)$ and our fitting parameters is:

$$|V_{cb}|\eta_{ew}\mathcal{F}(1) = \frac{1}{2\sqrt{m_Bm_{D^*}}}\left(\frac{|\tilde{b}_0|}{P_f(0)\phi_f(0)}\right). \tag{23}$$

Given the statistical ensemble of $\tilde{b}_0$ provided by the MCMC, we can estimate the likelihood, and thus the variance in the χ^2 about its minimum, i.e., $\Delta\chi^2$, for the value of $|V_{cb}|\eta_{ew}\mathcal{F}(1)$ extracted from the Belle data, as shown in Fig. 3. This results in:

$$|V_{cb}|\eta_{ew}\mathcal{F}(1) = (38.2^{+1.7}_{-1.6}) \times 10^{-3}. \tag{24}$$

Using the $\eta_{ew} = 1.0066$ and the FNAL/MILC value of $\mathcal{F}(1) = 0.906 \pm 0.013$ from Ref. [20], the final result for $|V_{cb}|$ is

$$|V_{cb}| = (41.9^{+2.0}_{-1.9}) \times 10^{-3}. \tag{25}$$

This result is nearly identical to that of the recent analysis by the authors of Ref. [14], and systematically larger than the value extracted by the Belle experiment using the CLN parameterization [7]. Our result is comparable to a recent analysis by the authors of Ref. [15], which reevaluate the CLN parameterization, using relations from heavy quark symmetry, but not ignoring important sources of theoretical uncertainty when fitting to the same data.

As explained above, extracting the value of $|V_{cb}|$ using the BGL parameterization is robust to changes in the numerical inputs. We explore further the effect of Blaschke factors on the extraction of $|V_{cb}|$ by squaring the Blaschke factors for f and $\mathcal{F}_1$. One might, a priori, expect to produce radical variations in the z dependence of the form factors. Surprisingly, while we fit very different values for the free parameters, the extracted value of $|V_{cb}|\eta_{ew}\mathcal{F}(1)$ is nearly unchanged (after updating the expression in Eq. (23) to account for the different definition of the form factors). This gives further evidence that the BGL parameterization is robust when extrapolating to zero recoil. On the other hand, the precise values of the theoretical inputs may be essential when extracting the functional dependence of the form factors over the entire kinematic range.

4. Summary and conclusions

The long-standing tension between the exclusive and inclusive determinations of $|V_{cb}|$ in semileptonic B decays could have been exacerbated by underestimated uncertainties in the exclusive analyses when using the CLN parameterization [6,15,14]. As it stands, the data has become too precise to use the original CLN proposal in the case of $\bar{B} \to D\ell\bar{\nu}$, as discussed in Ref. [6], and one might expect the same could be true for $\bar{B} \to D^*\ell\bar{\nu}$.

The Belle experiment recently released an unfolded dataset of $\bar{B}^0 \to D^{*+}\ell\bar{\nu}$, which resulted in the following value of $|V_{cb}|$ using the CLN parameterization [7]:

$$|V_{cb}| = (37.4 \pm 1.3) \times 10^{-3} \quad \text{(CLN)}. \tag{26}$$

Using this data, we fit to the BGL parameterization, which does not rely on any assumptions regarding heavy quark symmetry, and obtain:

$$|V_{cb}| = (41.9^{+2.0}_{-1.9}) \times 10^{-3} \quad \text{(BGL)}. \tag{27}$$

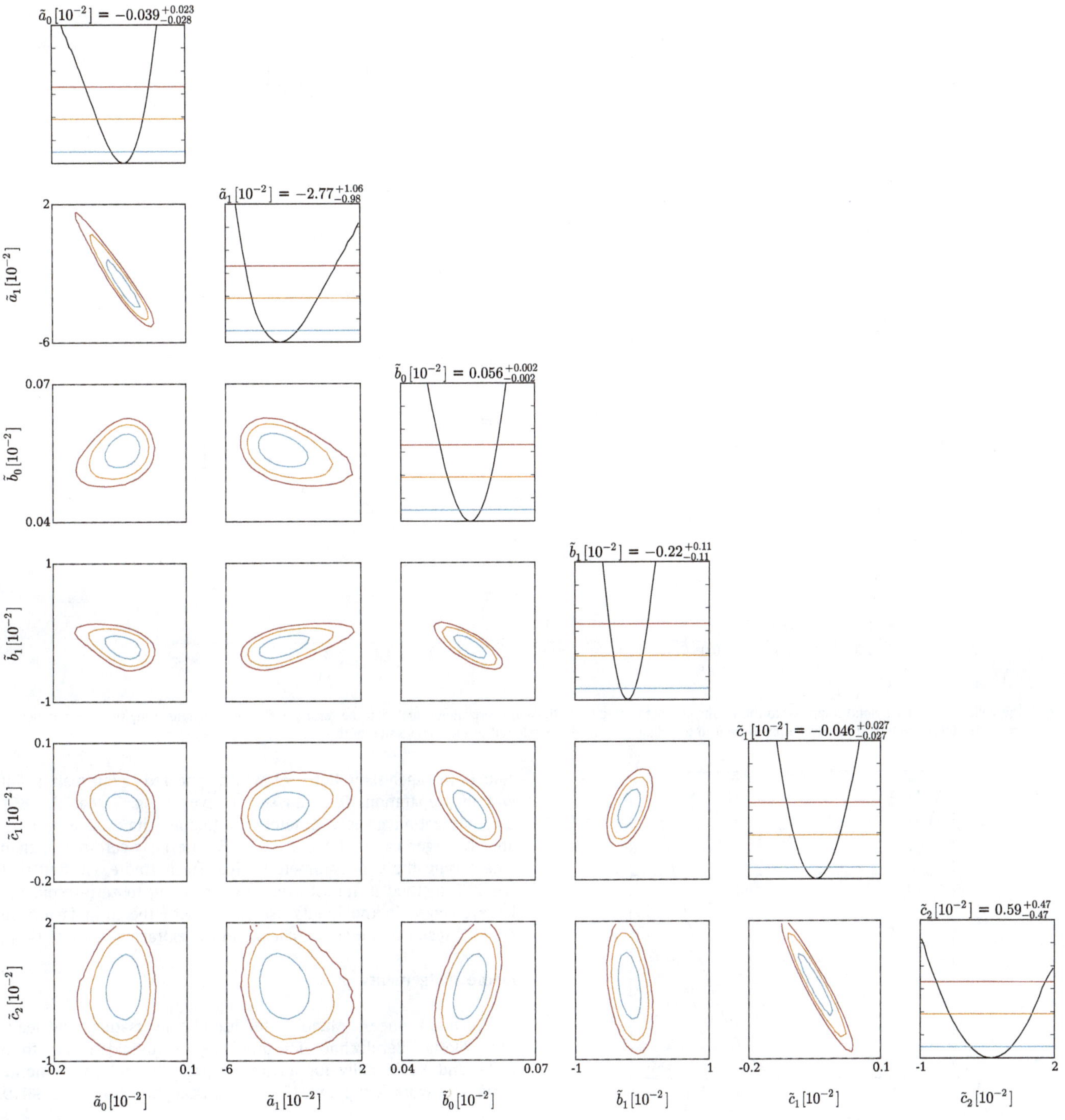

Fig. 1. The result of the fit to the form factors defined in Eq. (12), where $\tilde{a}_i$, $\tilde{b}_i$, and $\tilde{c}_i$ are defined as $\eta_{ew}|V_{cb}|a_i$, $\eta_{ew}|V_{cb}|b_i$, and $\eta_{ew}|V_{cb}|c_i$, respectively. The blue, yellow, and red lines correspond to the 68.27%, 95%, and 99% CL, respectively. Any jaggedness in the contours is from the finite size of the MCMC ensemble. (For interpretation of the references to color in this figure, the reader is referred to the web version of this article.)

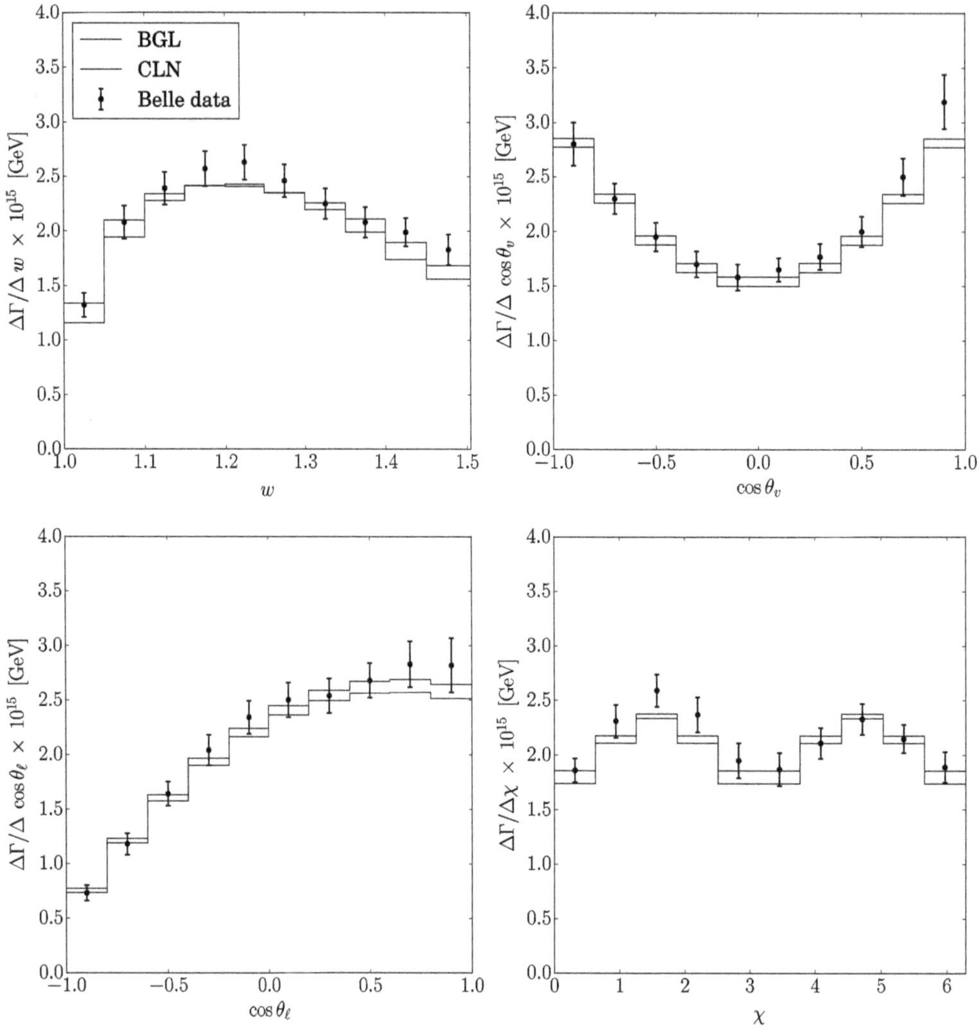

Fig. 2. The Belle data (black points) compared to the results of our fit using the BGL parameterization (blue), and the results of the Belle analysis using the CLN parameterization (red). (For interpretation of the references to color in this figure, the reader is referred to the web version of this article.)

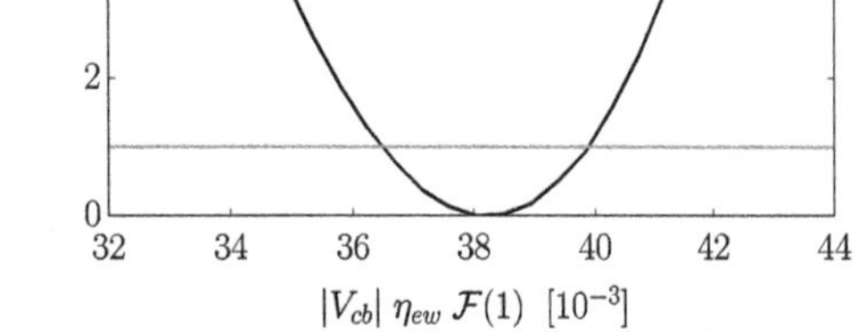

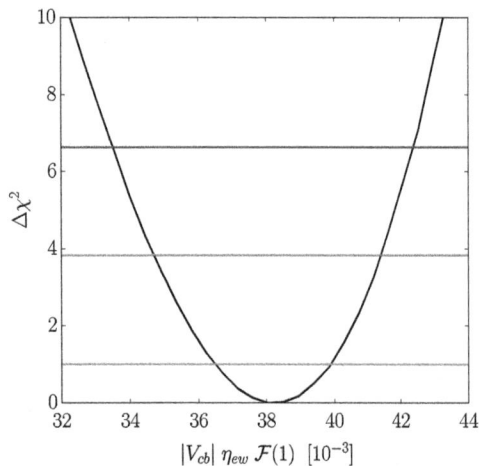

Fig. 3. The extracted value of $|V_{cb}|\eta_{ew}\mathcal{F}(1)$ from the Belle data. The blue, yellow, and red lines correspond to the 68.27%, 95%, and 99% CL, respectively. (For interpretation of the references to color in this figure, the reader is referred to the web version of this article.)

This result is consistent with the value of $|V_{cb}|$ measured using inclusive semileptonic B decays. We are able to corroborate the results of the analysis in Ref. [14], which appeared as our manuscript was in preparation. One can see by eye in Fig. 2 that the BGL parametrization gives a different fit to the data, and a systematically larger value of the $\bar{B} \to D^*\ell\bar{\nu}$ rate near zero-recoil, than when using the CLN parameterization. Furthermore, we find that the BGL method is robust when extrapolating to zero-recoil, i.e., large changes in the inputs associated with the form factors in Eq. (12) result in almost no change in the extracted value of $|V_{cb}|$.

Acknowledgements

We thank Aneesh Manohar for useful conversations and feedback, Florian Bernlochner for providing the unfolded data from Belle, and Kevin Kelly for advice on graphics and using Monte Carlo. This work is supported in part by DOE grant #DE-SC0009919.

References

[1] G. Ricciardi, Mod. Phys. Lett. A 32 (2017) 1730005, arXiv:1610.04387.
[2] Y. Amhis, et al., arXiv:1612.07233, 2016.
[3] C.W. Bauer, Z. Ligeti, M. Luke, A.V. Manohar, M. Trott, Phys. Rev. D 70 (2004) 094017, arXiv:hep-ph/0408002.
[4] I. Caprini, M. Neubert, Phys. Lett. B 380 (1996) 376, arXiv:hep-ph/9603414.
[5] I. Caprini, L. Lellouch, M. Neubert, Nucl. Phys. B 530 (1998) 153, arXiv:hep-ph/9712417.
[6] D. Bigi, P. Gambino, Phys. Rev. D 94 (2016) 094008, arXiv:1606.08030.

[7] A. Abdesselam, et al., Belle, arXiv:1702.01521, 2017.
[8] C.G. Boyd, B. Grinstein, R.F. Lebed, Phys. Rev. Lett. 74 (1995) 4603, arXiv:hep-ph/9412324.
[9] C.G. Boyd, B. Grinstein, R.F. Lebed, Phys. Lett. B 353 (1995) 306, arXiv:hep-ph/9504235.
[10] C.G. Boyd, B. Grinstein, R.F. Lebed, Nucl. Phys. B 461 (1996) 493, arXiv:hep-ph/9508211.
[11] C.G. Boyd, R.F. Lebed, Nucl. Phys. B 485 (1997) 275, arXiv:hep-ph/9512363.
[12] C.G. Boyd, B. Grinstein, R.F. Lebed, Phys. Rev. D 56 (1997) 6895, arXiv:hep-ph/9705252.
[13] B. Grinstein, R.F. Lebed, Phys. Rev. D 92 (2015) 116001, arXiv:1509.04847.
[14] D. Bigi, P. Gambino, S. Schacht, arXiv:1703.06124, 2017.
[15] F.U. Bernlochner, Z. Ligeti, M. Papucci, D.J. Robinson, arXiv:1703.05330, 2017.
[16] J.A. Bailey, et al., MILC, Phys. Rev. D 92 (2015) 034506, arXiv:1503.07237.
[17] A. Sirlin, Nucl. Phys. B 196 (1982) 83.
[18] E.J. Eichten, C. Quigg, Phys. Rev. D 49 (1994) 5845, arXiv:hep-ph/9402210.
[19] C. Patrignani, et al., Particle Data Group, Chin. Phys. C 40 (2016) 100001.
[20] J.A. Bailey, et al., Fermilab Lattice, MILC, Phys. Rev. D 89 (2014) 114504, arXiv:1403.0635.

15

Multi-parton interactions and rapidity gap survival probability in jet–gap–jet processes

Izabela Babiarz [a], Rafał Staszewski [b], Antoni Szczurek [b,*,1]

[a] Faculty of Mathematics and Natural Sciences, University of Rzeszów, ul. Pigonia 1, 35-310 Rzeszów, Poland
[b] Institute of Nuclear Physics, Polish Academy of Sciences, Radzikowskiego 152, 31-342 Kraków, Poland

ARTICLE INFO

Editor: A. Ringwald

ABSTRACT

We discuss an application of dynamical multi-parton interaction model, tuned to measurements of underlying event topology, for a description of destroying rapidity gaps in the jet–gap–jet processes at the LHC. We concentrate on the dynamical origin of the mechanism of destroying the rapidity gap. The cross section for jet–gap–jet is calculated within LL BFKL approximation. We discuss the topology of final states without and with the MPI effects. We discuss some examples of selected kinematical situations (fixed jet rapidities and transverse momenta) as distributions averaged over the dynamics of the jet–gap–jet scattering. The colour-singlet ladder exchange amplitude for the partonic subprocess is implemented into the Pythia 8 generator, which is then used for hadronisation and for the simulation of the MPI effects. Several differential distributions are shown and discussed. We present the ratio of cross section calculated with and without MPI effects as a function of rapidity gap in between the jets.

1. Introduction

Diffraction, *i.e.* strong interaction involving the exchange of the vacuum quantum numbers (the pomeron)[2] is a very broad field of research. In the recent years the understanding of diffraction and its connection to the microscopic picture of strong interactions has been greatly improved thanks to the studies of hard diffraction, *i.e.* diffraction involving a hard scale, like high p_T jets.

The jet–gap–jet process is an example of the diffractive jet production, in which the pomeron is exchanged between the produced jets. Contrary to other types of diffractive jet processes (*e.g.* single diffractive jets), the absolute value of the four momentum carried by the pomeron is large. This provides a unique possibility to apply perturbative calculation methods to fully describe the diffractive exchange. The jet–gap–jet processes were measured at Tevatron [2] and recently at the LHC [3].

One important ingredient in the calculations of the hard diffractive cross section is the rapidity gap survival probability. In many calculations, see *e.g.* [4,5], the gap survival factor is assumed to be constant with respect to the kinematics of the event, and depending only on the centre-of-mass energy. Recently, more detailed analyses were performed, in which kinematic dependence was taken into account. The calculations were done for exclusive processes (see *e.g.* [6] and references therein) and single diffraction [7]. Recently the gap survival in single diffractive processes was calculated dynamically by including MPIs [8].

In the present paper we study this topic for a different class of processes – the jet–gap–jet production. What is/are the process(es) responsible for destroying rapidity gap obtained in the pQCD calculation of colour-singlet exchange? In the present study we explore the role of multi-parton interactions that are the main mechanism responsible for understanding of underlying event topology, see *e.g.* [9–12]. In particular, we wish to address the problem how much the dynamical calculation changes the somewhat academic BFKL result.

2. Particle production in jet events

The difference in the underlying mechanism of the non-diffractive jet and jet–gap–jet production, the details of which are discussed in Section 4 and 5, affect not only the cross section and angular distribution of jets, but also the distributions of particles produced in the events. This difference originate from a different flow of the colour charges in the events, which affect the hadron formation process. These effects can be studied using Monte Carlo

* Corresponding author.
E-mail addresses: babiarz.i.m@gmail.com (I. Babiarz), rafal.staszewski@ifj.edu.pl (R. Staszewski), antoni.szczurek@ifj.edu.pl (A. Szczurek).
1 Also at University of Rzeszów, PL-35-959 Rzeszów, Poland.

2 The spin structure of the pomeron is a matter of current discussions [1].

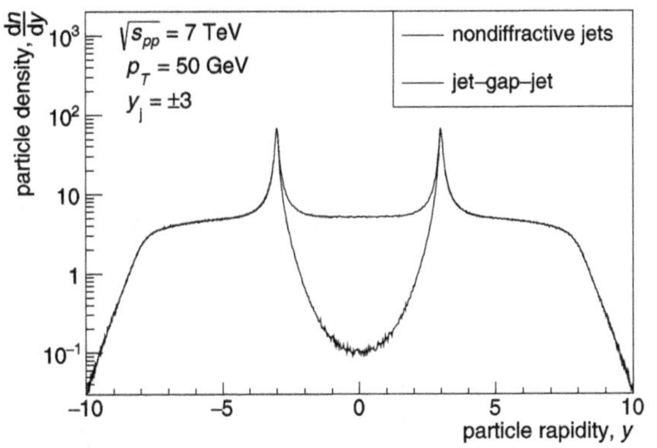

Fig. 1. Rapidity distributions of particles produced in non-diffractive jet (black) and jet–gap–jet (red, curve with a dip at y = 0) events, for the selected kinematical situation. (For interpretation of the references to colour in this figure legend, the reader is referred to the web version of this article.)

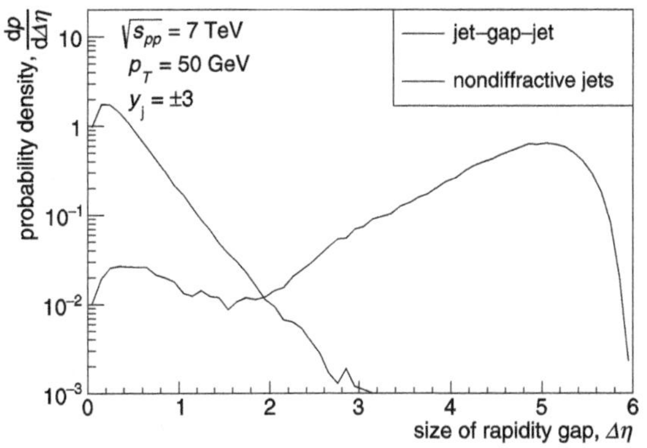

Fig. 2. Rapidity gap distributions for non-diffractive jet (red, with maximum at $\Delta\eta \sim 0$) and jet–gap–jet events (black, with maximum at $\Delta\eta \sim 5$), for our selected kinematical configuration. No MPI effects are included here. (For interpretation of the references to colour in this figure legend, the reader is referred to the web version of this article.)

event generators that simulate the hadronization process. The following results were obtained with Pythia 8.

First we wish to illustrate the situation for a selected kinematical situation. Fig. 1 presents the rapidity distribution of particles produced in a pp interaction at $\sqrt{s} = 7$ TeV for events obtained with the $gg \to gg$ hard subprocess, where the gluons are scattered with fixed transverse momentum $p_T = 50$ GeV at rapidities of $y = \pm 3$. The results were obtained with Pythia using the Les Houches Event File [13] interface. Two cases are studied: non-diffractive jets, when colour charges are exchanged between the scattered gluons, and jet–gap–jet production, in which interacting gluons keep their colours. One can clearly see that the rapidity density of produced particles is highest around rapidities of scattered gluons, which reflects the jet structure of the events. In this region one does not see a large difference between the two cases (colour structures). On the other hand, in the region between the jets the difference is quite dramatic. When no colour charge is transferred between the gluons, the density of produced particles is reduced by two orders of magnitude. The particles produced at rapidities outside the jet system originate from the hadronization of the proton remnants and from the fact that there is a colour transfer between the remnants and the scattered gluons.

The suppression of particles production in the region between the jets will lead to large rapidity regions devoid of particles – rapidity gaps. However, the actual values of particles rapidities, and thus the size of the rapidity gap, is to some extend random and will fluctuate from event to event. This is true both for non-diffractive jets as well as for jet–gap–jet events. On average, for the former ones one expects rather small gaps, and much bigger gaps for the latter ones. This is illustrated in Fig. 2, where the distributions of gap size are shown for jets with $p_T = 50$ GeV and $y = \pm 3$. One can see that even though the distributions are well separated, their tails are rather long and they have a non-negligible overlap. This shows that, even neglecting other effects discussed later, these two processes cannot be fully separated experimentally (at least based solely on the rapidity gap size).

3. Multiple parton interactions

An additional complication to the picture presented in the previous section comes from the fact that hadrons are complicated objects that consist of many partons. In a single hadron–hadron collision more than one parton–parton interaction can take place. This phenomenon is known as the multiple parton interactions (MPI) or the underlying event (UE) activity and it was extensively studied at Tevatron and the LHC [9,11,12].

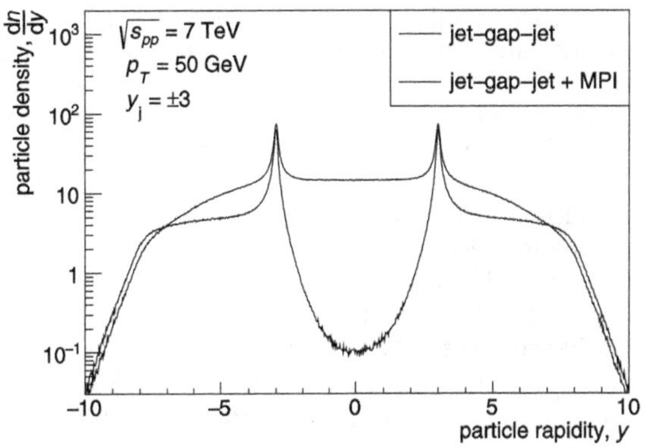

Fig. 3. Rapidity distributions of produced particles for jet–gap–jet events without (black) and with (red) multi parton interactions for our selected kinematical configuration. (For interpretation of the references to colour in this figure legend, the reader is referred to the web version of this article.)

The MPIs are modeled in Pythia with the help of minijets calculated in collinear factorization approach with a special treatment at low transverse momenta of minijets by multiplying standard cross section by a somewhat arbitrary suppression factor [14]

$$F_{sup}(p_t) = \frac{p_t^4}{(p_{t0}^2 + p_t^2)^2}\theta(p_t - p_{t,cut}) \,. \tag{3.1}$$

Typically, MPI effects are responsible for increasing the particle production in the events, but for diffraction they have particularly important consequences. If the $gg \to gg$ or another partonic subprocess with a colour-singlet exchange is accompanied by another independent parton–parton interaction, additional particles can be produced in the region where a gap was expected. This is presented in Fig. 3, where rapidity distributions of particles produces in jet–gap–jet events are shown for the MPI effects in pythia turned off (black) and turned on (red). It is crucial that even though the particle density in the region between the jets is greatly increased, it is possible to observe events with very large gap sizes. This is contrary to the case of non-diffractive jets, and it originates from the fact that it is possible to have events with no additional parton–parton interaction, in which a large gap can survive. The distributions of the gap size for events with and without MPI effects are presented in Fig. 4.

For events with MPI effects the gap distribution consists of two parts. The distributions for low gap sizes is steeply falling, similarly to the non-diffractive events. This comes from the fact that addi-

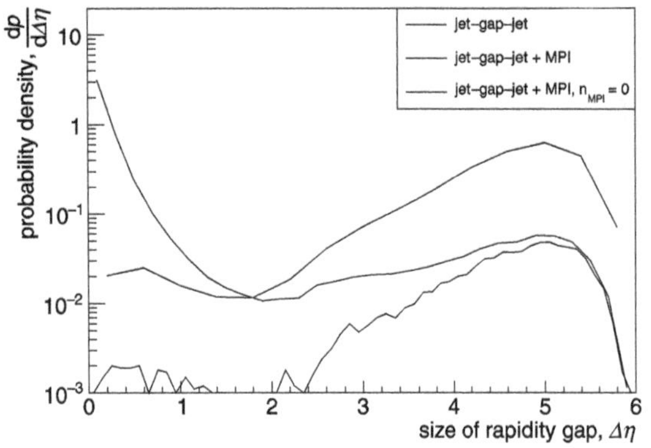

Fig. 4. Rapidity gap distributions for jet–gap–jet events without MPI effects (black, the highest curve at $\Delta\eta = 3$–6) and with MPI effects (red) and with MPI effects included, but for events in which no MPIs occurred *i.e.* $n_{MPI} = 0$ (blue, the lowest curve). (For interpretation of the references to colour in this figure legend, the reader is referred to the web version of this article.)

tional interactions produce particles between the jets. The large-gap part originates from events where no additional interactions occurred. The distribution is similar to the one obtained without MPI effects, but reduced by a factor of about one order of magnitude. Fig. 4 shows also the gap size distribution plotted with MPI effects included, but only for events that do not contain any additional interactions. For very large gap sizes this distribution agrees with the one for all events. However, for medium gap sizes it is not. On the other hand, the shape of this distribution is the same as for the distribution without MPI effects, but scaled down. The difference between the red and the blue curves comes from events in which the additional interactions produce very few particles. In these events the initial rapidity gap is reduced, but not completely filled.

Since experimentally the jet–gap–jet events are distinguished by the presence of large rapidity gaps, the MPI effects lead to a reduction of the measured cross sections with respect to the cross section of the actual colour-singlet exchange. This phenomenon is often called as absorptive corrections and the corresponding probability – gap survival probability.

In order to estimate its magnitude one can perform event simulation with Pythia and count in which fraction of events no additional parton interactions are present. For the jet–gap–jet processes it is possible to use the existing Monte Carlo generators that take into account modeling of multi-parton interactions. If such a generation correctly describes the MPI effects for standard jet production, it should also provide a correct description of the jet–gap–jet process.[3]

The MPI generation in Pythia is based on phenomenological models that contain several arbitrary parameters. Usually the values of these parameters are tuned in order to give the best description of the Tevatron and the LHC data for observables related to MPI. Therefore, it can be expected that the results presented in the present paper should also be close to reality. A big advantage of this approach is that it allows calculations of the cross section or gap survival probability as a function of the kinematical variables of the hard (sub)process.

Fig. 5 presents the dependence of a few kinematical variables of the gap survival probability, defined as a fraction of events that do not contain any additional parton–parton interactions apart from the hard one. Since Pythia assumes the initial partons to be collinear with the protons, the kinematics of an event can be described by four parameters. A sensible choice is: the centre-of-mass energy $\sqrt{s}$, the invariant mass of the hard subprocess M_{gg}, the difference of rapidities of the scattered gluons Δy and the rapidity of the gluon–gluon system y_{gg}.

The dependence of the gap survival probability as a function of the centre-of-mass energy is presented in Fig. 5a. Its value drops from about 15% at the Tevatron energy (2 TeV) to about 5% at the LHC nominal energy (14 TeV).

Fig. 5b shows the dependence on the M_{gg} for $\sqrt{s} = 7$ TeV. Here, the gap survival increases from about 7% for masses close to zero up to 30% for masses close to 6 TeV. The observed behaviour can be qualitatively explained by the energy conservation: when a bigger part of the proton energy is carried by the parton participating in the hard subprocess, less energy is available for additional interactions and they become less likely.

Fig. 5c presents the dependence of the survival probability on Δy_{gg}.

Fig. 5d presents the dependence on the rapidity of the gluon–gluon system. The dependence is rather flat for central values of rapidities and rapidly grows for $|y_{gg}| > 3$. For such strongly boosted events one of the incoming partons carries a large energy and the other one very small one. In this situation the possibility of additional interactions is also reduced, because additional partons from both protons are needed for extra MPI to occur.

In summary, there is a strong dependence of the gap survival probability on kinematical variables. However, not all kinematic configurations are equally probable. For example, the partonic distributions are larger at small values of Bjorken x, which favours small masses of the system. In addition, the dynamics of the colour-singlet exchange will also play some role.

4. Hard colour-singlet exchange

The calculation of the jet–gap–jet process is based on the QCD collinear factorisation, where the cross section for the hard subprocess, $\hat{\sigma}$, is convoluted with the appropriate parton densities. An example of the diagram of the full process is presented in Fig. 6. The cross section can be written in the simple form

$$\frac{d\sigma}{dp_T} = \int g_{eff}(x_1, \mu_F^2) g_{eff}(x_2, \mu_F^2) \frac{d\hat{\sigma}}{dp_T} dx_1 dx_2 ,$$

where

$$g_{eff}(x_k, \mu_F^2) = g(x_k, \mu_F^2) + \frac{16}{81} \sum_f (q_f(x_k, \mu_F^2) + \bar{q}_f(x_k, \mu_F^2))$$

and $k = 1, 2$. The jet–gap–jet process differs from the typical jet production in the colour structure of the subprocess. Here, the colour-singlet ladder is exchanged between partons.

To a first approximation, the colour-singlet exchange can be described in perturbative QCD as an exchange of a pair of gluons that carry opposite colour charges [15]. A better approach is to describe it as a gluon ladder, which can be performed within the BFKL framework. The first calculation of this type were done in [16]. This was followed *e.g.* by further studies, see *e.g.* [17–19].

In the present paper we shall use for illustration LL BFKL formalism used previously *e.g.* in [18,19].

$$A(\Delta\eta, p_T^2) = \frac{16 N_C \pi \alpha_s^2}{C_F p_T^2} \sum_{p=-\infty}^{\infty} \int \frac{d\gamma}{2i\pi} \times \frac{[p^2 - (\gamma - 1/2)^2] \exp(\bar{\alpha} \chi_{eff}[2p, \gamma, \bar{\alpha}] \Delta\eta)}{[(\gamma - 1/2)^2 - (p - 1/2)^2][(\gamma - 1/2)^2 - (p + 1/2)^2]} , \tag{4.1}$$

[3] This statement is not necessarily true for other diffractive jet processes, where the production mechanism is somewhat different and not so well understood.

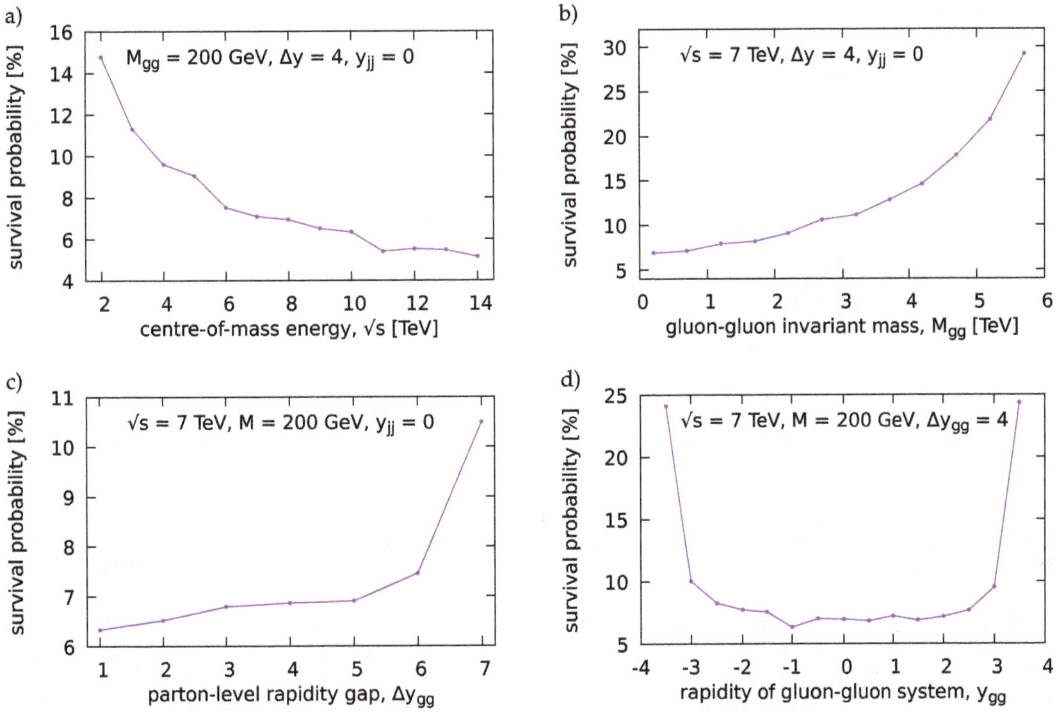

Fig. 5. Kinematic dependence of gap survival probability as a function of: a) centre-of-mass energy, b) invariant mass in the hard subprocess, c) rapidity distance between the scattered gluons, d) rapidity of the digluon system.

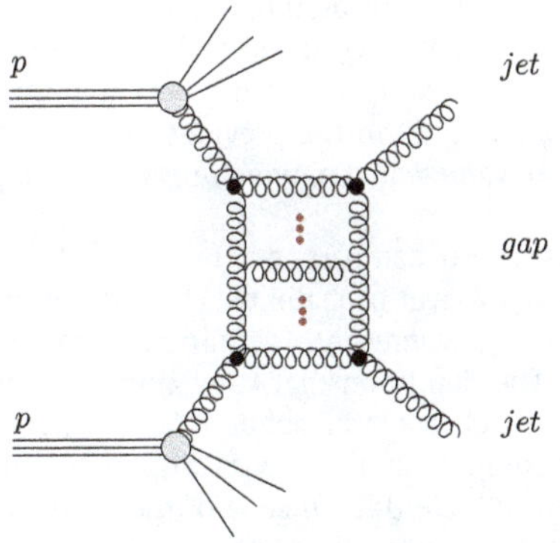

Fig. 6. A schematic QCD diagram of colour-singlet exchange for the jet–gap–jet process in a *pp* collision. Only gg initiated process is shown explicitly.

where p_t is jet transverse momentum and $\Delta\eta$ is rapidity distance between the partonic jets. The integral above runs along the imaginary axis from $\frac{1}{2} - i\infty$ to $\frac{1}{2} + i\infty$ and with only even conformal spins p in the sum. The χ_{LL} kernel reads

$$\chi_{LL} = 2\psi(1) - \psi\left(1 - \gamma + \frac{|p|}{2}\right)\psi\left(\gamma + \frac{|p|}{2}\right), \tag{4.2}$$

where $\psi(\gamma) = d\log\Gamma(\gamma)/d\gamma$. In the LL BFKL approach a constant value of α_s is used [18]. In Eq. (4.1) $\bar{\alpha} = \alpha_s N_c/\pi$.

In Fig. 7 we show the LL BFKL amplitude as a function of rapidity distance between jets for a few values of jet transverse momenta. The increase at large rapidity distance is a typical BFKL increase while the increase at low rapidity distances is "caused" by the presence of higher conformal spins. The calculation sketched above does not include gap survival factor which is a very important ingredient as discussed in the next section.

The leading-logarithm approximation of BFKL may be not sufficient to provide a reasonable description of the absolute cross section normalisation and the distributions shapes, but is sufficient

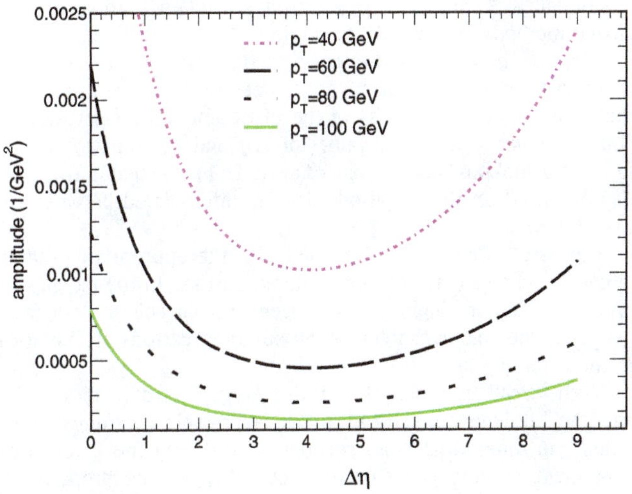

Fig. 7. Subprocess amplitude in the LL BFKL approach as a function of rapidity distance between jets for selected transverse momenta of the jets. In this calculation $\alpha_S = 0.17$.

for presentation of the MPI effects in suppressing rapidity gap discussed in the present paper, as will be explained below.

5. Realistic distributions of rapidity gap size

Here we wish to discuss our results for a broad range of jet rapidities when imposing only a minimal lower cut on jet transverse momenta. We fix transverse momenta of jets to be in the interval 40 GeV $< p_{1t}, p_{2t} <$ 200 GeV and impose no cuts on jet rapidities at all. The calculations presented in this section have been performed using the framework of the PYTHIA 8 Monte Carlo event generator. The BFKL leading-logarithm amplitude for the hard subprocess has been calculated following [18,19] and implemented

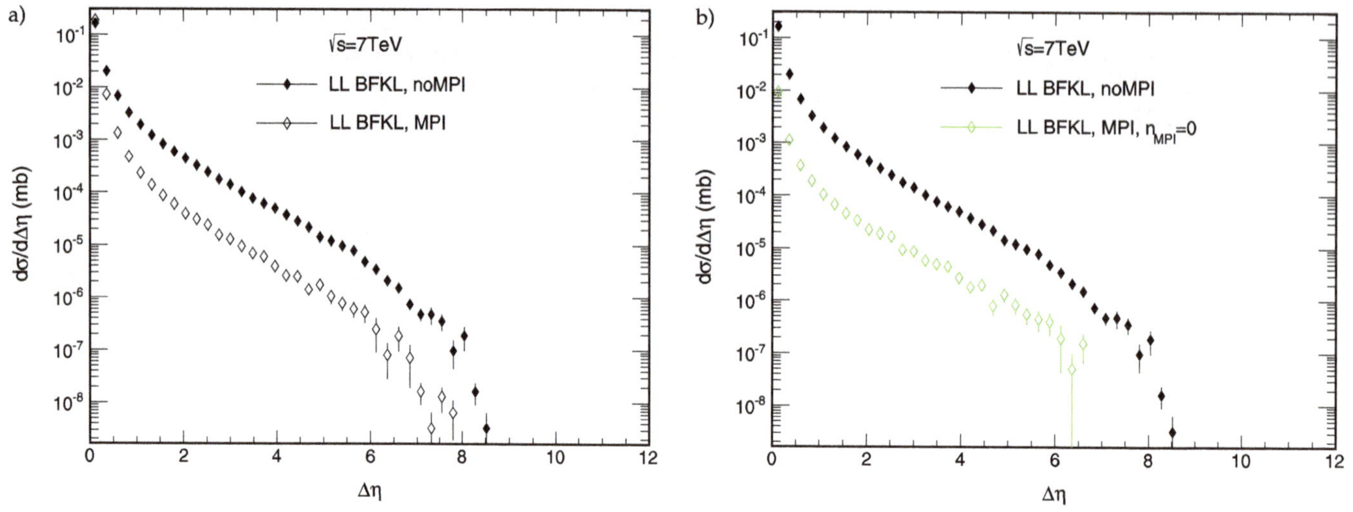

Fig. 8. Rapidity gap distributions for jet–gap–jet events generated with and without MPI effects as a function of rapidity gap size. All partons combinations are included here. The left panel shows the rapidity gap distribution when MPI effects are included while the right panel shows result with extra requirement of $n_{MPI} = 0$.

into PYTHIA as new $gg \to gg$, $qg \to qg$, $qq \to qq$, $q\bar{q} \to q\bar{q}$ subprocesses.

The implementation of jet–gap–jet processes into Pythia was done in analogy to the already existing parton–parton scattering processes. Technically, a dedicated class `Sigma2gg2jgj` has been created. It derives from PYTHIA's base class for hard $2 \to 2$ processes – `Sigma2Process`. In order to provide the appropriate functionality, `Sigma2gg2jgj` provides implementations of two virtual methods: `sigmaKin()` and `setIdColAcol()`.

`sigmaKin()` returns the cross section value for the jet–gap–jet process. In order to optimise the speed of the event generation, the time-consuming process of the BFKL amplitude calculation is done in advance for fixed values of $\Delta\eta$ and p_T and stored in a form of a look-up table. Then, `sigmaKin()` calculates the cross section based on the amplitude linearly interpolated between the stored values.

The `setIdColAcol()` method sets the appropriate transfer of colour charges between the scattered partons. In the jet–gap–jet process, a colour singlet is exchanged and no colour flow takes place, *i.e.* the colour charges of the outgoing partons are identical to the incoming ones.

With a realistic description of the jet–gap–jet dynamics (BFKL) and the MPI modeling by PYTHIA it is possible to study the rapidity gap differential cross sections. In this way the kinematics-dependent rapidity gap survival probability will be properly averaged over different kinematic configurations. In addition, the effects of rapidity gap reduction (see Fig. 4) is also taken into account.

Fig. 8 presents such distributions for events with and without MPI effects. Both distributions are rapidly falling, which originates predominantly from the shape of the parton densities. In the large-gap region the distribution with the MPI effects is reduced with respect to the one without MPI. However, for small $\Delta\eta$ the situation is opposite. This comes from the fact that the integrated cross section is in both cases the same, since it is given only by the hard partonic mechanism. The occurrence of MPI effects does not change the normalisation of the distributions, but shifts events to smaller rapidity gap sizes.

Fig. 9 presents the ratio of differential cross section for the gap distributions with and without MPI effects. This ratio can be treated as an effective gap survival factor, including all effects previously discussed and averaged over all configurations for the dynamics of the BFKL colour-singlet exchange. The occurrence of additional MPIs destroys large rapidity gaps and simultaneously increases of number of events with small rapidity gaps (see the left panels). Therefore the gap survival factor, defined in this way, depends on $\Delta\eta$ (gap size), see the left panels of Fig. 9. It is worth considering a different definition of the survival factor, namely the ratio between the number of events in which no additional events occur. This definition is similar to the one typically used in the literature, where it is assumed that any additional interaction destroys the rapidity gap. This assumption leads to a flat dependence of the gap survival with $\Delta\eta$, as seen in 9 (right panels). This is understandable, since here the events with additional interactions are not considered at all. In the previous case they were included, but with smaller values of $\Delta\eta$, which resulted in high ratio values in that region.

It is interesting to compare not only the shape, but also the value of the gap survival probability. This can be best done in the region of large $\Delta\eta$, where both definitions give an approximately flat behaviour. The definition that takes into account all events results in gap survival factor of about 10%. On the other hand, the definition that counts only the events with no additional MPIs give a value close to 6%. The difference is of the order of 40%, which is rather significant. It shows that latter definition may be too simplistic to provide a precise description of jet–gap–jet processes.

In addition, we also compare the situation where the jets can be created in the scattering of quarks and gluons, to the situation when only $gg \to gg$ process is included. The resulting gap survival probability in these two cases is the same within our present accuracy.

In the bottom panels of Fig. 9 we show for comparison also result obtained for the two-gluon exchange model with the regularisation parameter $m_g = 0.75$ GeV (see [15] for the cross section formula and [20] for the value of the nonperturbative parameter). The gap survival factor for the case of $n_{MPI} = 0$ is (at $\sqrt{s} = 7$ TeV) about 6–7 %, independent of modeling colour-singlet exchange.

In Fig. 10 we show similar ratios but now as a function of jet transverse momentum for a few selected rapidity gap intervals. No obvious dependence on the jet transverse momentum can be observed in the left panel where we show the ratio of the distribution with MPI effect included to the corresponding distribution without MPI effects in contrast to the dependence on rapidity gap size observed in the previous figure. For smallest gaps ($0 < \Delta\eta < 1$) the ratio is with a good precision equal to 1. This seems accidental and is connected with bin size. This could be better understood by inspection of the left panels of Fig. 9 at $\Delta\eta \sim 0$. In the right panel we show the ratio with extra academic condition

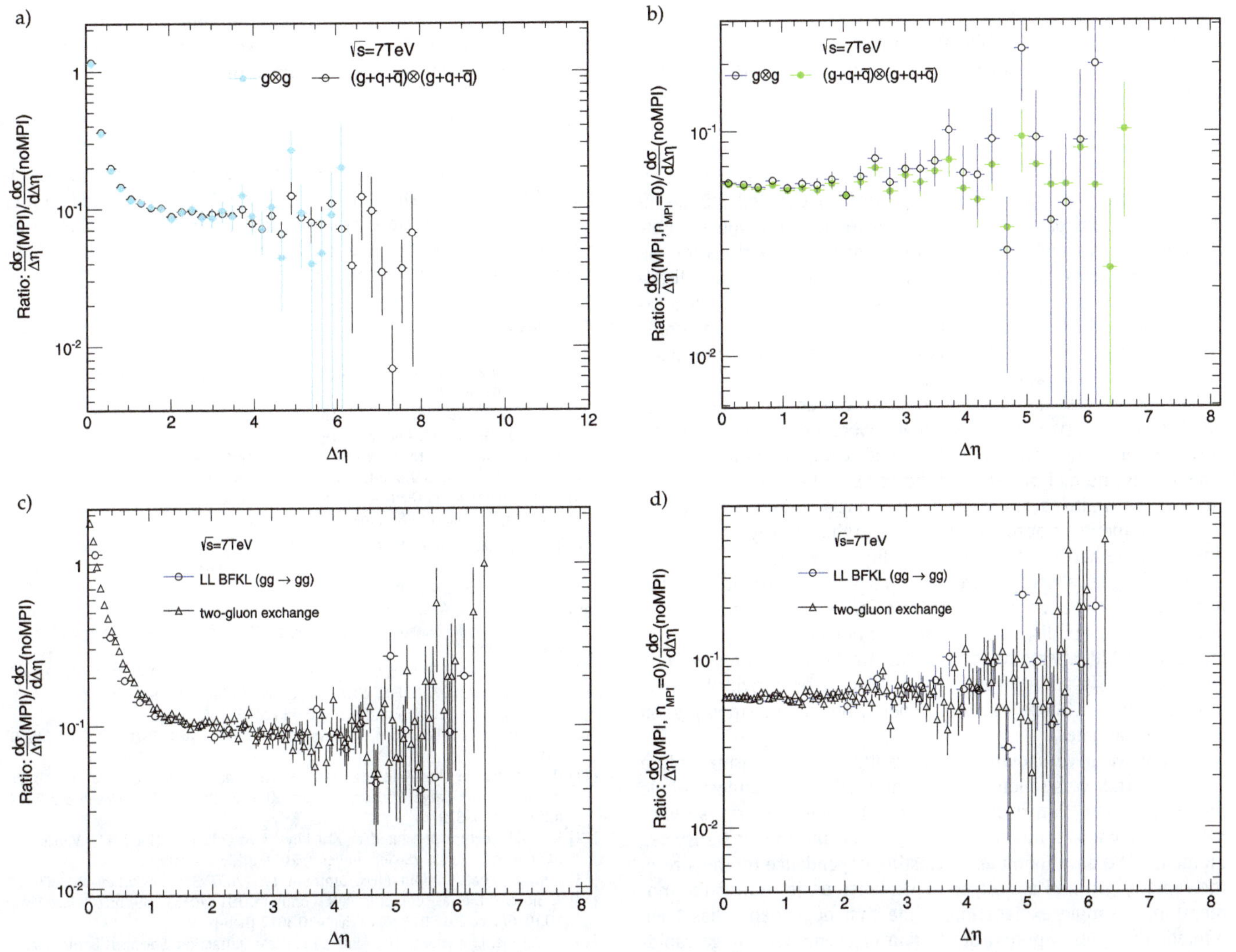

Fig. 9. Ratio of rapidity gap distributions for jet–gap–jet events generated with and without MPI effects as a function of rapidity gap size. All partons are included here. For comparison result for only $gg \to gg$ is shown by the dark blue line (a, b). Results for colour-singlet two-gluon exchange are shown in panels (c, d). (For interpretation of the references to colour in this figure legend, the reader is referred to the web version of this article.)

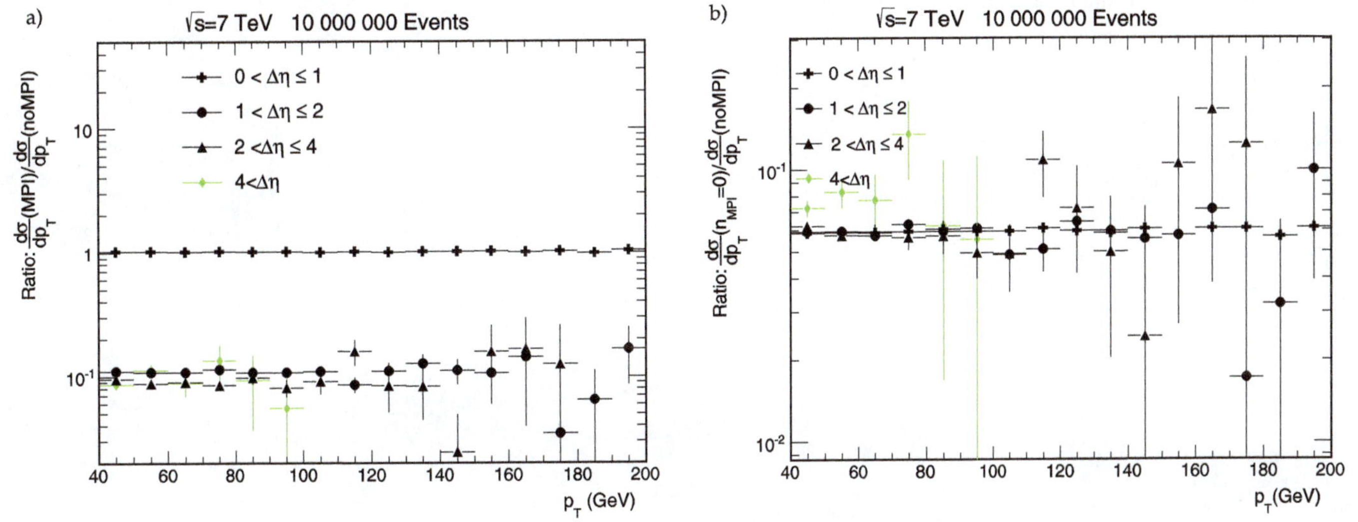

Fig. 10. Ratio of rapidity gap distributions for jet–gap–jet events generated with and without MPI effects (left panel) and with extra requirement $n_{MPI} = 0$ (right panel) as a function of jet transverse momentum for different intervals of rapidity gaps.

$n_{MPI} = 0$. Here we can observe that the result for all rapidity gap size intervals coincide within the limited Monte Carlo statistics. The ratios on the right panel are clearly smaller than those on the left panels.

6. Conclusions

In the present paper we have performed detailed studies of the role of multi-parton interactions in reduction of the theoretical cross section and/or different differential distributions for the jet–gap–jet processes. The cross section and the differential distributions for jet–gap–jet processes have been calculated for illustration in the LL BFKL framework. We have also tried to use a simple two-gluon exchange model regularised by the effective gluon mass to describe the jet–gap–jet process.

The subprocess amplitudes for the colour-singlet exchange (BFKL ladder or two-gluon exchange) were implemented to the PYTHIA 8 generator, which was then used to simulate multi-parton interactions and hadronisation of the generated events. The parameters of the multi-parton interactions models in PYTHIA are tuned to measurements of observables related to the underlying event. In this sense we have no freedom to modify the MPIs.

For pedagogical reasons we have first studied particle (hadron) final states for the jet–gap–jet process for fixed kinematic configurations (fixed rapidity and transverse momenta of the jets). After inclusion of MPI effects we have shown fractions of events with no extra activity (in addition to the hard jets) or no activity in some rapidity interval. Those fractions depend, but rather smoothly, on kinematic variables.

Finally we have shown similar results when imposing only a cut on jet transverse momenta and integrating over almost whole phase space (full range of jet rapidities). Again, the gap survival factor is shown as a function of the gap size and the jet transverse momenta. We have found an interesting dependence on the size of the rapidity gap and almost no dependence on jet transverse momentum. A simple explanation of the first dependence has been offered. The MPIs suppress production of events with large rapidity gaps but create events with smaller rapidity gaps. The resulting rapidity gap survival factor depends on the gap size. On the other hand, when imposing the requirement of occurring no additional MPIs, the corresponding gap survival factor is almost constant, independent of gap size. However, there is a sizeable dependence on the collision energy. The ratios obtained for colour-singlet two-gluon exchange and for the BFKL ladder are almost the same.

Summarising in one sentence, the MPI effects lead to a dependence on kinematical variables of the so-called gap survival factor, in contrast to what is usually assumed in the literature.

Acknowledgements

This study was partially supported by the Polish National Science Center grant DEC-2014/15/B/ST2/02528 and by the Center for Innovation and Transfer of Natural Sciences and Engineering Knowledge in Rzeszów. We are indebted to Cyrille Marquet for a discussion of their BFKL calculations.

References

[1] P. Lebiedowicz, O. Nachtmann, A. Szczurek, Phys. Rev. D 93 (2016) 054015, arXiv:1601.04537 [hep-ph];
C. Ewerz, P. Lebiedowicz, O. Nachtmann, A. Szczurek, Phys. Lett. B 763 (2016) 382–387, arXiv:1606.08067 [hep-ph];
C. Ewerz, M. Maniatis, O. Nachtmann, arXiv:1309.3478 [hep-ph].
[2] D0 Collaboration, B. Abbott, et al., Phys. Lett. B 440 (1998) 189.
[3] CMS Collaboration, Dijet production with a large rapidity gap between the jets, CMS Physics Analysis Summary FSQ-12-001.
[4] M. Łuszczak, R. Maciuła, A. Szczurek, Phys. Rev. D 91 (2015) 054024.
[5] A. Chuinard, C. Royon, R. Staszewski, J. High Energy Phys. 2016 (2016) 92, http://dx.doi.org/10.1007/JHEP04(2016)092.
[6] P. Lebiedowicz, A. Szczurek, Phys. Rev. D 92 (2015) 054001.
[7] M. Łuszczak, R. Maciuła, A. Szczurek, M. Trzebiński, J. High Energy Phys. 02 (2017) 089.
[8] C.O. Rasmussen, T. Sjöstrand, J. High Energy Phys. 1602 (2016) 142, http://dx.doi.org/10.1007/JHEP02(2016)142, arXiv:1512.05525 [hep-ph].
[9] V. Khachatryan, et al., Eur. Phys. J. C 76 (2016) 155.
[10] K. Akiba, et al., LHC forward physics, J. Phys. G, Nucl. Part. Phys. 43 (2016) 110201.
[11] Proceedings of the Sixth International Workshop on Multiple Partonic Interactions at the Large Hadron Collider, Krakow, Poland, 3–7 November 2014, arXiv:1506.05829.
[12] R. Field, Lecture presented at XXI Physics in Collisions (PIC2011), Vancouver, BC, Canada, August 28–September 1, 2011, arXiv:1202.0901.
[13] J. Alwall, et al., Comput. Phys. Commun. 176 (2007) 300, arXiv:hep-ph/0609017.
[14] T. Sjöstrand, et al., Comput. Phys. Commun. 191 (2015) 159, http://dx.doi.org/10.1016/j.cpc.2015.01.024, arXiv:1410.3012 [hep-ph].
[15] V. Barone, E. Predazzi, High-Energy Particle Diffraction, Springer, Berlin, 2002.
[16] A.H. Mueller, W.K. Tang, Phys. Lett. B 284 (1992) 123, http://dx.doi.org/10.1016/0370-2693(92)91936-4.
[17] L. Motyka, A.D. Martin, M.G. Ryskin, Phys. Lett. B 524 (2002) 107, http://dx.doi.org/10.1016/S0370-2693(01)01380-6, arXiv:hep-ph/0110273.
[18] O. Kepka, C. Marquet, C. Royon, Phys. Rev. D 83 (2011) 034036.
[19] F. Chevallier, O. Kepka, C. Marquet, C. Royon, Phys. Rev. D 79 (2009) 094019, http://dx.doi.org/10.1103/PhysRevD.79.094019, arXiv:0903.4598 [hep-ph].
[20] E. Meggiolaro, Phys. Lett. B 451 (1999) 414.

16

Nonstandard interactions in solar neutrino oscillations with Hyper-Kamiokande and JUNO

Jiajun Liao [a,*], Danny Marfatia [a], Kerry Whisnant [b]

[a] *Department of Physics and Astronomy, University of Hawaii at Manoa, Honolulu, HI 96822, USA*
[b] *Department of Physics and Astronomy, Iowa State University, Ames, IA 50011, USA*

ARTICLE INFO

Editor: M. Cvetič

ABSTRACT

Measurements of the solar neutrino mass-squared difference from KamLAND and solar neutrino data are somewhat discrepant, perhaps due to nonstandard neutrino interactions in matter. We show that the zenith angle distribution of solar neutrinos at Hyper-Kamiokande and the energy spectrum of reactor antineutrinos at JUNO can conclusively confirm the discrepancy and detect new neutrino interactions.

1. Introduction

There is currently about a 2σ tension in measurements of the neutrino mass splitting, $\delta m^2_{21} \equiv m^2_2 - m^2_1$, from solar and reactor neutrino experiments [1]. The discrepancy mainly arises from the measurement of the day-night asymmetry in the Super-Kamiokande (SK) experiment [2]. The latest SK combined measurement of the day-night asymmetry is $A^{SK}_{DN} \equiv 2(\Phi_D - \Phi_N)/(\Phi_D + \Phi_N) = -3.3 \pm 1.0 \pm 0.5\%$ [3], where Φ_D (Φ_N) is the measured solar neutrino flux during the day (night).[1] This day-night asymmetry is extracted for $\delta m^2_{21} = 4.8 \times 10^{-5}$ eV^2, while the global best-fit value (which is dominated by KamLAND data [5]) is $\delta m^2_{21} = 7.5 \times 10^{-5}$ eV^2 [1], for which the day-night asymmetry is -1.7%.

Analyzing the SK data in Table XII of Ref. [3], and the KamLAND data in Fig. 1 of Ref. [5] in the standard model (SM), we find the preferred parameters shown in Fig. 1 (which are consistent with those shown in Fig. 2 of Ref. [1]). From Fig. 1, we see a tension between the allowed regions at SK and KamLAND. For $\sin^2\theta_{12}$ close to the global best fit value of 0.31, SK data prefer a smaller value of δm^2_{21} than KamLAND. However, due to large uncertainties at SK, as shown in Fig. 1, current experiments are not able to resolve the tension. Nevertheless, if this discrepancy is due to a new physical effect, future solar and reactor experiments that have better control of the systematic uncertainties and larger datasets will see a significant difference in their measurements of δm^2_{21}, and could provide conclusive evidence for the existence of new physics.

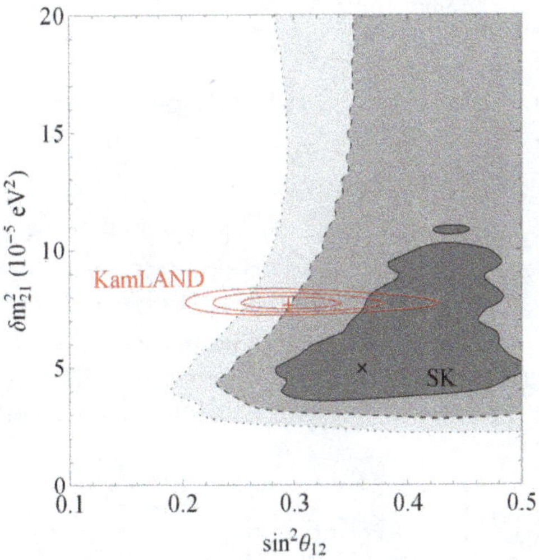

Fig. 1. 1σ, 2σ and 3σ allowed regions from SK (shaded regions) and KamLAND (ellipses) data in the SM. The cross (plus) sign marks the best-fit values at SK (KamLAND). We fix $\sin^2\theta_{13} = 0.023$, $\sin^2\theta_{23} = 0.43$ and $\delta m^2_{31} = 2.43 \times 10^{-3}$eV^2.

In this work, we explore this tension in the measurement of δm^2_{21} with future solar neutrino data at Hyper-Kamiokande (HK) [2] and future reactor antineutrino data at JUNO [6]. We use the framework of nonstandard interactions (NSI), which provides a model-independent way of studying new physics in neutrino oscillation experiments; for reviews see Ref. [7]. In particular, we focus on the day-night asymmetry and the zenith-angle distribution in solar neutrino experiments in the presence of matter NSI. Matter

* Corresponding author.
E-mail address: liaoj@hawaii.edu (J. Liao).

[1] SNO, on the other hand, does not disfavor a vanishing day-night asymmetry at more than 2σ [4].

NSI can be described by dimension-six four-fermion operators of the form [8,9]

$$\mathcal{L}_{\rm NSI} = -2\sqrt{2}G_F\epsilon_{\alpha\beta}^{fC}\left[\bar{\nu}_\alpha\gamma^\rho P_L\nu_\beta\right]\left[\bar{f}\gamma_\rho P_C f\right] + \text{h.c.}, \tag{1}$$

where $\alpha, \beta = e, \mu, \tau$, $C = L, R$, $f = u, d, e$, and $\epsilon_{\alpha\beta}^{fC}$ specifies the strength of the new interaction in units of G_F.

The JUNO experiment will measure δm_{21}^2 to the percent level [6], but it is not sensitive to the NSI parameters due to its short baseline and low neutrino energy. The HK solar neutrino oscillation probabilities are strongly dependent on the NSI parameters due to the MSW effect [8,10], but HK will not precisely measure δm_{21}^2 due to systematic uncertainties. A combination of the two experiments could provide direct evidence for the existence of new physics if the δm_{21}^2 discrepancy persists in HK and JUNO data.

The paper is organized as follows. In Section 2, we analyze the day–night asymmetry with NSI. In Section 3, we describe our simulations of HK and JUNO. We discuss our results in Section 4, and sum up in Section 5.

2. Day–night asymmetry with NSI

2.1. Formalism

The Hamiltonian in the three neutrino framework for neutrino propagation in the presence of matter NSI can be written in the flavor basis as

$$H = U\text{diag}\left(0, \frac{\delta m_{21}^2}{2E_\nu}, \frac{\delta m_{31}^2}{2E_\nu}\right)U^\dagger + V\,, \tag{2}$$

where U is the Pontecorvo–Maki–Nakagawa–Sakata mixing matrix [11],

$$U = R_{23}\Gamma_\delta R_{13}\Gamma_\delta^\dagger R_{12} = \tag{3}$$
$$\begin{pmatrix} c_{13}c_{12} & c_{13}s_{12} & s_{13}e^{-i\delta} \\ -s_{12}c_{23} - c_{12}s_{23}s_{13}e^{i\delta} & c_{12}c_{23} - s_{12}s_{23}s_{13}e^{i\delta} & c_{13}s_{23} \\ s_{12}s_{23} - c_{12}c_{23}s_{13}e^{i\delta} & -c_{12}s_{23} - s_{12}c_{23}s_{13}e^{i\delta} & c_{13}c_{23} \end{pmatrix}$$

where R_{ij} represents a real rotation by an angle θ_{ij} in the ij plane, $\Gamma_\delta = \text{diag}(1, 1, e^{i\delta})$, and s_{ij} and c_{ij} denote $\sin\theta_{ij}$ and $\cos\theta_{ij}$ respectively. The potential V originating from interactions of neutrinos in matter is

$$V = \sqrt{2}G_F N_e\begin{pmatrix} 1+\epsilon_{ee} & \epsilon_{e\mu} & \epsilon_{e\tau} \\ \epsilon_{e\mu}^* & \epsilon_{\mu\mu} & \epsilon_{\mu\tau} \\ \epsilon_{e\tau}^* & \epsilon_{\mu\tau}^* & \epsilon_{\tau\tau}\end{pmatrix}, \tag{4}$$

where $\epsilon_{\alpha\beta} \equiv \sum_f \epsilon_{\alpha\beta}^f \frac{N_f}{N_e}$ with $\epsilon_{\alpha\beta}^f \equiv \sum_C \epsilon_{\alpha\beta}^{fC}$, and N_f is the number density of fermion f at a given location.

Following Ref. [12], we work in the new basis $|\tilde{\nu}\rangle = \tilde{U}^\dagger|\nu_\alpha\rangle$, with $\tilde{U} = R_{23}\Gamma_\delta R_{13}$. The Hamiltonian in the new basis becomes

$$\tilde{H} = R_{12}\text{diag}\left(0, \frac{\delta m_{21}^2}{2E_\nu}, \frac{\delta m_{31}^2}{2E_\nu}\right)R_{12}^T + \tilde{V}\,, \tag{5}$$
$$= \begin{pmatrix} s_{12}^2\frac{\delta m_{21}^2}{2E_\nu} + \tilde{V}_{11} & s_{12}c_{12}\frac{\delta m_{21}^2}{2E_\nu} + \tilde{V}_{12} & \tilde{V}_{13} \\ s_{12}c_{12}\frac{\delta m_{21}^2}{2E_\nu} + \tilde{V}_{12}^* & c_{12}^2\frac{\delta m_{21}^2}{2E_\nu} + \tilde{V}_{22} & \tilde{V}_{23} \\ \tilde{V}_{13}^* & \tilde{V}_{23}^* & \frac{\delta m_{31}^2}{2E_\nu} + \tilde{V}_{33}\end{pmatrix},$$

where $\tilde{V} = \tilde{U}^\dagger V\tilde{U}$. For solar neutrinos, since $\frac{|\delta m_{31}^2|}{2E_\nu} \gg \tilde{V}_{ij}$, the third mass eigenstate decouples from the other mass eigenstates, and the evolution is governed by an effective 2×2 submatrix. After subtracting a constant diagonal matrix from the Hamiltonian, the effective matrix can be written as

$$\tilde{H}_{\rm eff} = \frac{\delta m_{21}^2}{4E_\nu}\times \tag{6}$$
$$\begin{pmatrix} -\cos 2\theta_{12} + 2\hat{A}(c_{13}^2 - \epsilon_D) & \sin 2\theta_{12} + 2\hat{A}\epsilon_N \\ \sin 2\theta_{12} + 2\hat{A}\epsilon_N^* & \cos 2\theta_{12} + 2\hat{A}\epsilon_D\end{pmatrix},$$

where $\hat{A} = 2\sqrt{2}G_F N_e E_\nu/\delta m_{21}^2$ and $\epsilon_X = \sum_f N_f\epsilon_X^f/N_e$ $(X = N, D)$ with [13]

$$\epsilon_D^f = -\frac{c_{13}^2}{2}(\epsilon_{ee}^f - \epsilon_{\mu\mu}^f) + \frac{s_{23}^2 - s_{13}^2c_{23}^2}{2}(\epsilon_{\tau\tau}^f - \epsilon_{\mu\mu}^f)$$
$$+ \text{Re}\left[c_{13}s_{13}e^{i\delta}(s_{23}\,\epsilon_{e\mu}^f + c_{23}\,\epsilon_{e\tau}^f)\right.$$
$$\left. -(1 + s_{13}^2)c_{23}s_{23}\epsilon_{\mu\tau}^f\right], \tag{7}$$

$$\epsilon_N^f = c_{13}(c_{23}\,\epsilon_{e\mu}^f - s_{23}\,\epsilon_{e\tau}^f) \tag{8}$$
$$+ s_{13}e^{-i\delta}\left[s_{23}^2\,\epsilon_{\mu\tau}^f - c_{23}^2\,\epsilon_{\mu\tau}^{f*} + c_{23}s_{23}(\epsilon_{\tau\tau}^f - \epsilon_{\mu\mu}^f)\right].$$

Solar neutrinos produced in the core of the Sun arrive at the surface of the Earth as an incoherent sum of the three mass-eigenstates ν_1, ν_2 and ν_3. During the day, the neutrinos only travel a few km in the Earth, and the ν_e survival probability from the source to the detector can be written as [12]

$$P_D = P_{e1}^S P_{1e}^V + P_{e2}^S P_{2e}^V + P_{e3}^S P_{3e}^V\,, \tag{9}$$

where the superscript V represents neutrinos propagating in vacuum, and we have

$$P_{1e}^V = c_{13}^2c_{12}^2\,, \qquad P_{2e}^V = c_{13}^2s_{12}^2\,, \qquad P_{3e}^V = s_{13}^2\,. \tag{10}$$

The superscript S represents neutrinos traveling in the Sun. We checked that the neutrino propagation in the Sun with NSI can be treated as adiabatic for the parameters we consider in this paper. For adiabatic propagation, we have

$$P_{e1}^S = c_{13}^2\cos^2\theta_{12}^m\,,$$
$$P_{e2}^S = c_{13}^2\sin^2\theta_{12}^m\,, \tag{11}$$
$$P_{e3}^S = s_{13}^2\,,$$

where θ_{12}^m can be found by diagonalizing the Hamiltonian in Eq. (6) at the production point, i.e.,

$$\tan 2\theta_{12}^m = \frac{|\sin 2\theta_{12} + 2\hat{A}_S\epsilon_N^S|}{\cos 2\theta_{12} - \hat{A}_S(c_{13}^2 - 2\epsilon_D^S)}\,. \tag{12}$$

Here $\hat{A}_S = 2\sqrt{2}G_F N_e^S E_\nu/\delta m_{21}^2$, and $\epsilon_X^S = \sum_f N_f^S\epsilon_X^f/N_e^S$ with N_f^S being the number density of fermion f at the production point.

During the night, the neutrinos travel a large distance through the Earth, and the survival probability becomes

$$P_N = P_{e1}^S P_{1e}^E + P_{e2}^S P_{2e}^E + P_{e3}^S P_{3e}^E\,, \tag{13}$$

where the superscript E represents neutrino propagation in the Earth. Since the third mass eigenstate decouples from the other mass eigenstates, $P_{3e}^E = s_{13}^2$. Also, from probability conservation, $P_{1e}^E = c_{13}^2 - P_{2e}^E$. However, the calculation of P_{2e}^E in the presence of NSI is nontrivial. Here we derive a simple expression of P_{2e}^E for a constant density profile.

For neutrino evolution in Earth matter, due to the decoupling of the third eigenstate, the effective Hamiltonian has the same form of Eq. (6) with $\hat{A}$ and ϵ_X replaced by $\hat{A}_E = 2\sqrt{2}G_F N_e^E E_\nu/\delta m_{21}^2$ and

$\epsilon_X^E = \sum_f N_f^E \epsilon_X^f / N_e^E$, respectively, where N_f^E is the number density of fermion f in the Earth. Then the effective Hamiltonian can be diagonalized by [14]

$$U' = \begin{pmatrix} \cos\tilde{\theta} & \sin\tilde{\theta}e^{-i\phi} \\ -\sin\tilde{\theta}e^{i\phi} & \cos\tilde{\theta} \end{pmatrix}, \tag{14}$$

where

$$\tan 2\tilde{\theta} = \frac{|\sin 2\theta_{12} + 2\hat{A}_E\epsilon_N^E|}{\cos 2\theta_{12} - \hat{A}_E(c_{13}^2 - 2\epsilon_D^E)}, \tag{15}$$

and

$$\phi = -\mathrm{Arg}\left(\sin 2\theta_{12} + 2\hat{A}_E\epsilon_N^E\right). \tag{16}$$

Then the evolution matrix in the new basis can be written as [12]

$$\tilde{S} = \begin{pmatrix} \tilde{\alpha} & \tilde{\beta} & 0 \\ -\tilde{\beta}^* & \tilde{\alpha}^* & 0 \\ 0 & 0 & \tilde{\gamma} \end{pmatrix}. \tag{17}$$

For a constant density profile,

$$\begin{aligned} \tilde{\alpha} &= \cos\omega L + i\cos 2\tilde{\theta}\sin\omega L, \\ \tilde{\beta} &= -i\sin 2\tilde{\theta}e^{-i\phi}\sin\omega L, \\ \tilde{\gamma} &= \exp(-i\frac{\delta m_{31}^2 L}{2E_\nu}), \end{aligned} \tag{18}$$

where

$$\omega = \frac{\delta m_{21}^2}{4E_\nu} \times \sqrt{(\cos 2\theta_{12} - \hat{A}_E(c_{13}^2 - 2\epsilon_D^E))^2 + |\sin 2\theta_{12} + 2\hat{A}_E\epsilon_N^E|^2}. \tag{19}$$

The evolution matrix in the neutrino flavor basis becomes

$$S = \tilde{U}\tilde{S}\tilde{U}^\dagger, \tag{20}$$

and we have

$$\begin{aligned} P_{2e}^E &= |\langle\nu_e|(SU)|\nu_2\rangle|^2 \\ &= c_{13}^2\left[s_{12}^2|\tilde{\alpha}|^2 + c_{12}^2|\tilde{\beta}|^2 + \sin 2\theta_{12}\mathrm{Re}(\tilde{\alpha}^*\tilde{\beta})\right]. \end{aligned} \tag{21}$$

Plugging the expressions in Eq. (18) into the above equation, we get

$$\begin{aligned} P_{2e}^E &= c_{13}^2\sin 2\tilde{\theta}\sin^2\omega L \times \\ &\left(\cos 2\theta_{12}\sin 2\tilde{\theta} - \sin 2\theta_{12}\cos 2\tilde{\theta}\cos\phi\right) \\ &- c_{13}^2\sin 2\theta_{12}\sin 2\tilde{\theta}\sin\phi\sin\omega L\cos\omega L + c_{13}^2 s_{12}^2. \end{aligned} \tag{22}$$

For earth matter, $\hat{A}_E \ll 1$, so we expand the above equation to leading order in $\hat{A}_E$ and find

$$\begin{aligned} P_{2e}^E &\approx \hat{A}_E c_{13}^2\sin 2\theta_{12}\sin^2\frac{\delta m_{21}^2 L}{4E_\nu} \times \\ &\left[2\cos 2\theta_{12}\mathrm{Re}(\epsilon_N^E) + \sin 2\theta_{12}(c_{13}^2 - 2\epsilon_D^E)\right] \\ &+ \hat{A}_E c_{13}^2\sin 2\theta_{12}\sin\frac{\delta m_{21}^2 L}{2E_\nu}\mathrm{Im}(\epsilon_N^E) + c_{13}^2 s_{12}^2. \end{aligned} \tag{23}$$

Then from Eqs. (9) and (13), the day-night symmetry is

$$\begin{aligned} A_{DN} &\equiv \frac{2(P_D - P_N)}{P_D + P_N} \\ &\approx \frac{2\hat{A}_E c_{13}^4\sin 2\theta_{12}\cos 2\theta_{12}^m}{c_{13}^4(1 + \cos 2\theta_{12}\cos 2\theta_{12}^m) + 2s_{13}^4} \times \\ &\left[\sin^2\frac{\delta m_{21}^2 L}{4E_\nu}\left(2\cos 2\theta_{12}\mathrm{Re}(\epsilon_N^E)\right.\right. \\ &\left.\left.+\sin 2\theta_{12}(c_{13}^2 - 2\epsilon_D^E)\right) + \sin\frac{\delta m_{21}^2 L}{2E_\nu}\mathrm{Im}(\epsilon_N^E)\right]. \end{aligned} \tag{24}$$

Since the day–night asymmetry is generally measured by integrating over the zenith angle and the oscillations in the above equation are averaged out, we obtain

$$\begin{aligned} \langle A_{DN}\rangle(E_\nu) &\approx \frac{\hat{A}_E c_{13}^4\sin 2\theta_{12}\cos 2\theta_{12}^m}{c_{13}^4(1 + \cos 2\theta_{12}\cos 2\theta_{12}^m) + 2s_{13}^4} \times \\ &\left[2\cos 2\theta_{12}\mathrm{Re}(\epsilon_N^E) + \sin 2\theta_{12}(c_{13}^2 - 2\epsilon_D^E)\right]. \end{aligned} \tag{25}$$

We have checked that the above equation is consistent with Eq. (13) in Ref. [15] for the two-flavor case.

2.2. Numerical analysis

Using Eq. (25), we estimate the day–night asymmetry for different SM and NSI parameters. In Fig. 2, we show iso-$\langle A_{DN}\rangle$ contours in the $\sin^2\theta_{12} - \delta m_{21}^2$ plane in the SM. We fix $\sin^2\theta_{13} = 0.023$, $E_\nu = 7.0$ MeV and Earth density $\rho_E = 3.0$ g/cm^3. The day–night asymmetry depends strongly on δm_{21}^2, and its size decreases as $\sin^2\theta_{12}$ increases in the second octant. In particular, a smaller value of δm_{21}^2 yields a larger $|\langle A_{DN}\rangle|$.

Now we study the dependence of the day–night asymmetry on the NSI parameters. For the NSI parameters, we assume there are no nonstandard couplings to electrons since they would affect the electron-neutrino scattering cross section, yielding NSI at the SK and HK detectors. We also assume the NSI couplings to the up and down quarks are the same for simplicity.

We first examine the dependence on the diagonal NSI parameters. We consider the case in which only $\epsilon_{ee}^u = \epsilon_{ee}^d$ is nonzero. From Eqs. (7) and (8), we see that $\epsilon_N^E = 0$ and ϵ_D^E is linearly dependent on ϵ_{ee}^u. We show the iso-$\langle A_{DN}\rangle$ contours in the space of δm_{21}^2 and ϵ_{ee}^u in Fig. 3. We find that for fixed δm_{21}^2, a more positive value of ϵ_{ee}^u implies a larger $|\langle A_{DN}\rangle|$. This result can be understood from Eq. (25). Since the dominant contribution to the change in $\langle A_{DN}\rangle$ comes from the factor $[2\cos 2\theta_{12}\mathrm{Re}(\epsilon_N^E) + \sin 2\theta_{12}(c_{13}^2 - 2\epsilon_D^E)]$, and ϵ_D^E is proportional to $-\epsilon_{ee}^u$, as ϵ_{ee}^u increases, $|\langle A_{DN}\rangle|$ becomes larger. This yields a degeneracy between δm_{21}^2 and ϵ_{ee}^u in the measurement of the day–night asymmetry, i.e., a day–night asymmetry that is consistent with $\delta m_{21}^2 = 4.8 \times 10^{-5}$ eV^2 in the SM can also be obtained with $\delta m_{21}^2 = 7.5 \times 10^{-5}$ eV^2 and $\epsilon_{ee}^u = \epsilon_{ee}^d \sim 0.1$.

We also checked the dependence of the day–night asymmetry on the off-diagonal NSI parameters. Here we consider the case in which only $\epsilon_{e\tau}^u = \epsilon_{e\tau}^d$ is nonzero. As can be seen from Eqs. (7) and (8), ϵ_D^E is suppressed by $\sin\theta_{13}$ and ϵ_N^E is proportional to $-\epsilon_{e\tau}^u$ in this case. We first assume $\delta = 0$ and $\epsilon_{e\tau}^u$ is real for simplicity. In Fig. 4, we show the iso-$\langle A_{DN}\rangle$ contours in the space of δm_{21}^2 and $\epsilon_{e\tau}^u$. The results in Fig. 4 can be understood from Eq. (25). As $\epsilon_{e\tau}^u$ approaches 0.2, the factor $[2\cos 2\theta_{12}\mathrm{Re}(\epsilon_N^E) + \sin 2\theta_{12}(c_{13}^2 - 2\epsilon_D^E)]$ approaches zero. We also checked the complex case by varying δ and the phase of $\epsilon_{e\tau}^u$, and find that for $\delta m_{21}^2 = 7.5 \times 10^{-5}$ eV^2, the day–night asymmetry is always smaller than 2% for $|\epsilon_{e\tau}^u| < 0.4$. Since from Eq. (8) we know that the dominant contribution to ϵ_N^f

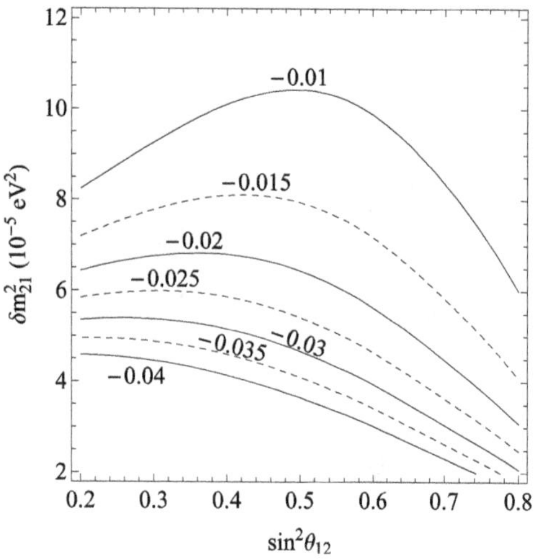

Fig. 2. Iso-$\langle A_{DN}\rangle$ contours in the $\sin^2\theta_{12} - \delta m^2_{21}$ plane in the SM. Here $\sin^2\theta_{13} = 0.023$, $E_\nu = 7.0$ MeV and $\rho_E = 3.0$ g/cm^3.

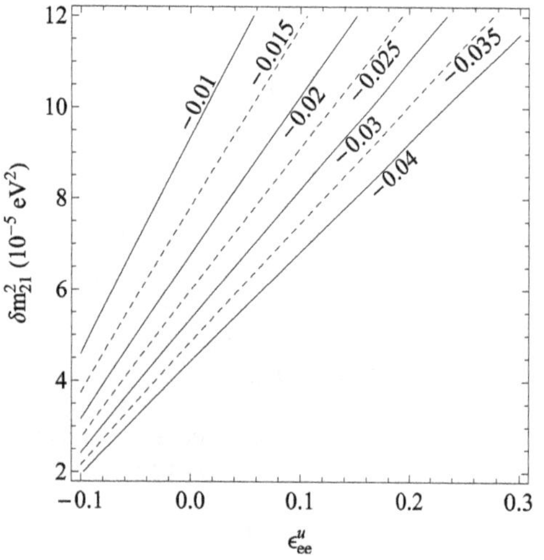

Fig. 3. Iso-$\langle A_{DN}\rangle$ contours in the $\epsilon^u_{ee} - \delta m^2_{21}$ plane. The parameters are the same as in Fig. 2, and $\sin^2\theta_{12} = 0.031$. For the NSI parameters we assume $\epsilon^u_{ee} = \epsilon^d_{ee}$ and all other NSI parameters are zero.

comes from $\epsilon^f_{e\mu}$ and $\epsilon^f_{e\tau}$, and the global-fit constraints on $\epsilon^f_{e\mu}$ are stronger than on $\epsilon^f_{e\tau}$ [13], an off-diagonal NSI parameter always gives a small day–night asymmetry for $\delta m^2_{21} = 7.5 \times 10^{-5}$ eV^2. We henceforth focus on the diagonal NSI parameters.

3. Experimental simulations

3.1. Hyper-Kamiokande

Solar neutrino experiments like HK detect neutrinos via the elastic scattering reaction,

$$\nu_x + e^- \to \nu'_x + e^- . \tag{26}$$

The expected event rate for the reconstructed electron kinetic energy of T is [16]

$$R(T) = \mathcal{N} \int dE_\nu \times \tag{27}$$
$$\left[\Phi_B(E_\nu) + 1.462 \times 10^{-3}\Phi_{\rm hep}(E_\nu)\right] S^{D,N}(E_\nu),$$

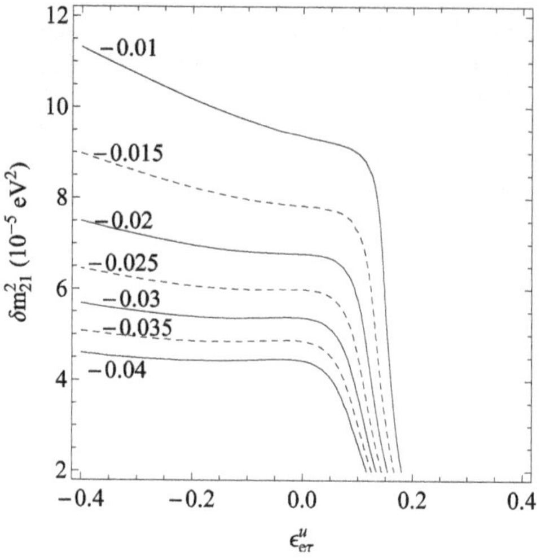

Fig. 4. Iso-$\langle A_{DN}\rangle$ contours in the $\epsilon^u_{e\tau} - \delta m^2_{21}$ plane. The parameters are the same as in Fig. 2, and $\sin^2\theta_{12} = 0.031$, $\sin^2\theta_{23} = 0.43$ and $\delta = 0$. For the NSI parameters, we assume $\epsilon^u_{e\tau} = \epsilon^d_{e\tau}$ is real, and all other NSI parameters are zero.

where $\mathcal{N}$ is the overall normalization that gives the expected event rate in the absence of oscillations, Φ_B ($\Phi_{\rm hep}$) is the normalized ^8B (hep) neutrino flux, and the factor 1.462×10^{-3} is the relative total flux of hep to ^8B neutrinos in the standard solar model (SSM) (B16-GS98) [17]. The effective cross section is

$$S^{D,N}(E_\nu) = P^{D,N}\sigma_e + (1 - P^{D,N})\sigma_\mu , \tag{28}$$

with

$$\sigma_i = \int dT \int dT' \frac{d\sigma_i}{dT'}(E_\nu, T')g(T, T') , \tag{29}$$

where $i = e, \mu$, T' is the true electron kinetic energy, $\frac{d\sigma_i}{dT'}(E_\nu, T')$ is the differential scattering cross section with radiative corrections taken from Ref. [18], and the energy resolution $g(T, T')$ is given by

$$g(T, T') = \frac{1}{\sqrt{2\pi}\sigma(T')}\exp\left[-\frac{(T - T')^2}{2\sigma(T')^2}\right] . \tag{30}$$

Since an energy resolution of 10% at 10 MeV is achievable at the HK experiment [19], we choose the energy resolution function,

$$\sigma(T') = (0.316\,\text{MeV})\sqrt{\frac{T'}{\text{MeV}}} . \tag{31}$$

Due to Earth matter effects, the electron neutrino survival probability at night is zenith angle dependent. Given a particular zenith angle, the relative amount of time that the detector is exposed to the Sun is determined by the latitude of the detector site. We use the exposure function at Kamiokande from Ref. [20], and weight each zenith angle by the exposure function. To obtain the survival probabilities, we adopt the average value of the production-point densities of the electron, up-quark and down-quark in the Sun from Ref. [21], and use the GLoBES software [22] with the new physics tools developed in Ref. [23] to calculate P^E_{2e} numerically.

We first simulate the detector with a fiducial volume of 0.56 Mt and the electron kinetic energy threshold of 7.0 MeV from the old HK design [24]. We normalize the number of events in our simulation to 200 events per day [24]. Since the HK collaboration has updated their design with a new two-tank configuration for the detector [2], we change the fiducial volume to 0.187 Mt per tank and assume the threshold energy is 5.0 MeV. Ergo, we expect 152 events per day per tank for the new design. The 2TankHK-staged

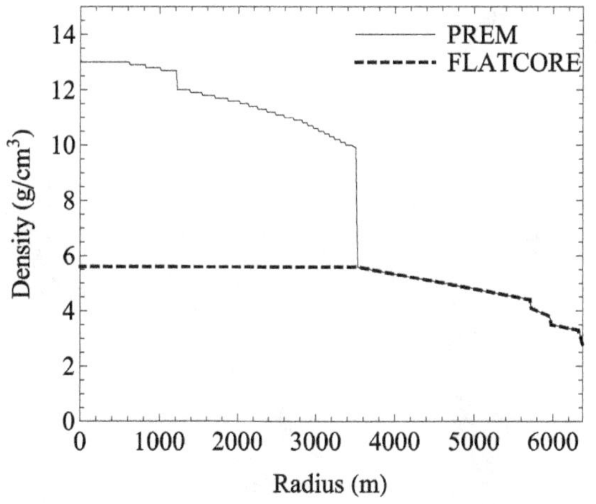

Fig. 5. The density profile of the Earth in two models.

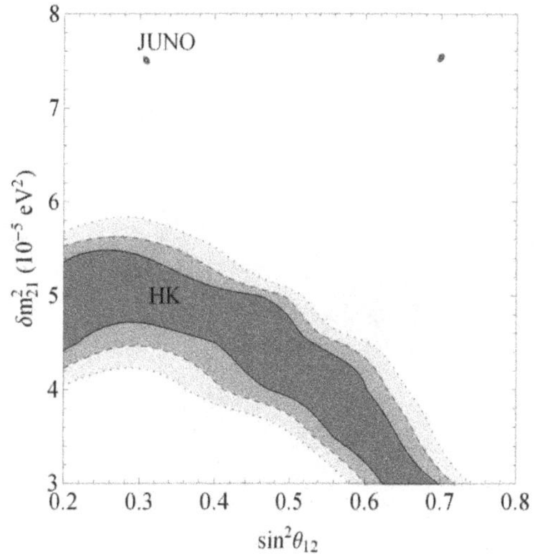

Fig. 6. 1σ, 2σ and 3σ allowed regions for HK and JUNO; for JUNO, note the second set of allowed regions at $\sin^2\theta_{12} = 0.69$. The data are simulated with $\delta m^2_{21} = 7.5 \times 10^{-5}$ eV^2, $\sin^2\theta_{12} = 0.31$, NH, $\epsilon^u_{ee} = \epsilon^d_{ee} = 0.1$, and fit with the SM allowing for both mass hierarchies. Here only the day–night asymmetry is used in the analysis, and the PREM model is used for the Earth density profile.

configuration [2] has one tank taking data for 6 years and a second tank is added for another 4 years. We checked that our sensitivity to the day–night asymmetry is consistent with Fig. 134 in Ref. [2] for a 6.5 MeV energy threshold.

In our simulation of the HK experiment, we also consider two Earth density profiles: the Preliminary Reference Earth Model (PREM) [25] and the FLATCORE model [26], in which the density of the core is a constant, as shown in Fig. 5. Note that the FLATCORE model does not match the Earth's mass, and we only use it as an example to study the effects of the Earth's density profile on our results.

3.2. *JUNO*

The 20-kt liquid scintillator JUNO experiment will detect reactor antineutrinos from two reactor complexes with a total power of 36 GW via the inverse beta-decay reaction,

$$\bar{\nu}_e + p \to e^+ + n\,. \tag{32}$$

Besides the primary goal of determining the neutrino mass hierarchy, JUNO will also provide a precise measurement of the solar neutrino oscillation parameters. We simulate the JUNO experiment using the GLoBES software with the tools developed in Refs. [23, 27]. The baseline of the experiment is 52.5 km and we take the detector energy resolution to be $3\%/\sqrt{E(\mathrm{MeV})}$. With 6 years running, the detector will collect a total of 1.52×10^5 events. An overall normalization error of 5% and a linear energy scale uncertainty of 3% is implemented in our simulation [28]. We consider 200 bins from 1.8 MeV to 8.0 MeV, and checked that the spectrum produced from our simulation is in good agreement with that in Fig. 2–15 of Ref. [6].

4. Results

4.1. *Resolving the tension in δm^2_{21}*

Since solar data are not sensitive to parameters related to m_3 or the CP phase for the case we are considering, we fix $\sin^2\theta_{13} = 0.023$, $\sin^2\theta_{23} = 0.43$ and $\delta = 0$. We simulate HK and JUNO data with $\delta m^2_{21} = 7.5 \times 10^{-5}$ eV^2, $\sin^2\theta_{12} = 0.31$, $\delta m^2_{31} = 2.43 \times 10^{-3}$ eV^2 for the normal mass hierarchy (NH), and $\epsilon^u_{ee} = \epsilon^d_{ee} = 0.1$, which gives a prediction for the day–night asymmetry that agrees with the current measurement at SK. We first perform a fit to only the SM parameters for each experiment separately to show how parameter degeneracies can occur with nonzero NSI, then perform a fit to the NSI parameters for the two experiments combined to study their ability to reject the SM. We always marginalize over the normal and inverted mass hierarchy (IH).

4.1.1. *Day–night asymmetry*

As an example, we first only use the day–night asymmetry in the HK analysis. The experimentally measured day–night asymmetry is defined as

$$A^{\mathrm{exp}}_{\mathrm{DN}} \equiv \frac{2(N_D - N_N)}{N_D + N_N}\,, \tag{33}$$

where N_D (N_N) denotes the total number of events detected in the day (night) time. We fit the SM to the simulated data with NSI. From Fig. 6, we see an allowed region for HK around $\delta m^2_{21} = 4.8 \times 10^{-5}$ eV^2. Note that the dependence of the HK allowed regions on $\sin^2\theta_{12}$ is consistent with the prediction of Eq. (25) shown in Figs. 2 and 3. The two sets of allowed regions for JUNO (shown for comparison) around $\sin^2\theta_{12} = 0.31$ and 0.69 are a consequence of the generalized mass-hierarchy degeneracy [29]. Although the exact generalized mass-hierarchy degeneracy requires $\epsilon_{ee} \to -\epsilon_{ee} - 2$, since JUNO is not sensitive to the NSI parameters, an approximate degeneracy holds.

4.1.2. *Zenith-angle distribution*

We now consider one bin with daytime data and six equisized (in the cosine of the zenith angle) nighttime bins, and define

$$\chi^2_{HK} = \sum_{i=1}^{7} \frac{(\alpha N^{\mathrm{fit}}_i - N^{\mathrm{data}}_i)^2}{N^{\mathrm{data}}_i} + \frac{(1-\alpha)^2}{\sigma^2_\alpha}\,, \tag{34}$$

where $\sigma_\alpha = 12\%$ is the flux uncertainty in the SSM (B16-GS98) [17]. The results of a SM parameter space scan are shown in Fig. 7. The allowed regions around $\delta m^2_{21} = 4.8 \times 10^{-5}$ eV^2 persist. Compared to Fig. 6, we see that the new analysis gives a better constraint on $\sin^2\theta_{12}$; however, the allowed regions in δm^2_{21} are similar to those obtained from the day–night asymmetry. Hence the sensitivity to δm^2_{21} at HK mainly comes from the day–night asymmetry, and HK alone cannot distinguish between the SM and NSI scenarios.

We also simulated data using the FLATCORE model for the Earth density profile. Then we fit the SM assuming the PREM model for

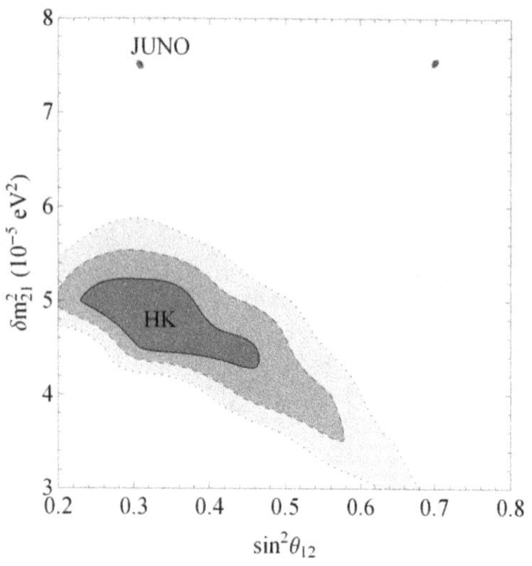

Fig. 7. Same as Fig. 6, except that the zenith-angle distribution is used in the analysis.

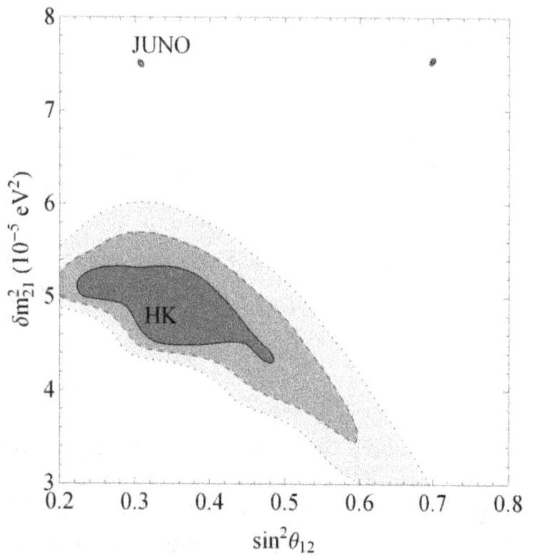

Fig. 8. Same as Fig. 7, except that the data are simulated with the FLATCORE model and fit with the PREM model.

the Earth density profile. The best-fit χ^2 to the HK data is 2.3, indicating that the PREM model provides a good fit to data simulated with the FLATCORE model. The allowed regions shown in Fig. 8 are similar to those in Fig. 7.

4.1.3. HK and JUNO combined analysis

Although HK data alone cannot distinguish between the SM and NSI scenarios, when combined with reactor data, measurements of the NSI parameters may be possible. We combine the data from JUNO and HK, and study their sensitivities to the NSI parameters. For the HK analysis, we use the zenith-angle distribution. Using simulated data with $\epsilon^u_{ee} = \epsilon^d_{ee} = 0.1$ at JUNO and HK, we scan over the range of ϵ^u_{ee} that is consistent with the global fit in Ref. [13]. After marginalizing over δm^2_{21}, θ_{12} and the mass hierarchy, we plot $\sqrt{\Delta\chi^2}$ as a function of ϵ^u_{ee} for the JUNO and HK combined analysis. The solid curves in Fig. 9 show that the SM (with $\epsilon^u_{ee} = 0$) is ruled out at 7.6σ, but large negative values of ϵ^u_{ee} are allowed at less than 3σ due to the generalized mass-hierarchy degeneracy.

In order to test the effect of the Earth density profile on our results, we also simulate the data with the FLATCORE model, and fit the data assuming the PREM model. The results are shown in Fig. 9 as the dashed curves. As expected, the sensitivity is reduced

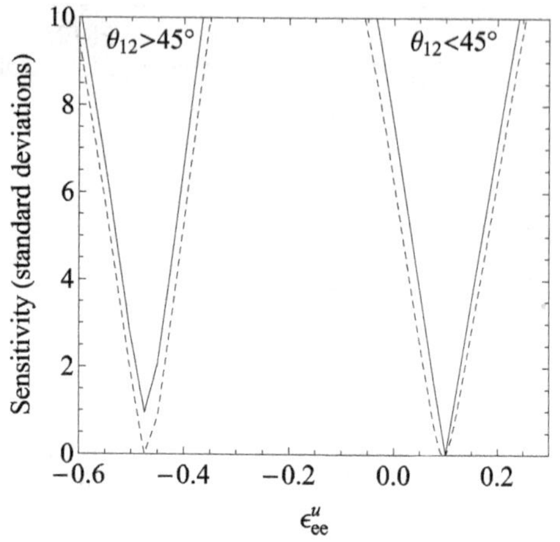

Fig. 9. The sensitivity to ϵ^u_{ee} for the HK and JUNO combined analysis when the true $\epsilon^u_{ee} = 0.1$. The data are simulated with $\delta m^2_{21} = 7.5 \times 10^{-5}$ eV^2, $\sin^2\theta_{12} = 0.31$ and NH. We assume $\epsilon^u_{ee} = \epsilon^d_{ee}$, and all other NSI parameters are zero. The solid (dashed) curves correspond to the case in which the data are simulated with the PREM (FLATCORE) model for the Earth density profile and fit with the PREM model.

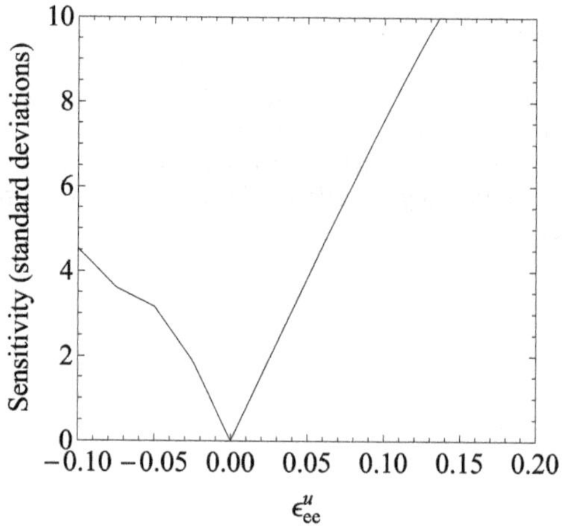

Fig. 10. The sensitivity to reject the SM as a function of true ϵ^u_{ee} for the HK and JUNO combined analysis. The data are simulated with $\delta m^2_{21} = 7.5 \times 10^{-5}$ eV^2, $\sin^2\theta_{12} = 0.31$, and NH.

if the Earth density profile employed is inaccurate. However, the SM is still excluded at 6.2σ.

4.2. Detecting NSI

We now study the significance with which $\epsilon^u_{ee} = \epsilon^d_{ee} \neq 0$ can be established by ruling out the SM. We simulate data with $\delta m^2_{21} = 7.5 \times 10^{-5}$ eV^2, NH, $\sin^2\theta_{12} = 0.31$ ($\sin^2\theta_{12} = 0.7$), and values of ϵ^u_{ee} that are roughly consistent with the global fit in Ref. [13] for the first (second) octant of θ_{12}. For each value of ϵ^u_{ee} we calculate the sensitivity to reject the SM allowing for both mass hierarchies. From Figs. 10 and 11, we see that the combination of HK and JUNO data can exclude the SM at high confidence for a range of ϵ^u_{ee} values.[2] The kink on the left side of the curve in Fig. 10 arises because the second octant of θ_{12} and IH provides a better fit than the first octant for $\epsilon^u_{ee} \sim -0.05$. If $\sin^2\theta_{12} = 0.7$, Fig. 11 shows that

[2] Guided by the generalized mass-hierarchy degeneracy, we also simulated data for the IH by fixing $\delta m^2_{31} = -2.355 \times 10^{-3}$ eV^2, thus yielding $(\delta m^2_{32})_{\rm IH} = -(\delta m^2_{31})_{\rm NH}$ for the simulated data. The sensitivity to reject the SM is identical to that in Figs. 10 and 11.

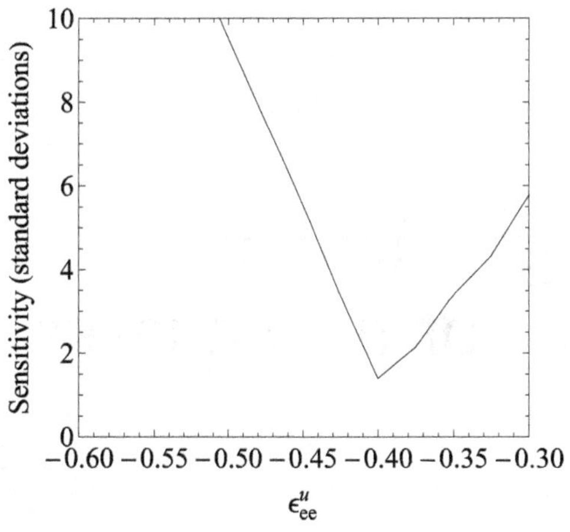

Fig. 11. Same as Fig. 10, except that the data are simulated with $\sin^2\theta_{12} = 0.70$.

for $\epsilon^{u}_{ee} \sim -0.4$, the SM is allowed at less than 3σ as a result of the generalized mass-hierarchy degeneracy.

5. Summary

We explored the discrepancy in the current measurements of δm^2_{21} from the SK solar neutrino and KamLAND reactor antineutrino experiments in the framework of NSI. Since the discrepancy mainly stems from the measurement of the day–night asymmetry, we first derived an analytic formula for the day–night asymmetry in the presence of NSI in the three-neutrino framework. We studied the dependence of the day–night asymmetry on both the diagonal and off-diagonal NSI parameters using the formula. We find that a diagonal NSI parameter could yield a large day–night asymmetry. In particular, for $\delta m^2_{21} = 7.5 \times 10^{-5}$ eV^2, the value preferred by KamLAND, $\epsilon^{u}_{ee} = \epsilon^{d}_{ee} = 0.1$ could give a day–night asymmetry that agrees with the current measurement at SK. We also find that an off-diagonal NSI parameter always yields a small day–night asymmetry for $\delta m^2_{21} = 7.5 \times 10^{-5}$ eV^2.

Since current SK solar and KamLAND reactor experiments cannot resolve the tension we studied the potential of the future solar neutrino experiment at HK and the future reactor antineutrino experiment at JUNO to provide a resolution. We find that by combining HK and JUNO data, the SM scenario can be rejected at 7.6σ if $\epsilon^{u}_{ee} = \epsilon^{d}_{ee} = 0.1$. Due to the generalized mass-hierarchy degeneracy, larger negative values of ϵ^{u}_{ee} are also allowed at less than 3σ. We find our conclusions to be robust under reasonable variations of the Earth density profile. Further, we demonstrated that by combining HK and JUNO data, the SM can be excluded at high confidence for a range of ϵ^{u}_{ee} values.

Acknowledgements

We thank S.-H. Seo and M.B. Smy for helpful discussions regarding HK and SK. KW thanks the University of Hawaii at Manoa for its hospitality during part of this work. This research was supported in part by the U.S. DOE under Grant No. DE-SC0010504.

References

[1] M. Maltoni, A.Y. Smirnov, Eur. Phys. J. A 52 (4) (2016) 87, arXiv:1507.05287 [hep-ph].
[2] Hyper-Kamiokande Collaboration, vol. KEK-PREPRINT-2016-21, ICRR-REPORT-701-2016-1, https://lib-extopc.kek.jp/preprints/PDF/2016/1627/1627021.pdf.
[3] K. Abe, et al., Super-Kamiokande Collaboration, Phys. Rev. D 94 (5) (2016) 052010, arXiv:1606.07538 [hep-ex].
[4] B. Aharmim, et al., SNO Collaboration, Phys. Rev. C 88 (2013) 025501, arXiv:1109.0763 [nucl-ex].
[5] A. Gando, et al., KamLAND Collaboration, Phys. Rev. D 83 (2011) 052002, arXiv:1009.4771 [hep-ex].
[6] F. An, et al., JUNO Collaboration, J. Phys. G 43 (3) (2016) 030401, arXiv: 1507.05613 [physics.ins-det].
[7] T. Ohlsson, Rep. Progr. Phys. 76 (2013) 044201, arXiv:1209.2710 [hep-ph]; O.G. Miranda, H. Nunokawa, New J. Phys. 17 (9) (2015) 095002, arXiv: 1505.06254 [hep-ph].
[8] L. Wolfenstein, Phys. Rev. D 17 (1978) 2369.
[9] M.M. Guzzo, A. Masiero, S.T. Petcov, Phys. Lett. B 260 (1991) 154.
[10] S.P. Mikheev, A.Y. Smirnov, Sov. J. Nucl. Phys. 42 (1985) 913, Yad. Fiz. 42 (1985) 1441.
[11] K.A. Olive, et al., Particle Data Group Collaboration, Chin. Phys. C 38 (2014) 090001.
[12] E.K. Akhmedov, M.A. Tortola, J.W.F. Valle, J. High Energy Phys. 0405 (2004) 057, arXiv:hep-ph/0404083.
[13] M.C. Gonzalez-Garcia, M. Maltoni, J. High Energy Phys. 1309 (2013) 152, arXiv: 1307.3092.
[14] J. Liao, D. Marfatia, K. Whisnant, Phys. Rev. D 92 (7) (2015) 073004, arXiv: 1506.03013 [hep-ph].
[15] A. Friedland, C. Lunardini, C. Pena-Garay, Phys. Lett. B 594 (2004) 347, arXiv: hep-ph/0402266.
[16] V.D. Barger, D. Marfatia, K. Whisnant, B.P. Wood, Phys. Rev. D 64 (2001) 073009, arXiv:hep-ph/0104095.
[17] N. Vinyoles, et al., Astrophys. J. 835 (2) (2017) 202, arXiv:1611.09867 [astro-ph.SR].
[18] J.N. Bahcall, M. Kamionkowski, A. Sirlin, Phys. Rev. D 51 (1995) 6146, arXiv: astro-ph/9502003.
[19] S.-H. Seo, private communication.
[20] J.N. Bahcall, P.I. Krastev, Phys. Rev. C 56 (1997) 2839, arXiv:hep-ph/9706239.
[21] J.N. Bahcall, M.H. Pinsonneault, S. Basu, Astrophys. J. 555 (2001) 990, arXiv: astro-ph/0010346.
[22] P. Huber, M. Lindner, W. Winter, Comput. Phys. Commun. 167 (2005) 195, arXiv:hep-ph/0407333; P. Huber, J. Kopp, M. Lindner, M. Rolinec, W. Winter, Comput. Phys. Commun. 177 (2007) 432, arXiv:hep-ph/0701187.
[23] J. Kopp, M. Lindner, T. Ota, J. Sato, Phys. Rev. D 77 (2008) 013007, arXiv: 0708.0152 [hep-ph].
[24] K. Abe, et al., arXiv:1109.3262 [hep-ex].
[25] A.M. Dziewonski, D.L. Anderson, Phys. Earth Planet. Inter. 25 (1981) 297.
[26] K. Hoshina, H. Tanaka, EGU General Assembly Conference Abstracts 14 (2012) 3246.
[27] J. Kopp, T. Ota, W. Winter, Phys. Rev. D 78 (2008) 053007, arXiv:0804.2261 [hep-ph].
[28] M. Blennow, T. Schwetz, J. High Energy Phys. 1309 (2013) 089, arXiv:1306.3988 [hep-ph].
[29] P. Bakhti, Y. Farzan, J. High Energy Phys. 1407 (2014) 064, arXiv:1403.0744 [hep-ph]; P. Coloma, T. Schwetz, Phys. Rev. D 94 (5) (2016) 055005, arXiv:1604.05772 [hep-ph].

On a reinterpretation of the Higgs field in supersymmetry and a proposal for new quarks

D.E. López-Fogliani [a,b,*], C. Muñoz [c,d,*]

[a] IFIBA, UBA & CONICET, Departamento de Física, FCEyN, Universidad de Buenos Aires, 1428 Buenos Aires, Argentina
[b] Pontificia Universidad Católica Argentina, Buenos Aires, Argentina
[c] Departamento de Física Teórica, Universidad Autónoma de Madrid, Campus de Cantoblanco, E-28049 Madrid, Spain
[d] Instituto de Física Teórica UAM-CSIC, Campus de Cantoblanco, E-28049 Madrid, Spain

ARTICLE INFO

Editor: G.F. Giudice

Keywords:
Supersymmetry
Higgses
Right-handed neutrinos
Phenomenology
New quarks

ABSTRACT

In the framework of supersymmetry, when R-parity is violated the Higgs doublet superfield H_d can be interpreted as another doublet of leptons, since all of them have the same quantum numbers. Thus Higgs scalars are sleptons and Higgsinos are leptons. We argue that this interpretation can be extended to the second Higgs doublet superfield H_u, when right-handed neutrinos are assumed to exist. As a consequence, we advocate that this is the minimal construction where the two Higgs doublets can be interpreted in a natural way as a fourth family of lepton superfields, and that this is more satisfactory than the usual situation in supersymmetry where the Higgses are 'disconnected' from the rest of the matter and do not have a three-fold replication. On the other hand, in analogy with the first three families where for each lepton representation there is a quark counterpart, we propose a possible extension of this minimal model including a vector-like quark doublet representation as part of the fourth family. We also discuss the phenomenology of the associated new quarks.

1. Introduction

The Higgs particle in the framework of the standard model is intriguing, being the only elementary scalar in the spectrum, and introducing the hierarchy problem in the theory. Besides, whereas for the rest of the matter there is a three-fold replication, this does not seem to be the case of the Higgs since only one scalar/family has been observed. In the framework of supersymmetry, the presence of the Higgs is more natural: scalar particles exist by construction, the hierarchy problem can be solved, and the models predict that the Higgs mass must be $\lesssim 140$ GeV if perturbativity of the relevant couplings up to high-energy scales is imposed. In a sense, the latter has been confirmed by the detection of a scalar particle with a mass of about 125 GeV. However, in supersymmetry the existence of at least two Higgs doublets, H_d and H_u, is necessary, as in the case of the minimal supersymmetric standard model (MSSM) [1], and as a consequence new neutral and charged scalar particles should be detected in the future to confirm the theory. Similar to the standard model, no theoretical explanation is given for the existence of only one family of Higgs doublets.

In this work we want to contribute a new vision of the Higgs(es) in the framework of supersymmetry. We will argue that the well known fact that the Higgs doublet superfield H_d has the same gauge quantum numbers as the doublets of leptons L_i, where $i = 1, 2, 3$ is the family index, is a clue that the Higgses can be reinterpreted as a fourth family of lepton superfields. Thus Higgs scalars are sleptons and Higgsinos are leptons. This can be done only when R-parity (R_p) is violated, since the standard model particles and their superpartners have opposite R_p quantum numbers. Early attempts in this direction can be found in Refs. [2,3]. In particular, in the first paper it was pointed out that in theories with TeV scale quantum gravity, the scalar H_d can be a fourth family slepton. Since H_u is not present in that construction, with its role in the Lagrangian played by H_d through non-renormalizable couplings, H_d is proposed to be part of a complete standard model family in order to cancel anomalies. In the second paper, in the context of low-energy supersymmetry the scalar H_u was also included as a slepton as part of another complete family with opposite quantum numbers to the fourth family. Thus, four chiral families with standard model quantum numbers and one chiral family with opposite quantum numbers are present in that construction.

However, with the matter content of the MSSM, which is sufficient to cancel anomalies, this interpretation of H_d as another

* Corresponding authors.
E-mail addresses: daniel.lopez@df.uba.ar (D.E. López-Fogliani), c.munoz@uam.es (C. Muñoz).

lepton superfield in the case of R_p violation cannot be extended to H_u in a natural way, as we will show in Section 2. Fortunately, as we will discuss in Section 3, when right-handed neutrino superfields are allowed in the spectrum, not only the violation of R_p turns out to be natural solving the μ problem and reproducing easily current neutrino data, but also the interpretation of H_u as part of the fourth family of lepton superfields is straightforward. Finally, we will argue in Section 4 that, as a consequence, a vector-like quark doublet representation might also be part of the new fourth family, and we will briefly discuss its phenomenology. Our conclusions are left for Section 5.

2. Supersymmetry without right-handed neutrinos

Unlike the standard model where only one Higgs doublet scalar (together with its complex conjugated representation) is sufficient to generate Yukawas couplings for quarks and charged leptons at the renormalizable level, in supersymmetry we need a vector-like Higgs doublet representation, with their superfields usually denoted as:

$$H_d = \begin{pmatrix} H_d^0 \\ H_d^- \end{pmatrix}, \quad H_u = \begin{pmatrix} H_u^+ \\ H_u^0 \end{pmatrix}. \tag{1}$$

In addition, the matter sector of the supersymmetric standard model, in the absence of right-handed neutrinos, contains also the following three families of superfields:

$$L_i = \begin{pmatrix} \nu_i \\ e_i \end{pmatrix}, \quad \begin{matrix} e_i^c \\ - \end{matrix}, \quad Q_i = \begin{pmatrix} u_i \\ d_i \end{pmatrix}, \quad \begin{matrix} d_i^c \\ u_i^c \end{matrix}, \tag{2}$$

where we have defined u_i, d_i, ν_i, e_i, and u_i^c, d_i^c, e_i^c, as the left-chiral superfields whose fermionic components are the left-handed fields of the corresponding quarks, leptons, and antiquarks, antileptons, respectively.

With this matter content, the most general gauge-invariant renormalizable superpotential is given by:

$$\begin{aligned} W = {} & \mu\, H_u\, H_d + Y_{ij}^e\, H_d\, L_i\, e_j^c + Y_{ij}^d\, H_d\, Q_i\, d_j^c - Y_{ij}^u\, H_u\, Q_i\, u_j^c \\ & + \mu_i H_u\, L_i + \lambda_{ijk}\, L_i\, L_j\, e_k^c + \lambda'_{ijk}\, L_i\, Q_j\, d_k^c + \lambda''_{ijk}\, u_i^c\, d_j^c\, d_k^c, \end{aligned} \tag{3}$$

where the summation convention is implied on repeated indexes, and our convention for the contraction of two $SU(2)$ doublets is e.g. $H_u\, H_d \equiv \epsilon_{ab} H_u^a\, H_d^b$, with ϵ_{ab} the totally antisymmetric tensor $\epsilon_{12} = 1$.

In the absence of the terms in the second line, the terms in the first line of Eq. (3) constitute the superpotential of the MSSM, where baryon (B) and lepton (L) numbers are conserved. This superpotential arises from imposing the Z_2 discrete symmetry R-parity [4], $R_p = (-1)^{2S}(-1)^{(3B+L)}$, which acts on the components of the superfields. Here S is the spin, and one obtains $R_p = +1$ for ordinary particles and -1 for their superpartners. Because of the different R_p quantum numbers, there can be no mixing between particles and superpartners.

If we allow the terms in the second line of Eq. (3) to be present, they violate R_p explicitly [4]. The first term $\mu_i H_u L_i$ which also violates lepton number, together with the superpotential of the MSSM, constitute the bilinear R-parity violation model (BRpV). This term contributes to the neutral scalar potential generating VEVs not only for the Higgses as in the MSSM, but also for the left sneutrinos, $\langle\widetilde{\nu}_{iL}\rangle \neq 0$. The other three terms are the conventional trilinear lepton- and baryon-number-violating couplings. The presence of the couplings $\mu_i, \lambda_{ijk}, \lambda'_{ijk}$, violating lepton number could have easily been argued, once the μ-term and the Yukawa couplings for d-type quarks and charged leptons are introduced in the first line of the superpotential (3), by noting that the superfields H_d and L_i have the same gauge quantum numbers. Actually, the latter fact might lead us to interpret the Higgs superfield H_d as a fourth family of lepton superfields L_4, in addition to the three families L_i of Eq. (2):

$$L_4 = \begin{pmatrix} \nu_4 \\ e_4 \end{pmatrix} = \begin{pmatrix} H_d^0 \\ H_d^- \end{pmatrix} = H_d\,. \tag{4}$$

Notice that this is not possible in the case of the MSSM because the components of the superfields H_d and L_i have opposite quantum numbers under R_p. Unfortunately, we cannot interpret naturally the other Higgs superfield H_u in a similar way, given that it has no leptonic counterpart, in particular its neutral component. We will see in the next section that this counterpart is present when we enter right-handed neutrinos in our supersymmetric framework.

On the other hand, it is well known that the simultaneous presence of the couplings λ'_{ijk} and λ''_{ijk}, violating lepton and baryon number respectively, can be dangerous since they would produce fast proton decay. The usual assumption in the literature of the MSSM of invoking R_p to avoid the problem is clearly too stringent, since then the other couplings λ_{ijk}, and μ_i in the superpotential (3), which are harmless for proton decay, would also be forbidden. A less drastic solution, taking into account that the choice of R_p is *ad hoc*, is to use other Z_N discrete symmetries to forbid only λ''_{ijk}. This is the case e.g. of Z_3 Baryon-parity [5] which also prohibits dimension-5 proton decay operators, unlike R_p. In addition, this strategy seems reasonable if one expects all discrete symmetries to arise from the breaking of gauge symmetries of the underlying unified theory [6], because Baryon-parity and R_p are the only two generalized parities which are 'discrete gauge' anomaly free [5]. Discrete gauge symmetries are also not violated [6] by potentially dangerous quantum gravity effects [7].

Given the relevance of string theory as a possible underlying unified theory, a robust argument in favour of the above mechanism is that, in string compactifications such as e.g. orbifolds, the matter superfields have several extra $U(1)$ charges broken spontaneously at high energy by the Fayet–Iliopoulos D-term, and as a consequence residual Z_N symmetries are left in the low-energy theory. As pointed out in Ref. [8], the same result can be obtained by the complementary mechanism that stringy selection rules can naturally forbid the λ''_{ijk} couplings discussed above, since matter superfields are located in general in different sectors of the compact space. As a whole, some gauge invariant operators violating R_p can be forbidden, but others are allowed [9].

Let us finally remark that although the BRpV has the interesting property of generating through the bilinear terms μ_i that mix the left-handed neutrinos ν_{iL} and the neutral Higgsino $\tilde{H}_u^0$, one neutrino mass at tree level (and the other two masses at one loop), the μ problem [10] is in fact augmented with the three new supersymmetric mass terms which must be $\mu_i \lesssim 10^{-4}$ GeV, in order to reproduce the correct values of neutrino masses. This extra problem can be avoided imposing a Z_3 symmetry in the superpotential, which implies that only trilinear terms are allowed. Actually, this is what one would expect from a high-energy theory where the low-energy modes should be massless and the massive modes of the order of the high-energy scale. As pointed out in Ref. [11], this is what happens in string constructions, where the massive modes have huge masses of the order of the string scale and the massless ones have only trilinear terms at the renormalizable level. Thus one ends up with an accidental Z_3 symmetry in the low-energy theory.

To summarize the discussion, instead of the superpotential of Eq. (3), a more natural superpotential (in the sense of free of problems) with the minimal matter content of Eqs. (1) and (2) seems to be

$$W = Y^e_{ij} H_d L_i e^c_j + Y^d_{ij} H_d Q_i d^c_j - Y^u_{ij} H_u Q_i u^c_j + \lambda_{ijk} L_i L_j e^c_k + \lambda'_{ijk} L_i Q_j d^c_k . \quad (5)$$

However, this implies that not only the bilinear terms μ_i are forbidden, but also the crucial μ term generating Higgsino masses. In the next section we will discuss a solution through the presence of right-handed neutrinos which will also allow us to interpret the two Higgs doublet superfields as a fourth family of leptons.

3. Right-handed neutrinos and reinterpretation of the Higgs superfields

Right-handed neutrinos are likely to exist in order to generate neutrino masses. Thus we will add these superfields to the minimal matter content of Eq. (2), allowing us also to write it in a more symmetric way:

$$L_i = \begin{pmatrix} \nu_i \\ e_i \end{pmatrix}, \ \begin{matrix} e^c_i \\ \nu^c_i \end{matrix}, \ Q_i = \begin{pmatrix} u_i \\ d_i \end{pmatrix}, \ \begin{matrix} d^c_i \\ u^c_i \end{matrix} . \quad (6)$$

This spectrum together with the Higgs superfields in Eq. (1) give rise to the following gauge invariant superpotential proposed in Refs. [11,8]:

$$W = Y^e_{ij} H_d L_i e^c_j + Y^d_{ij} H_d Q_i d^c_j - Y^u_{ij} H_u Q_i u^c_j - Y^\nu_{ij} H_u L_i \nu^c_j + \lambda_{ijk} L_i L_j e^c_k + \lambda'_{ijk} L_i Q_j d^c_k + \frac{1}{3}\kappa_{ijk} \nu^c_i \nu^c_j \nu^c_k + \lambda_i H_u H_d \nu^c_i . \quad (7)$$

This superpotential expands the one in Eq. (5) with the right-handed neutrinos, and is built using the same arguments of Section 2 in order to forbid the bilinear terms μ and μ_i, and the couplings violating baryon number λ''_{ijk} of superpotential (3). The first line corresponds to Yukawa couplings which conserve R_p, whereas the couplings in the second line violate R_p explicitly. In the absence of λ_{ijk} and λ'_{ijk}, R_p is restored in the limit $Y^\nu \to 0$.

The three terms with couplings Y^ν, λ_i and κ are characteristics of the 'μ from ν' supersymmetric standard model ($\mu\nu$SSM) [11, 12], and are harmless for proton decay. They contribute to the neutral scalar potential generating VEVs not only for the Higgses and the left sneutrinos as in the BRpV, but also for the right sneutrinos. As a consequence of the electroweak symmetry breaking, a μ term of the order of the electroweak scale is generated dynamically with $\mu = \lambda_i \langle \widetilde{\nu}_{iR} \rangle^*$. Let us also remark that the term with Y^ν contains the Dirac Yukawa couplings for neutrinos, and besides generates effectively bilinear couplings, $\mu_i = Y^\nu_{ij} \langle \widetilde{\nu}_{jR} \rangle^*$, as those discussed in Section 2 for the BRpV. The κ term produces Majorana masses for the right-handed neutrinos, $M_{ij} = \sqrt{2}\kappa_{ijk} \langle \widetilde{\nu}_{kR} \rangle^*$, instrumental in the generation of correct neutrino masses and mixing through a generalized electroweak-scale seesaw [11,8,13–17].

Because of the VEVs acquired by the neutral scalars and the violation of R_p, all fields in the spectrum with the same colour, electric charge and spin mix together contributing to the rich phenomenology of the $\mu\nu$SSM. For example, the neutral (scalar and pseudoscalar) Higgses mix with the left and right sneutrinos, the charged Higgses with the charged sleptons, the neutralinos of the MSSM with the left- and right-handed neutrinos, and the charginos with the charged leptons. Besides, in the $\mu\nu$SSM the scale of the breaking is set up by the soft terms, which is in the ballpark of a TeV. This nice features give rise to realistic signatures of this model at colliders [13,14,18–22], well verifiable at the LHC or at upcoming accelerator experiments. For example, prompt and/or displaced multi-leptons/taus/jets/photons final states.

Concerning cosmology in the $\mu\nu$SSM, as a consequence of R_p violation the lightest supersymmetric particle (LSP) is no longer a valid candidate for cold dark matter. Nevertheless, embedding the model in the context of supergravity, one can accommodate the gravitino [23,12] as an eligible decaying dark matter candidate with a lifetime greater than the age of the Universe. Its detection is also possible through the observation of a gamma-ray line in the Fermi satellite [23–26]. In Ref. [27], the generation of the baryon asymmetry of the universe was analysed in the $\mu\nu$SSM, with the interesting result that electroweak baryogenesis can be realised.

Similarly to the discussion for λ_{ijk} and λ'_{ijk} in Section 2, the presence of the λ_i term which violates lepton number in the superpotential (7), could have been deduced from the presence of the couplings Y^ν because of the same quantum numbers for the superfields H_d and L_i. Actually, the simultaneous presence of both terms in order to solve the μ problem and generate correct neutrino masses implies that all the charges of H_d and L_i must be the same even if extra $U(1)$'s are present. Thus the argument of the extra $U(1)$ charges used in Section 2 to forbid the couplings λ''_{ijk} is unlikely that can be used in typical string models to forbid the couplings λ_{ijk} and λ'_{ijk}. This makes more robust the superpotential (7). Besides, as discussed in Ref. [8], even if λ_{ijk} and λ'_{ijk} are set to zero, they are generated by loop corrections (although with very small values) due to the presence in the superpotential of couplings like Y^d, Y^ν, λ_i.

In Section 2, the fact that the superfields H_d and L_i have the same gauge quantum numbers led us to discuss in Eq. (4) the possibility of interpreting H_d as a fourth family of lepton superfields L_4. However, we were not able to interpret naturally the Higgs superfield H_u in a similar way, given that it has no leptonic counterpart in the spectrum of Eq. (2). On the contrary, for the spectrum of Eq. (6) it is possible to interpret H_u as another lepton superfield L^c_4:

$$L^c_4 = \begin{pmatrix} e^c_4 \\ \nu^c_4 \end{pmatrix} = \begin{pmatrix} H^+_u \\ H^0_u \end{pmatrix} = H_u . \quad (8)$$

Thus, at the level of weak eigenstates the superfield H_d/L_4 contains the fourth-family left sneutrino and the H_u/L^c_4 the fourth-family right sneutrino, as shown in Eqs. (4) and (8). In the limit were the others sneutrinos are decoupled in our model, the Higgs discovered at the LHC is described by a mixture of H_u and H_d as in the case of the MSSM. In addition, also as in the latter case, for reasonable values of $\tan\beta$ the standard model-like Higgs is mainly H_u. Therefore, in this supersymmetric framework the first scalar particle discovered at the LHC is mainly a right sneutrino belonging to a fourth-family vector-like lepton doublet representation.

To complete the argument, we must take into account what was mentioned above, that once the electroweak symmetry is broken the first three families of sneutrinos turn out to be mixed with the fourth one. Nevertheless, the left sneutrinos of the first three families are decoupled in all cases, since the mixing occurs through terms proportional to neutrino Yukawas or left sneutrino VEVs which are very small [8]. Concerning the right sneutrinos of the first three families, they are singlets of $SU(2)$ and can mix in general with the doublets H_u and H_d, similarly to the case of the Next-to-MSSM (NMSSM) [28] where one extra singlet is present. As a consequence, the decoupling limit is not necessarily a good approximation. For our model, where three singlets are present, discussions about viable regions of the parameter space and the expected signals at colliders were carried out in Refs. [19] and [22]. In those works, where not only LHC constraints but also LEP and

Tevatron ones were applied to the parameter space, viable regions were obtained.

Summarizing, Eqs. (4) and (8) constitute our reinterpretation of Eq. (1), and therefore we can write the whole spectrum in the following way:

$$L_i = \begin{pmatrix} \nu_i \\ e_i \end{pmatrix}, \quad \begin{matrix} e_i^c \\ \nu_i^c \end{matrix}, \quad Q_i = \begin{pmatrix} u_i \\ d_i \end{pmatrix}, \begin{matrix} d_i^c \\ u_i^c \end{matrix},$$
$$L_4 = \begin{pmatrix} \nu_4 \\ e_4 \end{pmatrix}, \; L_4^c = \begin{pmatrix} e_4^c \\ \nu_4^c \end{pmatrix}. \tag{9}$$

With this notation, Eq. (7) can be written in a more compact way as:

$$W = Y^e_{IJk} L_I L_J e^c_k + Y^d_{Ijk} L_I Q_j d^c_k - Y^u_{4jk} L^c_4 Q_j u^c_k - Y^\nu_{4Jk} L^c_4 L_J \nu^c_k + \frac{1}{3}\kappa_{ijk} \nu^c_i \nu^c_j \nu^c_k, \tag{10}$$

where $I = i, 4$ and $J = j, 4$ are the new family indexes, with $i, j, k = 1, 2, 3$, and the notation for the Yukawa couplings is self-explanatory.

4. Proposal for new quarks

We have identified in the previous section the minimal model where the two Higgs superfields can be interpreted in a natural way as a fourth family of leptons. One might think that this is just an academic discussion, in the sense that superpotential (10) is equivalent from the operational viewpoint to superpotential (7). Nevertheless, in this framework where in principle vector-like matter can be added to the fourth family consistently with the experiments, we find it natural to make the following proposal. In analogy with the first three families in Eq. (9), where each lepton representation has its quark counterpart, we add to the spectrum of the fourth family a vector-like quark doublet representation as counterpart of the vector-like lepton/Higgs doublet representation, implying in superfield notation:

$$L_i = \begin{pmatrix} \nu_i \\ e_i \end{pmatrix}, \quad \begin{matrix} e_i^c \\ \nu_i^c \end{matrix}, \quad Q_i = \begin{pmatrix} u_i \\ d_i \end{pmatrix}, \quad \begin{matrix} d_i^c \\ u_i^c \end{matrix},$$
$$L_4 = \begin{pmatrix} \nu_4 \\ e_4 \end{pmatrix}, \; L_4^c = \begin{pmatrix} e_4^c \\ \nu_4^c \end{pmatrix}, \; Q_4 = \begin{pmatrix} u_4 \\ d_4 \end{pmatrix}, \; Q_4^c = \begin{pmatrix} d_4^c \\ u_4^c \end{pmatrix}, \tag{11}$$

where Q_4 has hypercharge $\frac{1}{6}$ as for the first three families, whereas Q_4^c has by construction hypercharge $-\frac{1}{6}$ allowing the cancellation of anomalies[1]. It is worth noticing here that the presence of extra vector-like matter is a common situation in string constructions[2] (see e.g. [9,31–33]).

The spectrum of Eq. (11) implies that the following terms associated to the presence of the new quarks must be added to the superpotential in Eq. (7):

$$W = \lambda'_{i4k} L_i Q_4 d^c_k + Y^d_{4k} H_d Q_4 d^c_k - Y^u_{4k} H_u Q_4 u^c_k + Y^Q_{j4k} Q_j Q^c_4 \nu^c_k + Y^Q_{44k} Q_4 Q^c_4 \nu^c_k, \tag{12}$$

where the first one corresponds to trilinear lepton-number-violating couplings, the second and third contribute to the Yukawa couplings with the Higgses, and the last two terms contribute to the quark masses once the right sneutrinos acquire VEVs.

Working in low-energy supersymmetry, these terms will induce the corresponding trilinear soft-supersymmetry breaking terms in the Lagrangian. Together with the soft masses for the squark doublets $\widetilde{Q}_4$ and $\widetilde{Q}^c_4$ they constitute the new terms in the soft Lagrangian. Notice that none of them contributes to the minimization of the tree-level neutral scalar potential.

Using now the 'new' notation for the Higgs superfields, Eq. (12) can be written as

$$W = Y^d_{I4k} L_I Q_4 d^c_k - Y^u_{44k} L^c_4 Q_4 u^c_k + Y^Q_{J4k} Q_J Q^c_4 \nu^c_k. \tag{13}$$

This equation together with Eq. (10) allow us to write the whole superpotential in the compact notation:

$$W = Y^e_{IJk} L_I L_J e^c_k + Y^d_{IJk} L_I Q_J d^c_k - Y^u_{4Jk} L^c_4 Q_J u^c_k - Y^\nu_{4Jk} L^c_4 L_J \nu^c_k + Y^Q_{J4k} Q_J Q^c_4 \nu^c_k + \frac{1}{3}\kappa_{ijk} \nu^c_i \nu^c_j \nu^c_k. \tag{14}$$

New phenomenology is expected from the presence of the new quarks (and squarks) of the fourth family. Here we will discuss the specially interesting case of the quarks, given their mixing with the standard model ones and therefore the modification of the usual couplings to the W, Z and Higgs boson. For example, although in this construction the Higgs mass is already enhanced at tree level due to the λ_i couplings [8], the presence of the new quarks could help to enhance it further through one-loop effects [34]. Besides, the presence of flavour changing neutral currents (FCNCs) leads to a wide range of final states that can be analysed. Notice that large enough masses for the new quarks to be beyond the present experimental bounds, but still accessible at the LHC, can be generated by the last term in Eq. (12) with a Yukawa coupling $Y^Q_{44k} \sim 1$ and typical VEVs of the right sneutrinos $\langle\widetilde{\nu}_{kR}\rangle \sim$ TeV as discussed in Section 3.

In the basis of 2-components spinors $(u^*_L)^T = (u^*_{IL})$, $(u_R)^T = (u_{JR})$, one obtains the following up-quark mass terms in the Lagrangian:

$$\mathcal{L}_{\rm mass} = -(u^*_L)^T m_u u_R + \text{h.c.}, \tag{15}$$

where, using a compact block notation,

$$m_u = \begin{pmatrix} Y^{u*}_{ij}\langle H^0_u\rangle^* & Y^{Q\,*}_{i4k}\langle\widetilde{\nu}_{kR}\rangle \\ Y^{u\,*}_{4j}\langle H^0_u\rangle^* & Y^{Q\,*}_{44k}\langle\widetilde{\nu}_{kR}\rangle \end{pmatrix}. \tag{16}$$

We can simplify further this matrix redefining the left-handed fields in such a way that the new entries $(m_u)_{i4}$ are vanishing and $(m_u)_{ij} = Y'^{u*}_{ij}\langle H^0_u\rangle^*$, with Y'^u the redefined Yukawa coupling. After these replacements, the 4×4 mass matrix is diagonalized by two unitary matrices U^u_L and U^u_R:

$$U^{u\dagger}_L m_u U^u_R = m^{\rm dia}_u, \tag{17}$$

with

$$u_R = U^u_R U_R, \; u_L = U^u_L U_L. \tag{18}$$

Here, the 4 entries of the matrices U_L, U_R are the 2-component up-quark mass eigenstate fields. After a phase redefinition of the d_{4R} field to recover the conventions for the non-supersymmetric standard model extensions with vector-like quarks [35], the same formulas apply to the down-quark sector with the replacements $Y^u \to Y^d$ and $\langle H^0_u\rangle \to \langle H^0_d\rangle$ in Eq. (16).

Taking the above mixing matrices into account, in the basis of 4-components spinors with the projectors $P_{L,R} = \frac{1}{2}(1 \mp \gamma_5)$, charged currents are modified in the following way:

[1] Other extensions of the $\mu\nu$SSM were discussed in Ref. [29] in the context of an extra $U(1)$ gauge symmetry.

[2] For a standard-like model containing only the Higgs doublets as vector-like representations, see however Ref. [30]. Remarkably, in that model the presence of three families of right-handed neutrinos is mandatory to achieve anomaly cancellation.

$$\mathcal{L}_W = -\frac{g}{\sqrt{2}} \left(\overline{U}\gamma^\mu V_L P_L D + \overline{U}\gamma^\mu V_R P_R D \right) W^+_\mu + \text{h.c.}, \tag{19}$$

where

$$V_L = U_L^{u\dagger} U_L^d, \quad V_R = U_R^{u\dagger} \delta_R^{\text{dia}} U_R^d, \tag{20}$$

and the matrix $\delta_R^{\text{dia}} = \text{dia}(0, 0, 0, 1)$. Here, the measured CKM matrix corresponds to the (non-unitary) 3×3 block $(V_L)_{ij}$, and another (non-unitary) CKM matrix V_R must be defined for the right-handed quarks because of the new doublet $Q_{4R}^T = (u_{4R}, d_{4R})$.

Tree-level FCNC also occur due to the mixing in the right-handed sector induced by the new doublet. In particular, the neutral current interactions of quarks in the Lagrangian are:

$$\mathcal{L}_Z = -\frac{g}{2\cos\theta_W} \left(\overline{U}\gamma^\mu P_L U + \overline{U}\gamma^\mu X^u P_R U - \overline{D}\gamma^\mu P_L D \right.$$
$$\left. - \overline{D}\gamma^\mu X^d P_R D - 2\sin^2\theta_W J^\mu_{em} \right) Z_\mu, \tag{21}$$

where the matrices

$$X^u = V_R V_R^\dagger = U_R^{u\dagger} \delta_R^{\text{dia}} U_R^u, \quad X^d = V_R^\dagger V_R = U_R^{d\dagger} \delta_R^{\text{dia}} U_R^d, \tag{22}$$

are hermitian and non-diagonal, mediating FCNCs.

Finally, the modified couplings between the neutral components of the Higgs doublets and quarks

$$\mathcal{L}_{H_u} = -\frac{1}{\langle H_u^0 \rangle^*} \overline{U} \left(m_u^{\text{dia}} - (m_u)_{44} U_L^{u\dagger} \delta_R^{\text{dia}} U_R^u \right) P_R U \, H_u^{0*} + \text{h.c.},$$
$$\mathcal{L}_{H_d} = -\frac{1}{\langle H_d^0 \rangle^*} \overline{D} \left(m_d^{\text{dia}} - (m_d)_{44} U_L^{d\dagger} \delta_R^{\text{dia}} U_R^d \right) P_R D \, H_d^{0*} + \text{h.c.}, \tag{23}$$

may change the production and decay of the standard model-like Higgs.

Numerous analyses of the phenomenology of vector-like quark singlet, doublet and triplet representations have been carried out in the literature in extensions of the standard model [36,37,35], studying limits on mixing between ordinary quarks and heavy partners, the allowed range of splitting between the heavy states, and the production at the LHC. In particular, the non-observation of FCNCs put stringent constraints on mixing, and only one light quark can have significant mixing with the vector-like quark. Since the vector-like quarks are usually expected to mix predominantly with the third generation [38], one can obtain upper limits on the corresponding mixing angles from new contributions to the oblique parameters S and T, and $Z \to b\bar{b}$ observables [35]. These limits for the case of a vector-like doublet (T, B), where T is a new up-type of quark with charge $+2/3$ and B is a new down-type of quark with charge $-1/3$, can be applied to our model and are given for the right-handed fields by $\sin\theta_R^u \lesssim 0.1$ and $\sin\theta_R^d \lesssim 0.06$. The mixing angles for the left-handed fields are not independent and must satisfy $\tan\theta_L^u = \frac{m_t}{m_T} \tan\theta_R^u$ and $\tan\theta_L^d = \frac{m_b}{m_B} \tan\theta_R^d$.

Concerning detection at the LHC, pair production processes dominated by QCD have the advantage of being model independent, with the new heavy quarks subsequently decaying into ordinary quarks and a gauge boson or a Higgs (see couplings in Eqs. (19), (21) and (23) for our case). A recent search [39] yields observed lower limits on T ranging between 715 and 950 GeV for all possible values of the branching ratios into the three decay modes $T \to Wb$, $T \to Zt$ and $T \to Ht$. Similarly, for B the $B\bar{B}$ production implies that the limits range between 575 and 813 GeV for all possible values of the branching ratios into the three decay modes $B \to Wt$, $B \to Zb$ and $B \to Hb$. In these analyses, the above limits on mixing angles are applied since it is assumed that the new quarks mainly couple to the third generation. If they are allowed to mix with all standard model families, dedicated searches may be necessary.

The above mass bounds can be applied to our supersymmetric case if the light Higgs is a standard model-like Higgs particle and the decays of the fourth-family quarks involving non-standard model particles (such as e.g. the heavier Higgses or squarks) are negligible. Otherwise, the new branching ratios should be taken into account implying a new phenomenology. In addition, if the lepton-number-violating couplings λ' of Eqs. (7) and (12), which also violate R_p, are not small enough, they also could give rise to new channels modifying the single production of the new quarks, as well as their decay processes. The analysis of these possibilities is beyond the scope of this work, and we plan to cover it in a forthcoming publication [40]. On the other hand, the new processes induced by the terms characteristics of the $\mu\nu$SSM, and violating also R_p, can be safely neglected because of the small value of Y^ν.

5. Conclusions

In this work, in the framework of supersymmetry with right-handed neutrinos we have been able to reinterpret in a natural way the Higgs superfields as a fourth family of lepton superfields. From the theoretical viewpoint, this seems to be more satisfactory than the situation in usual supersymmetric models, where the Higgses are 'disconnected' from the rest of the matter and do not have a three-fold replication. Inspired by this interpretation of the Higgs superfields, we have also proposed the possible existence of a vector-like quark doublet representation in the low-energy supersymmetric spectrum. These new quark superfields have the implication of a potentially rich phenomenology at the LHC.

Acknowledgements

We gratefully acknowledge J.A. Aguilar-Saavedra, A. Casas, J. Moreno and A. Uranga for useful discussions. The work of D.E. López-Fogliani was supported by the Argentinian CONICET. He acknowledges the hospitality of the IFT during whose stay this work was started. The work of C. Muñoz was supported in part by the Programme SEV-2012-0249 'Centro de Excelencia Severo Ochoa'. We also acknowledge the support of the Spanish grant FPA2015-65929-P (MINECO/FEDER, UE), and MINECO's Consolider-Ingenio 2010 Programme under grant MultiDark CSD2009-00064.

References

[1] For a review, see: S.P. Martin, A supersymmetry primer, Adv. Ser. Dir. High Energy Phys. 18 (2008) 1, arXiv:hep-ph/9709356.

[2] A.K. Grant, Z. Kakushadze, Higgs as a slepton, Phys. Lett. B 465 (1999) 108, arXiv:hep-ph/9906556.

[3] C. Liu, Supersymmetry and an extra vectorlike generation, Phys. Rev. D 80 (2009) 035004, arXiv:0907.3011 [hep-ph].

[4] For a review, see: R. Barbier, et al., R-parity-violating supersymmetry, Phys. Rep. 420 (2005) 1, arXiv:hep-ph/0406039.

[5] L.E. Ibanez, G.G. Ross, Discrete gauge symmetry anomalies, Phys. Lett. B 260 (1991) 291;
Discrete gauge symmetries and the origin of baryon and lepton number conservation in supersymmetric versions of the standard model, Nucl. Phys. B 368 (1992) 3.

[6] L. Krauss, F. Wilczek, Discrete gauge symmetry in continuum theories, Phys. Rev. Lett. 62 (1989) 90.

[7] G. Gilbert, Wormhole-induced proton decay, Nucl. Phys. B 328 (1989) 159, and references therein.

[8] N. Escudero, D.E. López-Fogliani, C. Muñoz, R.R. de Austri, Analysis of the parameter space and spectrum of the $\mu\nu$SSM, J. High Energy Phys. 12 (2008) 099, arXiv:0810.1507 [hep-ph].

[9] J.A. Casas, E.K. Katehou, C. Muñoz, $U(1)$ charges in orbifolds: anomaly cancellation and phenomenological consequences, Oxford preprint, Nov. 1987, Ref: 1/88;
Nucl. Phys. B 317 (1989) 171;
J.A. Casas, C. Muñoz, Yukawa couplings in $SU(3) \times SU(2) \times U(1)_Y$ orbifold models, Phys. Lett. B 212 (1988) 343.

[10] J.E. Kim, H.P. Nilles, The μ problem and the strong CP problem, Phys. Lett. B 138 (1984) 150.
[11] D.E. López-Fogliani, C. Muñoz, Proposal for a supersymmetric standard model, Phys. Rev. Lett. 97 (2006) 041801, arXiv:hep-ph/0508297.
[12] For a recent review, see C. Muñoz, Searching for SUSY and decaying gravitino DM at the LHC and Fermi-LAT with the $\mu\nu$SSM, in: Proceedings, 11th International Workshop on the Dark Side of the Universe (DSU2015), Kyoto, Japan, December 14–18, 2015, PoS (**DSU2015**) 034, 2015, arXiv:1608.07912 [hep-ph].
[13] P. Ghosh, S. Roy, Neutrino masses and mixing, lightest neutralino decays and a solution to the μ problem in supersymmetry, J. High Energy Phys. 04 (2009) 069, arXiv:0812.0084 [hep-ph].
[14] A. Bartl, M. Hirsch, A. Vicente, S. Liebler, W. Porod, LHC phenomenology of the $\mu\nu$SSM, J. High Energy Phys. 05 (2009) 120, arXiv:0903.3596 [hep-ph].
[15] J. Fidalgo, D.E. López-Fogliani, C. Muñoz, R.R. de Austri, Neutrino physics and spontaneous CP violation in the $\mu\nu$SSM, J. High Energy Phys. 08 (2009) 105, arXiv:0904.3112 [hep-ph].
[16] P. Ghosh, P. Dey, B. Mukhopadhyaya, S. Roy, Radiative contribution to neutrino masses and mixing in $\mu\nu$SSM, J. High Energy Phys. 05 (2010) 087, arXiv:1002.2705 [hep-ph].
[17] For a review, see D.E. López-Fogliani, The Seesaw mechanism in the $\mu\nu$SSM, in: CTP International Conference on Neutrino Physics in the LHC Era Luxor, Egypt, November 15–19, 2009, arXiv:1004.0884 [hep-ph].
[18] P. Bandhopadhyay, P. Ghosh, S. Roy, Unusual Higgs boson signal in R-parity violating nonminimal supersymmetric models at the LHC, Phys. Rev. D 84 (2011) 115022, arXiv:1012.5762 [hep-ph].
[19] J. Fidalgo, C. Muñoz, The Higgs sector of the $\mu\nu$SSM and collider physics, J. High Energy Phys. 10 (2011) 020, arXiv:1107.4614 [hep-ph].
[20] P. Ghosh, D.E. López-Fogliani, V. Mitsou, C. Muñoz, R. Ruiz de Austri, Probing the μ-from-ν supersymmetric standard model with displaced multileptons from the decay of a Higgs boson at the LHC, Phys. Rev. D 88 (2013) 015009, arXiv:1211.3177 [hep-ph].
[21] P. Ghosh, D.E. López-Fogliani, V. Mitsou, C. Muñoz, R. Ruiz de Austri, Hunting physics beyond the standard model with unusual $W^{\pm}$ and Z decays, Phys. Rev. D 91 (2015) 035020, arXiv:1403.3675 [hep-ph].
[22] P. Ghosh, D.E. López-Fogliani, V. Mitsou, C. Muñoz, R. Ruiz de Austri, Probing the $\mu\nu$SSM with light scalars, pseudoscalars and neutralinos from the decay of a SM-like Higgs boson at the LHC, J. High Energy Phys. 11 (2014) 102, arXiv:1410.2070 [hep-ph].
[23] K.-Y. Choi, D.E. López-Fogliani, C. Muñoz, Roberto Ruiz de Austri, Gamma-ray detection from gravitino dark matter decay in the $\mu\nu$SSM, J. Cosmol. Astropart. Phys. 03 (2010) 028, arXiv:0906.3681 [hep-ph].
[24] G.A. Gómez-Vargas, M. Fornasa, F. Zandanel, A.J. Cuesta, C. Muñoz, F. Prada, G. Yepes, CLUES on Fermi-LAT prospects for the extragalactic detection of $\mu\nu$SSM gravitino dark matter, J. Cosmol. Astropart. Phys. 02 (2012) 001, arXiv:1110.3305 [hep-ph].
[25] A. Albert, G.A. Gómez-Vargas, M. Grefe, C. Muñoz, C. Weniger, E.D. Bloom, E. Charles, M.N. Mazziotta, A. Morselli, Search for 100 MeV to 10 GeV γ-ray lines in the Fermi-LAT data and implications for gravitino dark matter in the $\mu\nu$SSM, J. Cosmol. Astropart. Phys. 10 (2014) 023, arXiv:1406.3430 [astro-ph.HE].
[26] G.A. Gómez-Vargas, D.E. López-Fogliani, C. Muñoz, A.D. Pérez, R. Ruiz de Austri, Search for sharp and smooth spectral signatures of $\mu\nu$SSM gravitino dark matter with Fermi-LAT, J. Cosmol. Astropart. Phys. 03 (2017) 047, arXiv:1608.08640 [hep-ph].
[27] D.J.H. Chung, A.J. Long, Electroweak phase transition in the $\mu\nu$SSM, Phys. Rev. D 81 (2010) 123531, arXiv:1004.0942 [hep-ph].
[28] For a review, see U. Ellwanger, C. Hugonie, A.M. Teixeira, The next-to-minimal supersymmetric standard model, Phys. Rep. 496 (2010) 1, arXiv:0910.1785 [hep-ph].
[29] J. Fidalgo, C. Muñoz, The $\mu\nu$SSM with an Extra U(1), J. High Energy Phys. 04 (2012) 090, arXiv:1111.2836 [hep-ph].
[30] L.E. Ibáñez, F. Marchesano, R. Rabadán, Getting just the standard model at intersecting branes, J. High Energy Phys. 11 (2001) 002, arXiv:hep-th/0105155.
[31] J.A. Casas, C. Muñoz, Three-generation $SU(3) \times SU(2) \times U(1)_Y \times U(1)$ orbifold models through Fayet–Iliopoulos terms, Phys. Lett. B 209 (1988) 214;
Three-generation $SU(3) \times SU(2) \times U(1)_Y$ models from orbifolds, Phys. Lett. B 212 (1988) 343.
[32] A. Font, L.E. Ibáñez, H.P. Nilles, F. Quevedo, Yukawa couplings in degenerate orbifolds: towards a realistic $SU(3) \times SU(2) \times U(1)$ superstring, Phys. Lett. B 210 (1988) 101; Erratum: Phys. Lett. B 213 (1988) 564;
A. Font, L.E. Ibáñez, F. Marchesano, A. Sierra, The construction of 'realistic' four-dimensional strings through orbifolds, Nucl. Phys. B 331 (1990) 421.
[33] M. Cvetic, G. Shiu, A.M. Uranga, Three-family supersymmetric standard-like models from intersecting brane worlds, Phys. Rev. Lett. 87 (2001) 201801, arXiv:hep-th/0107143.
[34] K. Nickel, F. Staub, Precise determination of the Higgs mass in supersymmetric models with vectorlike tops and the impact on naturalness in minimal GMSB, J. High Energy Phys. 07 (2015) 139, arXiv:1505.06077 [hep-ph].
[35] For a review, see J.A. Aguilar-Saavedra, R. Benbrik, S. Heinemeyer, M. Pérez-Victoria, A handbook of vector-like quarks: mixing and single production, Phys. Rev. D 88 (2013) 094010, arXiv:1306.0572 [hep-ph].
[36] For a review, see J.A. Aguilar-Saavedra, Identifying top partners at LHC, J. High Energy Phys. 11 (2009) 030, arXiv:0907.3155 [hep-ph].
[37] For a review, see Y. Okada, L. Panizzi, LHC signatures of vector-like quarks, Adv. High Energy Phys. 11 (2013) 2013, arXiv:1207.5607 [hep-ph].
[38] J.A. Aguilar-Saavedra, Mixing with vector-like quarks: constraints and expectations, EPJ Web Conf. 60 (2013) 16012, arXiv:1306.4432 [hep-ph].
[39] G. Aad, et al., ATLAS Collaboration, Search for production of vector-like quark pairs and of four top quarks in the lepton-plus-jets final state in pp collisions at $\sqrt{8}$ TeV with the ATLAS detector, J. High Energy Phys. 08 (2015) 105, arXiv:1505.04306 [hep-ex].
[40] J.A. Aguilar-Saavedra, D.E. López-Fogliani, C. Muñoz, Novel signatures for vector-like quakrs, arXiv:1705.02526.

18

On the difference of time-integrated CP asymmetries in $D^0 \to K^+K^-$ and $D^0 \to \pi^+\pi^-$ decays: Unparticle physics contribution

Hosein Bagheri, Mohammadmahdi Ettefaghi, Reza Moazzemi *

Department of Physics, University of Qom, Ghadir Blvd., Qom 371614-6611, I.R. Iran

ARTICLE INFO

Editor: J. Hisano

Keywords:
CP violation
D meson decay
Unparticle physics

ABSTRACT

The LHCb Collaboration has recently measured the difference of time-integrated CP asymmetry in $D^0 \to K^+K^-$ and $D^0 \to \pi^+\pi^-$ decays, more precisely. The reported value is $\Delta A_{CP} = -0.10 \pm 0.08(\text{stat}) \pm 0.03(\text{syst})\%$ which indicates no evidence for CP violation. We consider the possible unparticle physics contribution in this quantity and by using the LCHb data try to constrain the parameter space of unparticle stuff.

1. Introduction

The standard model (SM), being a very successful model in all tests carried out so far, has been completed with the discovery of the Higgs boson, the last major missing piece of the SM, during the first LHC run. However, some problems such as baryon asymmetry, dark matter, neutrino oscillation etc. call for new physics beyond the SM. Therefore, physicists are now looking for phenomena that do not conform the SM predictions. In particular, to have an appropriate mechanism for the explanation of baryon asymmetry, one expects deviations of the SM predictions for CP violation. Until now, CP violation signals have been observed in the quark sector of the SM *e.g.* in K mesons [1] and B mesons [2,3] decays which are consistent with the SM predictions. The SM prediction of CP violation for D mesons is very small, so that "a measurement of CP violation in D decays would be a signal of new physics". For instance, in singly Cabibbo suppressed (SCS) decays the CP asymmetry is estimated to be about $\mathcal{O}(0.05\text{–}0.1)\%$ [4]. In addition, there exist large uncertainties in the theoretical estimations of parameters so that the interpretation of measurements in the context of the SM encounters with some ambiguities. In fact, the c quark is not heavy enough to apply heavy quark effective theory (like in B physics), on the other hand it is not light enough to use chiral perturbation theory (like in kaon physics). Moreover, the rate of mixing in neutral charm mesons is extremely small. Hence, the indirect CP violations (CP violation through mixing and the interference of the mixing and decay) are negligible in charm processes. In particular, when the difference of time-integrated CP asymmetries in $D^0 \to K^+K^-$ and $D^0 \to \pi^+\pi^-$ decay is measured, the indirect contribution almost vanishes in the difference.

The LHCb Collaboration has recently measured the difference between the CP asymmetry in $D^0 \to K^+K^-$ and $D^0 \to \pi^+\pi^-$ to be $\Delta A_{CP} \equiv A_{CP}(D^0 \to K^+K^-) - A_{CP}(D^0 \to \pi^+\pi^-) = [-0.10 \pm 0.08(\text{stat}) \pm 0.03(\text{syst})]\%$ [5]. This measurement has superseded the previous result obtained using the same decay channels [6–8].

According to the SM, the tree level amplitudes of $D^0 \to K^+K^-$ and $D^0 \to \pi^+\pi^-$ decays involve only the first two quark generations, which cannot have the CP violating Kobayashi–Maskawa (KM) phase. Therefore, both the weak and strong phases needed for the direct CP violation come from the loop-induced gluon penguin diagrams. This implies that the SM prediction is loop suppressed as well as CKM suppressed; parametrically we have $\mathcal{O}((\alpha_s/\pi)(V_{ub}V_{cb}^*)/(V_{us}V_{cs}^*)) \sim 10^{-4}$. Several authors have tried to improve estimates for ΔA_{CP} in the SM [4,9,10]. Although there is large uncertainty in the SM value of ΔA_{CP}, it is important and exciting to consider the recent measurement of ΔA_{CP} as possible new physics and explore allowed parameter space [11–13].

It has been conjectured that there may exist "stuff" that does not necessarily have zero mass but is still scale-invariant. This stuff cannot be described as particle, so it has been called "unparticle" [14]. If the unparticle stuff exists, it must couple with normal matter weakly, since there is not yet any observed signal confirming it. However, several new physical results due to the existence of unparticle have been explored extensively, see for instance [15–21]. The unparticle effects are also involved in the study of various

* Corresponding author.
E-mail address: r.moazzemi@qom.ac.ir (R. Moazzemi).

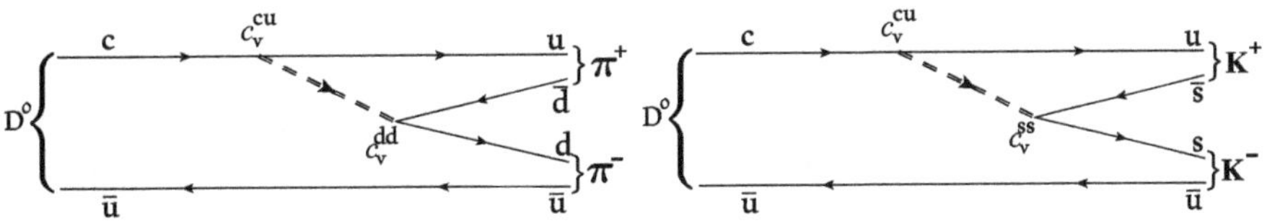

Fig. 1. Diagrams for decay of D^0 to $\pi^+\pi^-$ and K^+K^- final states via unparticle mediator.

decays, beyond the SM [22–26]. In particular, the CMS Collaboration has been recently searched for dark matter and unparticles in events with a Z boson and missing transverse momentum in proton–proton collisions at $\sqrt{s} = 13$ TeV [27]. Also, the peculiar CP conserving phases in unparticle propagators lead to a significant impact on CP violation [28–32]. Moreover, since the unparticle couplings to quarks can be complex in general, some CP violating phases, in addition to the SM weak CP phase, can arise from these new couplings. More recently, we study the CP violation in Cabibbo favored decays of D mesons via unparticle physics [33]. In this letter, we study this effect on singly Cabibbo suppressed (SCS) charm mesons decays. Namely, we obtain the unparticle contributions in the difference CP asymmetry of $D^0 \to \pi^+\pi^-$ and $D^0 \to K^+K^-$. Hereby, using the experimental measurement of this difference, we shall study the relevant parameter space of unparticle physics.

In the next two sections we first briefly review the unparticle physics then consider its contribution to the time-integrated CP asymmetries of $D^0 \to K^+K^-$ and $D^0 \to \pi^+\pi^-$ decays. In the last section we summarize our discussion and conclusions.

2. A brief review of unparticle physics

One can suppose that the very high energy physics contains a scale invariant sector with a nontrivial IR fixed point (Banks-Zaks theory) [34]. The properties and signatures of this sector are different from the SM particles, hence Georgi termed it "unparticle" [14]. Unparticles can interact with the SM particles through the exchange of particles with a large mass scale $M_{\mathcal{U}}$. Below this scale, one can write nonrenormalizable couplings involving both SM fields and Banks–Zaks (BZ) fields suppressed by powers of $M_{\mathcal{U}}$ as follows:

$$\frac{1}{M_{\mathcal{U}}^k}\mathcal{O}_{SM}\mathcal{O}_{BZ}, \tag{1}$$

where $\mathcal{O}_{SM}$ ($\mathcal{O}_{BZ}$) is an operator with mass dimension d_{SM} (d_{BZ}) built out of the SM (BZ) fields. The scale invariance of BZ sector emerges in an energy scale $\Lambda_{\mathcal{U}}$ where the BZ operators match onto unparticle operators. Here we have a dimensional transmutation due to the renormalizable coupling of BZ operators. Therefore, the effective interaction between unparticle and SM operators below $\Lambda_{\mathcal{U}}$ can be written as follows:

$$\frac{1}{M_{\mathcal{U}}^k}\mathcal{O}_{sm}\mathcal{O}_{BZ} \xrightarrow{\text{lower energy}} \frac{C_{\mathcal{U}}\Lambda_{\mathcal{U}}^{d_{BZ}-d_{\mathcal{U}}}}{M_{\mathcal{U}}^k}\mathcal{O}_{SM}\mathcal{O}_{\mathcal{U}}, \tag{2}$$

where $d_{\mathcal{U}}$ is the scaling dimension of the unparticle operator $\mathcal{O}_{\mathcal{U}}$, and the constant $C_{\mathcal{U}}$ is a coefficient function. The Lorentz structures of unparticle can be different, i.e. $\mathcal{O}_U \equiv \mathcal{O}_{\mathcal{U}}, \mathcal{O}_{\mathcal{U}}^{\mu}, \mathcal{O}_{\mathcal{U}}^{\mu\nu}$. Couplings of these operators to all possible gauge invariant SM ones, with dimensions less than or equal to 4 are listed in [35]. Here, we consider the following effective interactions of scalar and vector unparticle operators with operators composed of quarks,

$$\frac{c_V^{q'q}}{\Lambda_{\mathcal{U}}^{d_{\mathcal{U}}-1}}\bar{q}'\gamma_\mu(1-\gamma_5)q\mathcal{O}_{\mathcal{U}}^{\mu} + \frac{c_S^{q'q}}{\Lambda_{\mathcal{U}}^{d_{\mathcal{U}}}}\bar{q}'\gamma_\mu(1-\gamma_5)q\partial^\mu\mathcal{O}_{\mathcal{U}}, \tag{3}$$

where $c_V^{q'q}$ and $c_S^{q'q}$ are the dimensionless couplings. Moreover, the propagator of a scalar (vector) unparticle is given by [14]

$$\int d^4x e^{ip.x}\langle 0|T(\mathcal{O}_{\mathcal{U}}^{(\mu)}(x)\mathcal{O}_{\mathcal{U}}^{(\nu)}(0))|0\rangle = \Delta_{\mathcal{U}}^{s(v)}(p^2)e^{-i\phi_{\mathcal{U}}}, \tag{4}$$

with

$$\Delta_{\mathcal{U}}^{s}(p^2) = \frac{A_{d_{\mathcal{U}}}}{2\sin(d_{\mathcal{U}}\pi)}\frac{1}{(p^2+i\epsilon)^{2-d_{\mathcal{U}}}}, \tag{5}$$

$$\Delta_{\mathcal{U}}^{v}(p^2) = \frac{A_{d_{\mathcal{U}}}}{2\sin(d_{\mathcal{U}}\pi)}\frac{-g^{\mu\nu}+p^\mu p^\nu/p^2}{(p^2+i\epsilon)^{2-d_{\mathcal{U}}}}, \tag{6}$$

where $\phi_{\mathcal{U}} = (2-d_{\mathcal{U}})\pi$ and

$$A_{d_{\mathcal{U}}} = \frac{16\pi^{5/2}}{(2\pi)^{2d_{\mathcal{U}}}}\frac{\Gamma(d_{\mathcal{U}}+1/2)}{\Gamma(d_{\mathcal{U}}-1)\Gamma(2d_{\mathcal{U}})}. \tag{7}$$

Here, the transverse condition $\partial_\mu\mathcal{O}_{\mathcal{U}}^{(\mu)} = 0$ is used and the phase factor in Eq. (4) arises from $(-1)^{d_{\mathcal{U}}-2} = e^{-i\pi(d_{\mathcal{U}}-2)}$.

3. Time-integrated CP asymmetries of $D^0 \to K^+K^-$ and $D^0 \to \pi^+\pi^-$ decays with unparticle

The SM predicts very small CP violation effects in D-meson decays because, with an excellent approximation, only the first two quark generations are involved. The CP violation in neutral D decays could be direct or indirect (or both) which come from di-penguin and box diagrams being very small [36]. The time-dependent CP asymmetry of D^0 decay to final CP eigenstates $f = K^+K^-, \pi^+\pi^-$ can be approximated as [37,38]

$$A_{CP} = \frac{\Gamma(D^0(t)\to f) - \Gamma(\bar{D}^0(t)\to f)}{\Gamma(D^0(t)\to f) + \Gamma(\bar{D}^0(t)\to f)} \approx a_{CP}^{dir} - A_\Gamma\frac{t}{\tau_D}, \tag{8}$$

where

$$a_{CP}^{dir} = A_{CP}(t=0) = \frac{\Gamma(D^0\to f) - \Gamma(\bar{D}^0\to f)}{\Gamma(D^0\to f) + \Gamma(\bar{D}^0\to f)} \tag{9}$$

and τ_D is the average lifetime of the D^0 decay. Here, A_Γ is the asymmetry between the D^0 and $\bar{D}^0$ effective decay widths. The indirect contributions in D^0 decays, CP violation in mixing, are universal and negligible, hence A_Γ is mostly related to the decay. It is independent of final states and measured to be about 10^{-4} [38]. Consequently, we can write the ΔA_{CP} as

$$\Delta A_{CP} \approx a_{CP}^{dir}(K^+K^-) - a_{CP}^{dir}(\pi^+\pi^-). \tag{10}$$

Considering neutral unparticle such as Eq. (3) in addition to the SM contents (Fig. 1), one leads to write the total amplitude, $\mathcal{A}_{tot.}$, of the process $D \to \pi\pi(KK)$ as follows:

$$\mathcal{A}_{tot.} = \mathcal{A}_{SM} + \mathcal{A}_{\mathcal{U}}, \tag{11}$$

where $\mathcal{A}_{SM}$ and $\mathcal{A}_{\mathcal{U}}$ are the SM and unparticle contributions, respectively. At the tree level, $\mathcal{A}_{SM}$ is given by

$$\mathcal{A}_{SM}^{\pi\pi(KK)} = \frac{G_F}{\sqrt{2}}V_{cd(s)}^*V_{ud(s)}\mathcal{F}_{\pi(K)}, \tag{12}$$

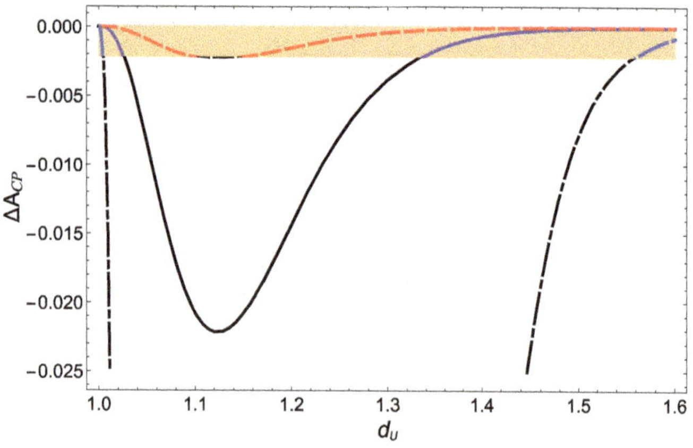

Fig. 2. ΔA_{CP} in terms of the scaling dimension $d_{\mathcal{U}}$ by fixing weak phases to $\gamma_\pi = -1$, $\gamma_K = 2.14$ and couplings to $|c_V^{dd(ss)} c_V^{cu}| \sim 10^{-4}$ (dot-dashed),10^{-6} (solid) and 10^{-7} (dashed), for the scale of unparticle $\Lambda_{\mathcal{U}} = 15$ TeV. The dark region shows the LHCb bounds.

where $\mathcal{F}_{\pi(K)}$ is a function which depends on the meson mass and QCD detail of the process, which will be finally removed in Eq. (9). The unparticle contribution in amplitude is

$$\mathcal{A}_{\mathcal{U}}^{\pi\pi(KK)} = |\mathcal{A}_{\rm SM}^{\pi\pi(KK)}| \chi_{\pi(K)} e^{-i\phi_{\mathcal{U}}} e^{-i\gamma_{\pi(K)}}, \tag{13}$$

with

$$\chi_{\pi(K)} = \frac{8\,|c_V^{dd(ss)} c_V^{cu}| A_{d_{\mathcal{U}}} m_W^2}{2\,g^2 a_1 N_c |V_{cd(s)} V_{ud(s)}| \sin(d_{\mathcal{U}}\pi) p^2} \left(\frac{p^2}{\Lambda_{\mathcal{U}}{}^2}\right)^{d_{\mathcal{U}}-1}, \tag{14}$$

where $a_1 = C_2 + C_1/N_C$ is the effective Wilson coefficient [39], N_C is the color number and $p^2 \sim m_D \bar{\Lambda}$ with $\bar{\Lambda} = m_D - m_c$. Here, $\gamma_{\pi(K)}$ is the weak phase related to the complex nature (in general) of the unparticle coupling coefficients, $c_V^{q'q}$. Furthermore, $\phi_{\mathcal{U}} = d_{\mathcal{U}}\pi$ is the unparticle phase which plays the role of strong phase in the corresponding direct CP violation. Note that, we have ignored the scalar unparticle contributions, since they are suppressed by $m_D^2/\Lambda_{\mathcal{U}}^2$. Consequently, the difference of time-integrated CP asymmetry in $D^0 \to K^+K^-$ and $D^0 \to \pi^+\pi^-$ decays from Eq. (9) becomes

$$\Delta A_{CP} = \frac{2\chi_\pi \sin(d_{\mathcal{U}}\pi)\sin\gamma_\pi}{1+\chi_\pi^2 + 2\chi_\pi \cos d_{\mathcal{U}}\pi \cos\gamma_\pi} - \frac{2\chi_K \sin(d_{\mathcal{U}}\pi)\sin\gamma_K}{1+\chi_K^2 + 2\chi_K \cos d_{\mathcal{U}}\pi \cos\gamma_K}. \tag{15}$$

We try to explore the relevant parameter space through some diagrams. We take the scale of unparticle ~ 15 TeV. First, the role of scaling dimension $d_{\mathcal{U}}$, which determines the strong phase, is illustrated in Fig. 2 for three different product of couplings s and fixed values of the weak phases. In this figure, we take $\gamma_\pi = -1$, $\gamma_K = 2.14$, which are corresponding to the approximate maximum of $|\Delta A_{CP}|$ (see Fig. 3), and $|c_V^{dd(ss)} c_V^{cu}| \sim 10^{-4}, 10^{-6}, 10^{-7}$. From this figure it is obvious that, with the current precisions of measurements, there is no exclusion region for $|c_V^{dd(ss)} c_V^{cu}| \lesssim 10^{-7}$. It is noticeable that, in the case of 10^{-4}, the vertex factor in Eq. (3) (the ratio of a coupling and $\Lambda_{\mathcal{U}}^{d_{\mathcal{U}}-1}$) is comparable to the least one chosen in [28]. In particular, note that while the unparticle has not been excluded by $B_d \to \pi^+\pi^-$ for this choice, here it is seriously excluded for $d_{\mathcal{U}}$ smaller than about 1.56. Second, in Fig. 3, ΔA_{CP} is plotted in terms of weak phases, γ_π and γ_K for a fixed value of $d_{\mathcal{U}} = 1.85$ and unity couplings. We find that, if two terms in Eq. (15) are in opposite phases, they are summed constructively. In this figure, the region between dashed lines is allowed and otherwise excluded. Here we should mention that, the future precise experiments, such as Phase-I and Phase-II upgrade of LHCb, or future analysis based on data collected of proton-proton collisions at the LHC, at a center-of-mass energy of 13 TeV, may lead to narrower allowed region, hence one can constrain the parameter space more strongly.

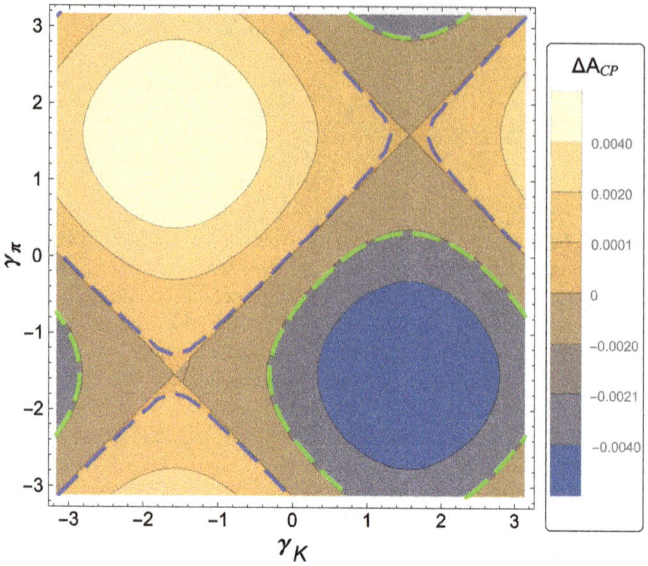

Fig. 3. Contourplot of ΔA_{CP} in terms of the weak phases with scaling dimension $d_{\mathcal{U}} = 1.85$ and unity couplings. Dashed lines correspond to LHCb upper (blue) and lower (green) bounds. The region between dashed lines is allowed and otherwise excluded.

4. Summary and conclusions

Recently, the difference of time-integrated CP asymmetry in $D^0 \to K^+K^-$ and $D^0 \to \pi^+\pi^-$ decays ΔA_{CP}, is measured by LHCb more precisely, which shows no evidence for CP violation. This can impose strong constraints on the parameter space of any proposed new physics. In this letter, we suppose unparticle physics to contribute in ΔA_{CP}. Here, there are seven parameters, $c^{dd}, c^{ss}, c^{cu}, d_{\mathcal{U}}, \gamma_\pi, \gamma_K$ and $\Lambda_{\mathcal{U}}$ which form our parameter space. Taking $\Lambda_{\mathcal{U}} \sim 15$ TeV, we have tried to illustrate the role of other parameters. According to Fig. 2, for $|c_V^{dd(ss)} c_V^{cu}| \gtrsim 10^{-7}$ there is an excluded region, due to the present precisions. In addition, for $|c_V^{dd(ss)} c_V^{cu}| \sim 10^{-4}$ we do not see an allowed region for $d_{\mathcal{U}} \lesssim 1.56$, while for the same order of couplings, all $d_{\mathcal{U}}$'s are allowed with $B_d \to \pi^+\pi^-$ [28]. Also, the dependence of A_{CP} on weak phases is shown via contourplot in Fig. 3. Considering the recent LHCb results, this figure illustrates the excluded regions for choosing of couplings ~ 1 and $d_{\mathcal{U}} = 1.85$.

References

[1] J. Christenson, J. Cronin, V. Fitch, Evidence for the 2π decay of the K_2^0 meson, Phys. Rev. Lett. 13 (1964) 138.
[2] A.B. Carter, A.I. Sanda, CP violation in B-meson decays, Phys. Rev. D 23 (1981) 1567.
[3] Belle Collaboration, K. Abe, et al., Observation of large CP violation in the neutral B meson system, Phys. Rev. Lett. 87 (2001) 091802.
[4] J. Brod, A.L. Kagan, J. Zupan, Size of direct CP violation in singly Cabibbo-suppressed D decays, Phys. Rev. D 86 (2012) 014023.
[5] LHCb Collaboration, R. Aaij, et al., Measurement of the difference of time-integrated CP asymmetries in $D^0 \to K^+K^-$ and $D^0 \to \pi^+\pi^-$ decays, Phys. Rev. Lett. 116 (2016) 191601.
[6] LHCb Collaboration, R. Aaij, et al., Evidence for CP violation in time-integrated $D^0 \to h^-h^+$ decay rates, Phys. Rev. Lett. 108 (2012) 111602, arXiv:1112.0938 [hep-ex].

[7] CDF Collaboration, T. Aaltonen, et al., Measurement of the difference in CP-violating asymmetries in $D^0 \to K^+K^-$ and $D^0 \to \pi^+\pi^-$ decays at CDF, Phys. Rev. Lett. 109 (2012) 111801, arXiv:1207.2158.
[8] Belle Collaboration, B.R. Ko, et al., Direct CP violation in charm at Belle, PoS ICHEP 2012 (2013) 353, arXiv:1212.1975.
[9] H.-Y. Cheng, C.-W. Chiang, Direct CP violation in two-body hadronic charmed meson decays, Phys. Rev. D 85 (2012) 034036, arXiv:1201.0785.
[10] D. Pirtskhalava, P. Uttayarat, CP violation and flavor SU(3) breaking in D-meson decays, Phys. Lett. B 712 (2012) 81, arXiv:1112.5451.
[11] Y. Grossman, A.L. Kagan, Y. Nir, New physics and CP violation in singly Cabibbo suppressed D decays, Phys. Rev. D 75 (2007) 036008, arXiv:hep-ph/0609178.
[12] W. Altmannshofer, R. Primulando, C.T. Yu, F. Yu, New physics models of direct CP violation in charm decays, J. High Energy Phys. 04 (2012) 049.
[13] Y. Grossman, A.L. Kagan, J. Zupan, Testing for new physics in singly Cabibbo suppressed D decays, Phys. Rev. D 85 (2012) 114036.
[14] H. Georgi, Unparticle physics, Phys. Rev. Lett. 98 (2007) 221601.
[15] H. Davoudiasl, Constraining unparticle physics with cosmology and astrophysics, Phys. Rev. Lett. 99 (2007) 141301.
[16] I. Lewis, Cosmological and astrophysical constraints on tensor unparticles, arXiv:0710.4147, 2007.
[17] J. McDonald, Cosmological constraints on unparticles as continuous mass particles, J. Cosmol. Astropart. Phys. 03 (2009) 019.
[18] F. Ayres, D. Wyler, Astro unparticle physics, J. High Energy Phys. 12 (2007) 033.
[19] H. Steen, G. Raffelt, Y.Y. Wong, Unparticle constraints from supernova 1987A, Phys. Rev. D 76 (2007) 121701.
[20] N.G. Deshpande, S.D.H. Hsu, J. Jiang, Long range forces and limits on unparticle interactions, Phys. Lett. B 659 (2008) 888.
[21] D. Boyanovsky, R. Holman, J.A. Hutasoit, Oscillation dynamics of active–unsterile neutrino mixing in a $2+1$ mixing scheme, Phys. Rev. D 81 (2010) 033009.
[22] H. Georgi, Another odd thing about unparticle physics, Phys. Lett. B 650 (2007) 275.
[23] G. Bhattacharyya, D. Choudhury, D.K. Ghosh, Unraveling unparticles through violation of atomic parity and rare beauty, Phys. Lett. B 655 (2007) 261–268.
[24] S.L. Chen, et al., Constraints on unparticle interactions from particle and antiparticle oscillations, Eur. Phys. J. C 59 (2009) 899.
[25] Z.T. Wei, H.W. Ke, X.F. Yang, Interpretation of the 'f_{D_s} puzzle' in the standard model and beyond, Phys. Rev. D 80 (2009) 015022.
[26] J.P. Lee, Constraints on unparticles from $B_s \to \mu^+\mu^-$, Phys. Rev. D 88 (2013) 116003.
[27] CMS Collaboration, Search for dark matter and unparticles in events with a Z boson and missing transverse momentum in proton-proton collisions at $\sqrt{s} =$ 13 TeV, J. High Energy Phys. 1703 (2017) 061.
[28] C.H. Chen, C.Q. Geng, Unparticle physics effects on direct CP violation, Phys. Rev. D 76 (2007) 115003.
[29] R. Zwicky, Unparticles at heavy flavor scales: CP violating phenomena, Phys. Rev. D 77 (2008) 036004.
[30] C.S. Huang, X.H. Wu, Direct CP violation of $B \to l\nu$ in unparticle physics, Phys. Rev. D 77 (2008) 075014.
[31] V. Bashiry, CP-conserving unparticle phase effects on the unpolarized and polarized direct CP asymmetry in $b \to dl^+l^-$ transition, Phys. Rev. D 77 (2008) 096005.
[32] J.K. Parry, Large CP phase in B_s–$\bar{B}_s$ mixing and unparticle physics, Phys. Rev. D 78 (2008) 114023.
[33] M.M. Ettefaghi, R. Moazzemi, M. Rousta, Constraining unparticle physics from CP violation in Cabibbo favored decays of D mesons, Phys. Rev. D 95 (2017) 095027, arXiv:1705.06330.
[34] T. Banks, A. Zaks, On the phase structure of vector-like gauge theories with massless fermions, Nucl. Phys. B 196 (1982) 189.
[35] S.L. Chen, X.G. He, Interactions of unparticles with standard model particles, Phys. Rev. D 76 (2007) 091702.
[36] J.F. Donoghue, E. Golowich, G. Valencia, New four-quark $\Delta S = 2$ local operator, Phys. Rev. D 33 (1986) 1387.
[37] CDF Collaboration, T. Aaltonen, et al., Measurement of CP-violating asymmetries in $D^0 \to K^+K^-$ and $D^0 \to \pi^+\pi^-$ decays at CDF, Phys. Rev. D 85 (2012) 012009, arXiv:1111.5023.
[38] LHCb Collaboration, R. Aaij, et al., Measurement of the CP violation parameter A_Γ in $D^0 \to K^+K^-$ and $D^0 \to \pi^+\pi^-$ decays, arXiv:1702.06490.
[39] G. Buchalla, A.J. Buras, M.E. Lautenbacher, Weak decays beyond leading logarithms, Rev. Mod. Phys. 68 (1996) 1125.

Perspectives of direct detection of supersymmetric dark matter in the NMSSM

C. Beskidt [a,*], W. de Boer [a,*], D.I. Kazakov [a,b], S. Wayand [a,*]

[a] Institut für Experimentelle Kernphysik, Karlsruhe Institute of Technology, P.O. Box 6980, 76128 Karlsruhe, Germany
[b] Bogoliubov Laboratory of Theoretical Physics, Joint Institute for Nuclear Research, 141980, 6 Joliot-Curie, Dubna, Moscow Region, Russia

ARTICLE INFO

Editor: L. Rolandi

Keywords:
Supersymmetry
NMSSM
Dark matter
Direct dark matter searches
Spin independent dark matter cross section
Spin dependent dark matter cross section

ABSTRACT

In the Next-to-Minimal-Supersymmetric-Standard-Model (NMSSM) the lightest supersymmetric particle (LSP) is a candidate for the dark matter (DM) in the universe. It is a mixture from the various gauginos and Higgsinos and can be bino-, Higgsino- or singlino-dominated. Singlino-dominated LSPs can have very low cross sections below the neutrino background from coherent neutrino scattering which is limiting the sensitivity of future direct DM search experiments. However, previous studies suggested that the combination of both, the spin-dependent (SD) and spin-independent (SI) searches are sensitive in complementary regions of parameter space, so considering both searches will allow to explore practically the whole parameter space of the NMSSM. In this letter, the different scenarios are investigated with a new scanning technique, which reveals that significant regions of the NMSSM parameter space cannot be explored, even if one considers both, SI and SD, searches.

1. Introduction

Experimental evidence shows that roughly 85% of the matter in the universe consists of dark matter (DM) [1], presumably made at least partially of Weakly Interacting Massive Particles (WIMPs). Supersymmetry (SUSY) [2–5] can provide a perfect WIMP candidate: the Lightest Supersymmetric Particle (LSP), in many models the lightest neutralino, has all the required WIMP properties: it is neutral, massive, stable and weakly interacting. The observed relic density is inversely proportional to the annihilation cross section [6] and indeed the LSP annihilation cross section can give the right amount of DM in the universe. This annihilation cross section is required to be some 10 orders of magnitude higher than the limits on the scattering cross section between WIMPs and nuclei, as found in the direct DM detection experiments, which try to detect WIMPs by measuring the recoil of a DM particle off a nucleus in deep underground experiments, see e.g. Refs. [7,8]. These many orders of magnitude between the scattering and annihilation cross section are easily explained in SUSY by a combination of the exchanged particle being a Higgs boson, which hardly couples to a nucleus because of the preponderance of light quarks inside a nucleus and the different kinematics from scattering and annihilation. The direct scattering can either be proportional to the spin (spin-dependent (SD)) or the scattering is coherent on the whole nucleus, in which case the cross section is enhanced by the square of the number of nuclei of the target material and independent of the spin (spin-independent (SI)).

In the Minimal-Supersymmetric-Standard-Model (MSSM) the LSP is a mixture of gauginos and Higgsinos, with the bino admixture typically being dominant. In this case the present limit of the SI cross section of $2 \cdot 10^{-10}$ pb from the LUX 2016 experiment starts to eliminate a significant fraction of the parameter space [9, 10]. Limits on the SD cross section are weaker and therefore neglected in the MSSM. With future expected sensitivity on the SI cross section of 10^{-13} pb [11] almost the whole parameter space will be accessible in the MSSM, so one would expect to either discover WIMP scattering or exclude the MSSM as the origin of DM.

However, in the Next-to-Minimal-Supersymmetric-Standard-Model (NMSSM) the situation is different, since the introduction of a Higgs singlet leads to an additional singlino. The Higgs singlet allows to avoid heavy stop masses and avoids the so-called μ-problem, see e.g. [12]. The LSP will mix with the singlino as well. So the LSP can become predominantly bino-, Higgsino- or

* Corresponding authors.
E-mail addresses: conny.beskidt@kit.edu (C. Beskidt), wim.de.boer@kit.edu (W. de Boer), KazakovD@theor.jinr.ru (D.I. Kazakov), stefan.wayand@kit.edu (S. Wayand).

singlino-like or be a mixture of them. The larger diversity of the LSP properties has led to many studies of direct DM detection in the NMSSM, see e.g. [13–29].

If the LSP is predominantly a singlino, it may hardly couple to any SM particle. In this case the non-observation of WIMP scattering may not exclude the NMSSM as the origin of DM, as was studied before in Ref. [22]. Here only the SI limits have been taken into account.

However, recently SD limits have become available [10,30], which have raised excitement, since they appeared to be complementary in that they exclude different regions of parameter space and it was suggested that in future the combination of SD and SI searches might be able to explore a large fraction of the NMSSM parameter space [24,25].

However, these papers relied on Markov Chain or random sampling of the NMSSM parameter space, in which case it is difficult to sample all regions of a multidimensional parameter space with highly correlated parameters [31,32]. The reason is simple: if 3 parameters are positively correlated, stepping through the parameter space with parameter 1 in one direction, one finds maximum likelihoods fastest, if the next steps of the other two parameters are in the same direction. In the constrained MSSM (NMSSM) the dimensionality of the parameter space is 5(9); in unconstrained models significantly larger. Without knowing the features of a likelihood function with its typical narrow features from correlated parameters, it is difficult to assure a complete sampling of the parameter space, as was demonstrated before for the 5-D parameter space of the MSSM [33–36] and the 10-D parameter space of the determination of the cosmological parameters of the CMB background [31, 37,38].

We therefore use a new sampling technique assuring that no regions of parameter space will be missed in the sampling. The main idea is to project the highly correlated parameter space of the couplings onto a space spanned by uncorrelated Higgs masses, which is only 3-D, if one considers one Higgs boson mass fixed to the measured 125 GeV and the heavy Higgs masses to be degenerate. In this space the couplings are marginalized over by a fit. Hence, the Higgs parameter space is reduced from 7-D to 3-D with largely uncorrelated parameters, which allows for an efficient sampling. An alternative way of explaining the sampling technique is as follows: suppose the LHC would have discovered all 7 Higgs bosons of the NMSSM. Would we be able to determine all couplings in the Higgs sector? The answer is: there is not a unique solution, but there are two preferred regions in the parameter space, which we called Scenario I and Scenario II in Ref. [39]. By repeating the fit to determine the couplings for each combination of Higgs boson masses in a 3-D grid of Higgs masses one can delineate the parameter regions of Scenario I and Scenario II.

It is the purpose of this letter to check if there are regions in the NMSSM parameter space, which evade exploration by a combination of SD and SI searches. We find that there are indeed regions of parameter space, which have cross sections below the "neutrino floor", both for the SD and SI searches. Below the "neutrino floor" direct detection will be difficult, because of the high background from the coherent scattering of neutrinos, which cannot be shielded in DM experiments. Only tails in the recoil spectrum, annual modulation or directional dependence of the events might allow to separate WIMP scattering from neutrino backgrounds given enough statistics, see Ref. [40] and references therein. Since in parameters regions near or below the neutrino floor the LSP is almost a pure singlino, these regions are not accessible at the LHC either.

After a short summary of the neutralino sector in the NMSSM and the elastic scattering processes, we discuss the fit strategy. We conclude by summarizing the impact of the DM constraints from future experiments on the NMSSM parameter space.

2. Semi-constrained NMSSM

Within the NMSSM the Higgs fields consist of the two Higgs doublets (H_u, H_d), which appear in the MSSM as well, but the NMSSM has an additional complex Higgs singlet S. The addition of a Higgs singlet yields more parameters in the Higgs sector to cope with the interactions between the singlet and the doublets and the singlet self interaction.

In the following we restrict the parameter space by assuming unification of couplings and masses at the GUT scale of about $2 \cdot 10^{16}$ GeV. Although this restricts the parameter space, it is a well motivated region of parameter space and it will be interesting to see if this region is within reach of the future experiments. In this case we have the GUT scale parameters of the Constrained-Minimal-Supersymmetric-Standard-Model (CMSSM): m_0 and $m_{1/2}$, where $m_0(m_{1/2})$ are the common mass scales at the GUT scale of the spin 0(1/2) SUSY particles, the trilinear coupling A_0 of the CMSSM Higgs sector and $\tan\beta$, the ratio of vacuum expectation values (vev) of the neutral components of the SU(2) Higgs doublets, i.e. $\tan\beta \equiv v_u/v_d$. For the NMSSM one has to add the coupling λ between the singlet and the doublets from the term $\lambda S H_u \cdot H_d$ and κ, the self-coupling of the singlet from the term $\kappa S^3/3$; A_λ and A_κ are the corresponding trilinear soft breaking terms; μ_{eff} represents an effective Higgs mixing parameter.

So in total the semi-constrained NMSSM has nine free parameters:

$$m_0,\ m_{1/2},\ A_0,\ \tan\beta,\ \lambda,\ \kappa,\ A_\lambda,\ A_\kappa,\ \mu_{eff}. \tag{1}$$

The effective Higgs mixing parameter is related to the vev of the singlet s via the coupling λ, i.e. $\mu_{eff} \equiv \lambda s$. Being proportional to a vev, μ_{eff} is naturally of the order of the electroweak scale, thus avoiding the μ-problem [12]. The supersymmetric partner of the singlet leads to an additional Higgsino, thus extending the neutralino sector from 4 to 5 neutralinos. This leads to modifications of the SI and SD cross sections, which are discussed in the following subsections.

2.1. The NMSSM neutralino sector

Within the NMSSM the singlino, the superpartner of the Higgs singlet, mixes with the gauginos and Higgsinos, leading to an additional fifth neutralino. The resulting mixing matrix reads [12,41]:

$$\mathcal{M}_0 = \begin{pmatrix} M_1 & 0 & -\frac{g_1 v_d}{\sqrt{2}} & \frac{g_1 v_u}{\sqrt{2}} & 0 \\ 0 & M_2 & \frac{g_2 v_d}{\sqrt{2}} & -\frac{g_2 v_u}{\sqrt{2}} & 0 \\ -\frac{g_1 v_d}{\sqrt{2}} & \frac{g_2 v_d}{\sqrt{2}} & 0 & -\mu_{\rm eff} & -\lambda v_u \\ \frac{g_1 v_u}{\sqrt{2}} & -\frac{g_2 v_u}{\sqrt{2}} & -\mu_{\rm eff} & 0 & -\lambda v_d \\ 0 & 0 & -\lambda v_u & -\lambda v_d & 2\kappa s \end{pmatrix} \tag{2}$$

with the gaugino masses M_1, M_2, the gauge couplings g_1, g_2 and the Higgs mixing parameter μ_{eff} as parameters. Furthermore, the vacuum expectation values of the two Higgs doublets v_d, v_u, the singlet s and the Higgs couplings λ and κ enter the neutralino mass matrix.

The upper left 4×4 submatrix of the neutralino mixing matrix corresponds to the MSSM neutralino mass matrix, see e.g. Ref. [4].

The neutralino mass eigenstates are obtained from the diagonalization of $\mathcal{M}_0$ in Eq. (2) and are linear combinations of the gaugino and Higgsino states:

$$\tilde{\chi}_i^0 = \mathcal{N}(i,1)\left|\tilde{B}\right\rangle + \mathcal{N}(i,2)\left|\tilde{W}^0\right\rangle + \mathcal{N}(i,3)\left|\tilde{H}_u^0\right\rangle + \mathcal{N}(i,4)\left|\tilde{H}_d^0\right\rangle + \mathcal{N}(i,5)\left|\tilde{S}\right\rangle. \tag{3}$$

Typically, the diagonal elements in Eq. (2) dominate over the off-diagonal terms, so the neutralino masses are of the order of M_1, M_2, the Higgs mixing parameter μ_{eff} for the Higgsinos and in case of the NMSSM $2\kappa s = 2(\kappa/\lambda)\mu_{eff}$ for the singlino-like neutralino.

The mass spectrum at the low mass SUSY scales is calculated from the GUT scale input parameters via the renormalization group equations (RGEs), which results in correlated masses including the large radiative corrections from the GUT scale to the electroweak scale. The gaugino masses at the electroweak scale are proportional to $m_{1/2}$ [2–4,42]:

$$M_1 \approx 0.4m_{1/2},\ M_2 \approx 0.8m_{1/2},\ M_3 \approx M_{\tilde{g}} \approx 2.7m_{1/2}. \tag{4}$$

In the CMSSM the Higgs mixing parameter μ is typically much larger than $m_{1/2}$ to fulfill radiative electroweak symmetry breaking (EWSB) [2–4,42], which leads to a bino-like lightest neutralino. In the NMSSM μ_{eff} is an input parameter, which is naturally of the order of the electroweak scale. In such natural NMSSM scenarios the lightest neutralino is singlino- or Higgsino-like and its mass can be degenerate with the second and third neutralino, all of which have a mass of the order of μ_{eff}. Bino-like neutralinos are also possible within the NMSSM but they require large values of $\mu_{eff} >> M_1$. This is not excluded, but not expected in natural NMSSM models. However, if the LSP in the NMSSM is bino-like, the situation is similar to the MSSM, which has been studied in great detail previously [43]. So in this letter we will concentrate on LSPs being singlino- or Higgsino-like in the NMSSM.

The amount of the Higgsino and singlino content of the lightest neutralino depends on the ratio and the absolute value of κ and λ, as can be seen from the coefficient $\mathcal{M}_0(5,5) = 2\kappa s = 2(\kappa/\lambda)\mu_{eff}$. The Higgsino fraction, which determines the coupling to the Higgs, is crucial for the elastic scattering cross section, since this proceeds mainly via the exchange of a Higgs boson.

2.2. Elastic WIMP-nucleon scattering

A WIMP might be detected by measuring the recoil of a nucleus after an elastic scattering of a WIMP on a nucleus taking place. Since such collisions are non-relativistic, only two cases need to be considered [44]: the spin–spin interaction (SD), where the WIMP couples to the spin of the nucleus, and the scalar interaction (SI), where the WIMP couples to the mass of the nucleus.

The SI cross section is proportional to the Higgsino content of the lightest neutralino $\sigma_{SI} \propto N_{13}^2 + N_{14}^2$ and to the mass of the nucleus squared, which leads to a substantial enhancement for heavy nuclei [45]. In addition, the cross section includes the effective quark form-factors which are similar for protons and neutrons and increase for large values of the strange quark content. However, the quark form-factors derived from pion–nucleon scattering measurements suffer from large uncertainties [46]. In addition, these measurements deviate from the form-factors resulting from lattice calculations. We calculate all DM cross sections with micrOMEGAs 3.6.9.2 [47]. The default form-factors given in micrOMEGAs are taken from the average of a variety of different measurements and lattice calculations [48]. The extreme values for the form factors lead to variations in the predicted cross section of about 20%.

The experimental best limit on the SI WIMP nucleon cross section is given by the LUX experiment [49]. It excludes discovery claims by DAMA/LIBRA [50] and CoGeNT [51]. The SI cross section is inversely proportional to the Higgs mass squared, so the prediction of two light scalar Higgs bosons can enhance the SI cross section in the NMSSM. However, a negative interference between them suppresses the SI cross section if the two lightest Higgs bosons are close in mass. In this case the predicted cross section is below the current LUX limits, which has been discussed in more detail in Refs. [23,25,43]. However, the SD cross section, which proceeds mainly by Z^0 exchange, does not suffer from such "blind" spots, which have a steep probability distribution in the parameter space, as demonstrated in Fig. 4c from Ref. [43].

The dominant diagram for the SD scattering is the Z^0 boson exchange. The corresponding cross section includes the difference of the Higgsino components $\sigma_{SD} \propto |N_{13}^2 - N_{14}^2|$. If the admixture of the two Higgsino components are large but similar, the SD cross section can become small. But then the SI cross section ($\propto N_{13}^2 + N_{14}^2$) will be large, so they do not become small simultaneously. The calculation of the nuclear matrix elements is at zero momentum transfer equivalent to the calculation of the average spins for neutrons and protons, while the corresponding coefficients can be extracted from data on polarized deep inelastic scattering [6]. Uncertainties in the experimental determination of these coefficients lead to variations in the predicted rates for WIMP detection as already discussed above for the SI cross section. The current best limit on the SD cross section is given by LUX for the WIMP-neutron interaction [10], as the majority of the nuclear spin is carried by the unpaired neutron in the Xenon isotopes. PICO-2L gives the best limit on the SD WIMP-proton cross section [30] because of the single unpaired proton in C_3F_8 providing a better sensitivity for SD WIMP-proton interactions. Naively, one would expect the SD cross sections to be the same for neutrons and protons. However, they are different because of the proton and neutron form factors which leads to $\sigma_n \approx 0.77\sigma_p$ [23]. If we take these different cross sections for proton and neutrons into account, the LUX experiment is more sensitive than the PICO experiment, so we continue to consider SD neutron cross sections, thus following Ref. [25].

The experimental limits require values for the exposure, which depends on the local DM density, which takes values between 0.3 and 1.3 GeV/cm^3, see e.g. [52]. This uncertainty leads to a variation of the limit of about a factor of 4. The limits given by different experiments are calculated for a local DM density of 0.3 GeV/cm^3, which leads to the most conservative limit.

3. Analysis

The additional particles and their interactions within the NMSSM lead to a large parameter space, even in the well-motivated subspace with unified masses and couplings at the GUT scale. We focus on the semi-constrained NMSSM and use the corresponding code NMSSMTools 4.6.0 [53,54] to calculate the SUSY mass spectrum from the NMSSM parameters. The Higgs masses depend on radiative corrections, which are calculated using the option 8-2 in NMSSMTools, which means that the full one loop and the full two loop corrections from top and bottom Yukawa couplings are taken into account. NMSSMTools has an interface to micrOMEGAs [47], which was used to calculate the relic density and LSP scattering cross sections.

As discussed in the introduction, we use a systematic sampling technique by considering a space spanned by the masses of the 3 scalar and 2 pseudo-scalar neutral Higgs boson masses m_{H_i} and m_{A_i}, as well as the two charged Higgs bosons $m_{H^\pm}$. This space reduces to a 3-D parameter space, if one requires m_{H1} or $m_{H2} \approx 125$ GeV with SM couplings and $m_{H3} \approx m_{A2} \approx m_{H^\pm}$ for $M_A >> M_Z$. We took the lightest [second-lightest] and heaviest neutral scalar Higgs boson masses and the lightest pseudo-scalar neutral Higgs mass as remaining masses, so after choosing these three "free" masses all Higgs masses are fixed. The "free" masses are distributed over a grid $m_{H1[H2]}$ vs. m_{H3} for different steps in m_{A1} [39]. These grid boundaries were chosen to lay between

$$5[125]\ \text{GeV} < m_{H1[H2]} < 125[500]\ \text{GeV}$$

$$100\,\text{GeV} < m_{H3} < 2\,\text{TeV} \quad (5)$$
$$5\,\text{GeV} < m_{A1} < 500\,\text{GeV}.$$

For each mass combination the allowed couplings were determined from a fit which minimizes the following χ^2 function with the parameters of Eq. (1) as free parameters:

$$\chi^2_{tot} = \chi^2_{H_S} + \chi^2_{H_{SM}} + \chi^2_{H_3} + \chi^2_{LEP}. \quad (6)$$

The χ^2 contributions are [39]

- $\chi^2_{H_S} = (m_{H_i} - m_{grid,H_i})^2/\sigma^2_{H_i}$: since one of the light Higgs bosons represent the observed SM Higgs, the other light Higgs boson H_i with $i = 1, 2$ has to be singlet-like. The term $\chi^2_{H_S}$ requires the NMSSM parameters to be adjusted such that the mass of the singlet-like light Higgs boson mass m_{H_i} with $i = 1, 2$ agrees with the chosen point in the 3-D mass space m_{grid,H_i}. The value of $\sigma^2_{H_i}$ is set to 2 GeV.
- $\chi^2_{H_{SM}} = (m_{H_i} - m_{obs})^2/\sigma^2_{SM} + \sum_i (c^i_{H_i} - c_{obs})^2/\sigma^2_{coup}$: the other light Higgs boson H_i with $i = 1, 2$ has to represent the observed Higgs boson with couplings close to the SM couplings, as required by the last term. $c^i_{H_i}$ represents the reduced couplings of H_i which is the ratio of the coupling of H_i to particle $i = f_u, f_d, W/Z, \gamma$ divided by the SM coupling. The observed couplings c_{obs} agree within 10% with the SM couplings, so $\sigma^2_{coup} = 0.1$. The first term is analogous to the term for m_{H_S}, except that the mass of the second light Higgs boson should have the observed Higgs boson mass, so m_{obs} is set to 125.4 GeV. The corresponding uncertainty σ^2_{SM} equals 1.9 GeV and results from the linear addition of the experimental and theoretical (1.5 GeV) uncertainties.
- $\chi^2_{H_3} = (m_{H_3} - m_{grid,H_3})^2/\sigma^2_{H_3}$: as $\chi^2_{H_S}$, but for the heavy scalar Higgs boson H_3.
- χ^2_{LEP}: includes the LEP constraints on the couplings of a light Higgs boson below 115 GeV and the limit on the chargino mass as discussed in Ref. [55].

The χ^2 function is insensitive to the SUSY mass parameters m_0 and $m_{1/2}$, so these were fixed to 1 TeV. This mass point leads to a sparticle spectrum consistent with the current limits of the direct SUSY searches from the LHC [56,57]. In addition to the LHC constraints, further constraints, like constraints from B-physics, are calculated and checked within NMSSMTools. The fit finds the best values of the couplings, but there is no unique solution, as can be seen already from the approximate expression for the Higgs mass [12]:

$$M^2_H \approx M^2_Z \cos^2 2\beta + \Delta_{\tilde{t}} + \lambda^2 v^2 \sin^2 2\beta - \frac{\lambda^2}{\kappa^2}(\lambda - \kappa \sin 2\beta)^2. \quad (7)$$

The first two terms are identical to the expression in the CMSSM, where the first tree level term can become as large as M^2_Z for large $\tan\beta$, but in the CMSSM the difference between M_Z and 125 GeV has to originate mainly from the logarithmic stop mass correction $\Delta_{\tilde{t}}$ [58]. The two remaining terms originate from the mixing with the singlet of the NMSSM. The SM Higgs boson in the NMSSM is fulfilled within two regions of the parameter space, as was determined in a previous paper [39]. The first region has large values of λ and κ and small values of $\tan\beta$ which we call *Scenario I*. Here the tree level mass of the Higgs is large due to the mixing with the singlet. Another possibility, which we call *Scenario II*, are small values for λ and κ, which requires large values of $\tan\beta$ in order to reach a Higgs mass of 125 GeV. Within these two scenarios either the lightest or the second lightest Higgs can be the discovered 125 GeV Higgs boson. The range of the couplings allowed by the fit in both scenarios has been given before [39].

The specific scenarios have distinctly different features since the range of the couplings differ. However, the ratio of λ and κ, which determines the Higgsino-singlino mixture of the LSP, can be the same in both scenarios, so the singlino content can be the same in both scenarios, as can be seen from the $\mathcal{M}_0(5,5)$ element in Eq. (2).

Since the Higgsino content is crucial for the SD and SI cross section, we divide the two scenarios further into singlino- and Higgsino-dominated scenarios. This means that either the Higgsino elements $\sqrt{N^2_{13} + N^2_{14}}$ or the singlino element $\sqrt{N^2_{15}}$ are above 0.8. All cases can be either fulfilled for the lightest or the second lightest Higgs boson being the SM Higgs boson, which gives in total 8 scenarios (I and II with Higgsino/Singlino LSP and either m_{H_1} or $m_{H_2} = 125$ GeV) be tested against the current SD and SI limits. Besides the direct DM detection limits the relic density Ωh^2 can be considered, either as an upper limit, if one assumes other particles contributing to the DM abundance in the universe as well, or one assumes that the LSPs saturate the relic density from the Planck data [59]. However, if the relic density is too high, this would overclose the universe and such points are excluded. For the sampled points the predicted value of the relic density is plotted versus $2\kappa/\lambda$ in Fig. 1. The dark grey (dark blue) points correspond to the Higgsino- and singlino-dominated points. The top/bottom row shows Scenario I/II. For the Higgsino-dominated (left) LSPs the relic density is usually below the experimental value because of the large annihilation cross section into ZZ and W^+W^-. In contrast, the singlino-dominated (right) LSP can cover a large range of relic densities, since many co-annihilation channels can contribute. Co-annihilation is important, because the lightest neutralinos all have similar masses of the order of μ_{eff}. If co-annihilation is not possible, large relic densities are obtained, because the singlinos hardly couple to SM particles leading to small annihilation cross sections. As mentioned before, such points over-close the universe and are rejected for further analyses. The correct relic density can also be fulfilled for resonant annihilation via Z^0 or H boson leading to narrow allowed regions around $m_{\tilde{\chi}^1_0} \approx 45$ and $m_{\tilde{\chi}^1_0} \approx 60$ GeV. Resonant annihilation via the light pseudo-scalar Higgs boson A_1 is possible for light neutralino masses of the order of a few GeV.

3.1. *Reach of direct DM searches*

The sampled points from the parameter space spanned by the Higgs masses for Scenario I and II with the lowest χ^2 are assumed to be representative for the NMSSM, so these points are compared with the relic density and DM scattering cross section limits.

As shown in Fig. 1, many sampled points have an expected relic density Ω_{theo} below the observed relic density, which is allowed if the DM has additional contributions from other particles, like axions. In this case the sensitivity of direct DM experiments will be reduced by the factor $\zeta = \Omega_{theo}/\Omega_{obs}$. If $\Omega_{theo} > \Omega_{obs}$ the points are excluded, so ζ cannot be above 1. In order to calculate the reach of direct DM search experiments we multiply the expected cross section with $min(1, \zeta)$ to obtain, what we call the reduced cross section. The sampled points can be projected into the WIMP mass – reduced cross section plane as shown in Fig. 2 for Scenario I/II for the SI and SD cross sections separately, as indicated. Here the second lightest Higgs boson is the SM Higgs boson. The results for $m_{H1} = 125$ GeV are similar. The left/right plots represent the Higgsino/singlino-dominated LSPs. The dark grey (dark blue) points fulfill the SM Higgs constraint, while the light grey (light blue) points also yield the correct relic density, which is mostly possible for singlino-dominated LSPs, as shown before in

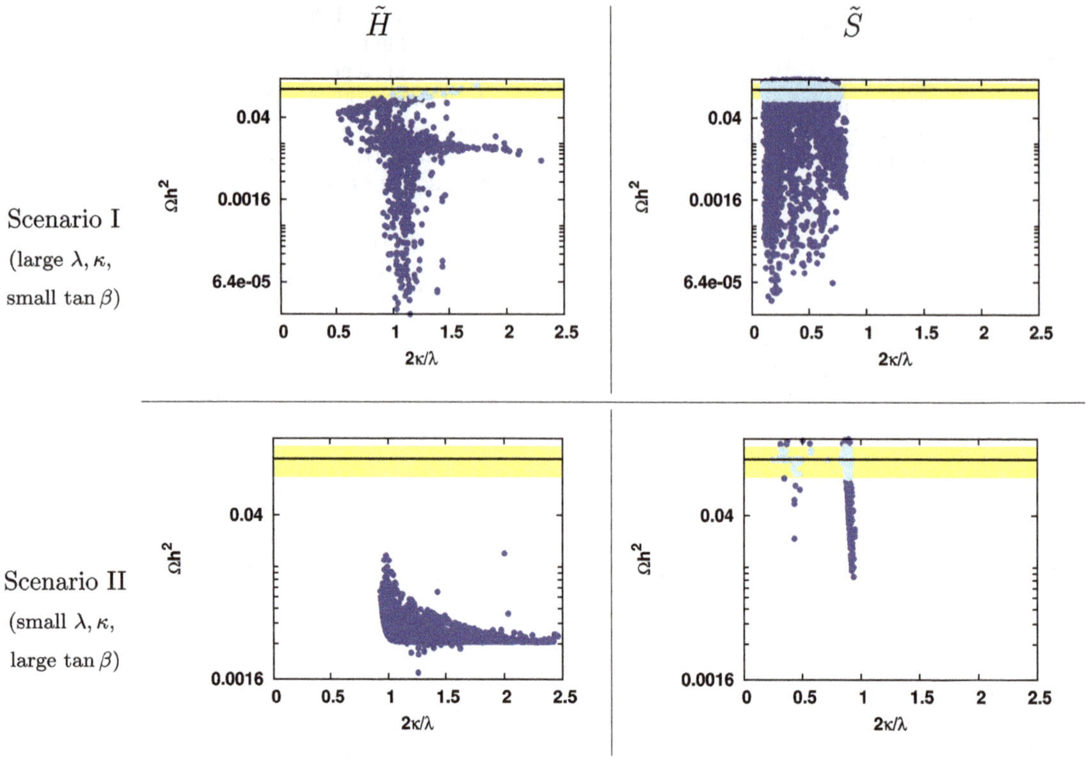

Fig. 1. Top row: The relic density versus $2\kappa/\lambda$ for Scenario I and $m_{H_2} = 125$ GeV for the Higgsino (left) and singlino-dominated points (right). Bottom row: as top row, but for Scenario II. The experimental value of the relic density is represented by the black solid line with the shaded (colored) band corresponding to the 95% C.L. region resulting from the linear addition of the experimental and theoretical error. The dark grey (dark blue) points represent Higgsino/singlino-type neutralinos with a corresponding neutralino content above 0.8. The light grey (light blue) points saturate the relic density at 95% C.L. The correct value of the relic density is easily fulfilled for the singlino-dominated LSPs, while for the Higgsino-type LSPs the predicted relic density is usually below the measured value due to the large coupling to Higgs bosons.

Fig. 1. For the Higgsino-dominated LSPs the relic density is usually too low. The red solid lines represent the current limits on the SI and SD cross sections, while the red dotted lines are the expectations from the future direct DM experiments XENON1T [60] and DARWIN [11]. The orange area below is the coherent neutrino scattering cross section of solar, atmospheric and diffuse supernova neutrinos on nuclei, which limits the sensitivity of direct detection experiments [61]. Points within this area are expected to be challenging to access in the future [40]. We choose not to give the percentage of the excluded points, since this number varies strongly with the size of the initial parameter space.

The predicted neutralino mass ranges differ for the different scenarios. For the Higgsino-dominated LSP the mass range starts at around 100 GeV, which is determined by the lowest value of μ_{eff} chosen around 100 GeV. For a singlino-dominated LSP the mass can be below μ_{eff}, since the mass is proportional to the ratio of κ and λ. The light neutralino masses in the order of a few GeV results from low values of the lightest pseudo-scalar Higgs boson A_1.

Most of the sampled points for the chosen scenarios will be within reach of the future direct DM searches. The comparison of the reduced cross section with the expected future sensitivity of DM experiments on the cross section, for which we take the proposed DARWIN experiment as an example, shows in Fig. 2 that in parts both, singlino- and Higgsino-dominated LSPs can be out of reach of future experiments. The Higgsino-dominated LSPs can be out of reach mainly because of the high coupling to Higgs bosons, which reduces the relic density, thus leading to a small reduced scattering cross section by the small value of $\zeta \approx 10^{-4}$, as shown in Fig. 1.

Singlino-dominated LSPs can be out of reach because of the small coupling to SM particles and thus small scattering cross section, which may be reduced even further by the factor ζ for a relic density being below the observed relic density (dark grey (dark blue) points). Points with a low SI and SD cross section have a large singlino component, as demonstrated in Fig. 3. Here the singlino-dominated points for Scenario I/II (left/right) for either $m_{H_1} = 125$ GeV or $m_{H_2} = 125$ GeV are shown in the reduced SI–SD cross section plane. The shaded (color) coding corresponds to the singlino content of the lightest neutralino. The vertical and horizontal dashed line show the lower limit on the SI and SD cross section from the future experiment DARWIN for a WIMP mass of about 100 GeV, which is close to the "neutrino floor". Scenario I will be fully covered by future direct dark matter experiments, while for Scenario II points with a singlino purity of about 99% will evade detection in the future. Such high purities are only possible for small values of λ/κ below $\sim$0.03/0.01, in which case the lightest neutralino is decoupled and interacts weakly with SM particles. At the same time the annihilation cross section for the correct relic density is still fulfilled by the sum of many co-annihilation channels with the second-lightest and third neutralino $\tilde{\chi}_2^0/\tilde{\chi}_3^0$ and lightest chargino $\tilde{\chi}_1^\pm$ for neutralino masses around 100 GeV. In this case the lightest chargino, the second-lightest and third neutralino are all of the order of $\mu_{eff} \approx 100$ GeV.

4. Conclusion

We surveyed the cross sections for the SI and SD dark matter searches in the semi-constrained NMSSM. The parameter space was sampled by considering a space spanned by the 7 Higgs masses, which reduces to a 3-D space, if one takes into account that one Higgs mass has to be equal to the observed Higgs mass of 125 GeV and the heavy Higgs bosons are practically mass-

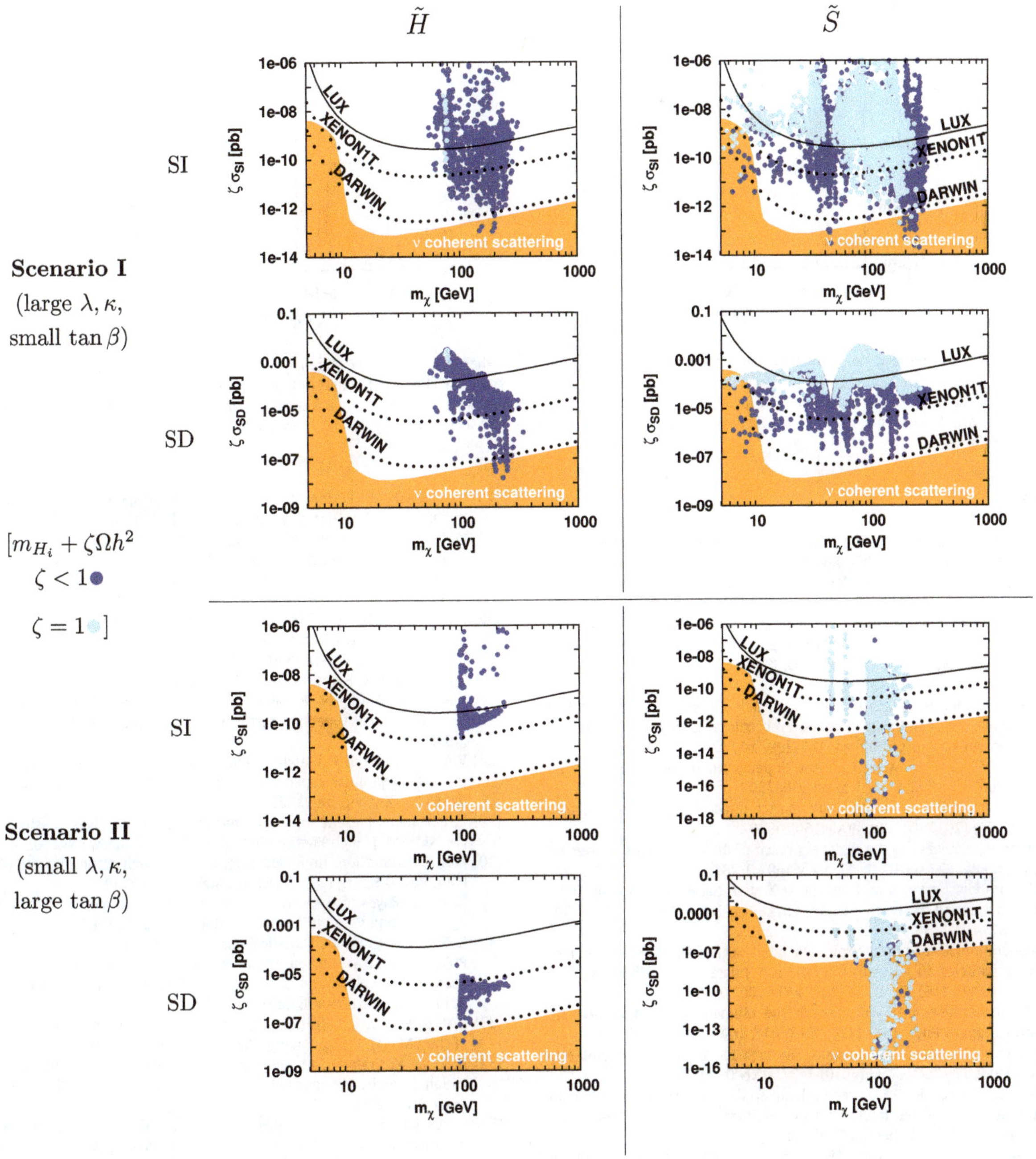

Fig. 2. Plots of the reduced scattering cross sections (SI and SD, as indicated) versus the WIMP mass. The left/right plots represent the Higgsino/singlino-dominated LSPs for Scenario I (top rows) and II (bottom rows). The Higgs mass m_{H_2} was chosen to be 125 GeV, but the case with $m_{H_1} = 125$ GeV looks similar. The dark grey (dark blue) points fulfill the SM Higgs constraint, while the light grey (light blue) points also yield the correct relic density. The dark grey (dark blue) points have a cross section multiplied by the sensitivity factor $\zeta = \Omega_{theo}/\Omega_{obs}$. The red solid/dotted lines represent the current/future sensitivities for various experiments. The orange area is below the neutrino coherent scattering cross section from solar, atmospheric and diffuse supernova neutrinos on nuclei, thus providing a high background for future DM searches, which makes this region challenging for future experiments [40].

degenerate. The advantage of projecting on the space spanned by masses is that the masses are largely uncorrelated and one can marginalize over the highly correlated couplings. From the sampling in the mass space we obtained the range of the neutralino masses and the corresponding SI and SD cross sections, as shown in Fig. 2 for two different ranges of allowed couplings corresponding to Scenarios I and II.

While Scenario I with large λ/κ couplings can be explored by the SI and SD searches, the new scanning technique reveals also that significant regions of the NMSSM parameter space in Scenario II cannot be explored with projected experiments, even if one considers both, SI and SD, searches. Such scenarios, which cannot be explored, are not evident from previous investigations [24,25, 29]. Since the singlino content of the LSP in these scenarios, displayed in the left bottom quadrant of the right panel of Fig. 3, is above 99%, they cannot be explored by the LHC either.

Acknowledgements

Support from the Heisenberg-Landau program and the Deutsche Forschungsgemeinschaft (DFG, Grant BO 1604/3-1) is warmly acknowledged.

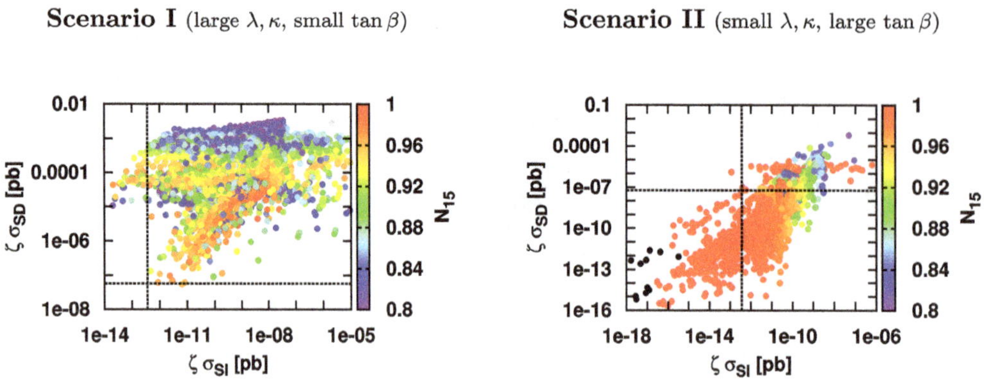

Fig. 3. Sampled points for the singlino-dominated points for Scenario I/II (left/right) for either $m_{H_1} = 125$ GeV or $m_{H_2} = 125$ GeV in the reduced SI-SD cross section plane. The shaded (color) coding corresponds to the singlino content of the lightest neutralino. The vertical and horizontal dashed lines show the lower limit on the SI and SD cross section expected for the future experiment DARWIN for a neutralino mass of about 100 GeV. Points in the lower left quadrant are below the "neutrino floor", which are only possible within Scenario II (right-hand side), since they require a singlino purity above 99%. Such pure singlinos are only possible for values of λ/κ below $\sim$0.03/0.01, as we discussed before [39].

References

[1] Planck Collaboration, Planck 2015 results. XIII. Cosmological parameters, Astron. Astrophys. 594 (2016) A13, arXiv:1502.01589.

[2] H.E. Haber, G.L. Kane, The search for supersymmetry: probing physics beyond the standard model, Phys. Rep. 117 (1985) 75–263.

[3] W. de Boer, Grand unified theories and supersymmetry in particle physics and cosmology, Prog. Part. Nucl. Phys. 33 (1994) 201–302, arXiv:hep-ph/9402266.

[4] S.P. Martin, A supersymmetry primer, in: G. Kane (Ed.), Perspectives on Supersymmetry II, 1997, arXiv:hep-ph/9709356.

[5] D.I. Kazakov, Beyond the standard model: in search of supersymmetry, in: 2000 European School of High-Energy Physics, Proceedings, Caramulo, Portugal, 20 August–2 September 2000, 2000, pp. 125–199, arXiv:hep-ph/0012288.

[6] G. Jungman, M. Kamionkowski, K. Griest, Supersymmetric dark matter, Phys. Rep. 267 (1996) 195–373, arXiv:hep-ph/9506380.

[7] T. Marrodn Undagoitia, L. Rauch, Dark matter direct-detection experiments, J. Phys. G 43 (1) (2016) 013001, arXiv:1509.08767.

[8] F. Mayet, et al., A review of the discovery reach of directional dark matter detection, Phys. Rep. 627 (2016) 1–49, arXiv:1602.03781.

[9] LUX Collaboration, First results from the LUX dark matter experiment at the Sanford Underground Research Facility, Phys. Rev. Lett. 112 (2014) 091303, arXiv:1310.8214.

[10] LUX Collaboration, Results on the spin-dependent scattering of weakly interacting massive particles on nucleons from the run 3 data of the LUX experiment, Phys. Rev. Lett. 116 (16) (2016) 161302, arXiv:1602.03489.

[11] DARWIN Collaboration, DARWIN: towards the ultimate dark matter detector, J. Cosmol. Astropart. Phys. 1611 (11) (2016) 017, arXiv:1606.07001.

[12] U. Ellwanger, C. Hugonie, A.M. Teixeira, The next-to-minimal supersymmetric standard model, Phys. Rep. 496 (2010) 1–77, arXiv:0910.1785.

[13] D.G. Cerdeno, C. Hugonie, D.E. Lopez-Fogliani, et al., Theoretical predictions for the direct detection of neutralino dark matter in the NMSSM, J. High Energy Phys. 12 (2004) 048, arXiv:hep-ph/0408102.

[14] J.F. Gunion, D. Hooper, B. McElrath, Light neutralino dark matter in the NMSSM, Phys. Rev. D 73 (2006) 015011, arXiv:hep-ph/0509024.

[15] D.G. Cerdeno, E. Gabrielli, D.E. Lopez-Fogliani, et al., Phenomenological viability of neutralino dark matter in the NMSSM, J. Cosmol. Astropart. Phys. 0706 (2007) 008, arXiv:hep-ph/0701271.

[16] C. Hugonie, G. Belanger, A. Pukhov, Dark matter in the constrained NMSSM, J. Cosmol. Astropart. Phys. 0711 (2007) 009, arXiv:0707.0628.

[17] G. Belanger, C. Hugonie, A. Pukhov, Precision measurements, dark matter direct detection and LHC Higgs searches in a constrained NMSSM, J. Cosmol. Astropart. Phys. 0901 (2009) 023, arXiv:0811.3224.

[18] D. Das, U. Ellwanger, Light dark matter in the NMSSM: upper bounds on direct detection cross sections, J. High Energy Phys. 09 (2010) 085, arXiv:1007.1151.

[19] G. Barenboim, G. Panotopoulos, Direct neutralino searches in the NMSSM with gravitino LSP in the degenerate scenario, J. High Energy Phys. 08 (2011) 027, arXiv:1102.0189.

[20] D. Albornoz Vasquez, G. Belanger, J. Billard, et al., Probing neutralino dark matter in the MSSM & the NMSSM with directional detection, Phys. Rev. D 85 (2012) 055023, arXiv:1201.6150.

[21] J. Kozaczuk, S. Profumo, Light NMSSM neutralino dark matter in the wake of CDMS II and a 126 GeV Higgs boson, Phys. Rev. D 89 (9) (2014) 095012, arXiv:1308.5705.

[22] U. Ellwanger, C. Hugonie, The semi-constrained NMSSM satisfying bounds from the LHC, LUX and Planck, J. High Energy Phys. 08 (2014) 046, arXiv:1405.6647.

[23] M. Badziak, M. Olechowski, P. Szczerbiak, Blind spots for neutralino dark matter in the NMSSM, J. High Energy Phys. 03 (2016) 179, arXiv:1512.02472.

[24] Q.-F. Xiang, X.-J. Bi, P.-F. Yin, et al., Searching for singlino–Higgsino dark matter in the NMSSM, Phys. Rev. D 94 (5) (2016) 055031, arXiv:1606.02149.

[25] J. Cao, Y. He, L. Shang, et al., Strong constraints of LUX-2016 results on the natural NMSSM, J. High Energy Phys. 10 (2016) 136, arXiv:1609.00204.

[26] V.A. Bednyakov, H.V. Klapdor-Kleingrothaus, Possibilities of directly detecting dark-matter particles in the next-to-minimal supersymmetric standard model, Phys. At. Nucl. 62 (1999) 966–974;
V.A. Bednyakov, H.V. Klapdor-Kleingrothaus, Yad. Fiz. 62 (1999) 1033.

[27] V.A. Bednyakov, A.N. Kovalev, Invisible Z boson width and restrictions on next-to-minimal supersymmetric standard model, JINR Rapid Commun. 4 (90) (1998) 5–14.

[28] V.A. Bednyakov, H.V. Klapdor-Kleingrothaus, About direct dark matter detection in next-to-minimal supersymmetric standard model, Phys. Rev. D 59 (1999) 023514, arXiv:hep-ph/9802344.

[29] U. Ellwanger, Present status and future tests of the Higgsino-singlino sector in the NMSSM, J. High Energy Phys. 02 (2017) 051, arXiv:1612.06574.

[30] PICO Collaboration, Improved dark matter search results from PICO-2L run 2, Phys. Rev. D 93 (6) (2016) 061101, arXiv:1601.03729.

[31] R. Trotta, Bayes in the sky: Bayesian inference and model selection in cosmology, Contemp. Phys. 49 (2008) 71–104, arXiv:0803.4089.

[32] R. Trotta, K. Cranmer, Statistical challenges of global SUSY fits, in: Proceedings, PHYSTAT 2011 Workshop on Statistical Issues Related to Discovery Claims in Search Experiments and Unfolding, CERN, Geneva, Switzerland 17–20 January 2011, 2011, pp. 170–176, arXiv:1105.5244.

[33] R. Ruiz de Austri, R. Trotta, L. Roszkowski, A Markov chain Monte Carlo analysis of the CMSSM, J. High Energy Phys. 05 (2006) 002, arXiv:hep-ph/0602028.

[34] F. Feroz, K. Cranmer, M. Hobson, et al., Challenges of profile likelihood evaluation in multi-dimensional SUSY scans, J. High Energy Phys. 06 (2011) 042, arXiv:1101.3296.

[35] L. Roszkowski, R. Ruiz de Austri, R. Trotta, Efficient reconstruction of CMSSM parameters from LHC data: a case study, Phys. Rev. D 82 (2010) 055003, arXiv:0907.0594.

[36] C. Beskidt, W. de Boer, D. Kazakov, et al., Constraints on supersymmetry from LHC data on SUSY searches and Higgs bosons combined with cosmology and direct dark matter searches, Eur. Phys. J. C 72 (2012) 2166, arXiv:1207.3185.

[37] A. Slosar, M. Hobson, An improved markov-chain Monte Carlo sampler for the estimation of cosmological parameters from cmb data, Mon. Not. R. Astron. Soc. (2003), submitted for publication, arXiv:astro-ph/0307219.

[38] C.M. Mueller, Cosmological Markov chain Monte Carlo simulation with CMBEASY, in: 39th Rencontres de Moriond Workshop on Exploring the Universe: Contents and Structures of the Universe, La Thuile, Italy, March 28–April 4, 2004, 2004, arXiv:astro-ph/0406206.

[39] C. Beskidt, W. de Boer, D.I. Kazakov, et al., Higgs branching ratios in constrained minimal and next-to-minimal supersymmetry scenarios surveyed, Phys. Lett. B 759 (2016) 141–148, arXiv:1602.08707.

[40] C.A. O'Hare, Dark matter astrophysical uncertainties and the neutrino floor, Phys. Rev. D 94 (6) (2016) 063527, arXiv:1604.03858.

[41] F. Staub, W. Porod, B. Herrmann, The electroweak sector of the NMSSM at the one-loop level, J. High Energy Phys. 1010 (2010) 040, arXiv:1007.4049.

[42] D.I. Kazakov, Introduction to supersymmetry, PoS CORFU2014 (2015) 024.

[43] C. Beskidt, W. de Boer, D.I. Kazakov, The impact of a 126 GeV Higgs on the neutralino mass, Phys. Lett. B 738 (2014) 505–511, arXiv:1402.4650.

[44] M.W. Goodman, E. Witten, Detectability of certain dark matter candidates, Phys. Rev. D 31 (1985) 3059.

[45] M. Drees, M. Nojiri, Neutralino–nucleon scattering revisited, Phys. Rev. D 48 (1993) 3483–3501, arXiv:hep-ph/9307208.
[46] J.R. Ellis, K.A. Olive, C. Savage, Hadronic uncertainties in the elastic scattering of supersymmetric dark matter, Phys. Rev. D 77 (2008) 065026, arXiv:0801.3656.
[47] G. Belanger, F. Boudjema, A. Pukhov, et al., micrOMEGAs-3: a program for calculating dark matter observables, Comput. Phys. Commun. 185 (2014) 960–985, arXiv:1305.0237.
[48] G. Belanger, F. Boudjema, A. Pukhov, et al., micrOMEGAs: a tool for dark matter studies, arXiv:1005.4133.
[49] LUX Collaboration, LZ Collaboration, M. Szydagis, The present and future of searching for dark matter with LUX and LZ, in: 38th International Conference on High Energy Physics (ICHEP 2016), Chicago, IL, USA, August 03–10, 2016, 2016, arXiv:1611.05525.
[50] DAMA Collaboration, LIBRA Collaboration, New results from DAMA/LIBRA, Eur. Phys. J. C 67 (2010) 39–49, arXiv:1002.1028.
[51] CoGeNT Collaboration, CoGeNT: a search for low-mass dark matter using p-type point contact germanium detectors, Phys. Rev. D 88 (2013) 012002, arXiv:1208.5737.
[52] M. Weber, W. de Boer, Determination of the local dark matter density in our Galaxy, Astron. Astrophys. 509 (2010) A25, arXiv:0910.4272.
[53] U. Ellwanger, J.F. Gunion, C. Hugonie, NMHDECAY: a Fortran code for the Higgs masses, couplings and decay widths in the NMSSM, J. High Energy Phys. 02 (2005) 066, arXiv:hep-ph/0406215.
[54] U. Ellwanger, C. Hugonie, NMHDECAY 2.0: an updated program for sparticle masses, Higgs masses, couplings and decay widths in the NMSSM, Comput. Phys. Commun. 175 (2006) 290–303, arXiv:hep-ph/0508022.
[55] C. Beskidt, Supersymmetry in the Light of Dark Matter and a 125 GeV Higgs Boson, PhD thesis, KIT, Karlsruhe, EKP, 2014.
[56] ATLAS Collaboration, Further Searches for Squarks and Gluinos in Final States with Jets and Missing Transverse Momentum at $\sqrt{s} = 13$ TeV with the ATLAS Detector, Technical Report ATLAS-CONF-2016-078, CERN, Geneva, August 2016.
[57] CMS Collaboration, Search for Supersymmetry in Events with Jets and Missing Transverse Momentum in Proton–Proton Collisions at 13 TeV, Technical Report CMS-PAS-SUS-16-014, CERN, Geneva, 2016.
[58] J.R. Ellis, G. Ridolfi, F. Zwirner, Radiative corrections to the masses of supersymmetric Higgs bosons, Phys. Lett. B 257 (1991) 83–91.
[59] Planck Collaboration, Planck 2013 results. XVI. Cosmological parameters, Astron. Astrophys. 571 (2014) A16, arXiv:1303.5076.
[60] XENON Collaboration, Physics reach of the XENON1T dark matter experiment, J. Cosmol. Astropart. Phys. 1604 (04) (2016) 027, arXiv:1512.07501.
[61] P. Cushman, et al., Working group report: WIMP dark matter direct detection, in: Proceedings, 2013 Community Summer Study on the Future of U.S. Particle Physics: Snowmass on the Mississippi (CSS2013), Minneapolis, MN, USA, July 29–August 6, 2013, 2013, p. 2013, arXiv:1310.8327.

Probing leptophilic dark sectors with hadronic processes

Francesco D'Eramo [a,b,*], Bradley J. Kavanagh [c], Paolo Panci [d,e]

[a] Department of Physics, University of California Santa Cruz, 1156 High St., Santa Cruz, CA 95064, USA
[b] Santa Cruz Institute for Particle Physics, 1156 High St., Santa Cruz, CA 95064, USA
[c] Laboratoire de Physique Théorique et Hautes Energies, CNRS, UMR 7589, 4 Place Jussieu, F-75252 Paris, France
[d] CERN Theoretical Physics Department, CERN, Case C01600, CH-1211 Genève, Switzerland
[e] Institut d'Astrophysique de Paris, UMR 7095 CNRS, Université Pierre et Marie Curie, 98 bis Boulevard Arago, F-75014 Paris, France

ARTICLE INFO

Editor: G.F. Giudice

ABSTRACT

We study vector portal dark matter models where the mediator couples only to leptons. In spite of the lack of tree-level couplings to colored states, radiative effects generate interactions with quark fields that could give rise to a signal in current and future experiments. We identify such experimental signatures: scattering of nuclei in dark matter direct detection; resonant production of lepton–antilepton pairs at the Large Hadron Collider; and hadronic final states in dark matter indirect searches. Furthermore, radiative effects also generate an irreducible mass mixing between the vector mediator and the Z boson, severely bounded by ElectroWeak Precision Tests. We use current experimental results to put bounds on this class of models, accounting for both radiatively induced and tree-level processes. Remarkably, the former often overwhelm the latter.

1. Introduction

Weakly Interacting Massive Particles (WIMPs) are motivated dark matter (DM) candidates testable with multiple and complementary methods [1–3]. An attractive class of WIMP models consists of those in which leptons are the only Standard Model (SM) degrees of freedom coupled to the DM [4–9]. Much interest in *leptophilic* models was ignited by the excess in the positron fraction at high energy observed by the PAMELA experiment [10,11], later confirmed by Fermi [12] and AMS-02 [13,14]. No associated excess in the antiproton flux has been observed. DM particles annihilating to leptons can provide a good fit to the positron excess, although complementary bounds from gamma-rays (see e.g. Refs. [15–17]) and Cosmic Microwave Background (see e.g. Refs. [18,19]) challenge this hypothesis and astrophysical explanations for this excess (e.g. pulsars [20–22]) remain plausible.

The expected experimental signals for leptophilic models are quite distinct. The absence of couplings to quarks and gluons leads (naïvely) to vanishing rates for direct detection and for DM production at hadron colliders. Furthermore, DM annihilation would lead to indirect detection spectra from only lepton final states. For these reasons nearly all phenomenological studies have focused on bounds coming from tree-level processes. These include mono-photon and lepton production at lepton colliders, four-lepton events at hadron colliders and diffuse gamma rays from DM annihilations. There are two noteworthy exceptions. On one hand, loop-induced contributions to DM scattering of target nuclei give rates observable in direct detection, as first pointed out by Ref. [5]. On the other hand, loop diagrams with a virtual leptophilic mediator give substantial corrections to lepton anomalous magnetic moments [23].

In this work, we identify new loop-induced signals coming from leptophilic dark sectors. We focus on the broad class of models where interactions between DM and leptons are mediated by a heavy vector particle. Radiative effects are accounted for by solving the renormalization group (RG) equations describing the evolution of the couplings with the energy scale [24,25], a procedure recently automated with the public code RUNDM [26,27]. We use these tools to explore the phenomenology of leptophilic models by rigorously accounting for the different energy scales. Crucially, different search strategies probe dark sector couplings at different energy scales, and RG flow induces new couplings through operator mixing. Despite the radiatively-induced couplings being suppressed by at least one loop factor, they can still give significant bounds, particularly in the case of loop-induced couplings to quarks (for related studied of RG effects see Refs. [28–35]).

Our work extends the previous literature on vector portal leptophilic models by identifying new processes arising from such

* Corresponding author.
E-mail addresses: fderamo@ucsc.edu (F. D'Eramo), bradley.kavanagh@lpthe.jussieu.fr (B.J. Kavanagh), paolo.panci@cern.ch (P. Panci).

loop-induced couplings and imposing the associated experimental constraints. As a result, we get novel bounds on the parameter space from the following signals:

- *LEP-II compositeness bound for μ and/or τ.* For models where the mediator couples only to μ and/or τ, we evaluate the RG-induced contribution to the coupling between the mediator and the electron. We constrain it by using LEP-II data for production of leptons in the final state.
- *Mass mixing with the Z boson.* We identify the RG-induced contribution to the mass mixing between the leptophilic vector mediator and the Z boson. We then place bounds using ElectroWeak Precision Tests (EWPT) data.
- *Dilepton resonance at the LHC.* We find the RG-induced contributions to the coupling between the mediator and light quarks. We constrain this coupling using data from LHC searches for dilepton resonances.
- *Direct detection.* Still using the RG-induced couplings between the mediator and light quarks, but this time at a much lower energy, we compute the DM elastic scattering rate of target nuclei and we impose direct detection bounds. If the mediator couples to the DM vector current, we improve the result of Ref. [5] by applying the RG analysis developed in Ref. [25]. On the contrary, if the mediator couples to the DM axial-vector current, we point out a contribution to the scattering rate proportional to the Yukawa coupling of the lepton under consideration. This effect was first pointed out in Ref. [25], and has not appeared in any previous analysis of leptophilic DM.

As a remarkable result of our analysis, we show that the most stringent constraints come from the RG-induced processes described above in a large region of the parameter space.

We describe in Sec. 2 the simplified model framework we adopt in this study. After setting up the notation, we give a simplified analytical solution for the RG evolution of the couplings in Sec. 3. While the analysis in this work is performed by employing the code RUNDM, these analytical solutions correctly capture the order of magnitude size of the full numerical results. The experimental constraints are enumerated in Sec. 4, where we highlight which bounds are present at the tree-level and which arise from RG-induced couplings. We present these bounds in Fig. 2 and we compare them for couplings to different lepton flavors in Sec. 5. In Sec. 6, we discuss in more detail the possible hadronic contamination of indirect detection spectra, an effect which has not previously been pointed out within this class of models. Finally, we present our conclusions in Sec. 7.

2. A leptophilic vector mediator

We augment the field content of the SM with two singlets: a fermion DM candidate χ assumed to be stabilized by a Z_2 symmetry, and a vector mediator V. We do not specify the microscopic origin of the mediator, but rather we employ a simplified model framework with interaction

$$\mathcal{L}_{\rm int} = \left(J_\mu^{(\rm DM)} + J_\mu^{(\rm ch.\ leptons)} + J_\mu^{(\rm neutrinos)} \right) V^\mu \, . \tag{1}$$

First, the mediator has tree-level couplings with the DM field through the fermion current

$$J_\mu^{(\rm DM)} = \overline{\chi} \gamma_\mu \left(g_{V\chi} + g_{A\chi} \gamma^5 \right) \chi \, . \tag{2}$$

We perform our phenomenological study for a Dirac DM field. The analysis for a Majorana DM particle would be very similar, with the absence of the DM vector current in Eq. (2) and appropriate factors of 2 in different observables. We find it convenient to parameterize the charged lepton current in terms of the vector and axial-vector pieces

$$J_\mu^{(\rm ch.\ leptons)} = \sum_{l=e,\mu,\tau} \bar{l} \gamma_\mu \left(g_{Vl} + g_{Al} \gamma^5 \right) l \, . \tag{3}$$

Although not electroweak gauge invariant, the connection with gauge invariant currents is straightforward. A manifestly gauge invariant lepton current would have two separate terms, associated to the left-handed weak isospin doublet l_L and the right-handed weak isospin singlet e_R. These gauge invariant currents can be unambiguously translated into the ones in Eq. (3) by using the relation $g_{V,Al} = (\pm c_{l_L} + c_{e_R})/2$. As a consequence of electroweak gauge invariance, the vector mediator has to couple also to SM neutrinos

$$J_\mu^{(\rm neutrinos)} = \sum_{l=e,\mu,\tau} \left(\frac{g_{Vl} - g_{Al}}{2} \right) \overline{\nu_l} \gamma_\mu \left(1 - \gamma^5 \right) \nu_l \, . \tag{4}$$

The only way to avoid neutrino couplings is for a mediator coupled to both vector and axial-vector currents of charged leptons with identical coefficient. This happens if the mediator couples only to the right-handed weak isospin singlet e_R, which is why the couplings to neutrinos vanish in this limit.

The simplified model with interactions as in Eq. (1) can only be valid up to some cut-off scale $\Lambda_{\rm UV}$. Above such a scale, we expect new degrees of freedom UV-completing the simplified model and protecting the mediator couplings to other SM fields. The dimensionless couplings in the DM and lepton currents in Eqs. (2) and (3), and the gauge invariant couplings to neutrino currents in Eq. (4), are always defined at the renormalization scale $\Lambda_{\rm UV}$. RG effects driven by SM couplings, mainly through operator mixing, induce new interactions at energy scales lower than $\Lambda_{\rm UV}$ that can have a phenomenological relevance, as we will extensively discuss in this work. For concreteness, we fix $\Lambda_{\rm UV} = 10$ TeV, though typically the induced couplings depend only logarithmically on the cut-off scale, so small changes to this value would not affect our results substantially.

3. RG analytical solutions

The couplings to SM particles other than leptons induced by RG evolutions at energy scales below $\Lambda_{\rm UV}$ are numerically evaluated by using the code RUNDM. In this Section, we provide the reader with simple analytical solutions to the RG equations derived in Ref. [25], which are helpful to understand the size of the effects. The analytical results are obtained by a fixed order calculation, whereas RUNDMalso accounts for the evolution of SM couplings (see Appendix A of Ref. [26]). For leptophilic models such a difference is pretty moderate, hence the solutions given here capture the RG effects with very good accuracy.

At renormalization scales μ between the cutoff and the electroweak scale, $\Lambda_{\rm UV} \gtrsim \mu \gtrsim m_Z$, all SM fields are in the spectrum. Radiative corrections driven by hypercharge interactions (Feynman diagram on the left of Fig. 1) induce flavor universal couplings between the mediator and SM fermions

$$c_i(\mu) = c_i(\Lambda_{\rm UV}) + \frac{2}{3} \frac{\alpha_Y}{\pi} y_i \left(\sum_{l=e,\mu,\tau} g_{Vl} \right) \log\left(\Lambda_{\rm UV}/\mu\right) \, . \tag{5}$$

Here, $\alpha_Y = g_Y^2/(4\pi)$ the hypercharge fine structure constant. The index i runs over all possible 5 SM fermions with well defined electroweak quantum numbers and hypercharge y_i. Crucially, the result in Eq. (5) accounts for both charged leptons and neutrinos

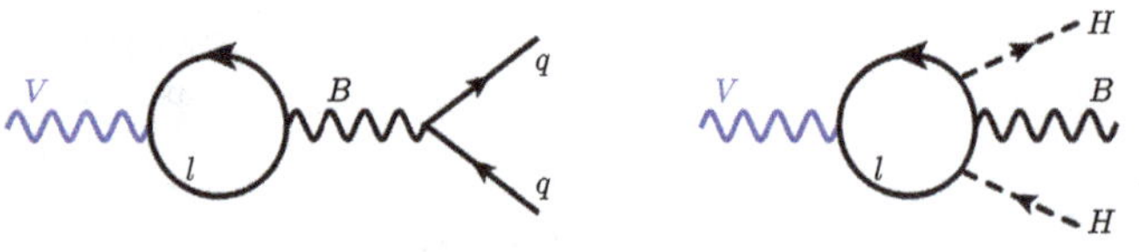

Fig. 1. Feynman diagrams for the operator mixing investigated in this work. Left: kinetic mixing between V and the hypercharge gauge boson B (and with the photon below the EWSB scale) induces mediator couplings to quarks. Right: loop-induced coupling between the mediator and the Higgs current that in turn generates mass mixing between V and the Z boson (plus other diagrams with only H external legs and with the $SU(2)_L$ gauge bosons).

in the loop, as dictated by the gauge invariant analysis in Ref. [25]. Using the same notation in Eq. (3), we employ the general result in Eq. (5) to identify the induced vector and axial-vector currents of up- and down-type quarks

$$\{g_{Vu}(\mu), g_{Vd}(\mu), g_{Au}(\mu), g_{Ad}(\mu)\} = \left\{\frac{5}{18}, -\frac{1}{18}, \frac{1}{6}, -\frac{1}{6}\right\} \frac{\alpha_Y}{\pi} \left(\sum_{l=e,\mu,\tau} g_{Vl}\right) \log\left(\Lambda_{\rm UV}/\mu\right) , \tag{6}$$

The same Feynman diagrams generate loop-induced currents to leptons as well, relevant only in the absence of the associated tree-level term. Hypercharge interactions generate flavor universal lepton currents

$$\{\Delta g_{Vl}(\mu), \Delta g_{Al}(\mu)\}_Y = \left\{-\frac{1}{2}, -\frac{1}{6}\right\} \frac{\alpha_Y}{\pi} \left(\sum_{l=e,\mu,\tau} g_{Vl}\right) \log\left(\Lambda_{\rm UV}/\mu\right) , \tag{7}$$

where we use the symbol Δ to emphasize that these are radiative corrections with respect to the tree-level terms in Eq. (3). The last operator we discuss above the weak scale is the mediator coupling with the Higgs current

$$\Delta\mathcal{L}_H \equiv g_H\, H^\dagger i \overleftrightarrow{D}_\mu H\, V^\mu , \tag{8}$$

generated by the Feynman diagram on the right of Fig. 1. Upon neglecting the electron and muon Yukawa couplings, the associated coefficient reads

$$g_H(\mu) = \left[\frac{1}{3}\frac{\alpha_Y}{\pi}\left(\sum_{l=e,\mu,\tau} g_{Vl}\right) - \frac{\alpha_\tau}{\pi} g_{A\tau}\right] \log\left(\Lambda_{\rm UV}/\mu\right) , \tag{9}$$

where we define $\alpha_\tau = \lambda_\tau^2/(4\pi)$ as the effective Yukawa coupling for the τ.

Heavy SM degrees of freedom are integrated-out below the weak scale. Integrating out the Z boson gives rise to threshold corrections to the vector and axial-vector currents of SM fermions coupled to the mediator, as a consequence of the induced g_H in Eq. (9) (for details of integrating out heavy SM states see the analysis in Ref. [25]). DD rates are set by the Wilson coefficients of the contact interactions

$$\mathcal{L}_{\rm DD} = -\frac{J_\mu^{(\rm DM)}}{m_V^2}\left[\sum_{q=u,d} c_V^{(q)}\, \overline{q}\gamma^\mu q + \sum_{q=u,d,s} c_A^{(q)}\, \overline{q}\gamma^\mu\gamma^5 q\right] . \tag{10}$$

Here, we distinguish between coupling to vector and axial-vector currents of light quarks, and we only keep the ones relevant for direct detection rates. The approximate expressions for the Wilson coefficients read

$$c_V^{(q)} = \frac{2\alpha_{\rm e.m.}}{3\pi} Q_q \left(\sum_{l=e,\mu,\tau} g_{Vl}\right) \log\left(\Lambda_{\rm UV}/\mu_N\right) + - \frac{\alpha_\tau}{\pi}\left(T_q^{(3)} - 2 s_w^2 Q_q\right) g_{A\tau} \log\left(\Lambda_{\rm UV}/\mu_N\right) . \tag{11}$$

Here, $\alpha_{\rm e.m.} = e^2/(4\pi)$ is the electromagnetic fine structure constant, whereas the nuclear scale relevant for DD is $\mu_N \simeq 1$ GeV. The third component of the weak isospin and the electric charge for the up and down quarks are $(T_u^{(3)}, T_d^{(3)}) = (+1/2, -1/2)$ and $(Q_u, Q_d) = (+2/3, -1/3)$, respectively. Likewise, RG-induced interactions with quark axial currents read

$$c_A^{(q)} = T_q^{(3)} \frac{\alpha_\tau}{\pi} g_{A\tau} \log\left(\Lambda_{\rm UV}/\mu_N\right) . \tag{12}$$

4. Phenomenological constraints

We list the experimental bounds on leptophilic models by dividing them into three different categories. First, we consider tree-level processes. An intermediate case is for experimental signals that could be present at the tree-level, depending on which specific lepton flavor is coupled to the mediator. Lastly, we list genuinely loop-induced constraints that cannot be obtained by considering only the operators in Eq. (1). All these constraints are implemented in the phenomenological analysis performed in Sec. 5, where we consider the mediator coupled to a single lepton flavor at a time. As explicitly shown in Fig. 2, the loop-induced bounds frequently dominate over the tree-level ones.

4.1. Bounds from tree-level processes

Perturbative unitarity The simplified model defined in Sec. 2 may violate perturbative unitarity at high energies. For pure vector couplings ($g_{Al} = g_{A\chi} = 0$) there is no unitarity issue, since in this case the mediator mass can be generated via a Stueckelberg mechanism [36] without any further low-energy degrees of freedom. Bounds from unitarity arise then only when axial couplings are present, putting a lower bound on the mediator mass [37]. The leptons are much lighter than the mediator masses considered in this work, so they are massless for our purposes and we do not get any bound in the presence of the axial coupling g_{Al}. On the contrary, an axial-vector coupling to DM particles gives a lower bound for the mediator mass

$$m_V \gtrsim \sqrt{2/\pi}\, g_{A\chi}\, m_\chi . \tag{13}$$

Relic density We numerically solve the Boltzmann equation for the DM number density, with the thermally averaged cross section computed as prescribed by Ref. [38]. In particular, we fully account for relativistic corrections and for resonance effects, the latter evaluated by computing the mediator width self-consistently with the simplified model field content. The non-relativistic limit for the annihilation cross sections are collected in Appendix A. We include in our analysis Sommerfeld corrections [39] to the annihilation cross section. As it turns out, they are relevant only for annihilation to mediators. For a DM particle coupled with the vector current, we use a standard analytic expression of the Sommerfeld factor obtained by replacing the Yukawa with the Hulthén potential (see e.g. [40–42]). In the phenomenological analysis of Sec. 5, we do not need to include such effects for DM coupled with the axial-vector current, since we always fix $g_{A\chi} = 1$; the region where Sommerfeld effects would be relevant ($m_\chi \gtrsim m_V$) is then excluded by perturbative unitarity (see Eq. (13)).

Constraints from relic density are not treated on an equal footing with the other experimental bounds listed in this Section.

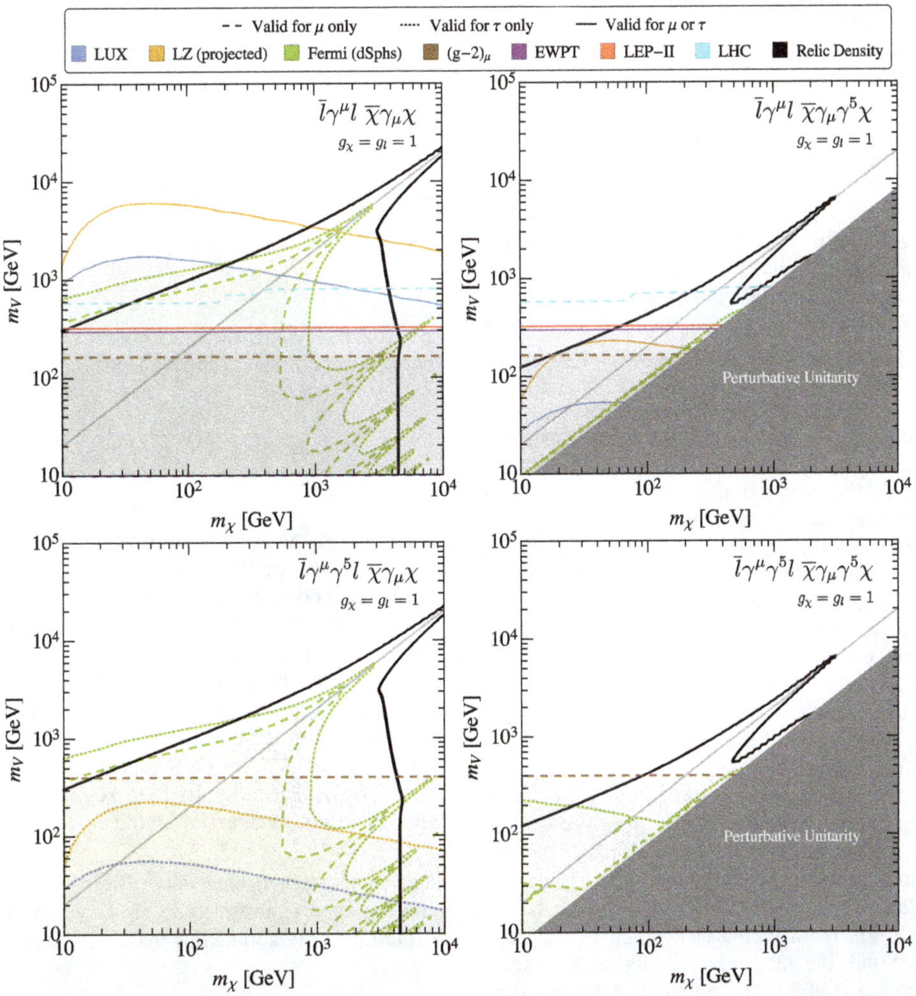

Fig. 2. Constraints on leptophilic DM coupling to muons and taus through a heavy vector mediator, detailed in Sec. 4. Dashed (dotted) lines show constraints which are valid for mediators coupling only to μ (τ) leptons. Solid lines are constraints valid in both cases. The solid gray area in the case of axial-vector couplings to DM is excluded as it violates perturbative unitarity [37]. The faint gray diagonal line is included to guide the eye and denotes the boundary $m_V = 2m_\chi$. Note that a vector mediator coupling to electrons (limits not shown) is strongly constrained by the LEP-II compositeness bound [55] for all types of interaction, with limits of $m_V \gtrsim \mathcal{O}$(3–4 TeV). All constraints are at the 95% confidence level. Note that for LZ we show projected (rather than current) constraints.

The computed dark matter abundance relies upon the extrapolation of the universe snapshot at the time of Big Bang Nucleosynthesis ($T_{\rm BBN} \simeq 1$ MeV), where the universe was a thermal bath of relativistic particles, to higher temperatures. We have no hint about the energy content of the universe for higher temperatures, relevant for dark matter freeze-out. Thus we remain agnostic about the thermal history at temperatures above $T_{\rm BBN}$, and we always assume that the dark matter is produced with the observed relic abundance when we impose limits such as direct and indirect detection. Our relic density lines serve only as benchmark values. Upon considering modified cosmological histories, a correct relic density can be achieved with annihilation cross sections much smaller [43–46] or much larger [47–49] than the thermal relic value.

Indirect detection We impose bounds from observations of diffuse gamma rays from dwarf spheroidal galaxies (dSphs). In the $m_\chi < m_V$ regime, DM can only annihilate to leptons and we impose the limits obtained by the Fermi collaboration in Ref. [50]. The Fermi Collaboration only report limits on the cross section for annihilation into single channels ($e^\pm$, $\mu^\pm$, $\tau^\pm$, etc.). For flavor universal interactions, annihilation proceeds into a mixture of channels. In that case, the limits can be calculated using the publicly available Fermi likelihoods [51], combined with the annihilation spectra from PPPC4DMID [52,53]. Once annihilations to mediators open up, we use the constraints of Ref. [54] for cascade annihilation into leptons. Given the small DM velocity, the inclusion of Sommerfeld corrections [39] is of importance, and we account for them as in the relic density calculation described above.

In Sec. 6, we explore the effects which loop-induced couplings (in particular to quarks) may have on annihilation spectra and therefore on indirect detection limits.

4.2. Bounds that can be present at tree-level

LEP-II compositeness bound Measurements of cross sections and forward-backward asymmetries for the production of leptons at LEP-II have been used to set limits on possible 4-fermion contact interactions and, in turn, the mass of a new vector mediator which could mediate such interactions [55]. These limits are often referred to as the LEP-II compositeness bounds. We impose constraints from measurements of the processes $e^+e^- \to e^+e^-$, $\mu^+\mu^-$, $\tau^+\tau^-$. The strongest constraints are therefore obtained when the mediator couples to electrons at tree-level, in which case couplings of order unity require a mediator heavier than 3–4 TeV. For mediators coupling primarily to μ and τ leptons at tree-level, we use the RUNDMcode [27] to evaluate the electron couplings at an energy scale of 209 GeV and then apply the LEP-II constraints. While for axial-vector SM currents, this loop-induced coupling is small, it can be sizable for vector currents, leading to limits of $m_V \gtrsim 300$ GeV.

Collider searches LHC collisions probe leptophilic dark sectors through tree-level processes. Lepton pair production with a V boson radiated by either of the particles in the final state give access to the dark sector. The reconstructed final state depends on the decay mode of the mediator. Drell–Yan processes pair produce charged leptons, and after V-radiation the final state can contain two leptons and missing energy or four leptons. The ATLAS four lepton cross section measurement [56] constrain a possible mediator to be heavier than $m_V \sim 8$ GeV, for $g_l = g_\chi = 1$ and for coupling to e or μ [6]. For coupling to τ the measurements are not good enough to give any meaningful constraint. Conversely, for neutrino-charged lepton pair production through s-channel W boson with the radiated mediator decaying invisibly, the final state contains one lepton and missing energy, as searched for in Refs. [57,58]. The analysis in Ref. [8] found that for a leptophilic vector mediator the most stringent bounds come from processes with at least three leptons in the final states. All of these constraints are typically weaker than the LEP-II compositeness bounds [55] and we therefore include only the latter in the phenomenological analysis performed in Sec. 5. However, we note that the relative importance of the different constraints depends on the assumed couplings and moving away from the case of $g_l = g_\chi = 1$ will shift the limits.

Mono-jet searches, usually a powerful DM probe at colliders, are not well suited for models where the DM does not couple to quarks or gluons at tree-level. Nevertheless, there exist collider processes capable of probing these models. Initial state photon radiation at leptonic colliders may yield mono-photon events, that are bounded by LEP [59,60]. For a mediator coupled to electrons with couplings of order one, the subsequent limits on the mediator mass are of order $m_V \gtrsim 400$–500 GeV for DM lighter than ~ 100 GeV [61]. Again, these limits are weaker than the LEP-II compositeness bound, though as above this statement holds only for $g_l = g_\chi = 1$.

4.3. Bounds absent at tree-level

EWPT RG flow generates a mediator coupling to the Higgs current. Upon expanding this interaction around the vacuum

$$\Delta\mathcal{L}_H = g_H\, H^\dagger i \overleftrightarrow{D}_\mu H\, V^\mu = -g_H \frac{2c_W}{g_2} m_Z^2\, Z_\mu V^\mu + \ldots\,, \tag{14}$$

we identify a mass mixing between V and the Z boson. Here, c_W and g_2 are the cosine of the weak mixing angle and the gauge coupling of the weak isospin gauge group, respectively. In the small mixing limit, this leads to a mixing angle between the two neutral vector particles

$$\theta_{\text{mix}} \simeq -g_H \frac{c_W}{g_2} \frac{m_Z^2}{m_V^2 - m_Z^2}\,. \tag{15}$$

This mixing angle is bounded to be $\theta_{\text{mix}} \lesssim 10^{-3}$ by ElectroWeak Precision Tests (EWPT) [62]. An approximate expression for g_H is given in Eq. (9). We emphasize that the EWPT limits we impose are the most conservative ones, since we assume that the mass mixing is vanishing at Λ_{UV}. This could easily not be the case in explicit microscopic realizations, and the mixing we account for is an irreducible contribution from loops of SM particles. Unless an unnatural cancellation takes place between UV and RG-induced terms, our analysis accounts for the minimum amount of mass mixing we expect in this class of models.

Dilepton resonances at the LHC The dilepton final state at the LHC is an excellent probe to look for new resonances. For the leptophilic models considered in this work, one may naively conclude that this channel is fruitless due to the lack of couplings to colored states. However, RG-induced couplings to quarks, although loop suppressed, still lead to meaningful bounds. The cross section in the narrow width limit reads

$$\sigma_{pp\to l^+l^-} = \frac{\pi\, BR_{V\to l^+l^-}}{3s} \sum_q C_{q\bar{q}}(m_V^2/s) \left(g_{Vq}^2 + g_{Aq}^2\right)\,, \tag{16}$$

where $BR_{V\to l^+l^-}$ is the branching ratio of the decay $V \to l^+l^-$. The parton luminosity $C_{q\bar{q}}(m_V^2/s)$ for the quark q reads

$$C_{q\bar{q}}(y) = \int_y^1 dx\, \frac{f_q(x)\, f_{\bar{q}}(y/x) + f_q(y/x)\, f_{\bar{q}}(x)}{x}\,, \tag{17}$$

with $f_{q,\bar{q}}(x)$ the quark and antiquark parton distribution function (PDF). We use the PDFs from Ref. [63] and we evaluate them using the public code available at the URL https://mstwpdf.hepforge.org/. The loop-induced couplings to quarks, whose approximate expressions can be found in Eq. (7), are evaluated with RUNDM. The cross section in Eq. (16) is then computed by evaluating the PDF and the couplings to quarks at the renormalization scale m_V. We impose the recent Run 2 bounds at $\sqrt{s} = 13$ TeV [64][1] (for earlier studies by the ATLAS and CMS collaborations see Refs. [65] and [66], respectively). For tau pairs [67], RG-induced signals are not strong enough to put meaningful bounds.

Anomalous magnetic moments The coupling between the mediator and the lepton current in Eq. (3) is responsible for a one-loop contribution to the anomalous magnetic moments [23]

$$\delta(g-2)_l \simeq \frac{1}{12\pi^2} \frac{m_l^2}{m_V^2} \left(g_{Vl}^2 - 5g_{Al}^2\right)\,. \tag{18}$$

The value of $(g-2)_\mu$ is measured to be larger than the Standard Model prediction at the 3σ level [68]. With the contribution of vector couplings, the agreement between the theory and measurement can be improved and we therefore set limits by requiring that the total theoretical prediction does not exceed the measured value at the 2σ level. On the other hand, axial-vector couplings reduce the theory prediction, worsening the agreement. In this case, we require that the loop-induced contribution not exceed the 2σ uncertainty on the measured value. While determination of $(g-2)_e$ is more precise, the contribution from New Physics is expected to be smaller by a factor of $m_e^2/m_\mu^2 \approx 2\times 10^{-5}$, leading to constraints on m_V which are weaker by about a factor of 10 [6]. In contrast, experimental limits on $(g-2)_\tau$ are currently too weak to give meaningful constraints on New Physics [69]. While in principle there are also loop-induced contributions to $(g-2)_\mu$ from mediators coupling to e or τ, the corresponding limits are negligible in our case. This is why in Sec. 5 we include only the $(g-2)_\mu$ bound, which applies to a mediator coupled to muons and generating an anomalous magnetic moment as given in Eq. (18).

[1] Note that the analysis of Ref. [64] places limits on the cross section using data from both the di-electron and di-muon channels, with roughly equal constraining power from each. In our analysis, we assume a coupling only to muons (not electrons) and we therefore rescale the limit by a factor of $\sqrt{2}$ to account for the reduced statistical power which would arise from using only the di-muon channel. We also assume a constant signal acceptance of 40%, which is typical for Z' models in the di-muon channel [64].

4.3.0.1. Direct Detection There is no direct detection at tree-level since the mediator couples only to leptons. However, RG mixing effects induce a coupling to light quarks [24,25], which can be compactly written by using the effective Lagrangian in Eq. (10). These loop-induced couplings can be evaluated by using the public code RUNDM [27]. We provided analytical solutions for the low-energy couplings in Sec. 3, which are a very accurate approximation to the ones derived by performing the full RG evolution. Previous works [5,6] focused on loop-induced direct detection rates for a mediator coupled to the lepton vector current. Here we point out a new effect, which follows from the general analysis of Ref. [25]. If the mediator couples to the axial-vector current, there is still a radiative contribution to direct detection rates arising from loops of the tau lepton (see Eqs. (11) and (12)). Although suppressed by the tau's Yukawa coupling, it still leads to meaningful constraints.

We present limits from the recent LUX WS2014-16 run [70] and projections for the upcoming LUX-ZEPLIN (LZ) experiment [71]. Calculation of the approximate 95% confidence limits from LUX is detailed in Appendix B of Ref. [72] and calculation of the LZ projection is described in Appendix D of Ref. [26], following the general procedure developed in Ref. [73].

5. Results for single lepton flavor

We analyze models where the vector mediator couples only to a single lepton flavor with dimensionless couplings of order one, namely $g_{\alpha\chi} = g_{\beta l} = 1$ (with $\alpha, \beta = V, A$). As emphasized at the end of Sec. 2, the charge assignments are made at the simplified model cutoff $\Lambda_{\mathrm{UV}} = 10$ TeV, and we assume no other interactions at such a scale. The case of a mediator coupled to electrons is not considered here, since the tree-level coupling to electrons is strongly constrained by the LEP-II compositeness bound. In this case, the mediator mass is constrained at the level of $m_V \gtrsim \mathcal{O}(3\text{–}4\ \text{TeV})$, which is much stronger than almost all other constraints and leaves only a small window of $m_\chi \gtrsim 1$ TeV in which DM could be produced by thermal freeze-out.

Constraints for coupling to μ or τ are presented in the (m_χ, m_V) plane in Fig. 2. Here, dashed (dotted) lines show bounds that are valid only in the case of couplings to muons (taus). Solid lines are constraints which apply in either case.

For a mediator having vector interactions with both DM and leptons (top-left panel), the dominant constraints come from direct and indirect detection. In the former case, this arises from a loop-induced interaction with quarks (and therefore nucleons) driven by the electromagnetic coupling, while the latter comes from tree-level annihilation to leptons. For couplings to muons, competitive constraints (particularly at high DM mass, $m_\chi \gtrsim 1$ TeV) come from searches for dilepton resonances at the LHC. The same loop-induced couplings with quarks which lead to direct detection constraints allow for the hadronic production of the vector mediator, which in turn decays into leptons. Since we consider coupling to a single lepton flavor, decays to both charged lepton and the associated neutrino of the same flavor are always open. Mediator decays to DM pairs are only allowed for $m_V > 2m_\chi$. Explicit expressions for the mediator width are provided in Appendix A. For couplings of order one as chosen in this Section, the mediator width is 6.6% (4%) if decays to DM pairs are (not) allowed. We impose the bounds from the recent analysis in Ref. [64], which provided limits on the production cross section for different mediator widths. For coupling to muons, we obtain a limit of $m_V \gtrsim 700$ GeV for the mass of the mediator. Limits on heavy mediators decaying to taus are weaker, owing to the relative difficulty of reconstructing tau leptons in the detector, and no significant constraints can be drawn for coupling to taus.

In the top-right panel we have the case of a mediator coupling to the vector current of leptons and the axial-vector current of DM. For couplings to muons, similar LHC constraints hold, which are the strongest ones. In contrast with the previous case, direct detection limits are much weaker, as the loop-induced interaction with nucleons is in this case velocity suppressed. Furthermore, indirect detection limits from Fermi are negligible as the annihilation cross section is p-wave suppressed (see Eq. (A.3)). For a mediator coupled to taus, the strongest bounds (at the level of $m_V \gtrsim 300$ GeV) come from EWPT and LEP-II compositeness bounds, which both arise from loop-induced couplings.

For axial-vector couplings to leptons and vector couplings to DM (bottom-left panel), the dominant constraint over much of the DM mass range is the Fermi limit on DM annihilation in dwarf spheroidal galaxies. In this case, the annihilation cross section is s-wave and proportional to the DM mass squared. At large DM masses, constraints on $(g-2)_\mu$ are also competitive when the mediator couples to muons. Direct detection proceeds through operator mixing. For axial-vector couplings to leptons, such a mixing is driven by the lepton Yukawa coupling, unlike the previous case of lepton vector current where it was driven by the EM coupling. As a result, these is no direct detection constraint for coupling to muons (and also no bounds from LHC and EWPT). In contrast, the tau Yukawa is large enough to induce an observable spin-independent direct detection signal (which is coherently enhanced), meaning that LUX can constrain such simplified models at the level of $m_V \gtrsim 50$ GeV. This effect has not previously been pointed out for leptophilic DM.

Finally, for axial-vector couplings to both leptons and DM (bottom-right panel), the parameter space is more poorly constrained. Though $(g-2)_\mu$ constraints still apply for tree-level couplings to muons, the annihilation cross section is suppressed by the mass of the lepton in the final state, leading to weaker indirect detection limits. For couplings to the tau, the loop-induced direct detection cross section is a combination of velocity-suppressed spin-independent and unsuppressed spin-dependent interactions. The latter receives no coherent enhancement of the rate and therefore gives no appreciable bounds in this case.

6. Hadronic contamination in indirect detection spectra

The gamma ray bounds in Sec. 5 were obtained by considering tree-level processes. Radiative corrections can potentially alter the spectral features of the predicted flux of cosmic rays (CRs). As an example, Refs. [74–77,53] considered leptophilic models and quantified these corrections for gamma rays arising from annihilations of DM particles with mass larger than the weak scale ($m_\chi \gtrsim m_Z$). In such a mass range, final state leptons have energies larger than the weak scale and the contamination of the spectra is due to electroweak bremsstrahlung.

In this section we discuss contamination of CR spectra within our framework and for DM masses smaller than the weak scale ($m_\chi \lesssim m_Z$). This effect is due to the mixing among the spin-one currents of SM particles coupled to the mediator [25], and it was not pointed out in previous studies. More specifically to our leptophilic framework, RG flow generates a coupling of the mediator to the Higgs current through operator mixing (see the EWPT discussion in Sec. 4.3). As a consequence, the DM acquires an effective vertex with the Z boson, which opens up new channels for indirect detection spectra. For example, it may be possible to have DM annihilations with hadronic final states mediated by a virtual Z boson exchanged in the s-channel.

The Z boson can always be integrated out for direct detection scattering. However, for DM annihilation the center of mass energy is approximately twice the DM mass, meaning that we cannot al-

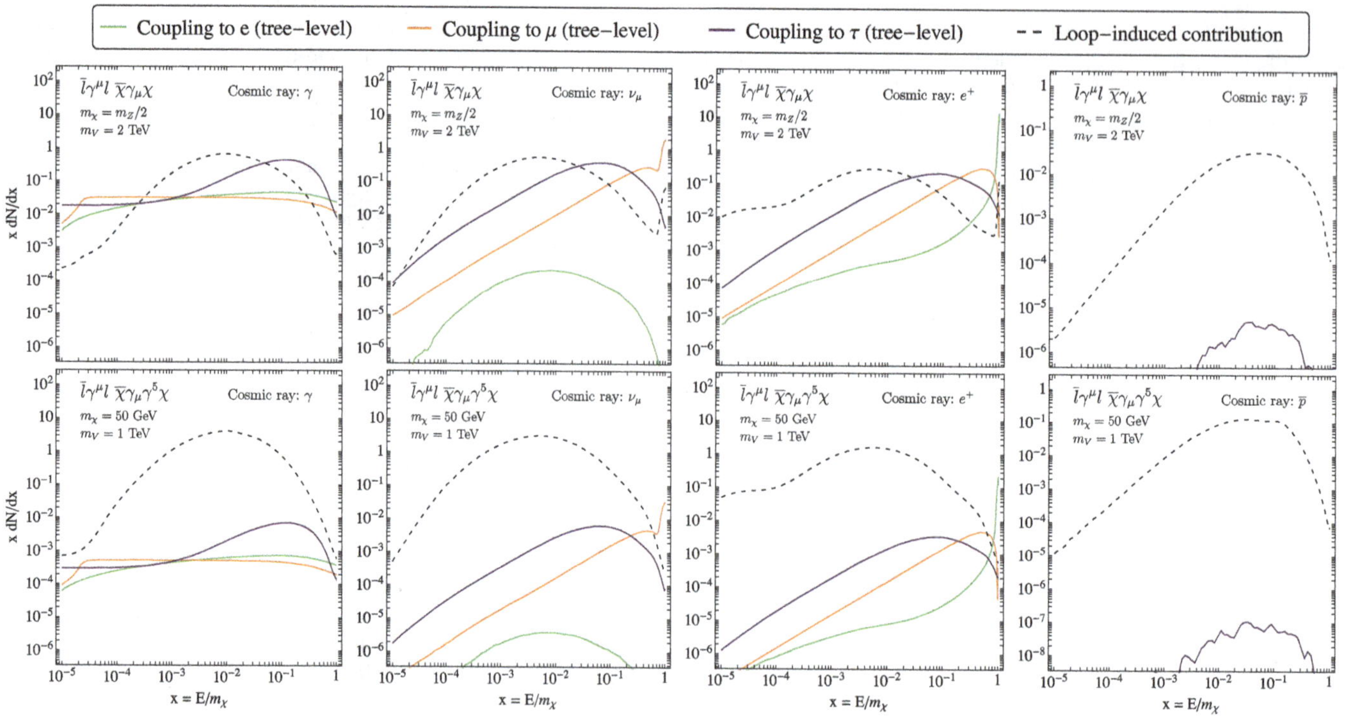

Fig. 3. Spectrum of cosmic rays produced (per annihilation) for leptophilic DM. The top row shows results for vector interactions to both leptons and DM, while the bottom row shows results for vector couplings to leptons and axial-vector couplings to DM. Each column shows the spectrum of a different cosmic ray species (from left to right): γ, ν_μ, e^+ and $\bar{p}$. The tree-level contribution to the spectrum is shown as a solid colored line for a mediator coupling to electron, muon or tau leptons. The dashed line shows the loop-induced contribution. The total spectrum is the sum of the tree-level and loop-induced contributions.

ways integrate out the Z. If the DM mass is close enough to the Z-resonance, we have to account for the full propagator of the intermediate Z boson, and the resonant enhancement may give rise to hadronic final states overtaking the leptonic ones already present at the tree-level. The total CR spectrum per annihilation is in general the sum of the tree-level and loop-induced contributions. Explicitly, the primary spectra at production of stable SM products i can be written

$$\frac{dN_i}{dx} = \left(\frac{\sigma_{\chi\chi\to Z'\to l^+l^-}}{\sigma_{\text{tot}}} \frac{dN_i^l}{dx} + \sum_f \frac{\sigma_{\chi\chi\to Z\to f\bar{f}}}{\sigma_{\text{tot}}} \frac{dN_i^f}{dx} \right), \tag{19}$$

where $x = E/m_\chi$ and $\sigma_{\text{tot}} = \sigma_{\chi\chi\to Z'\to l^+l^-} + \sum_f \sigma_{\chi\chi\to Z\to f\bar{f}}$ is the total annihilation cross section. Here $dN_i^{l(f)}/dx$ is the spectrum of products i arising solely from the primary $l(f)$ with energy E, obtained from PPPC4DMID [52,53].

In Fig. 3, we show the spectra at production of CRs from DM annihilation. We consider four different cases of stable SM products (different columns), and we take two benchmark scenarios: vector couplings to both leptons and DM (top row) and vector couplings to leptons with axial-vector coupling to DM (bottom row). For the former scenario, we show the special case where the DM can annihilate resonantly through a Z boson exchanged in the s-channel, $|m_\chi - m_Z/2| \ll \Gamma_Z$. We plot the tree-level CR spectra (first term in the parentheses of Eq. (19)) assuming annihilation to electrons, muons or taus (solid colored lines). We also plot the contribution to the spectrum arising from Z-mediated annihilation (dashed line), computed by summing the spectra for all possible final state annihilation products (second term in the parentheses of Eq. (19)).

For vector couplings to both DM and leptons (top row), mixing with the Z-boson leads to annihilation into quarks, which produces, via parton showering and hadronization, the broad low-energy bump in the CR spectra. In fact, in this case both the tree-level and loop-induced annihilations are s-wave processes, meaning that annihilation to quarks is loop-suppressed and so generally subdominant to annihilation to leptons. However, for DM masses around $m_\chi \sim m_Z/2$, Z-mediated annihilation to quarks is resonantly enhanced and can therefore become dominant at low energies (as in Fig. 3). The gamma-ray flux (top left-most panel) receives a substantial contribution from parton showering and hadronization of quarks. This means that the indirect detection limits presented in Fig. 2 should in principle be modified close to the resonance ($m_\chi \approx m_Z/2$) to account for this enhanced flux. Such an enhancement may be relevant for Simplified Model fits to the Galactic Centre Excess, where a DM particle with masses in the range 30–70 GeV has been proposed (see e.g. Refs. [78–80]). We note however that such a Simplified Model is likely to be strongly constrained by direct detection (as described in Sec. 5).

Loop-induced enhancements to the indirect detection spectra are particularly noticeable for antiprotons (top right-most panel). At the tree-level, annihilation to leptons produces very few antiprotons.[2] On the contrary, the antiproton flux from (loop-induced and resonant) annihilation to quarks is some 3 orders of magnitude larger. It is natural to investigate whether the antiproton flux measured by AMS-02 [81,82] can put bounds on leptophilic models. Assuming for simplicity universal antiproton spectra for all DM annihilations to quarks (and taking them to be the same as for b quarks), we answer this question in Fig. 4. For a DM mass at the Z resonance, the loop-induced antiproton flux leads to constraints on the mediator mass at the level of $m_V \gtrsim 750$ GeV. Remarkably, this is competitive with constraints from the Fermi gamma-ray search in dwarf Spheroidals already plotted in Fig. 2, which arise from tree-level, leptonic annihilations.

[2] The antiproton flux coming from the annihilation either to electrons or muons is zero for $m_\chi < m_Z$. For $m_\chi \gg m_Z$ one can get a sizable amount of antiprotons via electroweak gauge bosons bremsstrahlung [74–77,53].

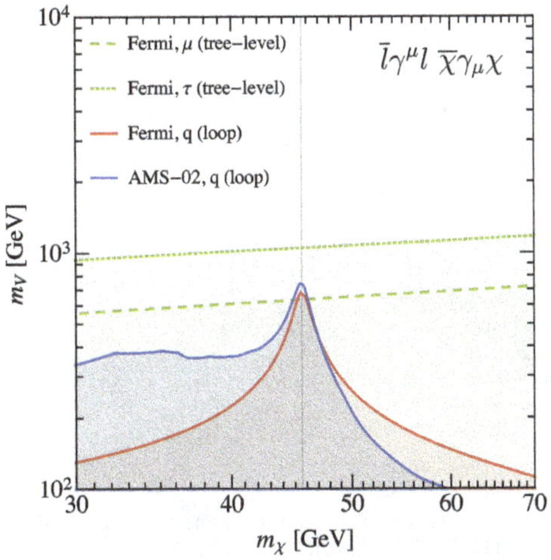

Fig. 4. Indirect detection constraints on leptophilic DM coupling to mu and tau leptons through the interaction $\bar{l}\gamma_{\mu}l\overline{\chi}\gamma^{\mu}\chi$. We show the tree-level constraints from Fermi dSphs arising from this tree-level coupling to muons and taus (dashed and dotted green, respectively). We also show the limits from loop-induced Z-mediated annihilation to quarks: gamma ray bounds from Fermi dSphs (solid red) and antiproton bounds from AMS-02 (solid blue). (For interpretation of the references to color in this figure legend, the reader is referred to the web version of this article.)

For vector couplings to leptons and axial-vector couplings to DM (bottom row), tree-level annihilation is p-wave suppressed. Mixing with the Z leads to an effective axial-vector coupling to leptons, which gives rise to s-wave annihilation (see Eq. (A.2)). Thus, the loop contribution can be substantially larger than the tree-level one, though the main contribution in this case is from annihilation to b quarks only (as the cross section is now suppressed by the mass of the final state particle). In this scenario, the loop-induced correction is generally expected to be large compared to the tree-level spectrum, regardless of the DM mass. However, we note that even including the large loop contribution, the CR fluxes are not large enough to be constrained by current indirect detection searches (limits from Fermi are negligible in the top right panel of Fig. 2).

We do not consider the impact of hadronic contamination for axial-vector couplings to leptons. As already mentioned in Sec. 5, the mixing for axial-vector currents is typically smaller than for the vector current, being driven now by the tau Yukawa. Contamination of the spectra through Z-mediated annihilation is therefore typically small in that scenario.

7. Conclusions

WIMPs remain among the best motivated candidates for particle DM. Models where the DM particle is coupled only to SM leptons alleviate the tension with experimental constraints, since the most severe bounds come from experimental processes involving hadrons. However, hadronic processes are still able to probe *leptophilic* dark sectors due to the RG-evolution of the couplings with the energy scale. More precisely, radiative effects generate couplings to quarks at higher-orders, which although suppressed still give sizable signals.

This work exploited the relevance of these effects for leptophilic dark sectors with a vector portal. We considered the massive mediator coupled to only a single lepton flavor at a time, and in each case we imposed the experimental bounds listed in Sec. 4. They arise both from tree-level and radiatively-induced processes. For a mediator coupled to electrons, the LEP-II compositeness bound is extremely severe. For comparable couplings to DM and electrons, it easily overtakes bounds from DM searches. For this reason, we have focused only on the cases of coupling to μ and τ.

Our main results are shown in Fig. 2. In the top panels we show the cases of vector couplings to muons and taus. A noteworthy feature of our results is that the most important constraints come from hadronic processes: direct detection in the case of vector couplings to DM and LHC dilepton resonances in the case of axial-vector couplings to DM (for which the direct detection rate is suppressed). In the bottom panels we show the cases of axial-vector couplings to leptons. In contrast, the dominant constraints here are from leptonic processes: indirect detection in the case of taus and $(g-2)_{\mu}$ in the case of muons. Only for axial-vector coupling to taus and vector couplings to DM do we find an appreciable hadronic constraint. In this case, mixing effects driven by the tau Yukawa lead to spin-independent DM-nucleon scattering, which is strongly constrained by LUX and other direct detection experiments. This effect was not pointed out in previous studies of leptophilic DM.

Indirect detection is the only canonical WIMP search with tree-level signals in leptophilic models. We studied higher order corrections to cosmic ray spectra in Sec. 6. While it was known that for high DM mass ($m_{\chi} \gtrsim m_Z$) electroweak bremsstrahlung may affect indirect detection spectra, no previous studies pointed out such a contamination in the low mass region. For vector coupling to leptons, RG flow induces an effective coupling of the DM particle to the Z boson, which in turn mediates annihilation to hadrons. For a DM mass sufficiently close to the Z resonance, these radiatively-induced contributions to the CRs fluxes may dramatically alter the predicted spectra. This is illustrated in Fig. 3, where for both vector and axial-vector DM couplings we compare the tree-level and loop-induced spectra of four different stable SM products. Final state antiprotons are a particular interesting case. At low DM masses ($m_{\chi} \lesssim m_Z$), the tree-level antiproton flux is either zero (coupling to electron and muon) or extremely suppressed (coupling to tau), thus the radiatively-induced contribution to the spectrum is significant. We illustrate this in Fig. 4, where we superimpose the recent AMS antiproton bounds over the other constraints used in this work. The bounds are competitive for a DM mass close enough to the Z resonance ($m_{\chi} \simeq m_Z/2$).

Leptophilic dark sectors are an attractive scenario. Motivated by anomalies in the observed CRs fluxes, they also alleviate the tension with experimental bounds from DM searches involving hadronic processes. In spite of expectations based on tree-level calculations, the constraints are not as mild when accounting for radiative corrections. This work quantified the magnitude of these effects and put novel bounds on the allowed parameter space. We also highlighted new prospects for probing leptophilic dark sector, such as future EWPT and dilepton resonance searches at hadron colliders. We find it intriguing that a leptophilic DM particle may ultimately be discovered in experimental searches not only probing dark sector couplings to leptons but also to quarks.

Acknowledgements

We thank Stefano Profumo for his valuable feedback on the manuscript and the Journal Referee for her/his very insightful and constructive comments. FD acknowledges very helpful discussions with Mike Hance about dilepton resonance searches. FD is supported by the U.S. Department of Energy grant number DE-SC0010107 (FD). BJK is supported by the European Research Council (ERC) under the EU Seventh Framework Programme (FP7/2007-2013)/ERC Starting Grant (agreement n. 278234–'NEWDARK' project).

Appendix A. Decay widths and cross sections

We collect results for the mediator partial decay widths and non-relativistic DM annihilation cross sections. The mediator width for the decay to a fermion/antifermion pair is

$$\Gamma_{V\to\bar{f}f} = \frac{g_{Vf}^2 + g_{Af}^2}{12\pi} m_V \,. \tag{A.1}$$

Here, g_{Vf} and g_{Af} are the mediator couplings to the vector and axial-vector fermion currents, respectively.

We expand DM annihilation cross sections in partial waves, $\sigma_{\chi\chi\to\text{final}} v_\text{rel} = \sigma^{(s)}_{\chi\chi\to\text{final}} + \sigma^{(p)}_{\chi\chi\to\text{final}} v_\text{rel}^2$. Annihilations to leptons, both charged and neutrinos, are always kinematically allowed

$$\sigma^{(s)}_{\chi\chi\to l^+l^-} = \frac{g_{V\chi}^2 (g_{Vf}^2 + g_{Af}^2)}{\pi} \frac{m_\chi^2}{(4m_\chi^2 - m_V^2)^2} + \frac{g_{A\chi}^2 g_{Af}^2}{2\pi} \frac{m_l^2}{m_V^4} \left(1 - \frac{m_l^2}{m_\chi^2}\right)^{1/2} , \tag{A.2}$$

$$\sigma^{(p)}_{\chi\chi\to l^+l^-} = \frac{g_{A\chi}^2 (g_{Vf}^2 + g_{Af}^2)}{6\pi} \frac{m_\chi^2}{(4m_\chi^2 - m_V^2)^2} \,. \tag{A.3}$$

Models with a vector (axial-vector) coupling to the DM have an s-(p-)wave cross section surviving the $m_l \to 0$ limit. If we keep terms suppressed by the lepton mass m_l, we also have a s-wave piece for theories where the couplings are both axial-vector. DM annihilations to mediators, if allowed, can be important since they are s-wave processes

$$\sigma^{(s)}_{\chi\chi\to VV} = \frac{(g_{V\chi}^4 - 6g_{V\chi}^2 g_{A\chi}^2 + g_{A\chi}^4) + \frac{8g_{V\chi}^2 g_{A\chi}^2}{\epsilon_V^2}}{16\pi\, m_\chi^2} f_{VV}(\epsilon_V) \,, \tag{A.4}$$

where $\epsilon_V \equiv m_V/m_\chi$ and $f_{VV}(x) = (1-x^2)^{3/2}(1-x^2/2)^{-2}$.

References

[1] G. Jungman, M. Kamionkowski, K. Griest, Phys. Rep. 267 (1996) 195–373, http://dx.doi.org/10.1016/0370-1573(95)00058-5, arXiv:hep-ph/9506380.
[2] G. Bertone, D. Hooper, J. Silk, Phys. Rep. 405 (2005) 279–390, http://dx.doi.org/10.1016/j.physrep.2004.08.031, arXiv:hep-ph/0404175.
[3] J.L. Feng, Annu. Rev. Astron. Astrophys. 48 (2010) 495–545, http://dx.doi.org/10.1146/annurev-astro-082708-101659, arXiv:1003.0904.
[4] P.J. Fox, E. Poppitz, Phys. Rev. D 79 (2009) 083528, http://dx.doi.org/10.1103/PhysRevD.79.083528, arXiv:0811.0399.
[5] J. Kopp, V. Niro, T. Schwetz, J. Zupan, Phys. Rev. D 80 (2009) 083502, http://dx.doi.org/10.1103/PhysRevD.80.083502, arXiv:0907.3159.
[6] N.F. Bell, Y. Cai, R.K. Leane, A.D. Medina, Phys. Rev. D 90 (2014) 035027, http://dx.doi.org/10.1103/PhysRevD.90.035027, arXiv:1407.3001.
[7] A. Freitas, S. Westhoff, J. High Energy Phys. 10 (2014) 116, http://dx.doi.org/10.1007/JHEP10(2014)116, arXiv:1408.1959.
[8] F. del Aguila, M. Chala, J. Santiago, Y. Yamamoto, J. High Energy Phys. 03 (2015) 059, http://dx.doi.org/10.1007/JHEP03(2015)059, arXiv:1411.7394.
[9] N. Chen, J. Wang, X.-P. Wang, arXiv:1501.04486, 2015.
[10] O. Adriani, et al., PAMELA Collaboration, Nature 458 (2009) 607–609, http://dx.doi.org/10.1038/nature07942, arXiv:0810.4995.
[11] O. Adriani, et al., PAMELA Collaboration, Phys. Rev. Lett. 111 (2013) 081102, http://dx.doi.org/10.1103/PhysRevLett.111.081102, arXiv:1308.0133.
[12] M. Ackermann, et al., Fermi-LAT Collaboration, Phys. Rev. Lett. 108 (2012) 011103, http://dx.doi.org/10.1103/PhysRevLett.108.011103, arXiv:1109.0521.
[13] M. Aguilar, et al., AMS Collaboration, Phys. Rev. Lett. 110 (2013) 141102, http://dx.doi.org/10.1103/PhysRevLett.110.141102.
[14] L. Accardo, et al., AMS Collaboration, Phys. Rev. Lett. 113 (2014) 121101, http://dx.doi.org/10.1103/PhysRevLett.113.121101.
[15] M. Cirelli, P. Panci, P.D. Serpico, Nucl. Phys. B 840 (2010) 284–303, http://dx.doi.org/10.1016/j.nuclphysb.2010.07.010, arXiv:0912.0663.
[16] P. Meade, M. Papucci, A. Strumia, T. Volansky, Nucl. Phys. B 831 (2010) 178–203, http://dx.doi.org/10.1016/j.nuclphysb.2010.01.012, arXiv:0905.0480.
[17] L. Ackermann, et al., Astrophys. J. 761 (2012) 91, http://dx.doi.org/10.1088/0004-637X/761/2/91, arXiv:1205.6474.
[18] M.S. Madhavacheril, N. Sehgal, T.R. Slatyer, Phys. Rev. D 89 (2014) 103508, http://dx.doi.org/10.1103/PhysRevD.89.103508, arXiv:1310.3815.
[19] T.R. Slatyer, Phys. Rev. D 93 (2016) 023527, http://dx.doi.org/10.1103/PhysRevD.93.023527, arXiv:1506.03811.
[20] D. Hooper, P. Blasi, P.D. Serpico, J. Cosmol. Astropart. Phys. 0901 (2009) 025, http://dx.doi.org/10.1088/1475-7516/2009/01/025, arXiv:0810.1527.
[21] S. Profumo, Cent. Eur. J. Phys. 10 (2011) 1–31, http://dx.doi.org/10.2478/s11534-011-0099-z, arXiv:0812.4457.
[22] J. Feng, H.-H. Zhang, Eur. Phys. J. C 76 (2016) 229, http://dx.doi.org/10.1140/epjc/s10052-016-4092-y, arXiv:1504.03312.
[23] P. Agrawal, Z. Chacko, C.B. Verhaaren, J. High Energy Phys. 08 (2014) 147, http://dx.doi.org/10.1007/JHEP08(2014)147, arXiv:1402.7369.
[24] A. Crivellin, F. D'Eramo, M. Procura, Phys. Rev. Lett. 112 (2014) 191304, http://dx.doi.org/10.1103/PhysRevLett.112.191304, arXiv:1402.1173.
[25] F. D'Eramo, M. Procura, J. High Energy Phys. 04 (2015) 054, http://dx.doi.org/10.1007/JHEP04(2015)054, arXiv:1411.3342.
[26] F. D'Eramo, B.J. Kavanagh, P. Panci, J. High Energy Phys. 08 (2016) 111, http://dx.doi.org/10.1007/JHEP08(2016)111, arXiv:1605.04917.
[27] F. D'Eramo, B.J. Kavanagh, P. Panci, runDM (Version 1.0), https://github.com/bradkav/runDM/, 2016.
[28] R.J. Hill, M.P. Solon, Phys. Lett. B 707 (2012) 539–545, http://dx.doi.org/10.1016/j.physletb.2012.01.013, arXiv:1111.0016.
[29] M.T. Frandsen, U. Haisch, F. Kahlhoefer, P. Mertsch, K. Schmidt-Hoberg, J. Cosmol. Astropart. Phys. 1210 (2012) 033, http://dx.doi.org/10.1088/1475-7516/2012/10/033, arXiv:1207.3971.
[30] U. Haisch, F. Kahlhoefer, J. Cosmol. Astropart. Phys. 1304 (2013) 050, http://dx.doi.org/10.1088/1475-7516/2013/04/050, arXiv:1302.4454.
[31] R.J. Hill, M.P. Solon, Phys. Rev. Lett. 112 (2014) 211602, http://dx.doi.org/10.1103/PhysRevLett.112.211602, arXiv:1309.4092.
[32] J. Kopp, L. Michaels, J. Smirnov, J. Cosmol. Astropart. Phys. 1404 (2014) 022, http://dx.doi.org/10.1088/1475-7516/2014/04/022, arXiv:1401.6457.
[33] A. Crivellin, U. Haisch, Phys. Rev. D 90 (2014) 115011, http://dx.doi.org/10.1103/PhysRevD.90.115011, arXiv:1408.5046.
[34] R.J. Hill, M.P. Solon, Phys. Rev. D 91 (2015) 043505, http://dx.doi.org/10.1103/PhysRevD.91.043505, arXiv:1409.8290.
[35] F. D'Eramo, J. de Vries, P. Panci, J. High Energy Phys. 05 (2016) 089, http://dx.doi.org/10.1007/JHEP05(2016)089, arXiv:1601.01571.
[36] E.C.G. Stueckelberg, Helv. Phys. Acta 11 (1938) 225–244, http://dx.doi.org/10.5169/seals-110852.
[37] F. Kahlhoefer, K. Schmidt-Hoberg, T. Schwetz, S. Vogl, J. High Energy Phys. 02 (2016) 016, http://dx.doi.org/10.1007/JHEP02(2016)016, arXiv:1510.02110.
[38] P. Gondolo, G. Gelmini, Nucl. Phys. B 360 (1991) 145–179, http://dx.doi.org/10.1016/0550-3213(91)90438-4.
[39] J. Hisano, S. Matsumoto, M.M. Nojiri, O. Saito, Phys. Rev. D 71 (2005) 063528, http://dx.doi.org/10.1103/PhysRevD.71.063528, arXiv:hep-ph/0412403.
[40] S. Cassel, J. Phys. G 37 (2010) 105009, http://dx.doi.org/10.1088/0954-3899/37/10/105009, arXiv:0903.5307.
[41] T.R. Slatyer, J. Cosmol. Astropart. Phys. 1002 (2010) 028, http://dx.doi.org/10.1088/1475-7516/2010/02/028, arXiv:0910.5713.
[42] M. Cirelli, P. Panci, K. Petraki, F. Sala, M. Taoso, arXiv:1612.07295, 2016.
[43] J. McDonald, Phys. Rev. D 43 (1991) 1063–1068, http://dx.doi.org/10.1103/PhysRevD.43.1063.
[44] M. Kamionkowski, M.S. Turner, Phys. Rev. D 42 (1990) 3310–3320, http://dx.doi.org/10.1103/PhysRevD.42.3310.
[45] D.J.H. Chung, E.W. Kolb, A. Riotto, Phys. Rev. D 60 (1999) 063504, http://dx.doi.org/10.1103/PhysRevD.60.063504, arXiv:hep-ph/9809453.
[46] G.F. Giudice, E.W. Kolb, A. Riotto, Phys. Rev. D 64 (2001) 023508, http://dx.doi.org/10.1103/PhysRevD.64.023508, arXiv:hep-ph/0005123.
[47] S. Profumo, P. Ullio, J. Cosmol. Astropart. Phys. 0311 (2003) 006, http://dx.doi.org/10.1088/1475-7516/2003/11/006, arXiv:hep-ph/0309220.
[48] P. Salati, Phys. Lett. B 571 (2003) 121–131, http://dx.doi.org/10.1016/j.physletb.2003.07.073, arXiv:astro-ph/0207396.
[49] F. D'Eramo, N. Fernandez, S. Profumo, J. Cosmol. Astropart. Phys. 1705 (2017) 102, http://dx.doi.org/10.1088/1475-7516/2017/05/012, arXiv:1703.04793.
[50] M. Ackermann, et al., Fermi-LAT Collab, Phys. Rev. Lett. 115 (2015) 231301, http://dx.doi.org/10.1103/PhysRevLett.115.231301, arXiv:1503.02641.
[51] Fermi-LAT Collaboration, https://www-glast.stanford.edu/pub_data/1048/, supplementary material to Phys. Rev. Lett. 115 (2015) 231301.
[52] M. Cirelli, G. Corcella, A. Hektor, G. Hutsi, M. Kadastik, P. Panci, M. Raidal, F. Sala, A. Strumia, J. Cosmol. Astropart. Phys. 1103 (2011) 051, http://dx.doi.org/10.1088/1475-7516/2011/03/051, arXiv:1012.4515, Erratum: J. Cosmol. Astropart. Phys. 1210 (2012) E01.
[53] P. Ciafaloni, D. Comelli, A. Riotto, F. Sala, A. Strumia, A. Urbano, J. Cosmol. Astropart. Phys. 1103 (2011) 019, http://dx.doi.org/10.1088/1475-7516/2011/03/019, arXiv:1009.0224.
[54] G. Elor, N.L. Rodd, T.R. Slatyer, W. Xue, J. Cosmol. Astropart. Phys. 1606 (2016) 024, http://dx.doi.org/10.1088/1475-7516/2016/06/024, arXiv:1511.08787.

[55] LEP, SLD Electroweak Group, SLD Heavy Flavor Group, DELPHI, LEP, ALEPH, OPAL, LEP Electroweak Working Group, L3 Collaborations, arXiv:hep-ex/0312023, 2003.
[56] G. Aad, et al., ATLAS Collaboration, Phys. Rev. Lett. 112 (2014) 231806, http://dx.doi.org/10.1103/PhysRevLett.112.231806, arXiv:1403.5657.
[57] G. Aad, et al., ATLAS Collaboration, J. High Energy Phys. 09 (2014) 037, http://dx.doi.org/10.1007/JHEP09(2014)037, arXiv:1407.7494.
[58] V. Khachatryan, et al., CMS Collaboration, Phys. Rev. D 91 (2015) 092005, http://dx.doi.org/10.1103/PhysRevD.91.092005, arXiv:1408.2745.
[59] J. Abdallah, et al., DELPHI Collaboration, Eur. Phys. J. C 38 (2005) 395–411, http://dx.doi.org/10.1140/epjc/s2004-02051-8, arXiv:hep-ex/0406019.
[60] J. Abdallah, et al., DELPHI Collaboration, Eur. Phys. J. C 60 (2009) 17–23, http://dx.doi.org/10.1140/epjc/s10052-009-0874-9, arXiv:0901.4486.
[61] P.J. Fox, R. Harnik, J. Kopp, Y. Tsai, Phys. Rev. D 84 (2011) 014028, http://dx.doi.org/10.1103/PhysRevD.84.014028, arXiv:1103.0240.
[62] K.A. Olive, et al., Particle Data Group, Chin. Phys. C 38 (2014) 090001, http://dx.doi.org/10.1088/1674-1137/38/9/090001.
[63] A. Martin, W. Stirling, R. Thorne, G. Watt, Eur. Phys. J. C 63 (2009) 189–285, http://dx.doi.org/10.1140/epjc/s10052-009-1072-5, arXiv:0901.0002.
[64] ATLAS Collaboration, ATLAS-CONF-2017-027 (2017).
[65] M. Aaboud, et al., ATLAS Collaboration, Phys. Lett. B 761 (2016) 372–392, http://dx.doi.org/10.1016/j.physletb.2016.08.055, arXiv:1607.03669.
[66] V. Khachatryan, et al., CMS Collaboration, Phys. Lett. B 768 (2017) 57–80, http://dx.doi.org/10.1016/j.physletb.2017.02.010, arXiv:1609.05391.
[67] V. Khachatryan, et al., CMS Collaboration, J. High Energy Phys. (2016), http://dx.doi.org/10.3204/PUBDB-2016-05677, arXiv:1611.06594.
[68] J. Beringer, et al., Particle Data Group, Phys. Rev. D 86 (2012) 010001, http://dx.doi.org/10.1103/PhysRevD.86.010001.
[69] S. Eidelman, M. Passera, Mod. Phys. Lett. A 22 (2007) 159–179, http://dx.doi.org/10.1142/S0217732307022694, arXiv:hep-ph/0701260.
[70] D.S. Akerib, et al., arXiv:1608.07648, 2016.
[71] D.S. Akerib, et al., LZ Collaboration, arXiv:1509.02910, 2015.
[72] B.J. Kavanagh, R. Catena, C. Kouvaris, arXiv:1611.05453, 2016.
[73] M. Cirelli, E. Del Nobile, P. Panci, J. Cosmol. Astropart. Phys. 1310 (2013) 019, http://dx.doi.org/10.1088/1475-7516/2013/10/019, arXiv:1307.5955.
[74] M. Kachelriess, P.D. Serpico, Phys. Rev. D 76 (2007) 063516, http://dx.doi.org/10.1103/PhysRevD.76.063516, arXiv:0707.0209.
[75] N.F. Bell, J.B. Dent, T.D. Jacques, T.J. Weiler, Phys. Rev. D 78 (2008) 083540, http://dx.doi.org/10.1103/PhysRevD.78.083540, arXiv:0805.3423.
[76] M. Kachelriess, P.D. Serpico, M.A. Solberg, Phys. Rev. D 80 (2009) 123533, http://dx.doi.org/10.1103/PhysRevD.80.123533, arXiv:0911.0001.
[77] P. Ciafaloni, A. Urbano, Phys. Rev. D 82 (2010) 043512, http://dx.doi.org/10.1103/PhysRevD.82.043512, arXiv:1001.3950.
[78] A. Alves, S. Profumo, F.S. Queiroz, W. Shepherd, Phys. Rev. D 90 (2014) 115003, http://dx.doi.org/10.1103/PhysRevD.90.115003, arXiv:1403.5027.
[79] M. Abdullah, A. DiFranzo, A. Rajaraman, T.M.P. Tait, P. Tanedo, A.M. Wijangco, Phys. Rev. D 90 (2014) 035004, http://dx.doi.org/10.1103/PhysRevD.90.035004, arXiv:1404.6528.
[80] F. Calore, I. Cholis, C. McCabe, C. Weniger, Phys. Rev. D 91 (2015) 063003, http://dx.doi.org/10.1103/PhysRevD.91.063003, arXiv:1411.4647.
[81] M. Aguilar, et al., AMS Collaboration, Phys. Rev. Lett. 117 (2016) 091103, http://dx.doi.org/10.1103/PhysRevLett.117.091103.
[82] A. Cuoco, M. Krämer, M. Korsmeier, arXiv:1610.03071, 2016.

21

Revisiting kaon physics in general Z scenario

Motoi Endo [a,b], Teppei Kitahara [c,d,*], Satoshi Mishima [a], Kei Yamamoto [a]

[a] Theory Center, IPNS, KEK, Tsukuba, Ibaraki 305-0801, Japan
[b] The Graduate University of Advanced Studies (Sokendai), Tsukuba, Ibaraki 305-0801, Japan
[c] Institute for Theoretical Particle Physics (TTP), Karlsruhe Institute of Technology, Engesserstraße 7, D-76128 Karlsruhe, Germany
[d] Institute for Nuclear Physics (IKP), Karlsruhe Institute of Technology, Hermann-von-Helmholtz-Platz 1, D-76344 Eggenstein-Leopoldshafen, Germany

ARTICLE INFO

Editor: J. Hisano

ABSTRACT

New physics contributions to the Z penguin are revisited in the light of the recently-reported discrepancy of the direct CP violation in $K \to \pi\pi$. Interference effects between the standard model and new physics contributions to $\Delta S = 2$ observables are taken into account. Although the effects are overlooked in the literature, they make experimental bounds significantly severer. It is shown that the new physics contributions must be tuned to enhance $\mathcal{B}(K_L \to \pi^0 \nu\bar{\nu})$, if the discrepancy of the direct CP violation is explained with satisfying the experimental constraints. The branching ratio can be as large as 6×10^{-10} when the contributions are tuned at the 10% level.

1. Introduction

A deviation of the standard model (SM) prediction from the experimental result is recently reported in the direct CP violation of the $K \to \pi\pi$ decays, which is called ϵ'. The latest lattice calculations of the hadron matrix elements significantly reduced the theoretical uncertainty [1–4] and yield [5,6]

$$\left(\frac{\epsilon'}{\epsilon}\right)_{\rm SM} = \begin{cases} (1.38 \pm 6.90) \times 10^{-4}, & \text{[RBC-UKQCD]} \\ (1.9 \pm 4.5) \times 10^{-4}, & \text{[Buras et al.]} \\ (1.06 \pm 5.07) \times 10^{-4}. & \text{[Kitahara et al.]} \end{cases} \tag{1.1}$$

They are lower than the experimental result [7–10],

$$\left(\frac{\epsilon'}{\epsilon}\right)_{\rm exp} = (16.6 \pm 2.3) \times 10^{-4}. \tag{1.2}$$

The deviations correspond to the 2.8–2.9σ level.

Several new physics (NP) models have been explored to explain the discrepancy [11–21]. In the literature, electroweak penguin contributions to ϵ'/ϵ have been studied.[1] In particular, the Z penguin contributions have been studied in detail [11,13,15,22]. The decay, $s \to dq\bar{q}$ $(q = u, d)$, proceeds by intermediating the Z boson, and its flavor-changing (s–d) interaction is enhanced by NP. Then, the branching ratios of $K \to \pi\nu\bar{\nu}$ are likely to be deviated from the SM predictions once the ϵ'/ϵ discrepancy is explained. This is because the Z boson couples to the neutrinos as well as the up and down quarks. They could be a signal to test the scenario.

Such a signal is constrained by the indirect CP violation of the K mesons. The flavor-changing Z couplings affect the indirect CP violation via the so-called double penguin diagrams; the Z boson intermediates the transition, both of whose couplings are provided by the flavor-changing Z couplings. Such a contribution is enhanced when there are both the left- and right-handed couplings because of the chiral enhancement of the hadron matrix elements. This is stressed by Ref. [15] in the context of the Z'-exchange scenario. In the Z-boson case, since the left-handed coupling is installed by the SM, the right-handed coupling must be constrained even without NP contributions to the left-handed one. Such interference contributions between the NP and the SM are overlooked in Refs. [11,13,15,22,23]. Therefore, the parameter regions allowed by the indirect CP violation will change significantly. In this letter, we revisit the Z-boson scenario.[2] It will be shown that the NP contributions to the right-handed s–d coupling are tightly constrained due to the interference, and thus, the branching ratio of $K_L \to \pi^0\nu\bar{\nu}$ is likely to be smaller than the SM predictions if the ϵ'/ϵ discrepancy is explained. We will discuss that NP parameters are necessarily tuned to enhance the ratio. A degree of

* Corresponding author.
E-mail address: teppei.kitahara@kit.edu (T. Kitahara).

1 QCD penguin contributions, e.g., through Kaluza–Klein gluons, have also been considered [11].

2 In this letter, we focus on the s–d transitions. The $\Delta F = 2$ transitions such as Δm_B generally involve the interference contributions.

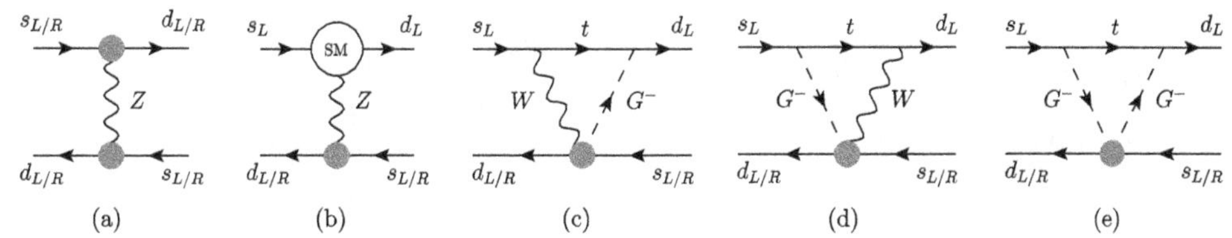

Fig. 1. The NP contributions to $\Delta S = 2$ process. The black bubble denotes the vertices in Eq. (2.4) originating from the dimension-6 effective operators: $\mathcal{O}_L$ and $\mathcal{O}_R$. The white bubble with "SM" denotes the SM flavor-changing Z interaction. Subfigures (b)–(e) correspond to the interference contributions between the NP and SM. A contribution from G^0-exchange diagram is negligible because it receives a suppression factor by the external momentum, so that we omit it here.

the parameter tuning will be investigated to estimate how large $\mathcal{B}(K_L \to \pi^0 \nu\bar{\nu})$ and $\mathcal{B}(K^+ \to \pi^+ \nu\bar{\nu})$ can become.

2. Z-penguin observables

In this section, we briefly review the Z-penguin contributions to $\Delta S = 2$ and $\Delta S = 1$ processes in the general Z scenario. Above the electroweak symmetry breaking scale, NP particles generate Wilson coefficients of the (dimension-6) effective operators,

$$\mathcal{O}_L = i(H^\dagger \overleftrightarrow{D}_\mu H)(\overline{Q}_L \gamma^\mu Q'_L), \tag{2.1}$$

$$\mathcal{O}_R = i(H^\dagger \overleftrightarrow{D}_\mu H)(\bar{d}_R \gamma^\mu d'_R), \tag{2.2}$$

$$\mathcal{O}_L^{(3)} = i(H^\dagger \sigma^a \overleftrightarrow{D}_\mu H)(\overline{Q}_L \gamma^\mu \sigma^a Q'_L). \tag{2.3}$$

They are gauge invariant under the SM gauge transformations. In this letter, we focus on the operators, $\mathcal{O}_L$ and $\mathcal{O}_R$, to demonstrate the impact of the interference between the SM and NP contributions.[3] After the electroweak symmetry breaking, they provide the flavor-changing (s–d) Z interactions,

$$\mathcal{L} = \Delta_L^{\rm NP} \left[Z_\mu + \frac{1}{m_Z} \partial_\mu G^0 - \frac{ig}{2m_W m_Z} G^- \overleftrightarrow{\partial_\mu} G^+ - \frac{g}{m_Z} \left(W_\mu^- G^+ + W_\mu^+ G^- \right) + \ldots \right] (\bar{s}\gamma^\mu P_L d) + (L \leftrightarrow R) + \text{H.c.}, \tag{2.4}$$

where the first term in the bracket is the Z-boson interaction, while the others are those of the Nambu–Goldstone boson, and we omitted the irrelevant terms for the interference effects. Here, the Wilson coefficients of $\mathcal{O}_L$ and $\mathcal{O}_R$ are normalized by the flavor-changing Z interactions. In the following, we omit the subscript "NP" in $\Delta_L^{\rm NP}$ and $\Delta_R^{\rm NP}$ for simplicity.

2.1. ϵ_K and Δm_K

In the $\Delta S = 2$ observables, there are the indirect CP violation ϵ_K and the mass difference Δm_K in the K^0–$\overline{K}^0$ mixing. Since ϵ_K has been measured precisely, and the SM prediction is accurate, it provides a severe constraint. The SM and NP contributions are shown as

$$\epsilon_K = e^{i\varphi_\epsilon} \left(\epsilon_K^{\rm SM} + \epsilon_K^{\rm NP} \right), \tag{2.5}$$

where $\varphi_\epsilon = (43.51 \pm 0.05)^\circ$. The NP contribution is given by the double penguin diagrams with the Z boson exchange (Fig. 1(a)),

$$\epsilon_K^{\rm NP} = \sum_{i=1}^{8} (\epsilon_K)_i^Z, \tag{2.6}$$

where the right-hand side is [15]

$$\begin{aligned}
(\epsilon_K)_1^Z &= -4.26 \times 10^7 \, \text{Im}\, \Delta_L \, \text{Re}\, \Delta_L, \\
(\epsilon_K)_2^Z &= -4.26 \times 10^7 \, \text{Im}\, \Delta_R \, \text{Re}\, \Delta_R, \\
(\epsilon_K)_3^Z &= 2.07 \times 10^9 \, \text{Im}\, \Delta_L \, \text{Re}\, \Delta_R, \\
(\epsilon_K)_4^Z &= 2.07 \times 10^9 \, \text{Im}\, \Delta_R \, \text{Re}\, \Delta_L.
\end{aligned} \tag{2.7}$$

In these expressions, renormalization group corrections and long-distance contributions are included [25]. In addition, one must take account of the interference terms between the SM and NP contributions (Figs. 1(b)–(e)),

$$\begin{aligned}
(\epsilon_K)_5^Z &= -4.26 \times 10^7 \, \text{Im}\, \Delta_L^{\rm SM} \, \text{Re}\, \Delta_L, \\
(\epsilon_K)_6^Z &= -4.26 \times 10^7 \, \text{Im}\, \Delta_L \, \text{Re}\, \Delta_L^{\rm SM}, \\
(\epsilon_K)_7^Z &= 2.07 \times 10^9 \, \text{Im}\, \Delta_L^{\rm SM} \, \text{Re}\, \Delta_R, \\
(\epsilon_K)_8^Z &= 2.07 \times 10^9 \, \text{Im}\, \Delta_R \, \text{Re}\, \Delta_L^{\rm SM}.
\end{aligned} \tag{2.8}$$

Here, the SM contribution, $\Delta_L^{\rm SM}$, is generated by radiative corrections. At the one-loop level, it is calculated as

$$\Delta_L^{\rm SM} = \frac{g^3 \lambda_t}{8\pi^2 c_W} \widetilde{C}\left(\frac{m_t^2}{m_W^2}, \mu_{\rm NP} \right), \quad \Delta_R^{\rm SM} = 0, \tag{2.9}$$

where $c_W = \cos\theta_W$, $\lambda_i \equiv V_{is}^* V_{id}$ with the CKM matrix V_{ij}, and $\mu_{\rm NP}$ corresponds to the NP scale.[4] In this letter, the CKMFITTER result [24] is used for the CKM elements, unless otherwise mentioned. The loop function is[5]

$$\widetilde{C}(x, \mu_{\rm NP}) = C(x) + \Delta C(x, \mu_{\rm NP}). \tag{2.10}$$

In the Feynman–'t Hooft gauge, the first term in the right-hand side corresponds to the Z-boson exchange diagram [22] (Fig. 1(b)),

$$C(x) = \frac{x}{8} \left[\frac{x-6}{x-1} + \frac{3x+2}{(x-1)^2} \ln x \right], \tag{2.11}$$

while the second term is obtained from the Nambu–Goldstone boson loops (Figs. 1(c)–(e)),

$$\Delta C(x, \mu_{\rm NP}) = -\frac{x}{16} \left[\frac{3x-17}{2(x-1)} - \frac{x(x-8)}{(x-1)^2} \ln x + \ln \frac{\mu_{\rm NP}^2}{m_W^2} \right]. \tag{2.12}$$

The explicit form of $\Delta C(x)$ depends on the effective operators above the electroweak symmetry breaking scale. Here, they are supposed to be $\mathcal{O}_L$ and $\mathcal{O}_R$. It is noted that the interference terms are gauge-independent.

[3] A similar discussion as follows is expected to hold for the effective operator $\mathcal{O}_L^{(3)}$.

[4] In order to introduce how significant the interference contributions are, we ignore the renormalization group corrections to the dimension-6 operators above the electroweak scale except for a first leading logarithmic contribution $\ln(\mu_{\rm NP}/m_W)$ which comes from Fig. 1(e). This approximation is valid when the NP scale is not so far from the electroweak scale.

[5] The loop function $\widetilde{C}(x, \mu_{\rm NP})$ is consistent with the result in Ref. [26].

The interference terms (2.8) have been overlooked in Refs. [11, 13,15,22]. They cannot be ignored, as we will see in the next section.

The latest estimation of the SM value is [27]

$$\epsilon_K^{\rm SM} = (2.24 \pm 0.19) \times 10^{-3}. \tag{2.13}$$

On the other hand, the experimental result is [10]

$$|\epsilon_K^{\rm exp}| = (2.228 \pm 0.011) \times 10^{-3}. \tag{2.14}$$

They are well consistent with each other, and $\epsilon_K^{\rm NP}$ must satisfy

$$-0.39 \times 10^{-3} < \epsilon_K^{\rm NP} < 0.37 \times 10^{-3}, \tag{2.15}$$

at the 2σ level.[6]

The kaon mass difference Δm_K consists of the SM and NP contributions:

$$\Delta m_K = \Delta m_K^{\rm SM} + \Delta m_K^{\rm NP}. \tag{2.16}$$

If we parameterize the NP contribution as

$$\frac{\Delta m_K^{\rm NP}}{\Delta m_K^{\rm exp}} = \sum_{i=1}^{8} R_i^Z, \tag{2.17}$$

the right-hand side is estimated as [15]

$$\begin{aligned} R_1^Z &= 6.43 \times 10^7 \left[({\rm Re}\,\Delta_L)^2 - ({\rm Im}\,\Delta_L)^2 \right], \\ R_2^Z &= 6.43 \times 10^7 \left[({\rm Re}\,\Delta_R)^2 - ({\rm Im}\,\Delta_R)^2 \right], \\ R_3^Z &= -6.21 \times 10^9\, {\rm Re}\,\Delta_L\, {\rm Re}\,\Delta_R, \\ R_4^Z &= 6.21 \times 10^9\, {\rm Im}\,\Delta_L\, {\rm Im}\,\Delta_R. \end{aligned} \tag{2.18}$$

Similarly to the case of ϵ_K, there are interference terms between the SM and NP contributions,

$$\begin{aligned} R_5^Z &= 12.9 \times 10^7\, {\rm Re}\,\Delta_L^{\rm SM}\, {\rm Re}\,\Delta_L, \\ R_6^Z &= -12.9 \times 10^7\, {\rm Im}\,\Delta_L^{\rm SM}\, {\rm Im}\,\Delta_L, \\ R_7^Z &= -6.21 \times 10^9\, {\rm Re}\,\Delta_L^{\rm SM}\, {\rm Re}\,\Delta_R, \\ R_8^Z &= 6.21 \times 10^9\, {\rm Im}\,\Delta_L^{\rm SM}\, {\rm Im}\,\Delta_R. \end{aligned} \tag{2.19}$$

Here, $\Delta_L^{\rm SM}$ is given by Eq. (2.9), and the result is gauge-independent. These terms have been overlooked in the literature.

The experimental result is [10]

$$\Delta m_K^{\rm exp} = (3.484 \pm 0.006) \times 10^{-15}\,{\rm GeV}. \tag{2.20}$$

Since the SM prediction involves sizable contributions of long-distance effects, the uncertainty is large.[7] Hence, we simply require that the NP contribution does not exceed the experimental value (with the 2σ uncertainty):

$$|\Delta m_K^{\rm NP}| < 3.496 \times 10^{-15}\,{\rm GeV}. \tag{2.21}$$

This constraint will turn out to be much weaker than $\epsilon_K^{\rm NP}$.

2.2. ϵ'/ϵ

The flavor-changing Z interaction also contributes to $\Delta S = 1$ observables. The direct CP violation ϵ'/ϵ is shown as

$$\frac{\epsilon'}{\epsilon} = \left(\frac{\epsilon'}{\epsilon}\right)_{\rm SM} + \left(\frac{\epsilon'}{\epsilon}\right)_{\rm NP}. \tag{2.22}$$

The NP contribution is estimated as [15]

$$\left(\frac{\epsilon'}{\epsilon}\right)_{\rm NP} = -2.64 \times 10^3\, B_8^{(3/2)} \left({\rm Im}\,\Delta_L + \frac{c_W^2}{s_W^2}\, {\rm Im}\,\Delta_R \right), \tag{2.23}$$

where $B_8^{(3/2)} = 0.76 \pm 0.05$ from the lattice calculation. Here, the terms which are not proportional to $B_8^{(3/2)}$ are omitted; the approximation is valid at the 10% accuracy. A factor in the parenthesis gives $c_W^2/s_W^2 \simeq 3.33$. Thus, the NP contribution can be enhanced easily by Δ_R.

As mentioned in Sec. 1, the SM prediction deviates from the experimental result at the 2.8–2.9σ level. In this letter, we require that the discrepancy of ϵ'/ϵ is explained at the 1σ level as

$$10.0 \times 10^{-4} < \left(\frac{\epsilon'}{\epsilon}\right)_{\rm NP} < 21.1 \times 10^{-4}, \tag{2.24}$$

where Ref. [6] is used for the SM prediction.

2.3. $K^+ \to \pi^+ \nu\bar\nu$ and $K_L \to \pi^0 \nu\bar\nu$

The (ultra-)rare kaon decay channels, $K^+ \to \pi^+ \nu\bar\nu$ and $K_L \to \pi^0 \nu\bar\nu$, are correlated with ϵ'/ϵ as well as ϵ_K and Δm_K in the general Z scenario.[8] They are represented as [15,32]

$$\begin{aligned} &\mathcal{B}(K^+ \to \pi^+ \nu\bar\nu) \\ &= \kappa_+ \left[\left(\frac{{\rm Im}\,X_{\rm eff}}{\lambda^5} \right)^2 + \left(\frac{{\rm Re}\,\lambda_c}{\lambda} P_c(X) + \frac{{\rm Re}\,X_{\rm eff}}{\lambda^5} \right)^2 \right], \end{aligned} \tag{2.25}$$

$$\mathcal{B}(K_L \to \pi^0 \nu\bar\nu) = \kappa_L \left(\frac{{\rm Im}\,X_{\rm eff}}{\lambda^5} \right)^2. \tag{2.26}$$

Here, $X_{\rm eff}$ is estimated as

$$X_{\rm eff} = \lambda_t\,(1.48 \pm 0.01) + 2.51 \times 10^2\,(\Delta_L + \Delta_R), \tag{2.27}$$

where the first term in the right-hand side is the SM contribution. Also, $\lambda = |V_{us}|$, $\kappa_+ = (5.157 \pm 0.025) \cdot 10^{-11} (\lambda/0.225)^8$, and $\kappa_L = (2.231 \pm 0.013) \cdot 10^{-10} (\lambda/0.225)^8$. The charm-quark contribution is $P_c(X) = (9.39 \pm 0.31) \cdot 10^{-4}/\lambda^4 + (0.04 \pm 0.02)$, where the first term in the right-hand side comes from short-distance effects, while the second one takes account of long-distance effects. Using the CKMFITTER result for the CKM elements, one obtains

$${\rm Re}\,X_{\rm eff} = -4.83 \times 10^{-4} + 2.51 \times 10^2\,({\rm Re}\,\Delta_L + {\rm Re}\,\Delta_R), \tag{2.28}$$

$${\rm Im}\,X_{\rm eff} = 2.12 \times 10^{-4} + 2.51 \times 10^2\,({\rm Im}\,\Delta_L + {\rm Im}\,\Delta_R). \tag{2.29}$$

The SM predictions become

$$\mathcal{B}(K^+ \to \pi^+ \nu\bar\nu)_{\rm SM} = (8.5 \pm 0.5) \times 10^{-11}, \tag{2.30}$$

$$\mathcal{B}(K_L \to \pi^0 \nu\bar\nu)_{\rm SM} = (3.0 \pm 0.2) \times 10^{-11}. \tag{2.31}$$

[6] The SM estimation $\epsilon_K^{\rm SM}$ is sensitive to the CKM elements. If one uses V_{cb} that is determined by the exclusive $B \to D^{(*)}\ell\nu$ decays [28], $\epsilon_K^{\rm SM} = (1.73 \pm 0.18) \cdot 10^{-3}$ is obtained [29]. Then, $\epsilon_K^{\rm NP} = (0.50 \pm 0.18) \cdot 10^{-3}$ is required at the 1σ level.

[7] The latest lattice simulation, which includes the long-distance contributions, provides $\Delta m_K^{\rm SM} = (3.19 \pm 1.04) \cdot 10^{-15}$ GeV [30]. However, it is performed on masses of unphysical pion, kaon and charmed quark.

[8] The branching ratios of $K \to \pi\ell^+\ell^-$ ($\ell = e, \mu$) are also affected in the general Z scenario. However, $K^+ \to \pi^+\ell^+\ell^-$ and $K_S \to \pi^0\ell^+\ell^-$ are dominated by a long-distance contribution through $K \to \pi\gamma^* \to \pi\ell^+\ell^-$ [31]. On the other hand, such a contribution to $K_L \to \pi^0\ell^+\ell^-$ is forbidden by the CP symmetry, but is dominated by an indirect CP-violating contribution, $K_L \to K_S \to \pi^0\ell^+\ell^-$ [31]. Therefore, it is challenging to discuss short-distance NP contributions in these channels.

On the other hand, the experimental results are [33,34]

$$\mathcal{B}(K^+ \to \pi^+ \nu\bar{\nu})_{\rm exp} = (17.3^{+11.5}_{-10.5}) \times 10^{-11}, \tag{2.32}$$

$$\mathcal{B}(K_L \to \pi^0 \nu\bar{\nu})_{\rm exp} \leq 2.6 \times 10^{-8}. \quad [90\% \text{ C.L.}] \tag{2.33}$$

Although the current constraints on the NP contributions are very weak, the experimental values will be improved significantly in the near future. The NA62 experiment at CERN, which already started the physics run at low beam intensity in 2015, has a potential to measure $\mathcal{B}(K^+ \to \pi^+ \nu\bar{\nu})$ at the 10% precision by 2018 [35]. The KOTO experiment at J-PARC is designed to improve the sensitivity for $\mathcal{B}(K_L \to \pi^0 \nu\bar{\nu})$, which enables us to measure it at the 10% level of the SM value [36,37]. As one can see from Eqs. (2.23) and (2.27), the NP contributions to $\mathcal{B}(K \to \pi\nu\bar{\nu})$ are correlated with those to ϵ'/ϵ in the general Z scenario. Thus, if the ϵ'/ϵ discrepancy is a signal of the scenario, these experiments would detect NP effects.

2.4. $K_L \to \mu^+\mu^-$

The branching ratio of $K_L \to \mu^+\mu^-$ is also sensitive to the NP contributions to the flavor-changing Z couplings. Theoretically, only the short-distance (SD) contributions can be calculated reliably. They are shown as [15,38,39]

$$\mathcal{B}(K_L \to \mu^+\mu^-)_{\rm SD} = \kappa_\mu \left(\frac{\mathrm{Re}\,\lambda_c}{\lambda} P_c(Y) + \frac{\mathrm{Re}\,Y_{\rm eff}}{\lambda^5} \right)^2, \tag{2.34}$$

where $\kappa_\mu = (2.01 \pm 0.02) \cdot 10^{-9} (\lambda/0.225)^8$. The charm-quark contribution is $P_c(Y) = (0.115 \pm 0.018) \cdot (0.225/\lambda)^4$. Using the CKMFITTER result, one obtains

$$\mathrm{Re}\,Y_{\rm eff} = -3.07 \times 10^{-4} + 2.51 \times 10^2 \,(\mathrm{Re}\,\Delta_L - \mathrm{Re}\,\Delta_R), \tag{2.35}$$

where the first term in the right-hand side is the SM contribution, and the minus sign between Δ_L and Δ_R is due to the axial-vector current. The SM value is obtained as

$$\mathcal{B}(K_L \to \mu^+\mu^-)_{\rm SD,\,SM} = (0.83 \pm 0.10) \times 10^{-9}. \tag{2.36}$$

On the other hand, it is challenging to extract a short-distance part in the experimental data $\mathcal{B}(K_L \to \mu^+\mu^-)_{\rm exp} = (6.84 \pm 0.11) \cdot 10^{-9}$ [10], because of huge long-distance contributions through $K_L \to \gamma^*\gamma^* \to \mu^+\mu^-$ [40]. An upper bound on the short-distance contribution is [40]

$$\mathcal{B}(K_L \to \mu^+\mu^-)_{\rm SD} < 2.5 \times 10^{-9}. \tag{2.37}$$

Since the constraint is much weaker than the SM uncertainties, we ignore them for simplicity and impose a bound on the Z couplings,

$$-1.08 \times 10^{-6} < \mathrm{Re}\,\Delta_L - \mathrm{Re}\,\Delta_R < 4.05 \times 10^{-6}. \tag{2.38}$$

The real parts of the NP contributions are constrained by $\mathcal{B}(K_L \to \mu^+\mu^-)$.

3. Analysis

In this section, we examine the general Z scenario. Although the discrepancy of ϵ'/ϵ could be explained by the scenario, the parameter regions would be constrained by ϵ_K, Δm_K and $K_L \to \mu^+\mu^-$. In particular, the interference between the SM and NP contributions, Eq. (2.8), affects ϵ_K significantly. In this section, we choose the NP scale, $\mu_{\rm NP} = 1\,\text{TeV}$, as a reference. As we will see, wide parameter regions are excluded. Thus, the discrepancy of ϵ'/ϵ will be explained by tuning the model parameters. Let us introduce a quantity which parameterizes the tuning[9]:

$$\xi = \max(\xi_1, \xi_2, \ldots, \xi_8), \quad \text{with } \xi_i = \left| \frac{(\epsilon_K)_i^Z}{\epsilon_K^{\rm NP}} \right|. \tag{3.1}$$

If $\epsilon_K^{\rm NP}$ is dominated by a single term, one obtains $\xi \simeq 1$ and there is no tuning among the model parameters. If the maximal value of $(\epsilon_K)_i^Z$ is about ten times larger than $\epsilon_K^{\rm NP}$, $\xi \sim 10$ is obtained; the model parameters are tuned such that there is a cancellation among $(\epsilon_K)_i^Z$ at the 10% level.

3.1. Simplified scenarios

First, we consider the following simplified scenarios (cf., Ref. [41]),

- left-handed scenario (LHS): $\Delta_R = 0$,[10]
- right-handed scenario (RHS): $\Delta_L = 0$,[11]
- pure imaginary scenario (ImZS): $\mathrm{Re}\,\Delta_L = \mathrm{Re}\,\Delta_R = 0$,
- left–right symmetric scenario (LRS): $\Delta_L = \Delta_R$.[12]

As shown below, these scenarios do not require large parameter tuning in $\epsilon_K^{\rm NP}$. However, $\mathcal{B}(K \to \pi\nu\bar{\nu})$ will turn out to be small.

In Fig. 2, the Z-penguin observables are shown as functions of $\Delta_{L,R}$ for LHS and RHS. In the green (light green) regions, the ϵ'/ϵ discrepancy is explained at $1\,(2)\,\sigma$. They depend only on the imaginary component of $\Delta_{L,R}$. Obviously, ϵ'/ϵ is enhanced by the right-handed Z coupling, Δ_R, more than Δ_L.

The blue regions are excluded by the ϵ_K, and the orange regions are by the $\mathcal{B}(K_L \to \mu^+\mu^-)$. The constraint from ϵ_K is much severer in RHS than LHS due to the interference contributions, Eq. (2.8). There is no constraint from Δm_K in the parameter regions of the plots.

The red and black dashed contours represent $\mathcal{B}(K_L \to \pi^0\nu\bar{\nu})/\mathcal{B}(K_L \to \pi^0\nu\bar{\nu})_{\rm SM}$ and $\mathcal{B}(K^+ \to \pi^+\nu\bar{\nu})/\mathcal{B}(K^+ \to \pi^+\nu\bar{\nu})_{\rm SM}$, respectively. Here and hereafter, $\mathcal{B}(K_L \to \pi^0\nu\bar{\nu})_{\rm SM}$ and $\mathcal{B}(K^+ \to \pi^+\nu\bar{\nu})_{\rm SM}$ denote the central values of the SM predictions, Eqs. (2.30) and (2.31). It is found that $\mathcal{B}(K_L \to \pi^0\nu\bar{\nu})$ cannot be as large as the SM value as long as ϵ'/ϵ is explained in LHS or RHS. On the other hand, if the ϵ'/ϵ discrepancy is explained by LHS, the NP contribution to $\mathcal{B}(K^+ \to \pi^+\nu\bar{\nu})$ is limited by $\mathcal{B}(K_L \to \mu^+\mu^-)$. In contrast, ϵ_K restricts RHS.

Next, we consider ImZS. Such a situation is often considered to amplify $(\epsilon'/\epsilon)_{\rm NP}$ but suppress $\epsilon_K^{\rm NP}$. In the left panel of Fig. 3, the Z-penguin observables are shown as functions of $\mathrm{Im}\,\Delta_{L,R}$. The most severe constraint is from ϵ_K due to the interference between the SM and NP. The other bounds are weak and absent in the plot. Since there are no real components of $\Delta_{L,R}$, $\mathcal{B}(K^+ \to \pi^+\nu\bar{\nu})$ is correlated with $\mathcal{B}(K_L \to \pi^0\nu\bar{\nu})$.

Finally, LRS is shown in the right panel of Fig. 3. Similarly to the cases of RHS and ImZS, most of the parameter regions are excluded by ϵ_K. The NP contributions to $\mathcal{B}(K_L \to \mu^+\mu^-)$ vanish because the process is the axial-vector current.

In Fig. 4, contours of the tuning parameter ξ are shown for the simplified scenarios: LHS, RHS, ImZS, and LRS on the plane of the branching ratios of $K \to \pi\nu\bar{\nu}$. We scanned the whole parameter space of $\Delta_{L,R}$ in each scenario and selected the parameters where ϵ'/ϵ is explained at the 1σ level, and the experimental bounds from ϵ_K, Δm_K, and $\mathcal{B}(K_L \to \mu^+\mu^-)$ are satisfied (see the previous

[9] Our definition is almost the same as that in Ref. [45], where the authors discuss correlations between the tuning parameter and flavor observables.

[10] This scenario is realized by chargino contributions to the Z penguin in the supersymmetric model [17,19,42–44].

[11] Randall–Sundrum models with custodial protection [45,46] can generate large Δ_R. However, there are additional effects, e.g., from KK-gluon diagrams for $\epsilon_K^{\rm NP}$.

[12] In axial-symmetric scenarios, $\Delta_L = -\Delta_R$, there are no NP contributions to $K \to \pi\nu\bar{\nu}$.

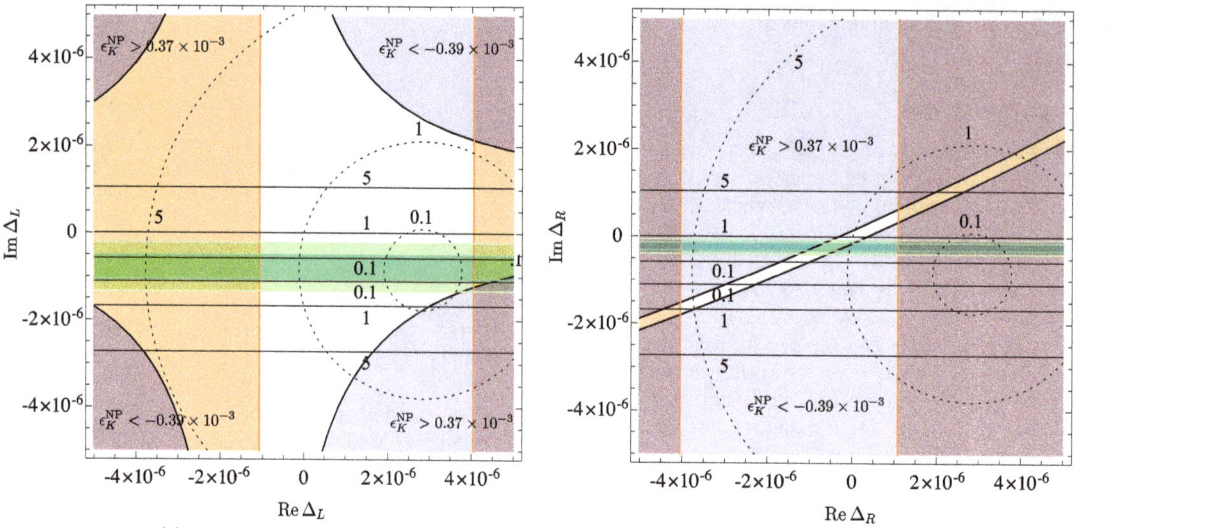

(a) Left-handed scenario (LHS) (b) Right-handed scenario (RHS)

Fig. 2. The Z-penguin observables are displayed in LHS (*left* panel) and RHS (*right*). In the green (light green) regions, the ϵ'/ϵ discrepancy is explained at 1 (2) σ. The blue and the orange shaded regions are excluded by ϵ_K and $\mathcal{B}(K_L \to \mu^+\mu^-)$, respectively. The ratios of $\mathcal{B}(K_L \to \pi^0\nu\bar{\nu})/\mathcal{B}(K_L \to \pi^0\nu\bar{\nu})_{SM}$ and $\mathcal{B}(K^+ \to \pi^+\nu\bar{\nu})/\mathcal{B}(K^+ \to \pi^+\nu\bar{\nu})_{SM}$ are shown by the red solid and black dashed contours, respectively. The NP scale is $\mu_{NP} = 1$ TeV. (For interpretation of the references to color in this figure, the reader is referred to the web version of this article.)

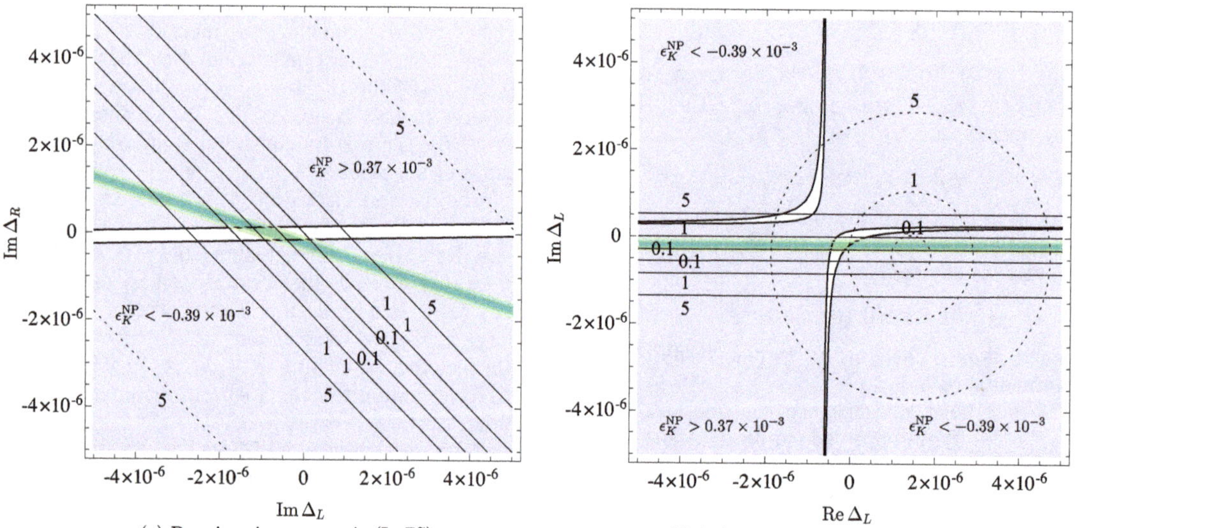

(a) Pure imaginary scenario (ImZS) (b) Left-right symmetric scenario (LRS)

Fig. 3. The Z-penguin observables are displayed in ImZS (*left* panel) and LRS (*right*). Notations of the lines and shaded regions are the same as in Fig. 2.

section for the experimental constraints). Then, ξ was estimated at each point. Several parameter sets predict the same $\mathcal{B}(K^+ \to \pi^+\nu\bar{\nu})$ and $\mathcal{B}(K_L \to \pi^0\nu\bar{\nu})$. Among them, the smallest ξ is chosen in Fig. 4 for each set of $\mathcal{B}(K^+ \to \pi^+\nu\bar{\nu})$ and $\mathcal{B}(K_L \to \pi^0\nu\bar{\nu})$. In most of the allowed parameter regions, $\xi = \mathcal{O}(1)$ is obtained. Thus, one does not require tight tunings in these scenarios.

In the figures, $\mathcal{B}(K_L \to \pi^0\nu\bar{\nu})$ is smaller than the SM value by more than 30%. Hence, the scenarios could be tested by the KOTO experiment. On the other hand, $\mathcal{B}(K^+ \to \pi^+\nu\bar{\nu})$ depends on the scenarios. In LHS, we obtain $0 < \mathcal{B}(K^+ \to \pi^+\nu\bar{\nu})/\mathcal{B}(K^+ \to \pi^+\nu\bar{\nu})_{SM} < 1.8$. In RHS, $\mathcal{B}(K^+ \to \pi^+\nu\bar{\nu})$ is comparable to or larger than the SM value, but cannot be twice as large. In ImZS, the branching ratios are perfectly correlated and displayed by a line in Fig. 4. Then, $\mathcal{B}(K^+ \to \pi^+\nu\bar{\nu})$ is not deviated from the SM one. The right panel of Fig. 4 is a result of the tuning parameter ξ in LRS. It is found that $\mathcal{B}(K_L \to \pi^0\nu\bar{\nu})$ does not exceed about a half of the SM value. On the other hand, $\mathcal{B}(K^+ \to \pi^+\nu\bar{\nu})$ is comparable to or larger than the SM value, but cannot be twice as large, as is similar to RHS.

3.2. General scenario

Let us consider the full parameter space in the general Z scenario. Both Δ_L and Δ_R are turned on. Then, $\mathcal{B}(K^+ \to \pi^+\nu\bar{\nu})$ and/or $\mathcal{B}(K_L \to \pi^0\nu\bar{\nu})$ can be enhanced if the tuning for ϵ_K^{NP} is allowed.

In Fig. 5, the branching ratios of $K \to \pi\nu\bar{\nu}$ and the tuning parameter are shown for the case of $(\epsilon'/\epsilon)_{NP} = 15.5 \cdot 10^{-4}$ and $\epsilon_K^{NP} = 0.37 \cdot 10^{-3}$. The flavor-changing Z couplings, and namely the NP contributions to $K \to \pi\nu\bar{\nu}$, are limited by $\mathcal{B}(K_L \to \mu^+\mu^-)$ and the tuning parameter.

In Fig. 6, contours of the tuning parameter ξ are shown on the plane of the branching ratios of $K \to \pi\nu\bar{\nu}$. The whole parameter space of the general Z scenario is scanned. In the colored regions, ϵ'/ϵ is explained at the 1σ level, and the experimental bounds of ϵ_K, Δm_K, and $\mathcal{B}(K_L \to \mu^+\mu^-)$ are satisfied (see the previous

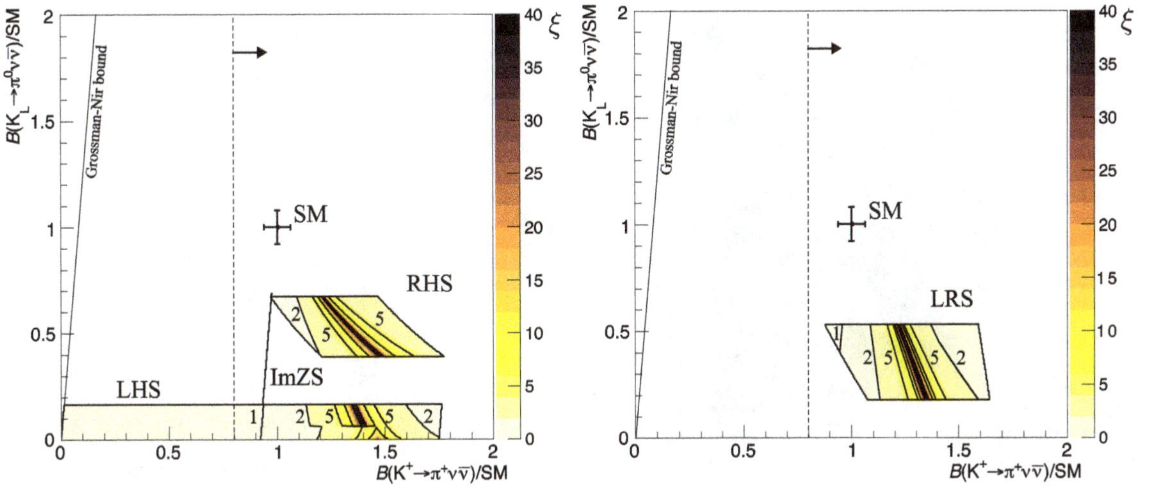

(a) LHS, RHS, and ImZS (b) LRS

Fig. 4. Contours of the tuning parameter ξ are shown in the simplified scenarios: LHS, RHS, and ImZS (*left* panel) and LRS (*right*). Here, "SM" in the axis labels denotes the central values of $\mathcal{B}(K_L \to \pi^0\nu\bar{\nu})_{\rm SM}$ and $\mathcal{B}(K^+ \to \pi^+\nu\bar{\nu})_{\rm SM}$, Eqs. (2.30) and (2.31). In the colored regions, ϵ'/ϵ is explained at 1σ, and the experimental bounds of ϵ_K, Δm_K, and $\mathcal{B}(K_L \to \mu^+\mu^-)$ are satisfied. The right region of the blue dashed line is allowed by the measurement of $\mathcal{B}(K^+ \to \pi^+\nu\bar{\nu})$ at 1σ. The Grossman–Nir bound [47] is shown by the blue solid line. The NP scale is set to be $\mu_{\rm NP} = 1\,$TeV. (For interpretation of the references to color in this figure legend, the reader is referred to the web version of this article.)

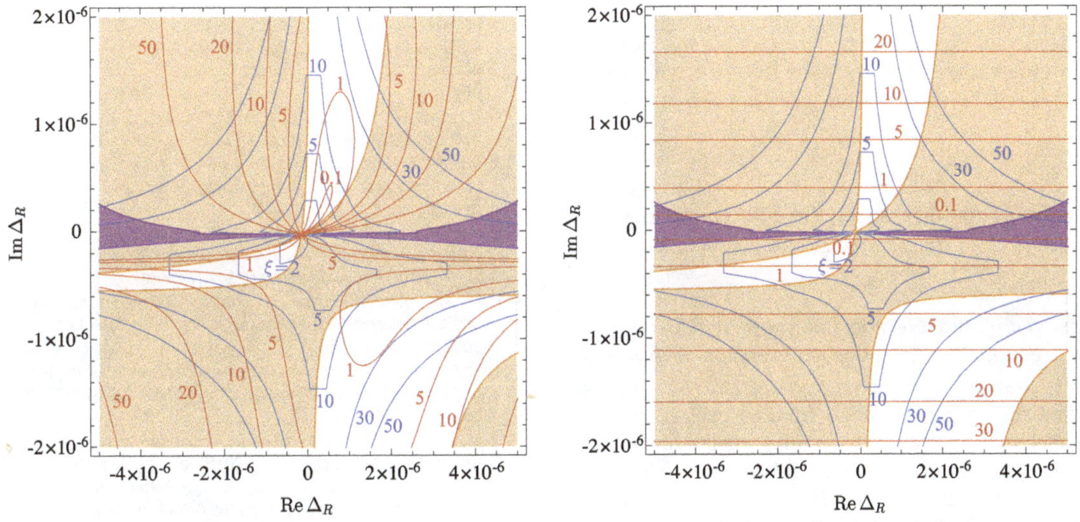

(a) $\mathcal{B}(K^+ \to \pi^+\nu\bar{\nu})/\mathcal{B}(K^+ \to \pi^+\nu\bar{\nu})_{\rm SM}$ (b) $\mathcal{B}(K_L \to \pi^0\nu\bar{\nu})/\mathcal{B}(K_L \to \pi^0\nu\bar{\nu})_{\rm SM}$

Fig. 5. $\mathcal{B}(K^+ \to \pi^+\nu\bar{\nu})/\mathcal{B}(K^+ \to \pi^+\nu\bar{\nu})_{\rm SM}$ (*left* panel) and $\mathcal{B}(K_L \to \pi^0\nu\bar{\nu})/\mathcal{B}(K_L \to \pi^0\nu\bar{\nu})_{\rm SM}$ (*right*) are shown by the red contours. The blue contours represent the tuning parameter ξ. The orange and purple shaded regions are excluded by $\mathcal{B}(K_L \to \mu^+\mu^-)$ and Δm_K, respectively. Here, $(\epsilon'/\epsilon)_{\rm NP} = 15.5 \cdot 10^{-4}$ and $\epsilon_K^{\rm NP} = 0.37 \cdot 10^{-3}$ as a reference. The NP scale is set to be $\mu_{\rm NP} = 1\,$TeV. (For interpretation of the references to color in this figure legend, the reader is referred to the web version of this article.)

section for the experimental constraints). For each set of $\mathcal{B}(K^+ \to \pi^+\nu\bar{\nu})$ and $\mathcal{B}(K_L \to \pi^0\nu\bar{\nu})$, the smallest ξ is chosen among the parameter sets which predict the same branching ratios.

Compared to the simplified cases in Fig. 4, $\mathcal{B}(K_L \to \pi^0\nu\bar{\nu})$ can be enhanced. The tuning parameter is not necessarily very large if only one of $\mathcal{B}(K_L \to \pi^0\nu\bar{\nu})$ and $\mathcal{B}(K_L \to \mu^+\mu^-)$ is enhanced. However, $\xi \gtrsim 30$–40 is required to amplify both of them. If $\xi \lesssim 10$ (5) is allowed, $\mathcal{B}(K_L \to \pi^0\nu\bar{\nu})$ can be as large as 6×10^{-10} (2×10^{-10}). In other words, $\mathcal{O}(10)\%$ tunings are required to enhance $\mathcal{B}(K_L \to \pi^0\nu\bar{\nu})$ by an order of magnitudes compared the SM prediction. The KOTO experiment can probe such large branching ratios in the near future.

4. Conclusion and discussion

The recent discrepancy of ϵ'/ϵ may be a sign of the NP contribution to the flavor-changing Z coupling. In this letter, we revisited the scenario with paying attention to the interference effects between the SM and NP contributions to the $\Delta S = 2$ observables. They affect ϵ_K significantly once the right-handed coupling is turned on. Consequently, $\mathcal{B}(K_L \to \pi^0\nu\bar{\nu})$ is smaller than the SM prediction in the simplified scenarios as long as ϵ'/ϵ is explained.

In the general Z scenario, $\mathcal{B}(K_L \to \pi^0\nu\bar{\nu})$ can be large if parameter tunings are allowed. It was found that the branching ratio can be enhanced by an order of magnitudes compared to the SM prediction if the NP contributions to ϵ_K are tuned at the $\mathcal{O}(10)\%$ level. It can be as large as 6×10^{-10} (2×10^{-10}) for $\xi \simeq 10$ (5), which implies that the NP contributions to ϵ_K are tuned at the 10% (20%) level. The KOTO experiment could probe such large branching ratios in the near future.

In the analysis, the NP scale was set to be 1 TeV. The NP contributions to ϵ_K as well as the tuning parameter depend on it through the interference terms of the SM and NP (see Eq. (2.12)). For $\mu_{\rm NP} \gtrsim 1\,$TeV, $\Delta_L^{\rm SM}$ is enhanced as the NP scale increases. Hence,

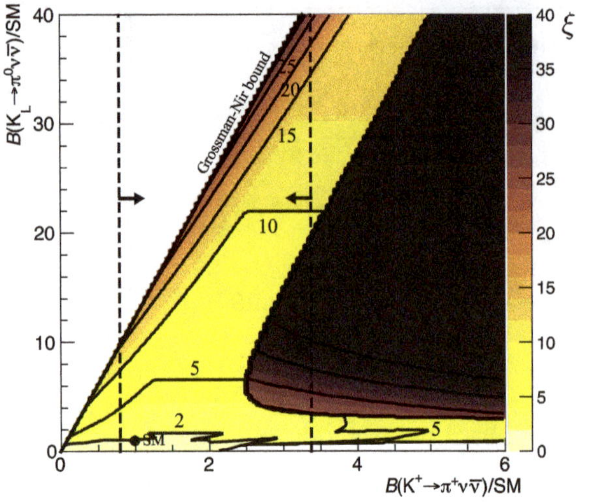

Fig. 6. Contours of the tuning parameter ξ are shown in the general Z scenario. In the colored regions, ϵ'/ϵ is explained at 1σ, and the experimental bounds of ϵ_K, Δm_K, and $\mathcal{B}(K_L\to\mu^+\mu^-)$ are satisfied. The region between the blue dashed lines is allowed by the measurement of $\mathcal{B}(K^+\to\pi^+\nu\bar{\nu})$ at 1σ. There are no available model parameters above the Grossman–Nir bound. The NP scale is set to be $\mu_{\rm NP}=1\,$TeV. (For interpretation of the references to color in this figure legend, the reader is referred to the web version of this article.)

one naively expects that tighter tuning is required in ϵ_K. However, renormalization group corrections could be larger in such a case. Such contributions will be studied elsewhere (see also Ref. [48]).

5. Note added

While resubmitting the manuscript, Ref. [48] appeared on the arXiv. In comparison with our analysis, the differences are as follows:

- we considered $\mathcal{O}_L$ and $\mathcal{O}_R$ operators with the first leading logarithmic renormalization group contribution $\ln(\mu_{\rm NP}/m_W)$ which comes from Fig. 1(e). Hence, the operator $\mathcal{O}_L^{(3)}$, the operator mixing among them through the renormalization group above the electroweak scale, nor running effects of the coupling constants are not considered in our analysis;
- a conservative constraint from $\mathcal{B}(K_L\to\mu^+\mu^-)$ [40] is imposed here, while Ref. [48] has also adopted an aggressive bound [49];
- the present analysis focuses on the parameter region where the ϵ'/ϵ discrepancy can be explained at $1\,\sigma$ level.

The loop function $\widetilde{C}(x,\mu_{\rm NP})$ in Eq. (2.9), which comes from $\mathcal{O}_L$ and $\mathcal{O}_R$, is in agreement with a result of Ref. [48], which comes from $\mathcal{O}_{Hq}^{(1)}$ and $\mathcal{O}_{Hd}$, at the first leading logarithmic approximation. Notice that $\mu_{\rm NP}$ corresponds to μ_Λ.

Acknowledgements

We would like to thank Andrzej J. Buras for useful discussions. We are indebted to the referee for important comments. This work is supported by JSPS KAKENHI No. 16K17681 (M.E.) and 16H03991 (M.E.).

References

[1] T. Blum, et al., The $K\to(\pi\pi)_{I=2}$ decay amplitude from lattice QCD, Phys. Rev. Lett. 108 (2012) 141601, arXiv:1111.1699 [hep-lat].

[2] T. Blum, et al., Lattice determination of the $K\to(\pi\pi)_{I=2}$ decay amplitude A_2, Phys. Rev. D 86 (2012) 074513, arXiv:1206.5142 [hep-lat].

[3] T. Blum, et al., $K\to\pi\pi$ $\Delta I=3/2$ decay amplitude in the continuum limit, Phys. Rev. D 91 (7) (2015) 074502, arXiv:1502.00263 [hep-lat].

[4] Z. Bai, et al., RBC and UKQCD Collaborations, Standard model prediction for direct CP violation in $K\to\pi\pi$ decay, Phys. Rev. Lett. 115 (21) (2015) 212001, arXiv:1505.07863 [hep-lat].

[5] A.J. Buras, M. Gorbahn, S. Jäger, M. Jamin, Improved anatomy of ϵ'/ϵ in the standard model, J. High Energy Phys. 1511 (2015) 202, arXiv:1507.06345 [hep-ph].

[6] T. Kitahara, U. Nierste, P. Tremper, Singularity-free next-to-leading order $\Delta S=1$ renormalization group evolution and ϵ'_K/ϵ_K in the standard model and beyond, J. High Energy Phys. 1612 (2016) 078, arXiv:1607.06727 [hep-ph].

[7] J.R. Batley, et al., NA48 Collaboration, A precision measurement of direct CP violation in the decay of neutral kaons into two pions, Phys. Lett. B 544 (2002) 97, arXiv:hep-ex/0208009.

[8] A. Alavi-Harati, et al., KTeV Collaboration, Measurements of direct CP violation, CPT symmetry, and other parameters in the neutral kaon system, Phys. Rev. D 67 (2003) 012005;
A. Alavi-Harati, et al., KTeV Collaboration, Phys. Rev. D 70 (2004) 079904 (Erratum), arXiv:hep-ex/0208007.

[9] E. Abouzaid, et al., KTeV Collaboration, Precise measurements of direct CP violation, CPT symmetry, and other parameters in the neutral kaon system, Phys. Rev. D 83 (2011) 092001, arXiv:1011.0127 [hep-ex].

[10] C. Patrignani, et al., Particle Data Group Collaboration, Review of particle physics, Chin. Phys. C 40 (10) (2016) 100001.

[11] A.J. Buras, F. De Fazio, J. Girrbach, $\Delta I=1/2$ rule, ϵ'/ϵ and $K\to\pi\nu\bar{\nu}$ in $Z'(Z)$ and G' models with FCNC quark couplings, Eur. Phys. J. C 74 (7) (2014) 2950, arXiv:1404.3824 [hep-ph].

[12] M. Blanke, A.J. Buras, S. Recksiegel, Quark flavour observables in the littlest Higgs model with T-parity after LHC run 1, Eur. Phys. J. C 76 (4) (2016) 182, arXiv:1507.06316 [hep-ph].

[13] A.J. Buras, D. Buttazzo, R. Knegjens, $K\to\pi\nu\bar{\nu}$ and ϵ'/ϵ in simplified new physics models, J. High Energy Phys. 1511 (2015) 166, arXiv:1507.08672 [hep-ph].

[14] A.J. Buras, F. De Fazio, ϵ'/ϵ in 331 models, J. High Energy Phys. 1603 (2016) 010, arXiv:1512.02869 [hep-ph].

[15] A.J. Buras, New physics patterns in ϵ'/ϵ and ϵ_K with implications for rare kaon decays and ΔM_K, J. High Energy Phys. 1604 (2016) 071, arXiv:1601.00005 [hep-ph].

[16] A.J. Buras, F. De Fazio, 331 models facing the tensions in $\Delta F=2$ processes with the impact on ϵ'/ϵ, $B_s\to\mu^+\mu^-$ and $B\to K^*\mu^+\mu^-$, J. High Energy Phys. 1608 (2016) 115, arXiv:1604.02344 [hep-ph].

[17] M. Tanimoto, K. Yamamoto, Probing the SUSY with 10 TeV stop mass in rare decays and CP violation of kaon, PTEP 2016 (12) (2016) 123B02, arXiv:1603.07960 [hep-ph].

[18] T. Kitahara, U. Nierste, P. Tremper, Supersymmetric explanation of CP violation in $K\to\pi\pi$ decays, Phys. Rev. Lett. 117 (9) (2016) 091802, arXiv:1604.07400 [hep-ph].

[19] M. Endo, S. Mishima, D. Ueda, K. Yamamoto, Chargino contributions in light of recent ϵ'/ϵ, Phys. Lett. B 762 (2016) 493, arXiv:1608.01444 [hep-ph].

[20] C. Bobeth, A.J. Buras, A. Celis, M. Jung, Patterns of flavour violation in models with vector-like quarks, arXiv:1609.04783 [hep-ph].

[21] V. Cirigliano, W. Dekens, J. de Vries, E. Mereghetti, An ϵ' improvement from right-handed currents, arXiv:1612.03914 [hep-ph].

[22] A.J. Buras, F. De Fazio, J. Girrbach, The anatomy of Z' and Z with flavour changing neutral currents in the flavour precision era, J. High Energy Phys. 1302 (2013) 116, arXiv:1211.1896 [hep-ph].

[23] A.J. Buras, private communication.

[24] J. Charles, et al., Current status of the standard model CKM fit and constraints on $\Delta F=2$ new physics, Phys. Rev. D 91 (7) (2015) 073007, arXiv:1501.05013 [hep-ph]. Updates on http://ckmfitter.in2p3.fr.

[25] A.J. Buras, D. Guadagnoli, G. Isidori, On ϵ_K beyond lowest order in the operator product expansion, Phys. Lett. B 688 (2010) 309, arXiv:1002.3612 [hep-ph].

[26] J. Aebischer, A. Crivellin, M. Fael, C. Greub, Matching of gauge invariant dimension-six operators for $b\to s$ and $b\to c$ transitions, J. High Energy Phys. 1605 (2016) 037, arXiv:1512.02830 [hep-ph].

[27] Z. Ligeti, F. Sala, A new look at the theory uncertainty of ϵ_K, J. High Energy Phys. 1609 (2016) 083, arXiv:1602.08494 [hep-ph].

[28] J.A. Bailey, et al., Fermilab Lattice and MILC Collaborations, Update of $|V_{cb}|$ from the $\bar{B}\to D^*\ell\bar{\nu}$ form factor at zero recoil with three-flavor lattice QCD, Phys. Rev. D 89 (11) (2014) 114504, arXiv:1403.0635 [hep-lat].

[29] J.A. Bailey, et al., SWME Collaboration, Standard model evaluation of ϵ_K using lattice QCD inputs for $\hat{B}_K$ and V_{cb}, Phys. Rev. D 92 (3) (2015) 034510, arXiv:1503.05388 [hep-lat].

[30] Z. Bai, N.H. Christ, T. Izubuchi, C.T. Sachrajda, A. Soni, J. Yu, K_L–K_S mass difference from lattice QCD, Phys. Rev. Lett. 113 (2014) 112003, arXiv:1406.0916 [hep-lat].

[31] V. Cirigliano, G. Ecker, H. Neufeld, A. Pich, J. Portoles, Kaon decays in the standard model, Rev. Mod. Phys. 84 (2012) 399, arXiv:1107.6001 [hep-ph].

[32] A.J. Buras, D. Buttazzo, J. Girrbach-Noe, R. Knegjens, $K^+ \to \pi^+\nu\overline{\nu}$ and $K_L \to \pi^0\nu\overline{\nu}$ in the standard model: status and perspectives, J. High Energy Phys. 1511 (2015) 033, arXiv:1503.02693 [hep-ph].

[33] A.V. Artamonov, et al., E949 Collaboration, New measurement of the $K^+ \to \pi^+\nu\bar{\nu}$ branching ratio, Phys. Rev. Lett. 101 (2008) 191802, arXiv:0808.2459 [hep-ex].

[34] J.K. Ahn, et al., E391a Collaboration, Experimental study of the decay $K_L^0 \to \pi^0\nu\bar{\nu}$, Phys. Rev. D 81 (2010) 072004, arXiv:0911.4789 [hep-ex].

[35] M. Moulson, NA62 Collaboration, Search for $K^+ \to \pi^+\nu\bar{\nu}$ at NA62, arXiv: 1611.04979 [hep-ex].

[36] Y.C. Tung, KOTO Collaboration, Status and prospects of the KOTO experiment, PoS CD 15 (2016) 068.

[37] Talk by H. Nanjo at International Workshop on Physics at the Extended Hadron Experimental Facility of J-PARC, KEK Tokai Campus, 2016.

[38] M. Gorbahn, U. Haisch, Charm quark contribution to $K_L \to \mu^+\mu^-$ at next-to-next-to-leading order, Phys. Rev. Lett. 97 (2006) 122002, arXiv:hep-ph/0605203.

[39] C. Bobeth, M. Gorbahn, E. Stamou, Electroweak corrections to $B_{s,d} \to \ell^+\ell^-$, Phys. Rev. D 89 (3) (2014) 034023, arXiv:1311.1348 [hep-ph].

[40] G. Isidori, R. Unterdorfer, On the short distance constraints from $K_{(L,S)} \to \mu^+\mu^-$, J. High Energy Phys. 0401 (2004) 009, arXiv:hep-ph/0311084.

[41] A.J. Buras, L. Silvestrini, Upper bounds on $K \to \pi\nu\bar{\nu}$ and $K_L \to \pi^0e^+e^-$ from ϵ'/ϵ and $K_L \to \mu^+\mu^-$, Nucl. Phys. B 546 (1999) 299, arXiv:hep-ph/9811471.

[42] A.J. Buras, A. Romanino, L. Silvestrini, $K \to \pi\nu\bar{\nu}$: a model independent analysis and supersymmetry, Nucl. Phys. B 520 (1998) 3, arXiv:hep-ph/9712398.

[43] G. Colangelo, G. Isidori, Supersymmetric contributions to rare kaon decays: beyond the single mass insertion approximation, J. High Energy Phys. 9809 (1998) 009, arXiv:hep-ph/9808487.

[44] A.J. Buras, G. Colangelo, G. Isidori, A. Romanino, L. Silvestrini, Connections between ϵ'/ϵ and rare kaon decays in supersymmetry, Nucl. Phys. B 566 (2000) 3, arXiv:hep-ph/9908371.

[45] M. Blanke, A.J. Buras, B. Duling, S. Gori, A. Weiler, $\Delta F = 2$ observables and fine-tuning in a warped extra dimension with custodial protection, J. High Energy Phys. 0903 (2009) 001, arXiv:0809.1073 [hep-ph].

[46] M. Blanke, A.J. Buras, B. Duling, K. Gemmler, S. Gori, Rare K and B decays in a warped extra dimension with custodial protection, J. High Energy Phys. 0903 (2009) 108, arXiv:0812.3803 [hep-ph].

[47] Y. Grossman, Y. Nir, $K_L \to \pi^0\nu\bar{\nu}$ beyond the standard model, Phys. Lett. B 398 (1997) 163, arXiv:hep-ph/9701313.

[48] C. Bobeth, A.J. Buras, A. Celis, M. Jung, Yukawa enhancement of Z-mediated new physics in $\Delta S = 2$ and $\Delta B = 2$ processes, arXiv:1703.04753 [hep-ph].

[49] G. D'Ambrosio, G. Isidori, J. Portoles, Can we extract short distance information from $\mathcal{B}(K_L \to \mu^+\mu^-)$?, Phys. Lett. B 423 (1998) 385, arXiv:hep-ph/9708326.

22

Scrutinizing the Higgs quartic coupling at a future 100 TeV proton–proton collider with taus and b-jets

Benjamin Fuks [a,b,c,*], Jeong Han Kim [d], Seung J. Lee [e,f]

[a] Sorbonne Universités, Université Pierre et Marie Curie (Paris 06), UMR 7589, LPTHE, F-75005 Paris, France
[b] CNRS, UMR 7589, LPTHE, F-75005 Paris, France
[c] Institut Universitaire de France, 103 boulevard Saint-Michel, 75005 Paris, France
[d] Department of Physics and Astronomy, University of Kansas, Lawrence, KS, 66045, USA
[e] Department of Physics, Korea University, Seoul 136-713, Republic of Korea
[f] School of Physics, Korea Institute for Advanced Study, Seoul 130-722, Republic of Korea

ARTICLE INFO

Editor: G.F. Giudice

ABSTRACT

The Higgs potential consists of an unexplored territory in which the electroweak symmetry breaking is triggered, and it is moreover directly related to the nature of the electroweak phase transition. Measuring the Higgs boson cubic and quartic couplings, or getting equivalently information on the exact shape of the Higgs potential, is therefore an essential task. However, direct measurements beyond the cubic self-interaction of the Higgs boson consist of a huge challenge, even for a future proton–proton collider expected to operate at a center-of-mass energy of 100 TeV. We present a novel approach to extract model-independent constraints on the triple and quartic Higgs self-coupling by investigating triple Higgs-boson hadroproduction at a center-of-mass energy of 100 TeV, focusing on the $\tau\tau b\bar{b}b\bar{b}$ channel that was previously overlooked due to a supposedly too large background. It is thrown into sharp relief that the assist from transverse variables such as m_{T2} and a boosted configuration ensures a high signal sensitivity. We derive the luminosities that would be required to constrain given deviations from the Standard Model in the Higgs self-interactions, showing for instance that a 2σ sensitivity could be achieved for an integrated luminosity of 30 ab^{-1} when Standard Model properties are assumed. With the prospects of combining these findings with other triple-Higgs search channels, the Standard Model Higgs quartic coupling could in principle be reached with a significance beyond the 3σ level.

1. Introduction

The discovery of a Higgs boson at the Large Hadron Collider (LHC) accomplished the long waited physics goals of getting hints on the nature of the electroweak symmetry breaking (EWSB) mechanism and understanding the generation of the fermion masses. While the discovered Higgs boson appears to be highly compatible with the Standard Model (SM) expectation [1], current data is still insufficient for revealing the true nature of the EWSB dynamics. Further pieces of information related to the shape of the Higgs potential are indeed needed, such as measurements of the Higgs cubic, quartic and even higher-order self-couplings. This would furthermore allow us to investigate whether the electroweak phase transition is of the first or second order, a fact related to the matter-antimatter asymmetry in the universe as a strong first order electroweak phase transition can potentially realize one of the Sakharov conditions for baryogenesis. Measuring the Higgs cubic and quartic self-couplings is consequently one of the major physics goals of the future high-energy physics program.

Di-Higgs production via gluon fusion offers the first playground to access the Higgs cubic coupling, in particular within the high-luminosity phase of the LHC expected to collect an integrated luminosity of 3 ab^{-1} of data at a center-of-mass energy of 14 TeV [2, 3]. The associated sizable (SM) production cross section of about 43 fb [4] allows one to make use of various final states to probe the Higgs cubic coupling, the two most promising signatures relying on final state systems made of four b-jets, or of a pair of photons and either a pair of b-jets or tau leptons [5–7]. At a future proton–proton collider aiming to operate at a center-of-mass energy of 100 TeV, the $b\bar{b}\gamma\gamma$ channel keeps its leading role and measurements at a precision of about 3–4% could be expected for

* Corresponding author.
E-mail address: fuks@lpthe.jussieu.fr (B. Fuks).

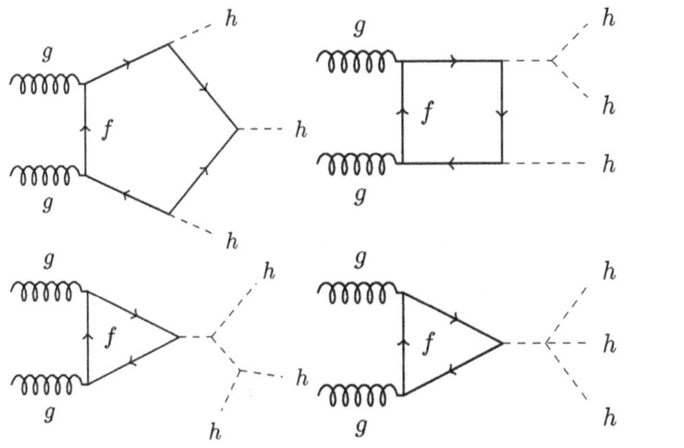

Fig. 1. Representative leading-order Feynman diagrams for triple Higgs production in proton–proton collisions.

a luminosity of 30 ab^{-1} [8,9]. None of these searches are, however, designed to probe the Higgs quartic coupling.

In the SM, triple-Higgs production mostly arises, at the leading-order in QCD, by gluon fusion (see Fig. 1). Such a process faces a rather grim prospect at the LHC, mainly because of a small signal rate of $\mathcal{O}(0.1)$ fb [10–12], so that the study of this process is left to the experimental program of the post-LHC era that is currently under discussion at CERN and IHEP [13]. Feasibility analyses have so far shown that the $b\bar{b}b\bar{b}\gamma\gamma$ channel can be used to constrain the size of the quartic Higgs coupling in a model-independent way, the interaction strength being allowed to deviate by a factor of at most $\mathcal{O}(10)$ from the SM after considering an integrated luminosity of 30 ab^{-1} [14–16]. The prospects of the $b\bar{b}WWWW$ decay mode have also been explored, and it was shown that a new physics triple-Higgs signal is in principle detectable [17].

In this article, we embark on reinvestigating triple Higgs production at a 100 TeV proton–proton collider to be more confident on the sensitivity of such a machine to the quartic Higgs self-coupling. We focus on the more challenging, branching-ratio-enhanced, $b\bar{b}b\bar{b}\tau^+\tau^-$ signature. Contrary to the $b\bar{b}b\bar{b}\gamma\gamma$ channel, it receives a severe background contamination that yields a weaker expected sensitivity [14]. However, with the effort of exploiting previously overlooked advantages of the ditau system and a boosted configuration, we show in this work that the $b\bar{b}b\bar{b}\tau\tau$ channel can be promoted to a leading discovery channel for triple-Higgs production.

This paper is organized as follows. In Sec. 2, we introduce the adopted simplified model parameterizing in a model-independent way any new physics effect on the Higgs self-interactions, and we present technical details related to our simulation setup. Sec. 3 is dedicated to our event selection strategy and exhibits details on its specificity. Our results are given in Sec. 4, together with prospects for a future 100 TeV proton–proton colliders.

2. Theoretical framework and technical details

In order to probe for possible new physics effects in multiple-Higgs interactions, we modify in a model-independent fashion the SM Higgs potential,

$$V_h = \frac{m_h^2}{2}h^2 + (1+\kappa_3)\lambda_{hhh}^{SM} v h^3 + \frac{1}{4}(1+\kappa_4)\lambda_{hhhh}^{SM} h^4 ,$$

by introducing two κ_i parameters that vanish in the SM. In our notation, h denotes the physical Higgs-boson field, m_h its mass and v its vacuum expectation value. The SM self-interaction strengths moreover read

$$\lambda_{hhh}^{SM} = \lambda_{hhhh}^{SM} = \frac{m_h^2}{2v^2} .$$

We simulate our triple Higgs signal and the associated backgrounds by implementing the above Lagrangian in the FeynRules package [18] that we use along with the NLoCT program [19] to generate a UFO library [20]. The latter allows for event generation for both tree-level and loop-induced processes within the MadGraph5_aMC@NLO [21,22] framework, that we use to convolute hard scattering matrix elements with the next-to-leading (NLO) set of NNPDF 2.3 parton densities [23] for a center-of-mass energy of $\sqrt{s} = 100$ TeV. The hard-scattering events are then decayed, showered and hadronized within the Pythia 6 environment [24] and reconstructed by using the anti-k_T algorithm [25] as implemented in FastJet [26], with a radius of $R = 1$ and 0.4 for a fat jet and slim jet definition, respectively.

Hadronic taus are defined as specific slim jets for which there is no hadronic object of $p_T > 1$ GeV and no photon with a $p_T > 1.5$ GeV at an angular distance of the jet axis greater than $r_{in} = 0.1$ and smaller than $r_{out} = 0.4$. The resulting tau-tagging efficiency is of about 50%, for a fake rate of mistagging a light-flavor jet as a tau of roughly 5%. Those performances can be compared to what could be expected from the high-luminosity phase of the LHC, for which an efficiency of 55% can be expected for a mistagging rate of 0.5% [7].

Our analysis relies on the reconstruction of boosted Higgs bosons. To this aim, we employ the template overlap method [27, 28] as embedded in the TemplateTagger program [29], and we use a new template observable derived from the $t\gamma$ quantity proposed in Ref. [30], which we here maximize over the different three-body Higgs templates. We make use of various two-body and three-body (NLO) Higgs templates featuring a sub-cone size of 0.1 to compute the discriminating overlaps Ov_2^h and Ov_3^h, respectively, that allow for a boosted Higgs boson identification. The performance of the method yields a tagging efficiency of 40% for a mistagging rate of 2%.

As suggested by the representative Feynman diagrams of Fig. 1, triple-Higgs production depends on both κ_i parameters as well as on the top Yukawa coupling. While in either an effective field theory framework or an ultraviolet-complete model building approach, the κ_i parameters are not independent, they will be varied independently in our study. Moreover, the top Yukawa coupling is assumed to be fixed to its SM value. The resulting production cross section is presented in Fig. 2 in the (κ_3, κ_4) plane after including a flat NLO K-factor of 2 [31]. The sign of the κ_3 parameter turns out to be crucial due to respective constructive and destructive interference patterns when κ_3 is negative and positive. As a consequence, the cross section can be reduced to below the fb level when both κ parameters are positive (and not too large), making this corner of the parameter space hard to probe. The variations in κ_4 are in addition mild for any fixed value of κ_3, so that only poor constraints could be expected from any potential measurement.

Among all triple-Higgs production signatures, we make use of the $b\bar{b}b\bar{b}\tau^+\tau^-$ channel with two hadronic tau decays to probe deviations in the Higgs self-interactions. Whilst the branching ratio is large ($\sim 6.3\%$), the background contamination is expected to be important [14]. We however demonstrate in the next sections that previously overlooked advantages stemming from the usage of specific kinematic properties of the ditau systems and the potentially boosted configuration of the b-jet pairs could largely increase the signal significance.

The various components of the SM background can be classified into three categories regarding their response to the basic selection criteria introduced in Sec. 3. We denote by t/W samples the ensemble of background processes featuring a top quark or a W-boson pair that decays into a tau-enriched final state, together

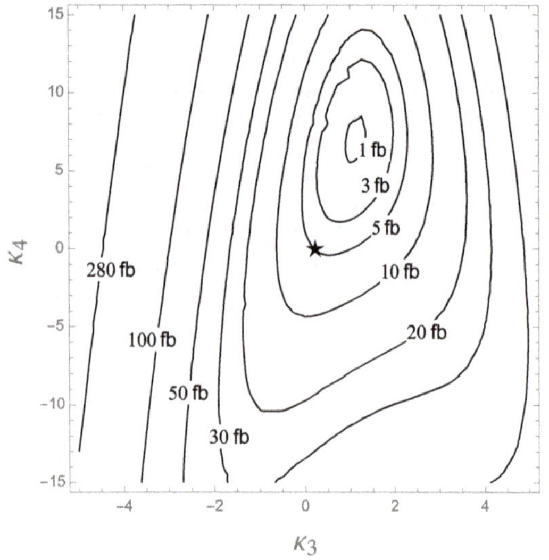

Fig. 2. Triple-Higgs production cross-section for a center-of-mass energy of $\sqrt{s} =$ 100 TeV presented as a function of the κ_3 and κ_4 parameters depicting the possible deviations from the SM (indicated by a black star). The results include a conservative NLO K-factor of 2.

Table 1
Fiducial cross sections of all components of the SM background after the baseline selection described in Sec. 3. The results include an NLO K-factor of 2, and the suffixes 'τ' and '$b\bar{b}$' respectively indicate decays into a tau-lepton and a $b\bar{b}$ pair, t_h denoting similarly a hadronically-decaying top quark.

Class	Backgrounds	Cross section [ab]
t/W samples	$t_\tau \bar{t}_\tau h_{b\bar{b}}$	2.3×10^4
	$t_\tau \bar{t}_\tau Z_{b\bar{b}}$	6.6×10^3
	$t_\tau \bar{t}_\tau b\bar{b}$	4.7×10^5
	$W_\tau^+ W_\tau^- b\bar{b}b\bar{b}$	4.7×10^5
	$t\bar{t}t\bar{t}$	6.6×10^4
$X_{\tau\tau}$ +jets	$X_{\tau\tau} b\bar{b}b\bar{b}$	6.9×10^4
	$X_{\tau\tau} b\bar{b}jj$	1.5×10^7
	$X_{\tau\tau} t_h \bar{t}_h$	1.6×10^5
	$X_{\tau\tau} Z_{b\bar{b}} b\bar{b}$	2.0×10^3
	$Z_{\tau\tau} h_{b\bar{b}} b\bar{b}$	300
	$X_{\tau\tau} Z_{b\bar{b}} Z_{b\bar{b}}$	23
	$Z_{\tau\tau} h_{b\bar{b}} Z_{b\bar{b}}$	15
	$h_{\tau\tau} h_{b\bar{b}} Z_{b\bar{b}}$ $h_{b\bar{b}} h_{b\bar{b}} Z_{\tau\tau}$	11
Di-Higgs	$h_{\tau\tau} h_{b\bar{b}}$ +jet	1.3×10^3

with the four-top background contributions. The second class of SM backgrounds consists of the $X_{\tau\tau}$ +jets category with X being a virtual photon, Higgs or Z-boson decaying into a pair of tau leptons. Di-Higgs production in association with jets finally forms the last class of background processes on its own. The full list of considered SM backgrounds is summarized in Table 1, where we additionally present the fiducial cross sections, multiplied by a conservative NLO K-factor of 2, obtained after requiring the presence of two hadronic taus and missing transverse energy (*cf.* the baseline selection described in Sec. 3).

3. Signal selection

Our triple-Higgs analysis relies for its baseline selection on the properties of the $b\bar{b}b\bar{b}\tau^+\tau^-$ final state. We preselect events featuring exactly two hadronic taus with a $p_T > 25$ GeV and a

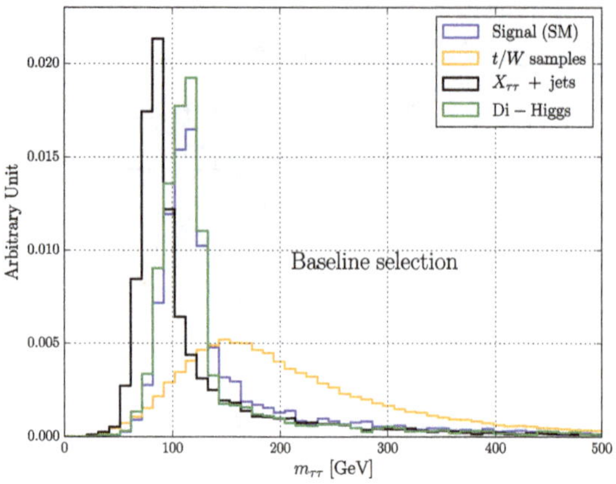

Fig. 3. Distribution in the $m_{\tau\tau}$ invariant mass (as defined in the text) after the baseline selection for the three SM background categories and for a SM triple-Higgs signal. (For interpretation of the references to color in this figure legend, the reader is referred to the web version of this article.)

pseudorapidity $|\eta| < 2.5$, as well as a missing transverse energy $\not{E}_T > 25$ GeV.

After this preselection, the two taus are enforced to be compatible with the decay of a Higgs boson by means of the $m_{\tau\tau}^{\text{Higgs-bound}}$ and m_T^{True} variables [32–34]. The former quantity is defined by minimizing, over all possible assignments for the neutrino four-momenta, the invariant mass of the system made of the two tau jets and the two invisible neutrinos. This minimization procedure however requires that each tau jet is matched with a neutrino and that the resulting two-body invariant mass is compatible with the tau mass. For cases for which there is no such a solution, the m_T^{True} variable is constructed instead in the same way, but without this last constraint. We present the resulting $m_{\tau\tau}$ distribution in Fig. 3, $m_{\tau\tau}$ generically denoting $m_{\tau\tau}^{\text{Higgs-bound}}$ when it can be constructed and m_T^{True} otherwise. Most signal events exhibit an $m_{\tau\tau}$ value lying between the Z and the Higgs boson masses, whereas background events from the $X_{\tau\tau}$ +jets category mainly feature smaller $m_{\tau\tau}$ values. We therefore impose that $m_{\tau\tau} \in [105, 135]$ GeV to ensure compatibility with a Higgs ditau decay and a very good discrimination from the $X_{\tau\tau}$ +jets background category.

We move on with the reconstruction of the two other Higgs bosons for which we rely on a configuration where one of them is boosted and the other one is resolved. We select events featuring at least one fat jet whose basic properties satisfy $p_T > 300$ GeV and $|\eta| < 2.5$. The fat jet invariant mass is moreover required to lie in the [105, 135] GeV window and the template overlaps are constrained to $Ov_3^h > 0.7$ and $Ov_2^h > 0.2$. We additionally require the presence of at least two slim jets and tag two of them as candidates for a non-boosted Higgs decay. This tagging is such that the dijet invariant mass $m_{jj} \in [105, 135]$ GeV minimizes $|m_{jj} - m_h|$. Furthermore, one of the two tagged slim jets must be b-tagged and the fat jet must contain a doubly-b-tagged substructure when we assume a b-tagging efficiency of 70% when a B-hadron is present in a cone of radius $R = 0.4$ around the jet direction, for a corresponding mistagging rate of 1%.

At this stage, the background is dominated by its t/W component (see Table 2). In contrast to the triple-Higgs signal in which the missing energy originates from the two neutrinos associated with the tau decays, most background events feature either more than two neutrinos, or a missing energy originating from a W-boson pair. This suggests to take advantage of the m_{T2} variable [35,36] to ensure an efficient background rejection. The m_{T2} spectrum is bounded from above and its shape depends both on a test mass and on the mass of the semi-invisibly decaying particle. Moreover, the upper bound sharply rises for increasing test

Table 2
Signal and background cross sections, in ab, at different stage of the analysis strategy depicted in Sec. 3. The signal to background ratio S/B and the significance σ for a luminosity of 30 ab^{-1} are also indicated.

Selection	Signal	t/W	$X_{\tau\tau}$	hh
Baseline	27	1.0×10^6	1.6×10^7	1.3×10^3
$m_{\tau\tau}$	12	1.4×10^5	2.6×10^6	670
Boosted Higgs	0.92	640	6.5×10^3	35
m_{jj}	0.47	180	81	4.1
b-tagging	0.15	15	0.20	0.034
m_{T2}	0.11	0.37	0.093	0.029
m_{hhh}	0.10	8.5×10^{-3}	0.012	0.026
S/B			2.1	
σ			2.0	

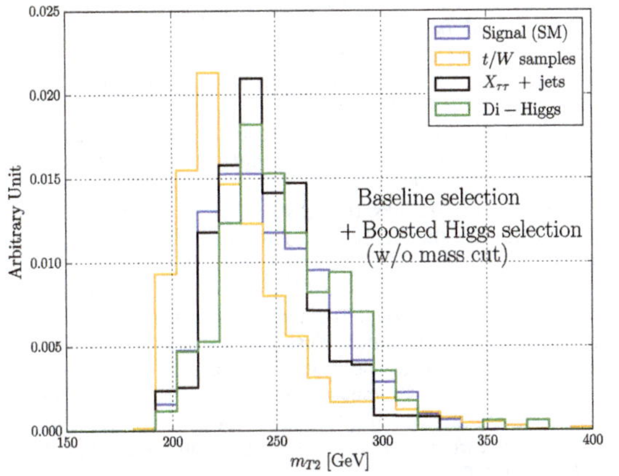

Fig. 4. m_{T2} spectra for the signal (in the case of a SM Higgs potential) and the various components of the background. (For interpretation of the references to color in this figure legend, the reader is referred to the web version of this article.)

masses above the true mass of the invisible particle [37]. As the true invisible mass is zero for the triple Higgs signal, the associated m_{T2} distribution is naturally broader than for the background, provided the test mass is taken large enough. This feature is illustrated in Fig. 4 for which we have chosen an optimized test mass of 190 GeV, which allows for a maximal background and signal separation.

After having reconstructed all three Higgs bosons, we derive the invariant mass of the triple-Higgs system m_{hhh} and constrain it to be smaller than 1.6 TeV.

4. Results and discussion

We present in Table 2 the fiducial cross sections resulting from the application of the various selections introduced in Sec. 3, both for the signal (assuming the SM case with $\kappa_3 = \kappa_4 = 0$) and the background. We can observe the complementarity of the various steps, the $m_{\tau\tau}$ and boosted Higgs requirements reducing the background by a factor of more than 2000, while the reconstruction of the resolved Higgs boson and the b-tagging conditions bring the signal over background (S/B) ratio down to the percent level. The background is at this stage dominated by t/W events and is further reduced to a manageable level by means of the m_{T2} selection. The selection on the triple-Higgs invariant mass finally brings the background rate to half the signal one for the considered benchmark.

In order to set limits and derive the future collider sensitivity in the (κ_3, κ_4) plane, we compute a significance σ defined as the likelihood ratio [38]

$$\sigma \equiv \sqrt{-2\ln\left(\frac{L(B|S+B)}{L(S+B|S+B)}\right)} \text{ with } L(x|n) = \frac{x^n}{n!}e^{-x},$$

where S and B are the expected number of signal and background events respectively. The signal sensitivity turns out to be of about 2σ in the SM case for a luminosity of 30 ab^{-1}, with a number of signal events $S \sim 3$ and background events $B \sim 1.4$. The number of signal events could however be increased by considering the strategic approach of including the contributions of a semi-leptonic $\tau_h\tau_l b\bar{b}b\bar{b}$ final state, as it has been recently proposed for di-Higgs searches at the LHC [7].

Scanning over the κ_i parameters, we show in Fig. 5 the luminosity goals of a 100 TeV proton–proton collider necessary for achieving a 2σ exclusion (left panel). Despite the dominance of destructive interferences on the upper-right-corner of the (κ_3, κ_4) plane, our analysis demonstrates that the SM expectation can in principle be excluded with 30 ab^{-1}. Conversely, we present in the right panel of the figure the significance contours obtained when

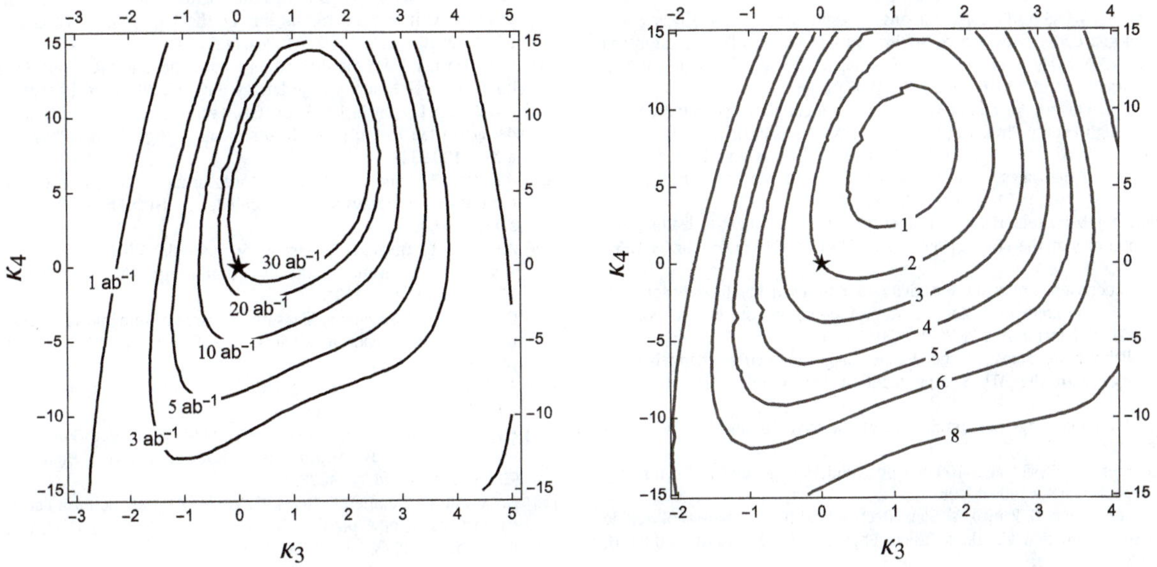

Fig. 5. Minimum luminosity of 100 TeV proton–proton collisions required to achieve a 2σ sensitivity to a triple-Higgs signal in the $b\bar{b}b\bar{b}\tau\tau$ channel shown in terms of the κ_3 and κ_4 parameters (left), and the corresponding sensitivity expected for a luminosity of 30 ab^{-1} (right).

considering a luminosity of 30 ab^{-1}. In order to access the sensitivity gap in the parameter space region limited by $\kappa_3 \in [0, 2]$ and $\kappa_4 \in [0, 14]$, one could combine our results with other channels, like the $\tau_h \tau_l b\bar{b}b\bar{b}$ mode that could enhance the sensitivity of the present analysis, and the $\gamma\gamma b\bar{b}b\bar{b}$ channel investigated in Refs. [14–16]. Our findings could moreover be merged with the more precise prospects on the κ_3 parameters that stem from di-Higgs probes expected to be produced at a large rate [8,9].

In this work, we have continued our investigation of the possibilities of a future proton–proton collider expected to run at $\sqrt{s} = 100$ TeV to unravel the true nature of the EWSB mechanism. We have shown that in addition to the $\gamma\gamma b\bar{b}b\bar{b}$ golden channel, the $b\bar{b}b\bar{b}\tau\tau$ mode is a complementary probe to the quartic Higgs self-interaction. Our results are comparable to those derived in other triple-Higgs channels, so that combinations of several searches could offer handles to parameter space regions featuring low cross sections and not accessible with a single triple-Higgs analysis. Such a combination also gives hope to access the SM couplings beyond the 3σ level.

Acknowledgements

We are very grateful to Minho Son for valuable help and discussions, in particular on tau reconstruction, as well as to K.C. Kong, Ian M. Lewis and Graham Wilson for useful comments and suggestions during the course of this project. We also thank the HTCaaS group of the Korea Institute of Science and Technology Information (KISTI) for providing the necessary computing resources and acknowledge the Korea Future Collider Study Group (KFCSG) for motivating us to proceed with this work. JHK is supported in part by US-DOE (DE-FG02-12ER41809) and by the University of Kansas General Research Fund allocation 2302091. SL is supported by the National Research Foundation of Korea (NRF) grant funded by the Korea government (MEST) (No. NRF-2015R1A2A1A15052408), and by the Korean Research Foundation (KRF) through the Korea-CERN collaboration program (NRF-2016R1D1A3B01010529). The work of BF is partly supported by French state funds managed by the Agence Nationale de la Recherche (ANR), in the context of the LABEX ILP (ANR-11-IDEX-0004-02, ANR-10-LABX-63), and by the FKPPL initiative of the CNRS.

References

[1] G. Aad, et al., ATLAS, CMS Collaborations, Measurements of the Higgs boson production and decay rates and constraints on its couplings from a combined ATLAS and CMS analysis of the LHC pp collision data at $\sqrt{s} = 7$ and 8 TeV, J. High Energy Phys. 08 (2016) 045, arXiv:1606.02266.

[2] M.J. Dolan, C. Englert, M. Spannowsky, Higgs self-coupling measurements at the LHC, J. High Energy Phys. 10 (2012) 112, arXiv:1206.5001.

[3] J. Baglio, A. Djouadi, R. Gröber, M.M. Mühlleitner, J. Quevillon, M. Spira, The measurement of the Higgs self-coupling at the LHC: theoretical status, J. High Energy Phys. 04 (2013) 151, arXiv:1212.5581.

[4] D. de Florian, J. Mazzitelli, Higgs pair production at next-to-next-to-leading logarithmic accuracy at the LHC, J. High Energy Phys. 09 (2015) 053, arXiv:1505.07122.

[5] ATLAS Collaboration, Projected sensitivity to non-resonant Higgs boson pair production in the $b\bar{b}b\bar{b}$ final state using proton–proton collisions at HL-LHC with the ATLAS detector, ATL-PHYS-PUB-2016-024.

[6] ATLAS Collaboration, Study of the double Higgs production channel $H(\to b\bar{b})H(\to \gamma\gamma)$ with the ATLAS experiment at the HL-LHC, ATL-PHYS-PUB-2017-001.

[7] CMS Collaboration, Higgs pair production at the High Luminosity LHC, CMS-PAS-FTR-15-002.

[8] R. Contino, et al., Physics at a 100 TeV pp collider: Higgs and EW symmetry breaking studies, arXiv:1606.09408.

[9] A. Azatov, R. Contino, G. Panico, M. Son, Effective field theory analysis of double Higgs boson production via gluon fusion, Phys. Rev. D 92 (3) (2015) 035001, arXiv:1502.00539.

[10] T. Plehn, M. Rauch, The quartic Higgs coupling at hadron colliders, Phys. Rev. D 72 (2005) 053008, arXiv:hep-ph/0507321.

[11] T. Binoth, S. Karg, N. Kauer, R. Ruckl, Multi-Higgs boson production in the Standard Model and beyond, Phys. Rev. D 74 (2006) 113008, arXiv:hep-ph/0608057.

[12] F. Maltoni, E. Vryonidou, M. Zaro, Top-quark mass effects in double and triple Higgs production in gluon–gluon fusion at NLO, J. High Energy Phys. 11 (2014) 079, arXiv:1408.6542.

[13] N. Arkani-Hamed, T. Han, M. Mangano, L.-T. Wang, Physics opportunities of a 100 TeV proton–proton collider, Phys. Rep. 652 (2016) 1–49, arXiv:1511.06495.

[14] B. Fuks, J.H. Kim, S.J. Lee, Probing Higgs self-interactions in proton–proton collisions at a center-of-mass energy of 100 TeV, Phys. Rev. D 93 (3) (2016) 035026, arXiv:1510.07697.

[15] C.-Y. Chen, Q.-S. Yan, X. Zhao, Y.-M. Zhong, Z. Zhao, Probing triple-Higgs productions via $4b2\gamma$ decay channel at a 100 TeV hadron collider, Phys. Rev. D 93 (1) (2016) 013007, arXiv:1510.04013.

[16] A. Papaefstathiou, K. Sakurai, Triple Higgs boson production at a 100 TeV proton–proton collider, J. High Energy Phys. 02 (2016) 006, arXiv:1508.06524.

[17] W. Kilian, S. Sun, Q.-S. Yan, X. Zhao, Z. Zhao, New physics in multi-Higgs boson final states, arXiv:1702.03554.

[18] A. Alloul, N.D. Christensen, C. Degrande, C. Duhr, B. Fuks, FeynRules 2.0 – a complete toolbox for tree-level phenomenology, Comput. Phys. Commun. 185 (2014) 2250–2300, arXiv:1310.1921.

[19] C. Degrande, Automatic evaluation of UV and R2 terms for beyond the Standard Model Lagrangians: a proof-of-principle, Comput. Phys. Commun. 197 (2015) 239–262, arXiv:1406.3030.

[20] C. Degrande, C. Duhr, B. Fuks, D. Grellscheid, O. Mattelaer, T. Reiter, UFO – the Universal FeynRules Output, Comput. Phys. Commun. 183 (2012) 1201–1214, arXiv:1108.2040.

[21] J. Alwall, R. Frederix, S. Frixione, V. Hirschi, F. Maltoni, O. Mattelaer, H.S. Shao, T. Stelzer, P. Torrielli, M. Zaro, The automated computation of tree-level and next-to-leading order differential cross sections, and their matching to parton shower simulations, J. High Energy Phys. 07 (2014) 079, arXiv:1405.0301.

[22] V. Hirschi, O. Mattelaer, Automated event generation for loop-induced processes, J. High Energy Phys. 10 (2015) 146, arXiv:1507.00020.

[23] R.D. Ball, et al., Parton distributions with LHC data, Nucl. Phys. B 867 (2013) 244–289, arXiv:1207.1303.

[24] T. Sjostrand, S. Mrenna, P.Z. Skands, PYTHIA 6.4 Physics and Manual J. High Energy Phys. 0605 (2006) 026, arXiv:hep-ph/0603175.

[25] M. Cacciari, G.P. Salam, G. Soyez, The anti-k(t) jet clustering algorithm, J. High Energy Phys. 0804 (2008) 063, arXiv:0802.1189.

[26] M. Cacciari, G.P. Salam, G. Soyez, FastJet user manual, Eur. Phys. J. C 72 (2012) 1896, arXiv:1111.6097.

[27] L.G. Almeida, S.J. Lee, G. Perez, G. Sterman, I. Sung, Template overlap method for massive jets, Phys. Rev. D 82 (2010) 054034, arXiv:1006.2035.

[28] L.G. Almeida, O. Erdogan, J. Juknevich, S.J. Lee, G. Perez, et al., Three-particle templates for a boosted Higgs boson, Phys. Rev. D 85 (2012) 114046, arXiv:1112.1957.

[29] M. Backović, J. Juknevich, TemplateTagger v1.0.0: a template matching tool for jet substructure, Comput. Phys. Commun. 185 (2014) 1322–1338, arXiv:1212.2978.

[30] J.H. Kim, K. Kong, S.J. Lee, G. Mohlabeng, Probing TeV scale top-philic resonances with boosted top-tagging at the high luminosity LHC, Phys. Rev. D 94 (3) (2016) 035023, arXiv:1604.07421.

[31] D. de Florian, J. Mazzitelli, Two-loop corrections to the triple Higgs boson production cross section, J. High Energy Phys. 02 (2017) 107, arXiv:1610.05012.

[32] A.J. Barr, S.T. French, J.A. Frost, C.G. Lester, Speedy Higgs boson discovery in decays to tau lepton pairs: h → tau, tau, J. High Energy Phys. 10 (2011) 080, arXiv:1106.2322.

[33] A.J. Barr, B. Gripaios, C.G. Lester, Measuring the Higgs boson mass in dileptonic W-boson decays at hadron colliders, J. High Energy Phys. 07 (2009) 072, arXiv:0902.4864.

[34] A.J. Barr, M.J. Dolan, C. Englert, M. Spannowsky, Di-Higgs final states augMT2ed – selecting hh events at the high luminosity LHC, Phys. Lett. B 728 (2014) 308–313, arXiv:1309.6318.

[35] C.G. Lester, D.J. Summers, Measuring masses of semiinvisibly decaying particles pair produced at hadron colliders, Phys. Lett. B 463 (1999) 99–103, arXiv:hep-ph/9906349.

[36] A. Barr, C. Lester, P. Stephens, m(T2): the truth behind the glamour, J. Phys. G 29 (2003) 2343–2363, arXiv:hep-ph/0304226.

[37] A.J. Barr, B. Gripaios, C.G. Lester, Weighing wimps with kinks at colliders: invisible particle mass measurements from endpoints, J. High Energy Phys. 02 (2008) 014, arXiv:0711.4008.

[38] G. Cowan, K. Cranmer, E. Gross, O. Vitells, Asymptotic formulae for likelihood-based tests of new physics, Eur. Phys. J. C 71 (2011) 1554, arXiv:1007.1727, Erratum: Eur. Phys. J. C 73 (2013) 2501.

23

Single vector-like T-quark search via the $T \rightarrow Wb$ decay channel at the LHeC

Lin Han [a], Yan-Ju Zhang [a], Yao-Bei Liu [b,*]

[a] School of Biomedical Engineering, Xinxiang Medical University, Xinxiang 453003, China
[b] Henan Institute of Science and Technology, Xinxiang 453003, PR China

ARTICLE INFO

Editor: A. Ringwald

ABSTRACT

In this letter, we analyze the prospects of observing the new vector-like T-quark in the leptonic $T \rightarrow bW$ decay channel at the Large Hadron Electron Collider (LHeC). The results show that the mixing between the T-quark with the first generation quarks can largely enhance the production cross section. We further study the observability of the single T through the process $e^+p \rightarrow T(\rightarrow bW^+)\bar{\nu}_e \rightarrow b\ell^+ + \not{E}_T^{miss}$ at the LHeC with the proposed 140 GeV electron beam (with 80% polarization) and 7 TeV proton beam. To be as model-independent as possible, a simplified model method with only two free parameters (g^* and R_L) has been applied. For three typical T-quark masses, the 3σ exclusion limits as well as the 5σ discovery region are respectively presented in terms of parameter space regions.

1. Introduction

To stabilize the discovered 125 GeV Higgs boson mass [1] and protect it from dangerous quadratic divergences [2], new heavy fermions are predicted in many physics models beyond the Standard Model (SM), such as little Higgs models [3], extra dimensions [4], twin Higgs models [5] and composite Higgs models [6]. In many cases, these new fermions are heavy vector-like top partners T, whose common feature is to decay into a SM quark and a gauge boson (W/Z), or a Higgs boson. Quite a few phenomenology studies for these new heavy fermions have been discussed in the literature, see for example [7–19]. Here we focus on the case of a $SU(2)_L$ singlet vector-like T-quark.

Up to now, previous studies of ATLAS and CMS Collaborations have established lower bounds on the T-quark mass in the range of 550–950 GeV [20–22], depending on the assumed branching ratios. On the other hand, the indirect searches for the heavy T-quarks through their contributions to the electroweak precision observables, such as the oblique parameters S and T [23], Z-pole observables [24] and various Higgs decay channels [25] have also been extensively investigated. As a simplifying assumption supported by theoretical expectations, most of these studies are based on the assumption that the heavy T-quarks only mix to the third quark generation. However, the new T-quarks can mix in a sizable way with the light generations by considering the constraints from flavor physics (see for example [26–31]). Due to the presence of valence quarks in the initial state, the crucial point is that even a small mixing to the first generation may have a severe impact on single T-quark production processes [32,33]. For example, the authors in Ref. [34] studied the discovery potential for $T \rightarrow tZ$ in the trilepton channel at the 13 TeV LHC for a singlet T-quark mixing with the first generation. Very recently, we have studied the observability of the heavy T-quark through the process $pp \rightarrow T(\rightarrow th)j \rightarrow t(\rightarrow b\ell\nu_\ell)h(\rightarrow \gamma\gamma)j$ at the high-luminosity (HL)-LHC [35].

From the upcoming experiment perspective, the proposed Large Hadron-Electron Collider (LHeC) [36] would be the next high energy e–p collider which is designed to collide an electron beam with a typical energy range, 60–150 GeV with a 7 TeV or higher proton beam from the LHC. The LHeC might deliver data samples of approximately 100 fb^{-1} and at the end of full data accumulation with 1000 fb^{-1} (with a higher detector coverage) [37]. Furthermore, the electron beam can be polarized [37] and has an enormous scope to probe electroweak and Higgs boson physics [38,39]. Although the $T \rightarrow tZ(\rightarrow \ell^+\ell^-)$ channel is a primary option for most experimental searches [40,41], it has small number of events even for a high luminosity [42]. Due to a larger expected cross section for the $T \rightarrow Wb$ decay mode, we expect that this channel will give a better constraint on the parameters of our model than the previously considered search $T \rightarrow Zt$ [42]. Therefore in this paper we mainly study the observability of a single T-quark

* Corresponding author.
E-mail address: liuyaobei@sina.com (Y.-B. Liu).

production at the LHeC for the bW decay channel. We consider the case that the vector-like T-quark has a small mixing to the first generation in the framework of a simplified model. We expect that such decay channels at the LHeC may become complementary to other production processes in the searches for the heavy vector-like T-quark at the LHC.

The rest of the paper is organized as follows. In Sec. 2 we briefly describe the main features of the simplified model and calculate the single T production at the LHeC and its decay channels that we consider in this paper. In Sec. 3 we turn to study the prospects of observing the single T production by performing a detailed analysis of the signal and backgrounds. Finally, we conclude in Sec. 4.

2. The simplified model and single T production at the LHeC

It is clear that the vector-like T-quarks with charge 2/3 share similar final state topologies, with different branching ratios and single production couplings depending on the particular underlying model. Thus, following [28], we here consider a simplified model where the heavy vector-like T quark is an $SU(2)$ singlet with charge 2/3, with couplings to both the first and the third generation of SM quarks. The top-partner sector of the model is described by the general effective Lagrangian (showing only the couplings relevant for our analysis) [28]

$$\mathcal{L}_T = \frac{gg^*}{\sqrt{2}}\left\{\sqrt{\frac{R_L}{1+R_L}}\,[\bar{T}_L W_\mu^+ \gamma^\mu d_L] + \sqrt{\frac{1}{1+R_L}}[\bar{T}_L W_\mu^+ \gamma^\mu b_L] + \frac{1}{\sqrt{2}\cos\theta_W}\sqrt{\frac{R_L}{1+R_L}}[\bar{T}_L Z_\mu^+ \gamma^\mu u_L] + \frac{1}{\sqrt{2}\cos\theta_W}\sqrt{\frac{1}{1+R_L}}[\bar{T}_L Z_\mu^+ \gamma^\mu t_L]\right\} + h.c., \quad (1)$$

where g is the $SU(2)_L$ gauge coupling constant and θ_W is the Weinberg angle, $v \simeq 246$ GeV, and the subscripts L and R label the chiralities of the fermions. There are two free parameters: g^* and R_L, which respectively denote the coupling strength and generation mixing coupling. The limits on these parameters come from the flavor physics and the oblique parameters, which have been studied in Refs. [26–28]. A full study of the precision bounds of this particular model is beyond the scope of this paper. Here we take a conservative range for these parameter [16,43]:

$$0.1 \le g^* \le 0.5, \quad 0 \le R_L \le 2. \quad (2)$$

The tree level Feynman diagrams for the process $e^+p \to \bar{\nu}_e T$ are plotted in Fig. 1, where the T-quark is produced due to the interaction with the b quark or due to the interaction with light quarks. From the Lagrangian in eq. (1), we know that the production cross sections are very sensitive to the strengths of the WTq couplings.

The model file [44] of the singlet vector-like T-quark is implemented via the `FeynRules` package [45]. The leading order cross sections are calculated using `MadGraph5-aMC@NLO` [46] and checked by CalcHEP [47] with CTEQ6L parton distribution function (PDF) [48] and the renormalization and factorization scales are set dynamically by default. The collider parameter is taken to be $E_e = 140$ GeV and $E_p = 7$ TeV, corresponding to a c.m. energy of approximately $\sqrt{s} = 1.98$ TeV. Here we take the positron (electron) beam being 0.8 (−0.8) polarized. The SM input parameters relevant in our study are taken from [49]. Considering the current bounds on the singlet T-quark masses, we take three typical values: 800, 900, and 1000 GeV.

In Fig. 2, we show the single production cross sections of the T (left) and $\bar{T}$ (right) depending on the mixing parameter R_L at the LHeC for $g^* = 0.1$ and several typical values of m_T. One can see that: (i) the production cross sections are very sensitive to R_L. This is because the mixing with the first generation can largely enhance the single production, especially due to the presence of valence quarks in the initial state. (ii) For the same nonzero value of R_L, the production cross section of $e^+p \to T\bar{\nu}_e$ is larger than that for the conjugate process $e^-p \to \bar{T}\nu_e$ due to the difference between the d-quark and $\bar{d}$-quark PDF of the proton.

In Fig. 3, we show the dependence of the cross sections $\sigma * Br(T \to XY)$ on the mixing parameter R_L at the LHeC. One can see that for the small range of R_L (i.e. $R_L \le 0.2$), the production cross section increases largely with the increase of R_L. While for the relatively large value of R_L (i.e. $R_L \ge 1$), the production cross section will become slightly smaller for increasing R_L. This is mainly because of the increased admixture of valence quarks in production, mitigated by a reduced $T \to Wb$ branching ratio with R_L increas-

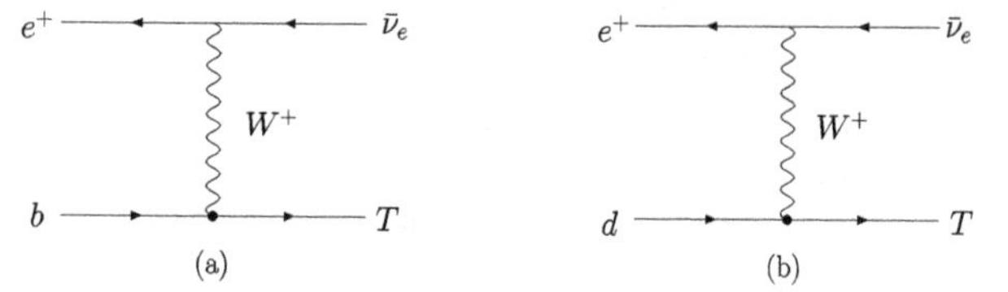

Fig. 1. The Feynman diagrams for single production of T at the LHeC for the process $e^+p \to \bar{\nu}_e T$.

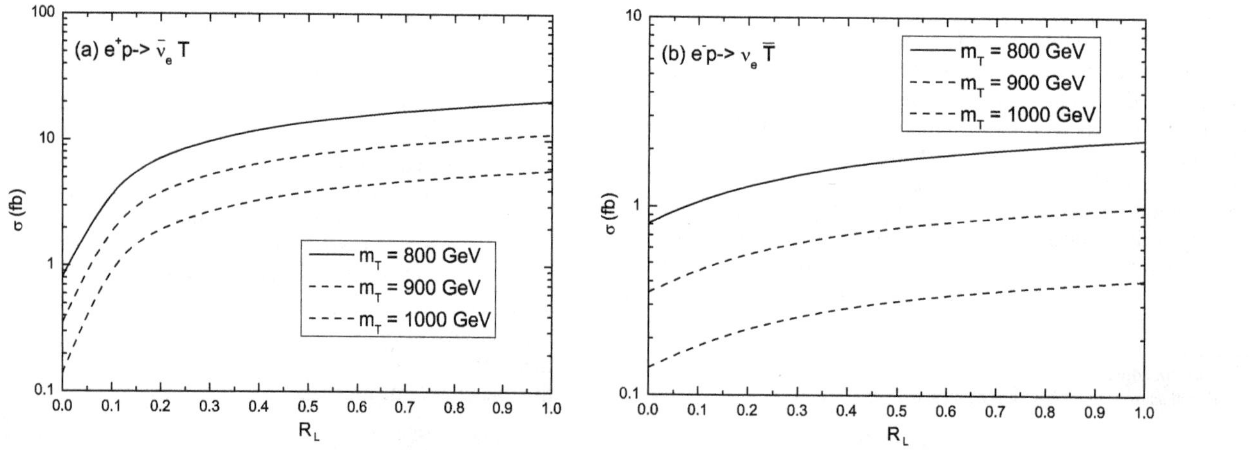

Fig. 2. The dependence of the cross sections σ on the mixing parameter R_L with $g^* = 0.1$ and three typical T quark masses for the processes (a) $e^+p \to T\bar{\nu}_e$ and (b) $e^-p \to \bar{T}\nu_e$, respectively.

ing. For instance, the value of $Br(T \to bW)$ is about 0.5 for $RL = 0$, and changed as about 0.35 for $R_L = 0.5$.

3. Event generation and analysis

In this section, we analyze the observation potential by performing a Monte Carlo simulation of the signal and background events and explore the sensitivity of single top partner at the LHeC through the channel

$$e^+p \to T(\to bW^+)\bar{\nu}_e \to bW^+(\to \ell^+\bar{\nu}_\ell)\bar{\nu}_e. \tag{3}$$

The Feynman diagram of production and decay chain is presented in Fig. 4.

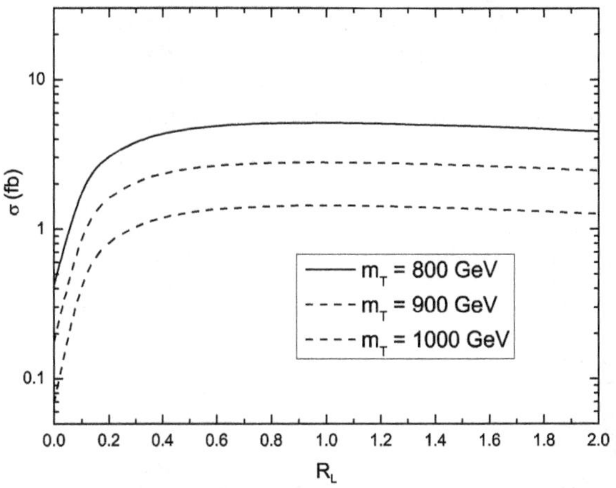

Fig. 3. The dependence of the cross sections σ on the mixing parameter R_L with $g^* = 0.1$ and three typical T quark masses for the process $e^+p \to T\bar{\nu}_e \to W^+b\bar{\nu}_e$.

For the fixed T-quark mass, the corresponding free parameters are the free parameters g^* and R_L. We take three typical values of the T quark mass: $m_T = 800, 900, 1000$ GeV with $g^* = 0.1$ and $R_L = 0.5$. We generate all event samples in this analysis at the leading order using MadGraph5-aMC@NLO with CTEQ6L PDF and PYTHIA [50] is used for parton showering and hadronization. For all the considered signals and backgrounds, the K-factors are taken to be 1 [51]. We apply jet and lepton energy smearing according to the following energy resolution formula [36]

$$\frac{\Delta E}{E} = \frac{a}{\sqrt{E}} \oplus b, \tag{4}$$

where $a = 0.45$ GeV$^{1/2}$, $b = 0.03$ for jets and $a = 0.085$ GeV$^{1/2}$, $b = 0.003$ for leptons and the symbol $\oplus$ represents a quadrature sum. Event analysis is performed by using the program of MadAnalysis5 [52].

For the bW channel, the typical signal is exactly one charged lepton, one b jet and missing energy. The main SM background are the processes containing a W boson in the final state, such as

$$e^+p \to t(\to bW^+)\bar{\nu}_e \to \ell^+ + b + \not{E}_T^{miss}, \quad (\bar{\nu}t)$$

$$e^+p \to W^+(\to \ell^+\bar{\nu}_\ell)j\bar{\nu}_e \to \ell^+ + j + \not{E}_T^{miss}, \quad (\bar{\nu}W^+j)$$

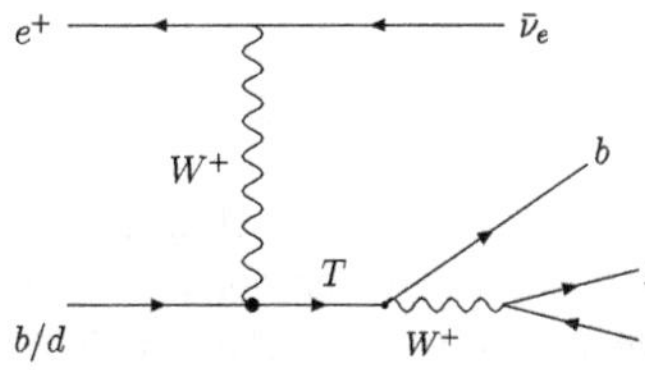

Fig. 4. The Feynman diagram for production of single T quark including the decay chain $T \to bW \to b\ell\nu$.

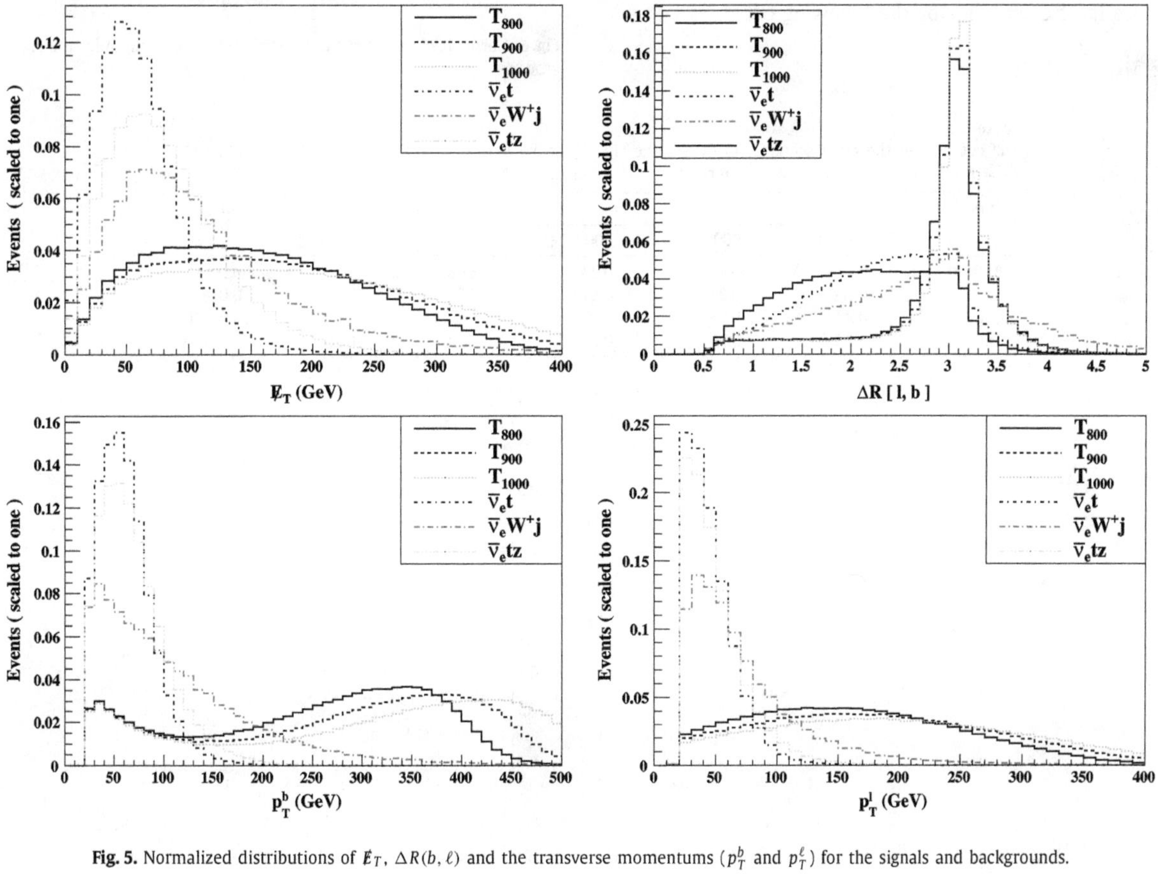

Fig. 5. Normalized distributions of $\not{E}_T$, $\Delta R(b, \ell)$ and the transverse momentums (p_T^b and p_T^ℓ) for the signals and backgrounds.

$$e^+p \to t(\to bW^+ \to b\ell^+\bar{\nu})Z(\to \nu\bar{\nu})\bar{\nu}_e$$
$$\to \ell^+ + b + \not{E}_T^{miss}, \quad (\bar{\nu}tZ) \tag{5}$$

where one light jet might be faked as b jet. We also checked that other background processes, such as the di-boson production are negligible with the selection cuts.

In our simulation, we first impose the following basic cuts to reduce the backgrounds:

- There is exactly one isolated lepton with $p_T^\ell > 20$ GeV and $|\eta_\ell| < 3$.
- There are exactly one b-tagged jet with $p_T^b > 20$ GeV and $|\eta_b| < 5$.
- The missing transverse momentum $\not{E}_T^{miss}$ is required to be larger than 20 GeV.

In Fig. 5, we show the normalized distributions of the missing transverse momentum $\not{E}_T$, the transverse momentums $p_T^{\ell,b}$ and the variable $\Delta R(b,\ell)$ for the signals and backgrounds. Based on these kinematical distributions, we impose the following cuts to get a high significance:

- Cut 1: $\not{E}_T > 130$ GeV.
- Cut 2: $p_T^\ell > 100$ GeV, $p_T^b > 200$ GeV and $2.8 < \Delta R(b,\ell) < 3.5$.

A very important selection cut, which is needed to suppress the background from the top quark production, is that the invariant mass of the lepton and b-jet pairs, $M_{b\ell}$, is above 170 GeV. This can forbid that the lepton and the b originate from the decay of a top quark. The invariant mass of the b-tagged jet and the lepton is plotted in Fig. 6 for the signals and the backgrounds. Thus we can further reduce the backgrounds by the following cut:

- Cut 3: $M_{b\ell} > 350$ GeV.

We present the cross sections of the signal and backgrounds after imposing the cuts in Table 1. From Table 1, one can see that all the backgrounds are suppressed very efficiently after imposing the selections. To estimate the observability quantitatively, we adopt the significance measurement [53]

$$SS = \sqrt{2\mathcal{L}_{int}[(\sigma_S + \sigma_B)\ln(1 + \sigma_S/\sigma_B) - \sigma_S]}, \tag{6}$$

where σ_S and σ_B are the signal and background cross sections and $\mathcal{L}_{int}$ is the integrated luminosity. Here we define the discovery significance as $SS = 5$ and exclusion limits as $SS = 3$.

In Figs. 7–8, we plot the excluded 3σ and 5σ discovery reaches in the plane of $g^* - R_L$ for three fixed typical T masses at the LHeC with 100 and 1000 fb^{-1} of integrated luminosity. As displayed in Fig. 3, the cross section for the final state reaches a maximum for $R_L \simeq 0.5$ due to the mixing effects. One can see that, for $m_T = 800, 900, 1000$ GeV and $R_L = 0.5$, the 5σ level discovery sensitivity of g^* is respectively about $0.1, 0.12, 0.16$ for $\mathcal{L}_{int} =$ 100 fb^{-1} and $R_L = 0.5$, and changed as about $0.05, 0.06, 0.08$ for

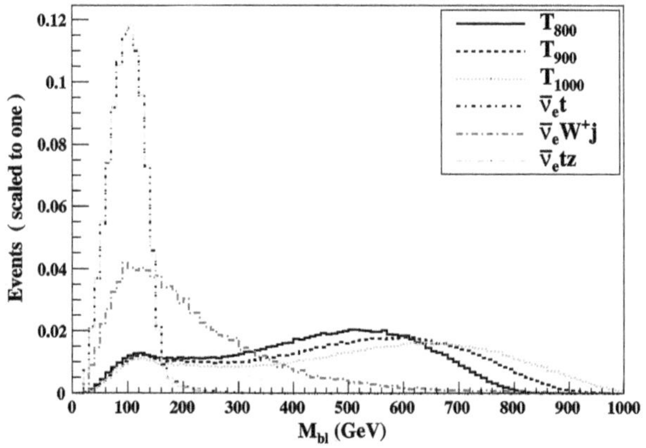

Fig. 6. Normalized invariant mass distribution of $b\ell$ system for the signals and backgrounds.

Table 1
The cut flow of the cross sections (in fb) for the signal and backgrounds at the LHeC with $E_e = 140$ GeV and $E_p = 7$ TeV. Here we take the parameters as $g^* = 0.1$ and $R_L = 0.5$.

Cuts	Signal			$\bar{\nu}_e t$	$\bar{\nu}_e W^+ j$	$\bar{\nu}_e tZ$
	800 GeV	900 GeV	1000 GeV			
Basic cuts	0.81	0.4	0.23	768	34	0.19
Cut 1	0.46	0.27	0.14	15.4	7.8	0.018
Cut 2	0.24	0.16	0.09	0.0036	0.3	4.3×10^{-6}
Cut 3	0.23	0.16	0.09	0.001	0.18	1.2×10^{-6}

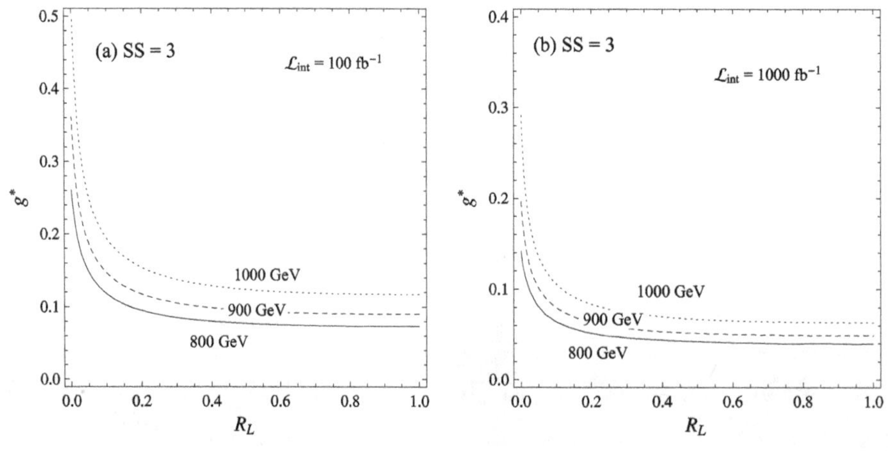

Fig. 7. 3σ contour plots for the signal in $g^* - R_L$ at the LHeC with 100 (left) and 1000 (right) fb^{-1} of integrated luminosity.

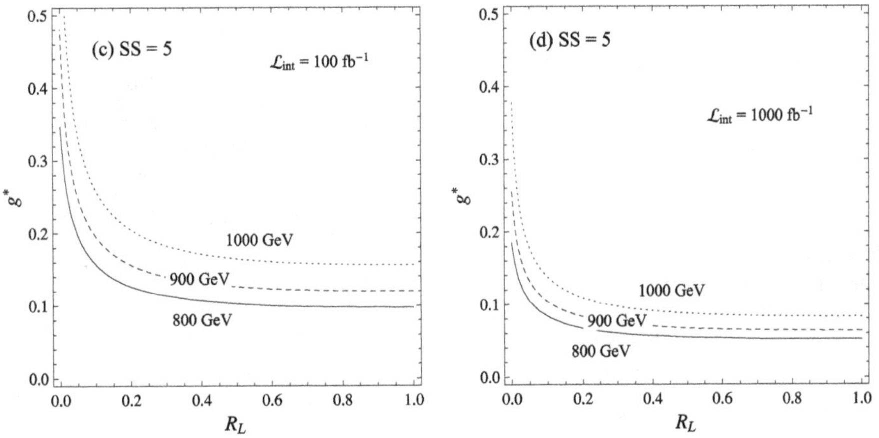

Fig. 8. 5σ contour plots for the signal in $g^* - R_L$ at the LHeC with 100 (left) and 1000 (right) fb^{-1} of integrated luminosity.

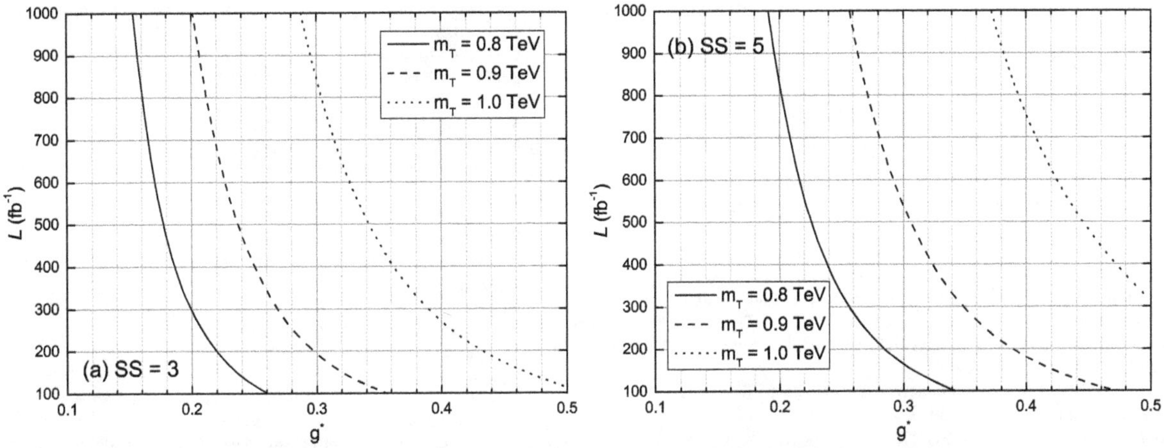

Fig. 9. 3σ (left) and 5σ (right) contour plots for the signal in $L - g^*$ at the LHeC for $R_L = 0$.

$\mathcal{L}_{int} = 1000$ fb^{-1}. If no signal is observed, it means that the coupling strength parameter g^* can not be too large. From the 3σ exclusion limits one can see that, for $m_T = 800, 900, 1000$ GeV and $\mathcal{L}_{int} = 1000$ fb^{-1}, the upper limits on the size of g^* are given as $g^* \leq 0.14, 0.20, 0.3$ for $R_L = 0$, and changed as $g^* \leq 0.04, 0.05, 0.06$ for the non-vanishing R_L.

Certainly, in some typical models such as the minimal $M1_5$ Composite Higgs (CH) models [15] and the littlest Higgs model with T-parity (LHT) [54], the heavy T-quark couplings only to the third generation of SM quarks. Due to the Goldstone-boson equivalence theorem, the branching fractions of T into bW, tZ and th are a good approximation given by ratios $2:1:1$ in the limit $m_T \gg m_t$. For comparison, we show in Fig. 9 the excluded 3σ and 5σ discovery reaches in the plane of the integrated luminosity and the coupling parameter g^* for $R_L = 0$. For $m_T = 800$ (900) GeV, we can see that the 5σ C.L. discovery sensitivity of g^* is 0.34 (0.48), 0.26 (0.35) and 0.19 (0.26) when the integrated luminosity is 100, 300 and 1000 fb^{-1}, respectively. Otherwise, the upper limits on the size of g^* are given as about 0.26 (0.36), 0.2 (0.27) and 0.15 (0.2) when the integrated luminosity is 100, 300 and 1000 fb^{-1}, respectively. Our results can be straightforwardly mapped within the context of the CH model and the LHT model, namely with [15,54]

$$g^* \simeq \frac{y m_W}{g m_T}, \quad \text{(CH)}$$

$$g^* \simeq \frac{R^2}{1+R^2}\frac{v}{f} + \mathcal{O}(\frac{v^2}{f^2}), \quad \text{(LHT)} \qquad (7)$$

where y, R and f are the model parameters (for more detail, see e.g. [15,54]).

We can now draw a comparison with other existing studies for searches via the leptonic $T \to Wb$ channel at the LHC. For example, the authors of Ref. [18] project at $\sqrt{s} = 13$ TeV and 100 fb^{-1} of integrated luminosity the exclusion potential with $Br(T \to Wb) = 0.5$, obtaining an exclusion reach up to 2.0 TeV for single production if $c_L^{Wb} \geq 0.4$. Analogously, in [19] the authors obtain an expected exclusion reach for masses up to 1.0 TeV for the leptonic $T \to Wb$ channel, including both pair and single production, with $\sqrt{s} = 14$ TeV and 30 fb^{-1} of integrated luminosity. Thus, our analysis can represent a viable and complementary candidate to pursue the search of a possible heavy T-quark.

4. Conclusion

In this letter we described the future LHeC potential to search for the heavy vector-like T-quark via the $T \to bW^+$ decay mode. We investigated the observability of the heavy vector-like top partner T production through the process $e^+p \to T(\to bW^+)\bar{\nu}_e \to bW^+(\to \ell^+\bar{\nu}_\ell)\bar{\nu}_e$ at the LHeC with $E_e = 140$ GeV (with 0.8 polarization) and $E_p = 7$ TeV. From our numerical calculations and the phenomenological analysis we found the following points:

1. Due to the presence of valence quarks in the initial state, the mixing between the T-quark with the first generation can largely enhance the production cross section. Furthermore, the production cross section of $e^+p \to T\bar{\nu}_e$ is more larger than

that for the conjugate process $e^- p \to \bar{T}\nu_e$ due to the difference between the d-quark and $\bar{d}$-quark PDF of the proton.

2. For the nonzero value of R_L, the future LHeC may observe the above signals at the 5σ confidence level. Otherwise, for $m_T = 800, 900, 1000$ GeV, the upper limits on the size of g^* are given as $g^* \le 0.09$ (0.04), 0.11 (0.05), 0.12 (0.06) with $\mathcal{L}_{int} = 100$ (1000) fb^{-1}.
3. In some typical models where the new vector-like T-quark only couplings to the third generation of SM quarks, for $m_T =$ 800 (900) GeV, the upper limits on the size of g^* are given as about 0.26 (0.36), 0.2 (0.27) and 0.15 (0.2) when the integrated luminosity is 100, 300 and 1000 fb^{-1}, respectively.

Acknowledgements

This work is supported by the Joint Funds of the National Natural Science Foundation of China (Grant No. U1304112), the Foundation of Henan Educational Committee (Grant No. 2015GGJS-059) and the Foundation of Henan Institute of Science and Technology (Grant No. 2016ZD01).

References

[1] G. Aad, et al., ATLAS collaboration, Phys. Lett. B 716 (2012) 1;
S. Chatrchyan, et al., CMS collaboration, Phys. Lett. B 716 (2012) 30.
[2] P.H. Frampton, P.Q. Hung, M. Sher, Phys. Rep. 330 (2000) 263;
A. De Simone, O. Matsedonskyi, R. Rattazzi, A. Wulzer, J. High Energy Phys. 1304 (2013) 004.
[3] N. Arkani-Hamed, A.G. Cohen, E. Katz, A.E. Nelson, J. High Energy Phys. 07 (2002) 034;
N. Arkani-Hamed, A.G. Cohen, E. Katz, A.E. Nelson, T. Gregoire, J.G. Wacker, J. High Energy Phys. 08 (2002) 021;
M. Schmaltz, D. Tucker-Smith, Annu. Rev. Nucl. Part. Sci. 55 (2005) 229.
[4] I. Antoniadis, Phys. Lett. B 246 (1990) 377;
D.B. Kaplan, Nucl. Phys. B 365 (1991) 259;
K. Agashe, G. Perez, A. Soni, Phys. Rev. D 75 (2007) 015002.
[5] Z. Chacko, H.-S. Goh, R. Harnik, Phys. Rev. Lett. 96 (2006) 231802;
Z. Chacko, Y. Nomura, M. Papucci, G. Perez, J. High Energy Phys. 01 (2006) 126.
[6] K. Agashe, R. Contino, A. Pomarol, Nucl. Phys. B 719 (2005) 165;
M. Low, A. Tesi, L.-T. Wang, Phys. Rev. D 91 (2015) 095012.
[7] L. Lavoura, J.P. Silva, Phys. Rev. D 47 (1993) 2046;
Csaba Balazs, Hong-Jian He, C.-P. Yuan, Phys. Rev. D 60 (1999) 114001;
H.-J. He, N. Polonsky, S. Su, Phys. Rev. D 64 (2001) 053004;
J.A. Aguilar-Saavedra, Phys. Lett. B 625 (2005) 234;
G. Cynolter, E. Lendvai, Eur. Phys. J. C 58 (2008) 463;
J.A. Aguilar-Saavedra, J. High Energy Phys. 0911 (2009) 030;
O. Cakir, A. Senol, A.T. Tasci, Europhys. Lett. 88 (2009) 11002;
A. Senol, A.T. Tasci, F. Ustabas, Nucl. Phys. B 851 (2011) 289.
[8] P. Meade, M. Reece, Phys. Rev. D 74 (2006) 015010;
R. Contino, G. Servant, J. High Energy Phys. 0806 (2008) 026;
M.M. Nojiri, M. Takeuchi, J. High Energy Phys. 0810 (2008) 025;
J. Alwall, J.L. Feng, J. Kumar, S. Su, Phys. Rev. D 81 (2010) 114027;
G. Cacciapaglia, A. Deandrea, D. Harada, Y. Okada, J. High Energy Phys. 1011 (2010) 159;
J. Berger, J. Hubisz, M. Perelstein, J. High Energy Phys. 1207 (2012) 016;
Y. Okada, L. Panizzi, Adv. High Energy Phys. 2013 (2013) 364936;
X.-F. Wang, C. Du, H.-J. He, Phys. Lett. B 723 (2013) 314;
H.-J. He, Z.-Z. Xianyu, J. Cosmol. Astropart. Phys. 1410 (2014) 019;
S.-F. Ge, H.-J. He, J. Ren, Z.-Z. Xianyu, Phys. Lett. B 757 (2016) 480.
[9] H.-C. Cheng, I. Low, L.-T. Wang, Phys. Rev. D 74 (2006) 055001;
S. Matsumoto, M.M. Nojiri, D. Nomura, Phys. Rev. D 75 (2007) 055006;
Y.-B. Liu, X.-L. Wang, Y.-H. Cao, Chin. Phys. Lett. 24 (2007) 57;
C.-X. Yue, L.-H. Wang, J. Wang, Chin. Phys. Lett. 25 (2008) 1613;
Q.-H. Cao, C.S. Li, C.-P. Yuan, Phys. Lett. B 668 (2008) 24;
C.-X. Yue, H.-D. Yang, W. Ma, Nucl. Phys. B 818 (2009) 1;
C.-X. Yue, X.-S. Su, W. Ma, T.-T. Zhang, Chin. Phys. Lett. 27 (2010) 101203;
C.-Y. Chen, A. Freitas, T. Han, K.S.M. Lee, J. High Energy Phys. 1211 (2012) 124;
J. Kearney, A. Pierce, J. Thaler, J. High Energy Phys. 1310 (2013) 230;
G. Burdman, Z. Chacko, R. Harnik, L. de Lima, C.B. Verhaaren, Phys. Rev. D 91 (2015) 055007;
A. Anandakrishnan, J.H. Collins, M. Farina, E. Kuflik, M. Perelstein, Phys. Rev. D 93 (2016) 075009;
N. Liu, L. Wu, B.-F. Yang, M.-C. Zhang, Phys. Lett. B 753 (2016) 664.
[10] G. Dissertori, E. Furlan, F. Moortgat, P. Nef, J. High Energy Phys. 1009 (2010) 019;
N. Chen, H.-J. He, J. High Energy Phys. 1204 (2012) 062;
O. Matsedonskyi, G. Panico, A. Wulzer, J. High Energy Phys. 1301 (2013) 164;
T. Flacke, J.H. Kim, S.J. Lee, S.H. Lim, J. High Energy Phys. 1405 (2014) 123;
J. Serra, J. High Energy Phys. 1509 (2015) 176;
T. DeGrand, Y. Shamir, Phys. Rev. D 92 (2015) 075039;
O. Matsedonskyi, G. Panico, A. Wulzer, J. High Energy Phys. 1604 (2016) 003;
S. Fichet, G. von Gersdorff, E. Ponton, R. Rosenfeld, J. High Energy Phys. 1701 (2017) 012.
[11] H.-S. Goh, C.A. Krenke, Phys. Rev. D 81 (2010) 055008;
Y.-B. Liu, X.-L. Wang, Int. J. Mod. Phys. A 25 (2010) 5885;
Y.-B. Liu, Z.-J. Xiao, Nucl. Phys. B 892 (2015) 63;
H.-C. Cheng, S. Jung, E. Salvioni, Y. Tsai, J. High Energy Phys. 1603 (2016) 074.
[12] C.-H. Chen, T. Nomura, Phys. Rev. D 94 (2016) 035001;
S. Moretti, D. O'Brien, L. Panizzi, H. Prager, arXiv:1603.09237 [hep-ph];
M. Endo, Y. Takaesu, Phys. Lett. B 758 (2016) 355;
E.L. Berger, Q.-H. Cao, Phys. Rev. D 81 (2010) 035006;
B. Holdom, Q.-S. Yan, Phys. Rev. D 83 (2011) 114031;
B. Holdom, Q.-S. Yan, Phys. Rev. D 84 (2011) 094012;
S. Yang, J. Jiang, Q.-S. Yan, X. Zhao, J. High Energy Phys. 1409 (2014) 035;
S. Gopalakrishna, T. Mandal, S. Mitra, G. Moreau, J. High Energy Phys. 1408 (2014) 079;
C. Han, A. Kobakhidze, N. Liu, L. Wu, B. Yang, Nucl. Phys. B 890 (2014) 388;
S.A.R. Ellis, R.M. Godbole, S. Gopalakrishna, J.D. Wells, J. High Energy Phys. 1409 (2014) 130;
M. Endo, K. Hamaguchi, K. Ishikawa, M. Stoll, Phys. Rev. D 90 (2014) 055027.
[13] N. Arkani-Hamed, T. Han, M. Mangano, L.-T. Wang, Phys. Rep. 652 (2016) 1;
S. Banerjee, D. Barducci, G. Bélanger, C. Delaunay, J. High Energy Phys. 1611 (2016) 154;
G. Brooijmans, et al., arXiv:1405.1617 [hep-ph];
G. Brooijmans, et al., arXiv:1605.02684 [hep-ph];
A. Azatov, M. Salvarezza, M. Son, M. Spannowsky, Phys. Rev. D 89 (2014) 075001;
M. Backović, T. Flacke, S.J. Lee, G. Perez, J. High Energy Phys. 1509 (2015) 022;
N. Vignaroli, J. High Energy Phys. 1207 (2012) 158;
N. Vignaroli, Phys. Rev. D 86 (2012) 075017.
[14] N.G. Ortiz, J. Ferrando, D. Kar, M. Spannowsky, Phys. Rev. D 90 (2014) 075009;
S. Beauceron, G. Cacciapaglia, A. Deandrea, J.D. Ruiz-Alvarez, Phys. Rev. D 90 (2014) 115008;
J. Li, D. Liu, J. Shu, J. High Energy Phys. 1311 (2013) 047;
M. Backovic, T. Flacke, J.H. Kim, S.J. Lee, J. High Energy Phys. 1604 (2016) 014.
[15] A. De Simone, O. Matsedonskyi, R. Rattazzi, A. Wulzer, J. High Energy Phys. 1304 (2013) 004.
[16] J. Reuter, M. Tonini, J. High Energy Phys. 1501 (2015) 088.
[17] C. Grojean, O. Matsedonskyi, G. Panico, J. High Energy Phys. 1310 (2013) 160;
D. Barducci, A. Belyaev, J. Blamey, S. Moretti, L. Panizzi, H. Prager, J. High Energy Phys. 1407 (2014) 142;
D. Barducci, A. Belyaev, M. Buchkremer, G. Cacciapaglia, A. Deandrea, S. De Curtis, J. Marrouche, S. Moretti, L. Panizzi, J. High Energy Phys. 1412 (2014) 080;
D. Pappadopulo, A. Thamm, R. Torre, A. Wulzer, J. High Energy Phys. 1409 (2014) 060.
[18] O. Matsedonskyi, G. Panico, A. Wulzer, J. High Energy Phys. 1412 (2014) 097.
[19] B. Gripaios, T. Mueller, M.A. Parker, D. Sutherland, J. High Energy Phys. 1408 (2014) 171.
[20] ATLAS collaboration, ATLAS-CONF-2013-018;
ATLAS collaboration, ATLAS-CONF-2014-036;
ATLAS collaboration, ATLAS-CONF-2013-060;
ATLAS collaboration, J. High Energy Phys. 1510 (2015) 150;
ATLAS collaboration, J. High Energy Phys. 1508 (2015) 105;
ATLAS collaboration, J. High Energy Phys. 1602 (2016) 110;
ATLAS collaboration, Eur. Phys. J. C 76 (2016) 442;
ATLAS collaboration, Phys. Lett. B 758 (2016) 249;
ATLAS collaboration, ATLAS-CONF-2016-072.
[21] CMS collaboration, CMS-PAS-JME-13-007;
CMS collaboration, Phys. Lett. B 729 (2014) 149;
CMS collaboration, Phys. Rev. D 93 (2016) 012003;
CMS collaboration, CMS-PAS-B2G-16-002;
CMS collaboration, CMS-PAS-B2G-15-008.
[22] ATLAS collaboration, ATLAS-CONF-2016-013;
ATLAS collaboration, ATLAS-CONF-2016-101.
[23] M.E. Peskin, T. Takeuchi, Phys. Rev. D 46 (1992) 381;
G. Altarelli, R. Barbieri, Phys. Lett. B 253 (1991) 161;
J. Hubisz, P. Meade, Phys. Rev. D 71 (2005) 035016;
H.-J. He, T.M.P. Tait, C.-P. Yuan, Phys. Rev. D 62 (2000) 011702;
H.-J. He, C.T. Hill, T.M.P. Tait, Phys. Rev. D 65 (2002) 055006;
J. Hubisz, P. Meade, A. Noble, M. Perelstein, J. High Energy Phys. 0601 (2006) 135.
[24] A. Djouadi, J.H. Kuhn, P.M. Zerwas, Z. Phys. C 46 (1990) 411;

F. Boudjema, A. Djouadi, C. Verzegnassi, Phys. Lett. B 238 (1990) 423;
R.S. Chivukula, B. Coleppa, S.D. Chiara, E.H. Simmons, H.-J. He, M. Kurachi, M. Tanabashi, Phys. Rev. D 74 (2006) 075011;
E.L. Berger, Q.-H. Cao, I. Low, Phys. Rev. D 80 (2009) 074020;
J.A. Aguilar-Saavedra, R. Benbrik, S. Heinemeyer, M. Perez-Victoria, Phys. Rev. D 88 (2013) 094010.

[25] A. Atre, M. Chala, J. Santiago, J. High Energy Phys. 1305 (2013) 099;
A. Djouadi, Eur. Phys. J. C 73 (2013) 2498;
C.-Y. Chen, S. Dawson, I.M. Lewis, Phys. Rev. D 90 (2014) 035016;
A. Djouadi, J. Quevillon, R. Vega-Morales, Phys. Lett. B 757 (2016) 412;
X.-F. Wang, C. Du, H.-J. He, Phys. Lett. B 723 (2013) 314;
T. Abe, M. Chen, H.-J. He, J. High Energy Phys. 1301 (2013) 082;
J. Baglio, A. Djouadi, J. High Energy Phys. 1103 (2011) 055;
S. Fichet, G. Moreau, Nucl. Phys. B 905 (2016) 391;
A. Angelescu, A. Djouadi, G. Moreau, Eur. Phys. J. C 76 (2016) 99.

[26] G. Cacciapaglia, A. Deandrea, D. Harada, Y. Okada, J. High Energy Phys. 11 (2010) 159.

[27] F. del Aguila, M. Perez-Victoria, J. Santiago, J. High Energy Phys. 09 (2000) 011;
G. Cacciapaglia, A. Deandrea, L. Panizzi, N. Gaur, D. Harada, Y. Okada, J. High Energy Phys. 03 (2012) 070;
F.J. Botella, G.C. Branco, M. Nebot, J. High Energy Phys. 12 (2012) 040.

[28] M. Buchkremer, G. Cacciapaglia, A. Deandrea, L. Panizzi, Nucl. Phys. B 876 (2013) 376.

[29] K. Ishiwata, Z. Ligeti, M.B. Wise, J. High Energy Phys. 10 (2015) 027;
A.K. Alok, S. Banerjee, D. Kumar, S.U. Sankar, D. London, Phys. Rev. D 92 (2015) 013002;
F.J. Botella, G.C. Branco, M. Nebot, M.N. Rebelo, J.I. Silva-Marcos, arXiv: 1610.03018.

[30] S.A.R. Ellis, R.M. Godbole, S. Gopalakrishna, J.D. Wells, J. High Energy Phys. 09 (2014) 130.

[31] G. Cacciapaglia, A. Deandrea, N. Gaur, D. Harada, Y. Okada, L. Panizzi, J. High Energy Phys. 09 (2015) 012.

[32] A. Atre, G. Azuelos, M. Carena, T. Han, E. Ozcan, J. Santiago, G. Unel, J. High Energy Phys. 1108 (2011) 080.

[33] S. Beauceron, G. Cacciapaglia, A. Deandrea, J.D. Ruiz-Alvarez, Phys. Rev. D 90 (2014) 115008.

[34] L. Basso, J. Andrea, J. High Energy Phys. 1502 (2015) 032.

[35] Y.-B. Liu, Phys. Rev. D 95 (2017) 035013.

[36] J. Abelleira Fernandez, et al., LHeC Study Group, J. Phys. G 39 (2012) 075001.

[37] O. Bruening, M. Klein, Mod. Phys. Lett. A 28 (2013) 1330011.

[38] T. Han, B. Mellado, Phys. Rev. D 82 (2010) 016009;
I.T. Cakir, O. Cakir, A. Senol, A.T. Tasci, Acta Phys. Pol. B 45 (2014) 1947;
I.A. Sarmiento-Alvarado, A.O. Bouzas, F. Larios, J. Phys. G 42 (2015) 085001;
S.P. Das, J. Hernandez-Sanchez, S. Moretti, A. Rosado, R. Xoxocotzi, Phys. Rev. D 94 (2016) 055003;
S.P. Das, M. Nowakowski, arXiv:1612.07241 [hep-ph];
G.R. Boroun, Phys. Lett. B 741 (2015) 197;
S. Mondal, S.K. Rai, Phys. Rev. D 94 (2016) 033008.

[39] A.O. Bouzas, F. Larios, Phys. Rev. D 88 (2013) 094007;
X.-P. Li, L. Guo, W.-G. Ma, R.-Y. Zhang, L. Han, M. Song, Phys. Rev. D 88 (2013) 014023;
W. Liu, H. Sun, X. Wang, X. Luo, Phys. Rev. D 92 (2015) 074015;
Y.-L. Tang, C. Zhang, S.-H. Zhu, Phys. Rev. D 94 (2016) 011702;
H. Sun, X.-J. Wang, arXiv:1602.04670 [hep-ph];
S. Liu, Y.-L. Tang, C. Zhang, S.-H. Zhu, arXiv:1608.08458 [hep-ph].

[40] G. Aad, et al., ATLAS collaboration, J. High Energy Phys. 1411 (2014) 104.

[41] M. Backović, T. Flacke, J.H. Kim, S.J. Lee, Phys. Rev. D 92 (2015) 011701.

[42] Y.-J. Zhang, L. Han, Y.-B. Liu, Phys. Lett. B 768 (2017) 241.

[43] S. Chatrchyan, et al., CMS collaboration, J. High Energy Phys. 1212 (2012) 035.

[44] https://feynrules.irmp.ucl.ac.be/wiki/VLQ_tsingletvl.

[45] A. Alloul, N.D. Christensen, C. Degrande, C. Duhr, B. Fuks, Comput. Phys. Commun. 185 (2014) 2250.

[46] J. Alwall, R. Frederix, S. Frixione, V. Hirschi, F. Maltoni, O. Mattelaer, H.-S. Shao, T. Stelzer, P. Torrielli, M. Zaro, J. High Energy Phys. 1407 (2014) 079.

[47] A. Belyaev, N.D. Christensen, A. Pukhov, Comput. Phys. Commun. 184 (2013) 1729.

[48] J. Pumplin, A. Belyaev, J. Huston, D. Stump, W.K. Tung, J. High Energy Phys. 0602 (2006) 032.

[49] K.A. Olive, et al., Particle Data Group, Chin. Phys. C 38 (2014) 090001.

[50] T. Sjostrand, S. Mrenna, P.Z. Skands, J. High Energy Phys. 0605 (2006) 026.

[51] B. Jager, Phys. Rev. D 81 (2010) 054018.

[52] E. Conte, B. Fuks, G. Serret, Comput. Phys. Commun. 184 (2013) 222.

[53] G. Cowan, K. Cranmer, E. Gross, O. Vitells, Eur. Phys. J. C 71 (2011) 1554.

[54] M. Blanke, A.J. Buras, A. Poschenrieder, S. Recksiegel, C. Tarantino, et al., J. High Energy Phys. 0701 (2007) 066.

24

Spin-dependent $\mu \to e$ conversion

Vincenzo Cirigliano [a], Sacha Davidson [b,*], Yoshitaka Kuno [c]

[a] Theoretical Division, Los Alamos National Laboratory, Los Alamos, NM 87545, USA
[b] IPNL, CNRS/IN2P3, Université Lyon 1, Univ. Lyon, 69622 Villeurbanne, France
[c] Department of Physics, Osaka University, 1-1 Machikaneyama, Toyonaka, Osaka 560-0043, Japan

ARTICLE INFO

Editor: J. Hisano

ABSTRACT

The experimental sensitivity to $\mu \to e$ conversion on nuclei is expected to improve by four orders of magnitude in coming years. We consider the impact of $\mu \to e$ flavour-changing tensor and axial-vector four-fermion operators which couple to the spin of nucleons. Such operators, which have not previously been considered, contribute to $\mu \to e$ conversion in three ways: in nuclei with spin they mediate a spin-dependent transition; in all nuclei they contribute to the coherent (A^2-enhanced) spin-independent conversion via finite recoil effects and via loop mixing with dipole, scalar, and vector operators. We estimate the spin-dependent rate in Aluminium (the target of the upcoming COMET and Mu2e experiments), show that the loop effects give the greatest sensitivity to tensor and axial-vector operators involving first-generation quarks, and discuss the complementarity of the spin-dependent and independent contributions to $\mu \to e$ conversion.

1. Introduction

New particles and interactions beyond the Standard Model of particle physics are required to explain neutrino masses and mixing angles. The search for traces of this New Physics (NP) is pursued on many fronts. One possibility is to look directly for the new particles implicated in neutrino mass generation, for instance at the LHC [1] or SHiP [2]. A complementary approach seeks new interactions among known particles, such as neutrinoless double beta decay [3] or Charged Lepton Flavour Violation (CLFV) [4].

CLFV transitions of charged leptons are induced by the observed massive neutrinos, at unobservable rates suppressed by $(m_\nu/m_W)^4 \sim 10^{-48}$. A detectable rate would point to the existence of new heavy particles, as may arise in models that generate neutrino masses, or that address other puzzles of the Standard Model such as the hierarchy problem. Observations of CLFV are therefore crucial to identifying the NP of the lepton sector, providing information complementary to direct searches.

From a theoretical perspective, at energy scales well below the masses of the new particles, CLFV can be parametrised with effective operators (see *e.g.* [5]), constructed out of the kinematically accessible Standard Model (SM) fields, and respecting the relevant gauge symmetries. In this effective field theory (EFT) description, information about the underlying new dynamics is encoded in the operator coefficients, calculable in any given model.

The experimental sensitivity to a wide variety of CLFV processes is systematically improving. Current bounds on branching ratios of τ flavour changing decays such as $\tau \to \mu\gamma$, $\tau \to e\gamma$ and $\tau \to 3\ell$ [6–8] are $\mathcal{O}(10^{-8})$, and Belle-II is expected to improve the sensitivity by an order of magnitude [9]. The bounds on the $\mu \leftrightarrow e$ flavour changing processes are currently of order $\sim 10^{-12}$ [10,11], with the most restrictive constraint from the MEG collaboration: $BR(\mu \to e\gamma) \leq 4.2 \times 10^{-13}$ [12]. Future experimental sensitivities should improve by several orders of magnitude, in particular, the COMET [13] and Mu2e [14] experiments aim to reach a sensitivity to $\mu \to e$ conversion on nuclei of $\sim 10^{-16}$, and the PRISM/PRIME proposal [15] could reach the unprecedented level of 10^{-18}.

In searches for $\mu \to e$ conversion, a μ^- from the beam is captured by a nucleus in the target, and tumbles down to the 1*s* state. The muon will be closer to the nucleus than an electron ($r \sim \alpha Z/m$), due to its larger mass. In the presence of a CLFV interaction with the quarks that compose the nucleus, or with its electric field, the muon can transform into an electron. This electron, emitted with an energy $E_e \simeq m_\mu$, is the signature of $\mu \to e$ conversion.

Initial analytic estimates of the $\mu \to e$ conversion rate were obtained by Feinberg and Weinberg [16], a wider range of nuclei were studied numerically by Shankar [17], and relativistic effects relevant in heavier nuclei were included in Ref. [18]. State of the art conversion rates for a broad range of nuclei induced by CLFV

* Corresponding author.
E-mail address: s.davidson@ipnl.in2p3.fr (S. Davidson).

operators which can contribute coherently to $\mu \to e$ conversion were obtained in Ref. [19], while some missing operators were included in Ref. [20].

The calculation has some similarities with dark matter scattering on nuclei [21–23], where the cross-section can be classified as spin-dependent (SD) or spin-independent (SI). Previous analyses of $\mu \to e$ conversion [19,20] focused on CLFV interactions involving a scalar or vector nucleon current, because, similarly to SI dark matter scattering, these sum coherently across the nucleus at the amplitude level, giving an amplification $\sim A^2$ in the rate, where A is the atomic number. However, other processes are possible, such as spin-dependent conversion on the ground state nucleus, which we explore here, or incoherent $\mu \to e$ conversion, where the final-state nucleus is in an excited state [17,24].

The upcoming exceptional experimental sensitivities motivate our study of new contributions to $\mu \to e$ conversion induced by tensor and axial vector operators,[1] which were not considered in Refs. [19,20]. These operators couple to the spin of the nucleus and can induce "spin-dependent" $\mu \to e$ conversion in nuclei with spin (such as Aluminium, the proposed target of COMET and Mu2e), not enhanced by A^2. In addition, the tensor and axial operators will contribute to "spin-independent" conversion via finite-momentum-transfer corrections [25,26], and Renormalisation Group mixing [27,28].[2] In an EFT framework, our analysis shows new sensitivities to previously unconstrained combinations of dimension-six operator coefficients, as we illustrate below. In the absence of CLFV, this gives new constraints on the coefficients, and when CLFV is observed, it could assist in determining its origin.

2. Estimating the $\mu \to e$ conversion rate

Our starting point is the effective Lagrangian [4]

$$\delta\mathcal{L} = -2\sqrt{2}G_F \sum_Y \Big(C_{D,Y}\mathcal{O}_{D,Y} + C_{GG,Y}\mathcal{O}_{GG,Y} + \sum_{q=u,d,s}\sum_{O} C^{qq}_{O,Y}\mathcal{O}^{qq}_{O,Y} + h.c.\Big) \quad (1)$$

where $Y \in \{L, R\}$ and $O \in \{V, A, S, T\}$ and the operators are explicitly given by ($P_{L,R} = 1/2(I \mp \gamma_5)$)

$$\begin{aligned}
\mathcal{O}_{D,Y} &= m_\mu(\bar{e}\sigma^{\alpha\beta}P_Y\mu)F_{\alpha\beta}\\
\mathcal{O}_{GG,Y} &= \frac{9}{32\pi^2 m_t}(\bar{e}P_Y\mu)\mathrm{Tr}[G_{\alpha\beta}G^{\alpha\beta}]\\
\mathcal{O}^{qq}_{V,Y} &= (\bar{e}\gamma^\alpha P_Y\mu)(\bar{q}\gamma_\alpha q)\\
\mathcal{O}^{qq}_{A,Y} &= (\bar{e}\gamma^\alpha P_Y\mu)(\bar{q}\gamma_\alpha\gamma_5 q)\\
\mathcal{O}^{qq}_{S,Y} &= (\bar{e}P_Y\mu)(\bar{q}q)\\
\mathcal{O}^{qq}_{T,Y} &= (\bar{e}\sigma^{\alpha\beta}P_Y\mu)(\bar{q}\sigma_{\alpha\beta}q)\,.
\end{aligned} \quad (2)$$

While our primary focus is on the tensor ($\mathcal{O}^{qq}_{T,Y}$) and axial ($\mathcal{O}^{qq}_{A,Y}$) operators, we include the vector, scalar, dipole and gluon operators because the first three are induced by loops, and the last arises by integrating out heavy quarks.

[1] We leave out the light-quark pseudoscalar operators and gluon operators such as $G\tilde{G}$ that can be induced by heavy-quark pseudoscalar operators at the heavy quark thresholds. The effect of this class of operators in a nucleus is suppressed both by spin and momentum transfer.

[2] The analogous mixing of SD to SI dark matter interactions was discussed in [29, 30].

At zero momentum transfer, the quark bilinears can be matched onto nucleon bilinears

$$\bar{q}(x)\Gamma_O q(x) \to G_O^{N,q}\bar{N}(x)\Gamma_O N(x) \quad (3)$$

where the vector charges are $G_V^{p,u} = G_V^{n,d} = 2$ and $G_V^{p,d} = G_V^{n,u} = 1$, and for the axial charges we use the results inferred in Ref. [22] by using the HERMES measurements [31], namely $G_A^{p,u} = G_A^{n,d} = 0.84(1)$, $G_A^{p,d} = G_A^{n,u} = -0.43(1)$, and $G_A^{p,s} = G_A^{n,s} = -.085(18)$. For the tensor charges we use the lattice QCD results [32] in the $\overline{\text{MS}}$ scheme at $\mu = 2$ GeV, namely $G_T^{p,u} = G_T^{n,d} = 0.77(7)$, $G_T^{p,d} = G_T^{n,u} = -0.23(3)$, and $G_T^{p,s} = G_T^{n,s} = .008(9)$. Finally, for the scalar charges induced by light quarks we use a precise dispersive determination [33], $G_S^{p,u} = \frac{m_N}{m_u}0.021(2)$, $G_S^{p,d} = \frac{m_N}{m_d}0.041(3)$, $G_S^{n,u} = \frac{m_N}{m_u}0.019(2)$, and $G_S^{n,d} = \frac{m_N}{m_d}0.045(3)$, and an average of lattice results [34] for the strange charge: $G_S^{p,s} = G_S^{n,s} = \frac{m_N}{m_s}0.043(11)$. In all cases, we take central values of the $\overline{\text{MS}}$ quark masses at $\mu = 2$ GeV, namely $m_u = 2.2$ MeV, $m_d = 4.7$ MeV, and $m_s = 96$ MeV [35].

Taking the above matching into account, the nucleon-level effective Lagrangian has the same structure of (1) with the replacements $\bar{q}\Gamma_O q \to \bar{N}\Gamma_O N$ and with effective couplings given by[3]

$$\tilde{C}^{NN}_{O,Y} = \sum_{q=u,d,s} G_O^{N,q} C^{qq}_{O,Y}\,. \quad (4)$$

However, we remove the tensor operators, because their effects can be reabsorbed into shifts to the axial-vector and scalar operator coefficients. In fact, to leading order in a non-relativistic expansion $\bar{N}\sigma^{ij}N = \epsilon^{ijk}\bar{N}\gamma^k\gamma_5 N$, so that the spin-dependent nucleon effective Lagrangian for $\mu \to e$ conversion reads

$$-2\sqrt{2}G_F\sum_N\sum_Y\Big(\tilde{C}^{NN}_{A,Y}(\bar{e}\gamma^\alpha P_Y\mu)(\bar{N}\gamma_\alpha\gamma_5 N) + h.c.\Big) \quad (5)$$

where $N \in \{n, p\}$, $Y \in \{L, R\}$, and

$$\tilde{C}^{NN}_{A,X} = \sum_q\Big(G_A^{N,q}C^{qq}_{A,X} + 2G_T^{N,q}C^{qq}_{T,X}\Big)\,. \quad (6)$$

Furthermore, at finite recoil the tensor operator induces a contribution to the SI amplitude, since $\overline{u_N}(p)\sigma^{0i}u_N(p-q)$ contains a term proportional to q^i/m_N [25,26], which contracts, in the amplitude, with the spin of the helicity-eigenstate electron. The net effect is tantamount to replacing the coefficient of the scalar operator with

$$\tilde{C}^{NN}_{S,Y} \to \tilde{C}^{NN}_{S,Y} + \frac{m_\mu}{m_N}\tilde{C}^{NN}_{T,Y}\,. \quad (7)$$

We write the conversion rate $\Gamma = \Gamma_{SI} + \Gamma_{SD}$, where Γ_{SI} is the A^2-enhanced rate occurring in any nucleus, and Γ_{SD} is only relevant in nuclei with spin. The usual SI branching ratio reads [4,19]

$$\begin{aligned}
\mathrm{BR}_{SI} = 2\mathrm{B}_0\Big|&[\tilde{C}^{pp}_{V,R} + \tilde{C}^{pp}_{S,L}]\,Z\,F_p(m_\mu)\\
&+ [\tilde{C}^{nn}_{V,R} + \tilde{C}^{nn}_{S,L}]\,[A-Z]\,F_n(m_\mu)\\
&+ 2C_{D,L}ZeF_p(m_\mu)\Big|^2 + \{L \leftrightarrow R\},
\end{aligned} \quad (8)$$

where $\mathrm{B}_0 = G_F^2 m_\mu^5(\alpha Z)^3/(\pi^2\Gamma_{cap})$, Γ_{cap} is the rate for the muon to transform to a neutrino by capture on the nucleus (0.7054 ×

[3] The gluon operators $\mathcal{O}_{GG,Y}$ induce a shift in the coefficient of the nucleon scalar density $\tilde{C}^{NN}_{S,Y}$, as discussed in Ref. [20]. We do not explicitly include this effect as it is not relevant to our discussion.

10^6/sec in Aluminium [36]), and the form factors $F_{p,n}(|\vec{k}|) = \int d^3x e^{-i\vec{k}\cdot\vec{x}} \rho_{p,n}(x)$ can be found in Eq. (30) of Ref. [19].

In the evaluation of Γ_{SD} from (5) we treat the muon as non-relativistic and the electron as a plane wave. Both are good approximations for low-Z nuclei; for definiteness we focus on Aluminium ($Z = 13, A = 27, J = 5/2$) the proposed target for the COMET and Mu2e experiments. After approximating the muon wavefunction in the nucleus to its value at the origin and taking it outside the integral over the nucleus [16], the nuclear part of the spin-dependent $\mu \to e$ amplitude corresponds to that of "standard" spin-dependent WIMP nucleus scattering. At momentum transfer $\vec{q}$, this is

$$\int d^3x e^{-i\vec{q}\cdot\vec{x}} \langle Al|\overline{N}(x)\gamma^k\gamma_5 N(x)|Al\rangle \,. \tag{9}$$

The $\mu \to e$ amplitude is then obtained by multiplying by the appropriate lepton current and coefficients.[4] By analogy with WIMP scattering [22,23,37], we obtain:

$$\mathrm{BR}_{SD} = 8\mathrm{B}_0 \frac{J_{Al}+1}{J_{Al}} \left| S_p^{Al} \widetilde{C}^{pp}_{A,L} + S_n^{Al} \widetilde{C}^{nn}_{A,L} \right|^2 \frac{S_A(m_\mu)}{S_A(0)} + \{L \leftrightarrow R\} \,. \tag{10}$$

The spin expectation values S_N^{Al} are defined as $S_N^{Al} = \langle J_{Al}, J_z = J_{Al}|S_N^z|J_{Al}, J_z = J_{Al}\rangle$, where S_N^z is the z component of the total nucleon spin, and the expectation value is over the nuclear ground state. They can be implemented in our QFT notation (with relativistic state normalisation for Al) by setting Eqn. (9) at $|\vec{q}| = 0$ to

$$2S_N^{Al} \frac{(J_{Al})^k}{|J_{Al}|} \times 2m_{Al}(2\pi)^3 \delta^{(3)}(p_{Al,out} - p_{Al,in}) \,.$$

The axial structure factor $S_A(|\vec{q}|)$ [23,37] reads

$$S_A(q) = a_{L,+}^2 S_{00}(q) + a_{L+}a_{L,-} S_{01}(q) + a_{L,-}^2 S_{11}(q)$$

where $a_{L,\pm} = \widetilde{C}^{pp}_{A,L} \pm \widetilde{C}^{nn}_{A,L}$. The S_N^{Al} and $S_{ij}(q)$ have been calculated in the shell model in Refs. [37,38]. At $|\vec{q}| \equiv q = 0$ the conversion rate is controlled by the spin expectation values; we use $S_n^{Al} = 0.030$ and $S_p^{Al} = 0.34$ [38]. At finite momentum transfer $q = m_\mu$, the structure factors provide a non-trivial correction. Using dominance of the proton contribution ($S_p^{Al} >> S_n^{Al}$) we find from Ref. [38] $S_{Al}(m_\mu))/S_{Al}(0) \simeq 0.29$.

3. Loop effects and the RGEs

QED and QCD loops change the magnitude of some operator coefficients, and QED loops can transform one operator into another. Such Standard Model loops are necessarily present, and their dominant (log-enhanced) effects are included in the evolution with scale of the operator coefficients, as described by the Renormalisation Group Equations (RGEs) of QED and QCD (see [5] for an introduction to the RG running of operators with the scale μ). If the New Physics scale is well above m_W, loops involving the W, Z, and h could also be relevant. However, we focus here on the RGE evolution from the experimental scale μ_N up to the weak scale m_W. Since any UV model can be mapped into a set of operator coefficients at $\mu = m_W$, our calculation does not lose generality while remaining quite simple.

We consider the one-loop RGEs of QED and QCD for $\mu \leftrightarrow e$ flavour-changing operators [27,28]. Defining $\lambda = \frac{\alpha_s(m_W)}{\alpha_s(\mu_N)}$, their solution can be approximated as

$$C_I(\mu_N) \simeq C_J(m_W)\lambda^{a_J} \left(\delta_{JI} - \frac{\alpha_e \widetilde{\Gamma}^e_{JI}}{4\pi} \log \frac{m_W}{\mu_N} \right) \tag{11}$$

where I, J represent the super- and subscripts which label operator coefficients. The a_I describe the QCD running and are only non-zero for scalars and tensors: for $N_f = 5$ one has $a_I = \frac{\Gamma^s_{II}}{2\beta_0} = \{-\frac{12}{23}, \frac{4}{23}\}$ for $I = S, T$,. We use this scaling to always give results in terms of coefficients at the low scale $\mu_N = 2$ GeV, where we match quarks to nucleons. Γ^e is the one-loop QED anomalous dimension matrix, rescaled [39,40] for $J, I \in T, S$ to account for the QCD running:

$$\tilde{\Gamma}^e_{JI} = \Gamma^e_{JI} f_{JI}, \quad f_{JI} = \frac{1}{1 + a_J - a_I} \frac{\lambda^{a_I - a_J} - \lambda}{1 - \lambda} \,. \tag{12}$$

In the estimates presented here, we focus on the effects of the off-diagonal elements of $\widetilde{\Gamma}^e_{JI}$, which mix one operator into another, and neglect the QED running of individual coefficients.

In RG evolution down to μ_N, photon exchange between the external legs of a tensor operator can mix it to a scalar operator. This contribution to the scalar coefficient is

$$\Delta\widetilde{C}^{NN}_{S,X}(\mu_N) \sim \sum_q G_S^{N,q} f_{TS} 24 Q_q \frac{\alpha_e}{\pi} \log \frac{m_W}{\mu_N} C^{qq}_{T,X}(\mu_N) \tag{13}$$

where f_{TS} is from Eq. (12).

The tensor operator also mixes to the dipole, when the quark lines are closed and an external photon is attached. This gives a contribution to the dipole coefficient

$$\left|\Delta C^{e\mu}_{D,X}(\mu_N)\right| \sim \frac{2Q_q N_c m_q}{e m_\mu} \frac{\alpha_e}{\pi} \log \frac{m_W}{\mu_N} C^{qq}_{T,X}(\mu_N) \tag{14}$$

which is suppressed by m_q/m_μ, due to a mass insertion on the quark line. For tensor operators involving u, d or s quark bilinears, the mixing to the scalar operator described in Eq. (13) gives a larger contribution to SI $\mu \to e$ conversion than this mixing to the dipole. So for the remainder of this letter, we do not discuss the contribution of Eq. (14) to $\mu \to e$ conversion. We will discuss heavier quarks[5] in a later publication [41].

Curiously, one-loop QED corrections to the axial operator generate the vector [28].[6] If a New Physics model induces a non-zero coefficient $C^{qq}_{A,Y}(m_W)$, then photon exchange between the external legs induces a contribution to the vector coefficient at the experimental scale:

$$\Delta C^{qq}_{V,Y}(\mu_N) \simeq -3Q_q \frac{\alpha_e}{\pi} \log \frac{m_W}{\mu_N} C^{qq}_{A,Y}(\mu_N) \tag{15}$$

As a result, the SI and SD processes will have comparable sensitivities to axial vector operators.

4. Results

To interpret our results, we first estimate the *sensitivity* of SD and SI $\mu \to e$ conversion to the coefficients of the tensor and axial operators of Eqn. (2). We allow a single operator coefficient to be non-zero at m_W, and consider its various contributions to SD

[4] At finite recoil, the vector or scalar operators can also contribute to the spin-dependent amplitude [26]. We neglect these contributions, because we estimate their interference with the axial vector is suppressed by $\mathcal{O}(m_\mu/m_N)$.

[5] The heavy quark scalar contribution to $\mu \to e$ conversion [20] is suppressed $\propto 1/m_Q$, so the tensor mixing to the dipole could dominate.

[6] If the lepton current contained γ^μ, rather than $\gamma^\mu P_Y$, this would not occur.

and SI $\mu \to e$ conversion (sometimes referred to as setting bounds "one-operator-at-a-time").

Suppose first that only the tensor coefficient $C^{uu}_{T,L}$ is present at m_W. Recall that $C^{uu}_{T,L}(m_W)$ can contribute to $\mu \to e$ conversion in three ways: to the SI rate via the finite momentum transfer effects of Eqn. (7), to the SI rate via the RG mixing to the scalar given in Eqn. (13), and directly to the SD rate as given in Eqn. (10). It is easy to check that the RG mixing contribution to $\widetilde{C}^{NN}_S(\mu_N)$ is an order of magnitude larger than the finite recoil contribution. Furthermore, the RG mixing effect is dominant contribution of $C^{uu}_{T,L}(m_W)$ to $\mu \to e$ conversion, as can been seen numerically by calculating the SD and SI contributions to the branching ratio:

$$BR(\mu\text{Al} \to e\text{Al}) \sim .12|1.54C^{uu}_{T,L}|^2 + .27|47C^{uu}_{T,L}|^2 \qquad (16)$$

where the coefficients are at the experimental scale, and the second term is the A^2-enhanced SI contribution.

The RG mixing is the largest contribution of $C^{uu}_{T,L}(m_W)$ to $\mu \to e$ conversion due to three enhancements: first, the anomalous dimension Γ^e_{TS} is large, and second, the $G^{N,q}_S$ coefficients of Eqn. (3) are an order of magnitude larger than $G^{N,q}_T$. The combination of these gives $\Delta\widetilde{C}^{NN}_{S,X}(\mu_N) \gtrsim \widetilde{C}^{pp}_{T,X}(\mu_N)$, which respectively contribute to the SI and SD rates. Finally, the scalar coefficient benefits from a further A^2 enhancement in the SI conversion rate. This shows that including the RG effects can change the branching ratio by orders of magnitude.

A similar estimate for the axial operator $\mathcal{O}^{uu}_{A,L}$ gives

$$BR(\mu\text{Al} \to e\text{Al}) \sim .12|0.84C^{uu}_{A,L}|^2 + .27|.69C^{uu}_{A,L}|^2 . \qquad (17)$$

We see that the RG mixing of $\mathcal{O}^{uu}_{A,L}$ into $\mathcal{O}^{uu}_{V,L}$, whose coefficient contributes to SI $\mu \to e$ conversion, also gives the best sensitivity to $C^{uu}_{A,L}$. However, the ratio of SI to SD contributions is smaller than in the tensor case, due to the smaller anomalous dimension in Eqn. (15).

SI $\mu \to e$ conversion will also give the best sensitivity to tensor and axial operators involving d quarks. However, in the case of strange quarks, the vector current vanishes in the nucleon, so $\mathcal{O}^{ss}_{A,Y}$ only contributes to SD $\mu \to e$ conversion. The largest contribution of the strange tensor operator is via its mixing to the scalar, with a sensitivity to $C^{ss}_{T,X}$ reduced by a factor $\sim G^{Ns}_T/2G^{Nu}_T$ with respect to $C^{uu}_{T,X}$. The strange tensor also mixes significantly to the dipole (see Eqn. (14)) which contributes to $\mu \to e\gamma$; we estimate that the sensitivity to $C^{ss}_{T,Y}$ of the MEG-II experiment with $BR \sim$ few $\times 10^{-14}$ (as expected after their upgrade), would be comparable to that of COMET or Mu2e with $BR \sim$ few $\times\, 10^{-16}$.

Let us now focus on the *complementarity* of SD and SI contributions to the $\mu \to e$ conversion rate, which depend on different combinations of operator coefficients. So once a signal is observed, measuring $\mu \to e$ conversion in targets with and without spin could assist in differentiating among operators or models. To illustrate this complementarity, we restrict to scalar and tensor operators involving u quarks, whose coefficients we would like to determine. Fig. 1 represents the allowed parameter space for $C^{uu}_{T,L}$ and $C^{uu}_{S,L}$ evaluated at μ_N (dotted blue) and m_W (solid red). We see that, irrespective of the operator scale, SD $\mu \to e$ conversion always gives an independent constraint. In its absence, there would be an unconstrained direction in parameter space, corresponding to $C^{uu}_{T,Y}$ at the experimental scale, or the diagonal red band at m_W. The figure also shows that the enhanced sensitivity of SI conversion illustrated in Eqn. (16) requires the (model-dependent) assumption that the model does not induce a scalar contribution which cancels the mixing of the tensor into the scalar, which would correspond to venturing along the red ellipse in the plot.

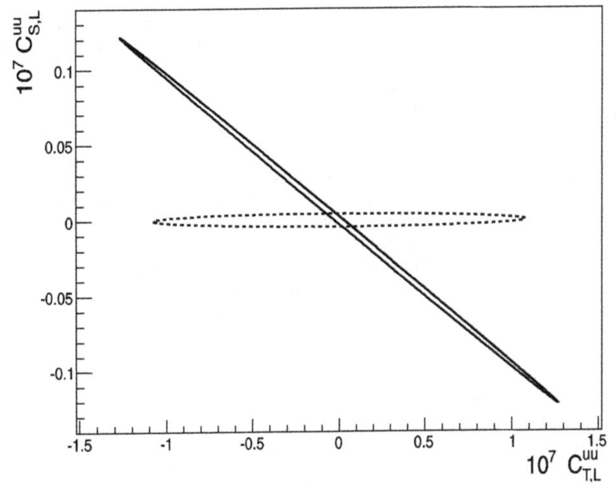

Fig. 1. The horizontal dotted blue (diagonal red) areas are the allowed parameter space at the experimental scale (at m_W), if $BR(\mu \to e \text{ conversion}) \leq 10^{-14}$. This plot assumes that CLFV only occurs in up-quark operators.

5. Prospects

In this letter, we followed the pragmatic low-energy perspective of parametrising charged Lepton Flavour Violating interactions with effective operators, and considered the contribution of axial vectors and tensors to $\mu \to e$ conversion. To our knowledge, this has not been studied previously. We found that the Spin-Dependent process depends on different operator coefficients from the Spin-Independent case, so comparing $\mu \to e$ conversion rates in targets with and without spin would give additional constraints, and could allow to identify axial or tensor operators coefficients. In future work [41], we plan to give rates for a complete set of operators, estimate their uncertainties due to higher order terms and neglected effects, and explore realistic prospects for distinguishing models/operators using targets with and without spin, such as different isotopes of Ti, a nucleus used for the past $\mu \to e$ conversion conversion searches.

Acknowledgements

VC and SD thank Martin Hoferichter for discussions, and acknowledge the partial support and hospitality of the Mainz Institute for Theoretical Physics (MITP) during the completion of this work. The work of Y.K. is supported in part by the Japan Society for the Promotion of Science (JSPS) KAKENHI Grant No. 25000004.

References

[1] V. Khachatryan, et al., CMS Collaboration, Search for heavy Majorana neutrinos in $e^\pm e^\pm$ + jets and $e^\pm$ $\mu^\pm$ + jets events in proton–proton collisions at $\sqrt{s} = 8$ TeV, J. High Energy Phys. 1604 (2016) 169, http://dx.doi.org/10.1007/JHEP04(2016)169, arXiv:1603.02248 [hep-ex];
V. Khachatryan, et al., CMS Collaboration, Search for heavy Majorana neutrinos in $\mu^\pm\mu^\pm$ jets events in proton–proton collisions at $\sqrt{s} = 8$ TeV, Phys. Lett. B 748 (2015) 144, http://dx.doi.org/10.1016/j.physletb.2015.06.070, arXiv:1501.05566 [hep-ex];
G. Aad, et al., ATLAS Collaboration, Search for heavy Majorana neutrinos with the ATLAS detector in pp collisions at $\sqrt{s} = 8$ TeV, J. High Energy Phys. 1507 (2015) 162, http://dx.doi.org/10.1007/JHEP07(2015)162, arXiv:1506.06020 [hep-ex].

[2] S. Alekhin, et al., A facility to search for hidden particles at the CERN SPS: the SHiP physics case, Rep. Prog. Phys. 79 (12) (2016) 124201, http://dx.doi.org/10.1088/0034-4885/79/12/124201, arXiv:1504.04855 [hep-ph].

[3] F.T. Avignone III, S.R. Elliott, J. Engel, Double beta decay, Majorana neutrinos, and neutrino mass, Rev. Mod. Phys. 80 (2008) 481, http://dx.doi.org/10.1103/RevModPhys.80.481, arXiv:0708.1033 [nucl-ex].

[4] Y. Kuno, Y. Okada, Muon decay and physics beyond the standard model, Rev. Mod. Phys. 73 (2001) 151, http://dx.doi.org/10.1103/RevModPhys.73.151, arXiv:hep-ph/9909265.
[5] H. Georgi, Effective field theory, Annu. Rev. Nucl. Part. Sci. 43 (1993) 209.
[6] B. Aubert, et al., BABAR Collaboration, Searches for lepton flavor violation in the decays $\tau^{\pm} \to e^{\pm}\gamma$ and $\tau^{\pm} \to \mu^{\pm}\gamma$, Phys. Rev. Lett. 104 (2010) 021802, arXiv:0908.2381 [hep-ex].
[7] K. Hayasaka, et al., Belle Collaboration, New search for $\tau \to e\gamma$ and $\tau \to \mu\gamma$, decays at Belle, Phys. Lett. B 666 (2008) 16–22, arXiv:0705.0650 [hep-ex].
[8] K. Hayasaka, et al., Search for lepton flavor violating tau decays into three leptons with 719 million produced $\tau^+\tau^-$ pairs, Phys. Lett. B 687 (2010) 139, http://dx.doi.org/10.1016/j.physletb.2010.03.037, arXiv:1001.3221 [hep-ex].
[9] T. Aushev, et al., Physics at Super B factory, arXiv:1002.5012 [hep-ex].
[10] W.H. Bertl, et al., SINDRUM II Collaboration, A search for muon to electron conversion in muonic gold, Eur. Phys. J. C 47 (2006) 337, http://dx.doi.org/10.1140/epjc/s2006-02582-x;
C. Dohmen, et al., SINDRUM II Collaboration, Test of lepton flavor conservation in $\mu \to e$ conversion on titanium, Phys. Lett. B 317 (1993) 631.
[11] U. Bellgardt, et al., SINDRUM Collaboration, Search for the Decay $\mu \to 3e$, Nucl. Phys. B 299 (1988) 1, http://dx.doi.org/10.1016/0550-3213(88)90462-2.
[12] A.M. Baldini, et al., MEG Collaboration, Search for the lepton flavour violating decay $\mu^+ \to e^+\gamma$ with the full dataset of the MEG experiment, Eur. Phys. J. C 76 (8) (2016) 434, http://dx.doi.org/10.1140/epjc/s10052-016-4271-x, arXiv:1605.05081 [hep-ex].
[13] Y. Kuno, COMET Collaboration, A search for muon-to-electron conversion at J-PARC: The COMET experiment, PTEP 2013 (2013) 022C01, http://dx.doi.org/10.1093/ptep/pts089.
[14] R.M. Carey, et al., Mu2e Collaboration, Proposal to search for $\mu^- N \to e^- N$ with a single event sensitivity below 10^{-16}, FERMILAB-PROPOSAL-0973.
[15] Y. Kuno, et al., PRISM collaboration, An experimental search for a $\mu N \to eN$ conversion at sensitivity of the order of 10^{-18} with a highly intense muon source: PRISM, unpublished, J-PARC LOI, 2006.
[16] S. Weinberg, G. Feinberg, Electromagnetic transitions between μ meson and electron, Phys. Rev. Lett. 3 (1959) 111, http://dx.doi.org/10.1103/PhysRevLett.3.111.
[17] O.U. Shanker, Z dependence of coherent μe conversion rate in anomalous neutrinoless muon capture, Phys. Rev. D 20 (1979) 1608, http://dx.doi.org/10.1103/PhysRevD.20.1608.
[18] A. Czarnecki, W.J. Marciano, K. Melnikov, Coherent muon electron conversion in muonic atoms, AIP Conf. Proc. 435 (1998) 409, http://dx.doi.org/10.1063/1.56214, arXiv:hep-ph/9801218.
[19] R. Kitano, M. Koike, Y. Okada, Detailed calculation of lepton flavor violating muon electron conversion rate for various nuclei, Phys. Rev. D 66 (2002) 096002, http://dx.doi.org/10.1103/PhysRevD.66.096002, arXiv:hep-ph/0203110; Erratum: Phys. Rev. D 76 (2007) 059902, http://dx.doi.org/10.1103/PhysRevD.76.059902.
[20] V. Cirigliano, R. Kitano, Y. Okada, P. Tuzon, On the model discriminating power of $\mu \to e$ conversion in nuclei, Phys. Rev. D 80 (2009) 013002, http://dx.doi.org/10.1103/PhysRevD.80.013002, arXiv:0904.0957 [hep-ph].
[21] G. Jungman, M. Kamionkowski, K. Griest, Supersymmetric dark matter, Phys. Rep. 267 (1996) 195, http://dx.doi.org/10.1016/0370-1573(95)00058-5, arXiv:hep-ph/9506380.
[22] G. Belanger, F. Boudjema, A. Pukhov, A. Semenov, Dark matter direct detection rate in a generic model with micrOMEGAs 2.2, Comput. Phys. Commun. 180 (2009) 747, arXiv:0803.2360 [hep-ph].
[23] J. Engel, S. Pittel, P. Vogel, Nuclear physics of dark matter detection, Int. J. Mod. Phys. E 1 (1992) 1, http://dx.doi.org/10.1142/S0218301392000023.
[24] H.C. Chiang, E. Oset, T.S. Kosmas, A. Faessler, J.D. Vergados, Coherent and incoherent (μ^-, e^-) conversion in nuclei, Nucl. Phys. A 559 (1993) 526;
T.S. Kosmas, G.K. Leontaris, J.D. Vergados, Lepton flavor nonconservation, Prog. Part. Nucl. Phys. 33 (1994) 397, http://dx.doi.org/10.1016/0146-6410(94)90047-7, arXiv:hep-ph/9312217.
[25] A.L. Fitzpatrick, W. Haxton, E. Katz, N. Lubbers, Y. Xu, The effective field theory of dark matter direct detection, J. Cosmol. Astropart. Phys. 1302 (2013) 004, http://dx.doi.org/10.1088/1475-7516/2013/02/004, arXiv:1203.3542 [hep-ph].
[26] M. Cirelli, E. Del Nobile, P. Panci, Tools for model-independent bounds in direct dark matter searches, J. Cosmol. Astropart. Phys. 1310 (2013) 019, http://dx.doi.org/10.1088/1475-7516/2013/10/019, arXiv:1307.5955 [hep-ph].
[27] S. Davidson, $\mu \to e\gamma$ and matching at m_W, arXiv:1601.07166 [hep-ph].
[28] A. Crivellin, S. Davidson, G.M. Pruna, A. Signer, Renormalisation-group improved analysis of $\mu \to e$ processes in a systematic effective-field-theory approach, arXiv:1702.03020 [hep-ph].
[29] U. Haisch, F. Kahlhoefer, On the importance of loop-induced spin-independent interactions for dark matter direct detection, J. Cosmol. Astropart. Phys. 1304 (2013) 050, http://dx.doi.org/10.1088/1475-7516/2013/04/050, arXiv:1302.4454 [hep-ph].
[30] A. Crivellin, F. D'Eramo, M. Procura, Phys. Rev. Lett. 112 (2014) 191304, http://dx.doi.org/10.1103/PhysRevLett.112.191304, arXiv:1402.1173 [hep-ph].
[31] A. Airapetian, et al., HERMES Collaboration, Precise determination of the spin structure function g(1) of the proton, deuteron and neutron, Phys. Rev. D 75 (2007) 012007, http://dx.doi.org/10.1103/PhysRevD.75.012007, arXiv:hep-ex/0609039.
[32] T. Bhattacharya, V. Cirigliano, R. Gupta, H.W. Lin, B. Yoon, Neutron electric dipole moment and tensor charges from lattice QCD, Phys. Rev. Lett. 115 (21) (2015) 212002, http://dx.doi.org/10.1103/PhysRevLett.115.212002, arXiv:1506.04196 [hep-lat].
[33] M. Hoferichter, J. Ruiz de Elvira, B. Kubis, U.G. Meißner, Phys. Rev. Lett. 115 (2015) 092301, http://dx.doi.org/10.1103/PhysRevLett.115.092301, arXiv:1506.04142 [hep-ph].
[34] P. Junnarkar, A. Walker-Loud, Phys. Rev. D 87 (2013) 114510, http://dx.doi.org/10.1103/PhysRevD.87.114510, arXiv:1301.1114 [hep-lat].
[35] K.A. Olive, et al., Particle Data Group, Chin. Phys. C 38 (2014) 090001, http://dx.doi.org/10.1088/1674-1137/38/9/090001.
[36] T. Suzuki, D.F. Measday, J.P. Roalsvig, Total nuclear capture rates for negative muons, Phys. Rev. C 35 (1987) 2212, http://dx.doi.org/10.1103/PhysRevC.35.2212.
[37] P. Klos, J. Menéndez, D. Gazit, A. Schwenk, Large-scale nuclear structure calculations for spin-dependent WIMP scattering with chiral effective field theory currents, Phys. Rev. D 88 (8) (2013) 083516, http://dx.doi.org/10.1103/PhysRevD.88.083516, arXiv:1304.7684 [nucl-th];
Erratum: Phys. Rev. D 89 (2) (2014) 029901, http://dx.doi.org/10.1103/PhysRevD.89.029901.
[38] J. Engel, M.T. Ressell, I.S. Towner, W.E. Ormand, Response of mica to weakly interacting massive particles, Phys. Rev. C 52 (1995) 2216, http://dx.doi.org/10.1103/PhysRevC.52.2216, arXiv:hep-ph/9504322.
[39] S. Bellucci, M. Lusignoli, L. Maiani, Nucl. Phys. B 189 (1981) 329, http://dx.doi.org/10.1016/0550-3213(81)90384-9.
[40] G. Buchalla, A.J. Buras, M.K. Harlander, Nucl. Phys. B 337 (1990) 313, http://dx.doi.org/10.1016/0550-3213(90)90275-I.
[41] V. Cirigliano, S. Davidson, Y. Kuno, A. Saporta, in preparation.

The method of arbitrarily large moments to calculate single scale processes in quantum field theory

Johannes Blümlein [a,*], Carsten Schneider [b]

[a] Deutsches Elektronen-Synchrotron, DESY, Platanenallee 6, D-15738 Zeuthen, Germany
[b] Research Institute for Symbolic Computation (RISC), Johannes Kepler University Linz, Altenbergerstraße 69, A-4040 Linz, Austria

ARTICLE INFO

Editor: A. Ringwald

ABSTRACT

We devise a new method to calculate a large number of Mellin moments of single scale quantities using the systems of differential and/or difference equations obtained by integration-by-parts identities between the corresponding Feynman integrals of loop corrections to physical quantities. These scalar quantities have a much simpler mathematical structure than the complete quantity. A sufficiently large set of moments may even allow the analytic reconstruction of the whole quantity considered, holding in case of first order factorizing systems. In any case, one may derive highly precise numerical representations in general using this method, which is otherwise completely analytic.

1. Introduction

Single scale higher order QED and QCD calculations in the massless [1,2] and massive [3] cases at fixed Mellin moment n are given by polynomials of rational numbers and a series of a few special constants, the multiple ζ-values, and possible generalizations thereof [4]. This is irrespectively the case, whether or not the functional representation for general values of n obeys an equation factorizing in first order, [1], or not [5]. If one has access[1] to an algorithm, through which a large number of moments e.g. $N = 8000$, can be calculated, the method of guessing, cf. [7], for holonomic problems, which often appear in physics applications, allows one to gain one large difference equation describing the corresponding problem [8].[2] If this equation is solvable in difference ring theory [9–18] one finds the solution for general values of n for this quantity without any further assumptions, e.g. made in [2,19]. In any case, significantly more moments will allow one to constrain the considered quantity much better numerically using approximation methods, e.g. through Chebyshev-polynomials or other interpolation methods.

In the following we describe an algorithm through which the system of differential equations, or associated to it, that of difference equations, available by the integration-by-parts relations [20], can be used to compute a large number of Mellin moments for the master integrals, and through them for the whole problem. The solution of the associated difference equations will need a relatively low number of initial values which have to be provided. The corresponding sequences of rational numbers mentioned above actually form the problematic part in gaining the general n result from the moments, since very involved, and in some cases yet unknown, functions span the corresponding sequences. In any case, the method allows either to find the one-dimensional distribution from a large but finite amount of moments, or at least to constrain it numerically at high accuracy.

We will give a brief illustration of the algorithm in case of a system of massive 3-loop master integrals, for which first a larger number of moments is generated, corresponding difference equations are found and are solved for general values of the Mellin variable n.

2. The algorithm

Single scale master integrals can be represented as analytic functions $\hat{I}_i(x) = \sum_{n=0}^{\infty} I_i(n)x^n$ with $1 \le i \le m$. Here m denotes the number of connected differential equations to be treated together, see Eq. (3). We aim at computing a large number of coefficients $I_i(0), I_i(1), \ldots, I_i(N)$ where N can has magnitudes like 8000. Usually the coefficients $I_i(n)$ depend on the dimensional parameter ε which itself can be expanded in a Laurent series in ε of a cer-

* Corresponding author.
E-mail address: Johannes.Bluemlein@desy.de (J. Blümlein).

[1] Standard algorithms like MINCER and MATAD [6] do only allow the calculation of a comparably low number of moments of up to $O(20-30)$ at 3-loop order.

[2] In calculating massive 3-loop integrals we have generated this number of moments using the present method. The master integrals to compute these moments amount to 114 Gbyte, leading to a final representation of the moments of ~ 1 Gbyte in size.

tain order $o \in \mathbb{Z}$. We are interested in calculating the coefficients $I_i^{(k)}(n) \in \mathbb{R}$ of the expansions

$$\hat{I}_i(x) = \sum_{n=0}^{\infty} I_i(n)x^n = \sum_{n=0}^{\infty}\left(\sum_{k=o}^{\infty} I_i^{(k)}(n)\varepsilon^k\right)x^n \tag{1}$$

up to a certain degree in ε, $t_i \in \mathbb{Z}$. More precisely, we want to compute for $1 \le i \le m$ the first $N+1$ initial values

$$I_i^{(k)}(0), I_i^{(k)}(1), I_i^{(k)}(2), \ldots, I_i^{(k)}(N) \in \mathbb{R} \tag{2}$$

for the ε-orders $o \le k \le t_i$.

In our approach we rely on the property that these unknown functions $\hat{I}_i(x)$ are usually described by a coupled system of first order linear differential equations

$$D_x \begin{pmatrix} \hat{I}_1(x) \\ \hat{I}_2(x) \\ \ldots \\ \hat{I}_m(x) \end{pmatrix} = A \begin{pmatrix} \hat{I}_1(x) \\ \hat{I}_2(x) \\ \ldots \\ \hat{I}_m(x) \end{pmatrix} + \begin{pmatrix} \hat{r}_1(x) \\ \hat{r}_2(x) \\ \ldots \\ \hat{r}_m(x) \end{pmatrix} \tag{3}$$

with $D_x = d/dx$ and where A is an $m \times m$ matrix with entries consisting of polynomials in ε and x and the entries $\hat{r}_i$ can be given in form of the expansions

$$\hat{r}_i(x) = \sum_{n=0}^{\infty} r_i(n)x^n = \sum_{n=0}^{\infty}\left(\sum_{k=o}^{\infty} r_i^{(k)}(n)\varepsilon^k\right)x^n \tag{4}$$

with $r_i^{(k)}(n) \in \mathbb{R}$ for $n \in \mathbb{N}$. Another important assumption is that the coefficients $r_i^{(k)}(n)$ can be determined efficiently for $n \in \{0, 1, \ldots, N\}$. Here one can use either the method under consideration in a recursive fashion or one can use, e.g., symbolic summation and integration techniques [9,21] if the arising functions in (4) are simple enough for a symbolic treatment. More precisely, we computed for various instances first symbolic representations of the $r_i^{(k)}(n)$ in terms of nested sums [29–31,33] and used afterwards this representation to compute the first $N+1$ initial values. Further, we assume that for reasonably small numbers s'_i the coefficients $I_i^{(k)}(n)$ with $0 \le n \le s'_i$ can be computed up to certain ε-degrees t'_i, i.e., $o \le k \le t'_i$ as a preprocessing step. Given this input we introduce the following efficient algorithm that computes the coefficients (2) for $n \in \{0, 1, 2, \ldots, N\}$.

(1) We are using first decoupling algorithms [22,23]. They operate both for systems of (inhomogeneous) differential and difference equations and transform these systems into a single scalar differential or difference equation, respectively, depending on one of the functions only. Furthermore they deliver equations for all the other functions, which can be expressed directly knowing the first function. We apply decoupling on the system (3) symbolically to obtain one scalar linear differential equation

$$\sum_{k=0}^{m} a_k(x,\varepsilon) D_x^{(k)} \hat{I}_1(x) = \sum_{i=1}^{m} d_i(x,\varepsilon)\hat{r}_i(x), \tag{5}$$

where the $a_i(x,\varepsilon)$ and $r_i(x)$ are polynomials in x and ε; in addition, one obtains identities of the form

$$\hat{I}_i(x) = \sum_{j=2}^{m}\sum_{k} e_{i,j,k}(x,\varepsilon) D_x^k \hat{I}_j(x) + \sum_{j=1}^{m}\sum_{k} f_{i,j,k}(x,\varepsilon) D_x^k \hat{r}_j(x) \tag{6}$$

for $2 \le i \le m$ where the $e_{i,j,k}(x,\varepsilon)$ and $f_{i,j,k}(x,\varepsilon)$ are rational functions in x and ε.

The algorithm proceeds now as follows: Compute in steps (2)–(4) the initial values for $\hat{I}_1(x)$ by using the scalar differential equation (5), and compute afterwards in step (5) the initial values for the remaining integrals with the given formulas (6).

(2) Plug $\hat{I}_1(x) = \sum_{n=0}^{\infty} I_1(n)x^n$ and $\hat{r}_i(x) = \sum_{n=0}^{\infty} r_i(n)x^n$ into (5) and eliminate D_x by using the property $D_x \sum_{n=0}^{\infty} h(n)x^n = \sum_{n=1}^{\infty} n\,h(n)x^{n-1}$ for a power series $\sum_{n=0}^{\infty} h(n)x^n$. Then by coefficient comparison w.r.t. x^n and using appropriate shifts one gets a linear recurrence of the form

$$\sum_{k=0}^{d} b_k(n,\varepsilon) I_1(n+k) = \rho(n,\varepsilon), \tag{7}$$

with $\rho(n,\varepsilon) = \sum_{j=1}^{m}\sum_{k=0}^{l} g_{j,k}(n,\varepsilon) r_j(n+k)$ for some $l \in \mathbb{N}$ where the $b_k(n,\varepsilon)$ and $g_{j,k}(n,\varepsilon)$ are polynomials in n and ε. Finally, divide the equation by a factor ε^u for some $u \ge 0$ in order to obtain updated polynomials $b_k(n,\varepsilon)$ in n, ε where not all $b_k(n,0)$ with $0 \le k \le d$ are zero; the $g_{j,k}(n,\varepsilon)$ are now polynomials in n and Laurent polynomials in ε.

(3) We write the right hand side of (7) in the expanded representation

$$\rho(n,\varepsilon) = \sum_{k=o}^{\infty} \rho^{(k)}(n)\varepsilon^k, \tag{8}$$

with $\rho^{(k)}(n) \in \mathbb{R}$. Since the $r_i^{(k)}(n) \in \mathbb{R}$ in (4) can be computed efficiently by assumption, the coefficients $\rho^{(k)}(n)$ for sufficiently large $k \in \mathbb{Z}$ and $n \in \{0, 1, \ldots, N\}$ can be obtained explicitly without any cost.

(4) We proceed as follows; compare [24], Lemma 1. Plugging

$$I_1(n) = \sum_{k=o}^{\infty} I_1^{(k)}(n)\varepsilon^k \tag{9}$$

into (7) and doing coefficient comparison w.r.t. ε^o yield the constraint

$$\sum_{k=0}^{d'} b_k(n,0) I_1^{(o)}(n+k) = \rho^{(o)}(n)$$

for some $d' \le d$ with $b_{d'}(n) \ne 0$. Choose $\delta \in \mathbb{N}$ such that $b_{d'}(n) \ne 0$ for all $n \in \mathbb{N}$ with $n \ge \delta$. Then we can compute with the first values $I_1^{(o)}(0), \ldots, I_1^{(o)}(n+d'+\delta-1)$ and the rule

$$I_1^{(o)}(n) \leftarrow \frac{\rho^{(o)}(n') - \sum_{k=0}^{d'-1} b_k(n',0) I_1^{(o)}(n'+k)}{b_{d'}(n',0)} \tag{10}$$

for $n' = n - d', n \ge d' + \delta$ all the other values in linear time. Now insert (9) with the explicitly computed values $I_1^{(o)}(n)$ with $0 \le n \le N$ into (7) and move these given values to the right hand side. This yields

$$\sum_{k=0}^{d'} b_k(n,\varepsilon) I'_1(n+k) = \sum_{k=o+1}^{\infty} \rho'^{(k)}(n)\varepsilon^k \tag{11}$$

for $I'_1(n) = \sum_{k=o+1}^{\infty} I_1^{(k)}(n)$ where the $\rho'^{(k)}(n)$ for $0 \le n \le N$ and sufficiently large k are given explicitly. Now we are in the position to repeat this tactic iteratively to compute the remaining coefficients: next $I_1^{o+1}(n)$ with $0 \le n \le N$, afterwards $I_1^{o+2}(n)$ with $0 \le n \le N$, and eventually $I_1^{t_i}(n)$ with $0 \le n \le N$.

(5) Now expand the rational functions $f_{i,j,k}(x,\varepsilon)$ and $g_{i,j,k}(x,\varepsilon)$ in (6) in a Laurent-series expansion w.r.t. ε and compute their coefficients as power series in x up to the necessary orders. More precisely, after the series expansion in ε is computed, we first cluster cleverly the resulting expressions with common denominators. Let $\frac{a(x)}{d(x)}$ be such an expression where $a(x) = \sum_{n=0}^{N} \alpha(n) x^n$ is already given in terms of $\alpha(n) \in \mathbb{R}$ (usually in form of polynomial expressions in certain parameters and special numbers like [4] whose coefficients are rational numbers) and $d(x) \in \mathbb{R}(x)$. Next, $d(x)^{-1} = \sum_{n=0}^{N} \delta_i(n) x^n$ is computed with $\delta_i(n) \in \mathbb{R}$ (usually in $\mathbb{Q}$). Then the Cauchy product of the truncated power series $a(x)$ and $d(x)^{-1}$ is computed in a recursive fashion. Finally, one combines all the explicitly given truncated expansions in (6) which yield the coefficients $I_i^{(k)}(n)$ for $2 \le i \le m$, $o \le k \le t_i$ and $0 \le n \le s$.

The following remarks are in order; related considerations have been applied also to our algorithms to solve coupled systems in terms of nested sums over hypergeometric products [21,25–27].

(i) Our algorithm requires that sufficiently many initial values/coefficients of $\hat{r}_i(x)$ and $\hat{I}_1(x)$ up to the right ε-order are computed as a preprocessing step. The necessary bounds for these numbers can be determined by analyzing the formulas (7) and (6) accordingly.

(ii) For simplicity, we assumed that only one scalar differential equation arises, as it happens in most examples. In general, several scalar differential equations might arise, i.e., steps (2)–(4) have to be executed several times.

(iii) One can choose any λ with $1 \le \lambda \le m$ to determine a scalar differential equation in $\hat{I}_\lambda(x)$. Different choices of λ might lead to different recurrences (7) with different orders d and formulae (6) of different size. In our calculations we analyze for each λ $(1 \le \lambda \le m)$ the obtained symbolic formulae and choose that λ for the calculation steps (3)–(5) that serves us best, e.g., to minimize the required ε-orders for the $\hat{r}_i(x)$ or minimize the recurrence order d.

(iv) Often it is a challenge to determine the first initial values of $I_i^k(n)$ to activate the recurrence formula (10). Thus it is highly desirable to keep the order d in (7) as small as possible. Since d gets smaller if the degrees of the $a_i(x,\varepsilon)$ w.r.t. x in (5) can be made smaller, the following refinement can be applied. Divide (5) by the polynomial $h = \gcd_x(a_0, \dots, a_n)$ in x. In most applications this reduces the degrees of the polynomials $a_i(x,\varepsilon)$ w.r.t. x heavily and thus produces recurrences with much lower order, e.g., $d = 4$ instead of $d = 20$ in (7). The price to be paid is that the right hand side of (5) contains now denominators which are formed by polynomials in x and ε. As a consequence it is now much harder to extract the nth coefficient of the right hand side of (5) in order to get the coefficients $\rho^{(k)}(n)$ in (8). Luckily, we can reuse our efficient tactics described in step (5) in order to obtain these coefficients $\rho^{(k)}(n)$.

All these aspects have been implemented within our package `SolveCoupledSystem` [21,26,27] which uses subroutines of the summation package `Sigma` [9] and the uncoupling package `OreSys` [23].

3. An example

Let us consider a set of three 3-loop master integrals $\{J_1(x,\varepsilon), J_2(x,\varepsilon), J_3(x,\varepsilon)\}$ which obey a 3×3 inhomogeneous linear coupled system, cf. (3), as an illustrative example. Our aim is to solve a case, which requires a number of moments normally not accessible using codes like `MINCER` and `MATAD` [6]. The requested number of moments will turn out to be $N = 89$. The example describes contributions to the massive operator matrix element $A_{gg}^{(3)}$ [28]. The system of differential equations is given by:

$$J_1'(x) = \frac{2+(1+x)\varepsilon}{2(-1+x)x} J_1(x) + \frac{2}{1-x} J_2(x) + \frac{R_1(x)}{x(x+\varepsilon)} \quad (12)$$

$$J_2'(x) = \frac{(1+\varepsilon)^2(4+3\varepsilon)}{4(-1+x)(\varepsilon+x)} J_1(x) + \frac{-2x(1+x)+(2-9x)\varepsilon+(1-2x(1+x))\varepsilon^2}{2(-1+x)x(\varepsilon+x)} \times J_2(x) + \frac{4\varepsilon}{\varepsilon+x} J_3(x) + \frac{R_2(x)}{x(x+\varepsilon)} \quad (13)$$

$$J_3'(x) = \frac{(1+\varepsilon)^2(4+3\varepsilon)(1-x+(1+x)\varepsilon)}{16(-1+x)x(\varepsilon+x)} J_1(x) + \frac{P_1(x,\varepsilon)}{8(-1+x)x(\varepsilon+x)} J_2(x) + \frac{P_2(x,\varepsilon)}{2(-1+x)x(\varepsilon+x))} J_3(x) + \frac{R_3(x)}{x(x+\varepsilon)} \quad (14)$$

with the inhomogeneities

$$R_1(x) = \frac{(2+\varepsilon)x(x+\varepsilon)}{2(1-x)} J_{16} \quad (15)$$

$$R_2(x) = \frac{(2+\varepsilon)^3 x}{8(-1+x)} J_{15} - \frac{(2+\varepsilon)(4+3\varepsilon)(2x+(1+x)\varepsilon)}{16(-1+x)} J_{16}$$

$$R_3(x) = \frac{(2+\varepsilon)^3(-2x-x\varepsilon+(3+2x)\varepsilon^2}{64\varepsilon(-1+x)} J_{15} - \frac{P_3(x,\varepsilon)}{128\varepsilon(-1+x)} J_{16} \quad (16)$$

and the polynomials

$$P_1(x,\varepsilon) = -4(1-x)+(-12+5x+x^2)\varepsilon + (-9-5x+x^2)\varepsilon^2 - (5x+2x^2)\varepsilon^3 \quad (17)$$

$$P_2(x,\varepsilon) = 2x-3x^2+x(1-x)\varepsilon + (x-2(1-x^2))\varepsilon^2 \quad (18)$$

$$P_3(x,\varepsilon) = (2+\varepsilon)(4+3\varepsilon)(-2x-3x\varepsilon+3(1+x)\varepsilon^2 + (5+2x)\varepsilon^3). \quad (19)$$

Here the integrals to be evaluated, $J_i = \hat{I}_i$, are given in Eqs. (3.62–3.64) in [28]. Moreover, the inhomogeneities $R_i = \hat{r}_i, i = 1,2,3$, cf. (5) are determined by the given expansions

$$J_{15}(\varepsilon) = \frac{8}{\varepsilon^3} - \frac{12}{\varepsilon^2} + \frac{12+3\zeta_2}{\varepsilon} - 10 - \frac{9\zeta_2}{2} + \zeta_3 + \left[\frac{15}{2} + \frac{9\zeta_2}{2} - \frac{3\zeta_3}{2} + \frac{57\zeta_2^2}{80}\right]\varepsilon + O(\varepsilon^2) \quad (20)$$

$$J_{16}(\varepsilon) = \frac{16}{\varepsilon^3} - \frac{92}{3\varepsilon^2} + \frac{35+6\zeta_2}{\varepsilon} - \frac{275}{12} - \frac{23\zeta_2}{2} + 2\zeta_3 + \left[-\frac{189}{16} + \frac{105\zeta_2}{8} + \frac{89\zeta_3}{6} + \frac{57\zeta_4}{16}\right]\varepsilon + O(\varepsilon^2). \quad (21)$$

In a first step, we will illustrate how one can compute with our algorithm from Section 2 a large set of moments for $J_{1,2,3}(x)$. In many application, this data might be already the desired result. For instance, one can use this large amount of moments to determine precise numerics using, e.g., Chebyshev polynomials. In particular, if an analytic result in terms of special functions seems unreachable, this result is the best that one can expect. Still, having these large moments at hand, one can try to compile more ambitious representations as follows.

Namely, in the second step we will take care that the number of moments are sufficiently high such that one can guess [7] from this data homogeneous linear recurrences for each of the $J_i(n)$ for $i = 1, 2, 3$.

In the last step, we will exploit summation algorithms in the setting of difference rings [9,10,12–18] and compute all d'Alembertian solutions [11]. This means that we look for all solutions of the found recurrences that are expressible in terms of nested sums over hypergeometric products. In our concrete example these solutions will establish a closed form representation of the integrals $J_{1,2,3}(x)$. In other instances, this approach might fail and proves that the class of nested sums over hypergeometric products is too small to express these integrals. Still, having these recurrences in hand might is rather useful to hunt for further classes of function spaces in which the integrals can be expressed.

With this motivation let us start to execute our proposed machinery.

Step 1: computing moments.

The system (12)–(21) is decoupled in x-space, cf. Eq. (5) in Section 2 using [22,23], yielding a scalar differential equation for $J_1(x)$:

$$\left\{\frac{d^3}{dx^3} + \frac{(-4+11x)}{2(-1+x)x}\frac{d^2}{dx^2} - \frac{Q_4(x,\varepsilon)}{4(1-x)^2x^2}\frac{d}{dx} - \frac{Q_3(x)}{8(1-x)^2x^3}\right\}J_1(x) + \frac{Q_1(x,\varepsilon)R_1(x,\varepsilon)}{4(1-x)^2x^3(\varepsilon+x)^3} + \frac{Q_2(x,\varepsilon)R_2(x,\varepsilon)}{(1-x)^2x^2(\varepsilon+x)^2} - \frac{8\varepsilon R_3(x,\varepsilon)}{(1-x)x(\varepsilon+x)^2} + \frac{Q_5(x,\varepsilon)R_1'(x,\varepsilon)}{2(1-x)x^2(\varepsilon+x)^2} + \frac{2R_2'(x,\varepsilon)}{(-1+x)x(\varepsilon+x)} - \frac{R_1''(x,\varepsilon)}{x(\varepsilon+x)} = 0, \quad (22)$$

with

$$Q_1(x,\varepsilon) = 6(-2+x)x^2 + x(-4-16x+11x^2)\varepsilon + x(-2-8x+13x^2)\varepsilon^2 + (4-9x+8x^2+6x^3)\varepsilon^3 + (2-5x+6x^2)\varepsilon^4 \quad (23)$$

$$Q_2(x,\varepsilon) = -x^2 - \varepsilon(2-x)(1-x) - \varepsilon^2(2-x-2x^2) \quad (24)$$

$$Q_3(x,\varepsilon) = -8x + (8-14x+6x^2)\varepsilon + (8-7x+2x^2)\varepsilon^2 + (2-x)\varepsilon^3 \quad (25)$$

$$Q_4(x,\varepsilon) = 8(1+x) - 20x^2 + (10-10x+3x^2)\varepsilon + (3-3x+x^2)\varepsilon^2 \quad (26)$$

$$Q_5(x,\varepsilon) = 3x(1+2x) + (2+8x+x^2)\varepsilon + (1+x)\varepsilon^2. \quad (27)$$

The solutions for $J_2(x)$ and $J_3(x)$ are finally obtained by inserting $J_1(x)$ into the following equations (see Eq. (6) in step (1) of Section 2),

$$J_2(x) = \frac{-1+x}{2x(\varepsilon+x)}R_1(x,\varepsilon) + \frac{2+(1+x)\varepsilon}{4x}J_1(x) + \frac{1}{2}(1-x)J_1'(x) \quad (28)$$

$$J_3(x) = \frac{6x+9\varepsilon x+(-1+2x+x^2)\varepsilon^2}{16\varepsilon x^2(\varepsilon+x)}R_1(x,\varepsilon) - \frac{R_2(x,\varepsilon)}{4\varepsilon x} + \frac{P_3(x,\varepsilon)}{32\varepsilon x^2}J_1(x) + \frac{-1+x}{8\varepsilon x}R_1'(x,\varepsilon) + \frac{-4x^2(4-10x+x^2)\varepsilon+(2-x-2x^2)\varepsilon^2}{16\varepsilon x}J_1'(x) - \frac{(-1+x)(\varepsilon+x)}{8\varepsilon}J_1''(x), \quad (29)$$

where

$$P_3(x,\varepsilon) = -8x - (18x-2x^2)\varepsilon + (2-7x)\varepsilon^2 + (1+2x^2)\varepsilon^3. \quad (30)$$

Using the serial Ansatz, cf. Section 2, (2),

$$J_1(x) = \sum_{n=0}^{\infty} f(n,\varepsilon)x^n \quad (31)$$

in (22) leads to the difference equation (see Eqs. (7)–(11) in step (2) of Section 2)

$$-2(4+\varepsilon-2n)(-1+n)(-1+\varepsilon+2n)f(n-2,\varepsilon) - (\varepsilon+2n)(6-18n+8n^2-(1+4n)\varepsilon-\varepsilon^2) \times f(n-1,\varepsilon) - 2(2+\varepsilon-n) \times (\varepsilon+2n)(2+\varepsilon+2n)f(n) = \rho(n,\varepsilon), \quad (32)$$

where $\rho(n,\varepsilon)$ denotes the inhomogeneity, being obtained taking the coefficient of the term x^n in the corresponding serial expansion of the inhomogeneity of (22). In our particular example, the first two coefficients boil down to the following very simple form: $\rho^{(-3)}(0) = 0$ and $\rho^{(-3)}(n) = -64$ for all $n \geq 1$, $\rho^{(-2)}(0) = \frac{448}{3}$ and $\rho^{(-2)}(n) = 0$ for $n \geq 1$.

Eq. (32) is of order $o = 2$ and needs two boundary conditions, implied by the respective physics case :

$$f(1,\varepsilon) = \frac{8}{\varepsilon^3} - \frac{46}{3\varepsilon^2} + \left[\frac{35}{2} + 3\zeta_2\right]\frac{1}{\varepsilon} - \frac{275}{24} - \frac{23\zeta_2}{4} + \zeta_3 + \left[-\frac{189}{32} + \frac{105\zeta_2}{16} + \frac{57\zeta_2^2}{80} + \frac{89\zeta_3}{12}\right]\varepsilon + O(\varepsilon^2) \quad (33)$$

$$f(2,\varepsilon) = \frac{56}{9\varepsilon^3} - \frac{298}{27\varepsilon^2} + \left[\frac{1873}{162} + \frac{7}{3}\zeta_2\right]\frac{1}{\varepsilon} - \frac{11009}{1944} - \frac{149}{36}\zeta_2 - \frac{7}{9}\zeta_3 + \left[-\frac{211991}{23328} + \frac{1873\zeta_2}{432} + \frac{1013}{108}\zeta_3 + \frac{16}{3}\mathrm{Li}_4\left(\frac{1}{2}\right) + \frac{2}{9}\ln^4(2) - \frac{4}{3}\ln^2(2)\zeta_2 - \frac{137}{80}\zeta_2^2\right]\varepsilon + O(\varepsilon^2). \quad (34)$$

Table 1
The number of moments needed to find the associated difference equations.

	ε^{-3}	ε^{-2}	ε^{-1}	ζ_2	ε^0	ζ_2	ζ_3
J_1	8	24	48	8	89	24	8
J_2	8	23	42	8	81	21	8
J_3	–	8	19	–	63	–	5

One now recursively solves (32) to derive the set of higher moments up to the required maximal value of N, also expanding to the lth power in ε

$$f(n,\varepsilon) = \sum_{k=-3}^{l} \varepsilon^k f(n,k). \tag{35}$$

The functions $f(n,k)$ still may contain different parameters, like color factors and special numbers, as ζ-values [4], and form otherwise rational number series. We first determine the highest pole contribution. Its corresponding difference equation is obtained by setting $\varepsilon = 0$, cf. (32). More precisely, we obtain the calculation rule

$$f(n,-3) \leftarrow \frac{4n^2-9n+3}{2(n-2)(n+1)} f(n-1,-3) \tag{36}$$

$$- \frac{(n-1)(2n-1)}{2n(n+1)} f(n-2,3) \tag{37}$$

$$+ \frac{1}{8(n-2)n(n+1)} \rho^{(-3)}(n). \tag{38}$$

Taking the first two initial values $f(1,-3) = 8$ and $f(2,-3) = \frac{56}{9}$ from (33) and (34) enables one to compute arbitrarily many moments in linear time. The first values of $f(n,-3)$ for $n = 1,2,3,\ldots$ are

$$8, \frac{56}{9}, \frac{16}{3}, \frac{24}{5}, \frac{40}{9}, \frac{88}{21}, 4, \frac{104}{27}, \frac{56}{15}, \frac{40}{11}, \frac{32}{9}, \frac{136}{39}, \frac{24}{7}, \ldots$$

To prepare for the next order, the former pole-solution has to be inserted repeating the process. In this way one obtains the corresponding series for the different contributions $\propto \varepsilon^k$. The latter has already been described in Ref. [21].

In this way one obtains the series of moments $f(n,\varepsilon)$ associated to the solution $J_1(x)$. Note that to obtain also the solutions $J_{2,3}(x)$ using the formulas (28) and (29) higher terms in ε may be requested, e.g. for solving the whole system to a certain power in ε.

Step 2: guess recurrences.

In the next step all these individual series are considered separately using the method of guessing [7], i.e. determining a recurrence from a large set of the corresponding moments. In the present example we consider the contributions $O(\varepsilon^{-3})$ to $O(\varepsilon^0)$. Contributions to higher powers in ε are accessible as well, however, they usually request a growing computational effort.

Since the number of moments N needed to guess a recurrence is not known a priori, the moments are computed up to a reasonable number. If the number turns out not to be sufficient, further moments can be computed reusing the previous calculations. E.g., using our algorithm the time to generate $N = 500$ moments amounts to 48 sec and $N = 2000$ moments require 569 sec in the present example.

The guessing method to find the minimal difference equation requires the numbers N, depending on ε and whether the respective set corresponds to the purely rational term or a respective contribution $\propto \zeta_i$. The values are summarized in Table 1. Here the highest number turns out to be $N = 89$. The guessing method [7] yields a difference equation

$$\sum_{k=0}^{O} a_k(n) F(n+k) = 0 \tag{39}$$

for each power in ε and factor in ζ_i. Here O and R denote the order and degree of the difference equation, i.e. the number of shift-operators and the maximal degree of the polynomials $a_k(n)$. These difference equations are usually of (much) higher order and degree then the recursion (32) with which the moments $f(n,k)$ were generated.

The orders and degrees of the different difference equations are listed in Table 2. The number of moments always includes a sufficiently large safety margin to validate the corresponding difference equations, which may even be enlarged. We generate additional moments according to their order to provide the needed initial values to solve the difference equations.

Step 3: solve the found recurrences.

In the present case, all difference equations can be solved by `Sigma.m` [9] using difference ring theory [9–18]. Besides rational terms in n the result for $J_1(n)$ and $J_2(n)$ is spanned by the harmonic sums [29] up to $S_{1,2}(n)$. $J_3(n)$, furthermore, contains a finite binomial sum. One obtains

$$\begin{aligned} J_3(n,\varepsilon) = {} & \frac{2(2n+5)}{3(n+1)} \frac{1}{\varepsilon^2} + \left[-\frac{8n^2+20n+15}{3(n+1)^2} \right. \\ & \left. + \frac{(2n-1)S_1}{3(n+1)} \right] \frac{1}{\varepsilon} + \frac{(1-2n)S_2}{4(n+1)} \\ & + \frac{24n^3+76n^2+84n+35}{6(n+1)^3} + \frac{(2n-1)S_1^2}{12(n+1)} \\ & - \frac{(8n^2+8n-3)S_1}{6(n+1)^2} + \frac{(2n+5)\zeta_2}{4(n+1)} \\ & + \frac{1}{4^n}\binom{2n}{n}\frac{2n+1}{2n+2}\left\{ \sum_{k=1}^{n} \frac{4^k S_1(k-1)}{\binom{2k}{k}k^2} - 7\zeta_3 \right\} \\ & + O(\varepsilon), \end{aligned} \tag{40}$$

where the harmonic sums [29] are defined by

$$S_{b,\vec{a}}(n) = \sum_{k=1}^{n} \frac{\mathrm{sign}(b)^k}{k^{|b|}} S_{\vec{a}}(k),\ b, a_i \in \mathbb{Z}\backslash\{0\}, n \in \mathbb{N}\backslash\{0\}. \tag{41}$$

The solutions $J_k(x,\varepsilon)$ in the present example have been obtained using a different method before [21] and are given up to $O(\varepsilon^0)$ in Eqs. (3.106–3.108) in Ref. [28]. All appearing letters building the nested sums forming the master integrals are found automatically. One may also give a closed form solution in x-space summing the infinite series for the functions J_i, expanded in the dimensional parameter ε, cf. (31). While one obtains harmonic polylogarithms [31] in case of $J_{1,2}(x)$, also root-valued iterated integrals appear for $J_3(x)$ [30].

4. Conclusion

The above example is a simpler one. Let us mention that, for comparison, the computational time for the complete pole terms of more involved massive 3-loop examples requires the knowledge of about ~ 1200 even moments [8] and is estimated to amount to several weeks [32]. In intermediary and also the final results, depending on the problem, all the different known types of nested sums and iterated integrals [29–31,33] may appear.

Table 2
The order and degree of the associated difference equations.

	ε^{-3}		ε^{-2}		ε^{-1}		ζ_2		ε^{0}		ζ_2		ζ_3	
	O	R	O	R	O	R	O	R	O	R	O	R	O	R
J_1	1	2	2	5	3	11	1	2	4	20	2	5	1	2
J_2	1	2	2	5	3	10	1	2	4	18	2	5	1	2
J_3	–	–	1	2	2	4	–	–	4	12	–	–	1	1

In case the difference equations would turn out not to be first order factorizable using the algorithm of Ref. [21], at least the first order factorizable terms will be gained analytically, leaving the non-factorizable part behind for further analysis using different methods, like those for elliptic integrals and their potential generalization [5,34].

Having the above number of moments, Eq. (1) also yields a first numerical approximation of $\hat{I}_k(x)$, up to the required power in ε.

The present method is suitable to obtain precise numerical representations in case of various current massless and massive calculations at 3- and 4-loop order. In case one may solve the associated difference equations algebraically, one will even obtain the complete analytic result without any prejudice, i.e. a sufficiently large set of scalar moments allows one to fully reconstruct a one-dimensional distribution. This automatic method is therefore suited to carry out many more higher loop calculations in contemporary elementary particle physics in a completely or at least widely analytic way.

Acknowledgements

We would like to thank J. Ablinger, A. Behring, A. De Freitas, M. Kauers, and P. Marquard for discussions. This work was supported in part by the Austrian Science Fund (FWF) grant SFB F50 (F5009-N15), the European Commission through contract PITN-GA-2012-316704 (HIGGSTOOLS).

References

[1] J.A.M. Vermaseren, A. Vogt, S. Moch, Nucl. Phys. B 724 (2005) 3, arXiv:hep-ph/0504242.
[2] B. Ruijl, T. Ueda, J.A.M. Vermaseren, J. Davies, A. Vogt, PoS (LL2016) 071, arXiv:1605.08408 [hep-ph].
[3] I. Bierenbaum, J. Blümlein, S. Klein, Nucl. Phys. B 820 (2009) 417, arXiv:0904.3563 [hep-ph].
[4] J. Blümlein, D.J. Broadhurst, J.A.M. Vermaseren, Comput. Phys. Commun. 181 (2010) 582, arXiv:0907.2557 [math-ph];
J. Ablinger, J. Blümlein, C. Schneider, J. Phys. Conf. Ser. 523 (2014) 012060, arXiv:1310.5645 [math-ph].
[5] J. Ablinger, A. Behring, J. Blümlein, A. De Freitas, M. van Hoeij, E. Imamoglu, C.G. Raab, C.-S. Radu, C. Schneider, Iterated Elliptic and Hypergeometric Integrals for Feynman Diagrams, DESY 16–147.
[6] S.A. Larin, F.V. Tkachov, J.A.M. Vermaseren, NIKHEF-H-91-18;
S.G. Gorishnii, S.A. Larin, L.R. Surguladze, F.V. Tkachov, Comput. Phys. Commun. 55 (1989) 381;
M. Steinhauser, Comput. Phys. Commun. 134 (2001) 335, arXiv:hep-ph/0009029.
[7] M. Kauers, Guessing Handbook, Version 0.32, 2009-04-07, RISC, Linz, 2009.
[8] J. Blümlein, M. Kauers, S. Klein, C. Schneider, Comput. Phys. Commun. 180 (2009) 2143, arXiv:0902.4091 [hep-ph].
[9] C. Schneider, Sémin. Lothar. Comb. 56 (2007) 1, article B56b;
C. Schneider, J. Phys. Conf. Ser. 523 (2014) 012037, arXiv:1310.0160 [cs.SC];
C. Schneider, in: C. Schneider, J. Blümlein (Eds.), Computer Algebra in Quantum Field Theory: Integration, Summation and Special Functions, in: Texts and Monographs in Symbolic Computation, Springer, Wien, 2013, p. 325, arXiv:1304.4134 [cs.SC].
[10] M. Karr, J. ACM 28 (1981) 305.
[11] M. Petkovšek, J. Symb. Comput. 14 (1992) 243;
S.A. Abramov, M. Petkovšek, in: J. von zur Gathen (Ed.), Proc. ISSAC'94, ACM Press, 1994, p. 169.
[12] C. Schneider, Symbolic Summation in Difference Fields, Ph.D. thesis RISC, Johannes Kepler University, Linz technical report 01-17, 2001.
[13] C. Schneider, Difference equations in $\Pi\Sigma$-extensions, An. Univ. Timiş. Ser. Mat.-Inform. 42 (2004) 163;
J. Differ. Equ. Appl. 11 (2005) 799;
Equations in $\Pi\Sigma$-fields, Appl. Algebra Eng. Commun. Comput. 16 (2005) 1.
[14] C. Schneider, J. Algebra Appl. 6 (2007) 415.
[15] C. Schneider, in: A. Carey, D. Ellwood, S. Paycha, S. Rosenberg (Eds.), Motives, Quantum Field Theory, and Pseudodifferential Operators, in: Clay Mathematics Proceedings, vol. 12, Amer. Math. Soc., 2010, p. 285, arXiv:0904.2323.
[16] C. Schneider, Ann. Comb. 14 (2010) 533, arXiv:0808.2596.
[17] C. Schneider, in: J. Gutierrez, J. Schicho, M. Weimann (Eds.), Computer Algebra and Polynomials, Applications of Algebra and Number Theory, in: Lecture Notes in Computer Science (LNCS), vol. 8942, 2015, p. 157, arXiv:cs.SC/1307788.
[18] C. Schneider, J. Symb. Comput. 43 (2008) 611, arXiv:0808.2543v1;
J. Symb. Comput. 72 (2016) 82, arXiv:1408.2776 [cs.SC];
J. Symb. Comput. 80 (2017) 616, arXiv:1603.04285 [cs.SC].
[19] V.N. Velizhanin, arXiv:1411.1331 [hep-ph].
[20] K.G. Chetyrkin, F.V. Tkachov, Nucl. Phys. B 192 (1981) 159.
[21] J. Ablinger, A. Behring, J. Blümlein, A. De Freitas, A. von Manteuffel, C. Schneider, Comput. Phys. Commun. 202 (2016) 33, arXiv:1509.08324 [hep-ph].
[22] A. Danilevskiĭ, Mat. Sb. 2 (1937) 169;
M.A. Barkatou, Appl. Algebra Eng. Commun. Comput. 4 (3) (1993) 185;
B. Zürcher, Rationale Normalformen von pseudo-linearen Abbildungen, Master's thesis, ETH Zürich, 1994;
M. Bronstein, M. Petkovšek, Theor. Comput. Sci. 157 (1) (1996) 3;
S.A. Abramov, E.V. Zima, Proc. Int. Conf. on Computational Modelling and Computing in Physics, Dubna, RU, Sept. 16–26, 1996, p. 16;
A. Bostan, F. Chyzak, E. de Panafieu, ISSAC 2013, Boston, arXiv:1301.5414 [cs.SC] and references therein.
[23] S. Gerhold, Uncoupling Systems of Linear Ore Operator Equations, Master's thesis, RISC, J. Kepler University, Linz, 2002.
[24] J. Blümlein, S. Klein, C. Schneider, F. Stan, J. Symb. Comput. 47 (2012) 1267, arXiv:1011.2656 [cs.SC].
[25] C. Schneider, A. De Freitas, J. Blümlein, PoS (LL2014) 017, arXiv:1407.2537 [cs.SC].
[26] J. Ablinger, J. Blümlein, A. De Freitas, C. Schneider, PoS (RADCOR2015) 060, arXiv:1601.01856 [cs.SC].
[27] J. Ablinger, A. Behring, J. Blümlein, A. De Freitas, C. Schneider, PoS (LL2016) 005, arXiv:1608.05376 [cs.SC].
[28] J. Ablinger, J. Blümlein, A. De Freitas, A. Hasselhuhn, A. von Manteuffel, M. Round, C. Schneider, Nucl. Phys. B 885 (2014) 280, arXiv:1405.4259 [hep-ph].
[29] J.A.M. Vermaseren, Int. J. Mod. Phys. A 14 (1999) 2037, arXiv:hep-ph/9806280;
J. Blümlein, S. Kurth, Phys. Rev. D 60 (1999) 014018, arXiv:hep-ph/9810241.
[30] J. Ablinger, J. Blümlein, C.G. Raab, C. Schneider, J. Math. Phys. 55 (2014) 112301, arXiv:1407.1822 [hep-th].
[31] E. Remiddi, J.A.M. Vermaseren, Int. J. Mod. Phys. A 15 (2000) 725, arXiv:hep-ph/9905237.
[32] J. Ablinger, A. Behring, J. Blümlein, A. De Freitas, A. von Manteuffel, C. Schneider, The three-loop splitting functions $P_{qg}^{(2)}$ and $P_{gg}^{(2,N_F)}$, arXiv:1705.01508 [hep-ph].
[33] J. Ablinger, J. Blümlein, C. Schneider, J. Math. Phys. 52 (2011) 102301, http://dx.doi.org/10.1063/1.3629472, arXiv:1105.6063 [math-ph];
J. Math. Phys. 54 (2013) 082301, arXiv:1302.0378 [math-ph];
J. Ablinger, J. Blümlein, arXiv:1304.7071 [math-ph].
[34] L. Adams, C. Bogner, A. Schweitzer, S. Weinzierl, arXiv:1607.01571 [hep-ph].

Three-body final state interaction in $\eta \to 3\pi$ updated

P. Guo [a], I.V. Danilkin [b,c,*], C. Fernández-Ramírez [d], V. Mathieu [e,f], A.P. Szczepaniak [e,f,g]

[a] Department of Physics and Engineering, California State University, Bakersfield, CA 93311, USA
[b] Institut für Kernphysik and PRISMA Cluster of Excellence, Johannes Gutenberg Universität, D-55099 Mainz, Germany
[c] SSC RF ITEP, Bolshaya Cheremushkinskaya 25, 117218 Moscow, Russia
[d] Instituto de Ciencias Nucleares, Universidad Nacional Autónoma de México, Ciudad de México 04510, Mexico
[e] Center for Exploration of Energy and Matter, Indiana University, Bloomington, IN 47403, USA
[f] Physics Department, Indiana University, Bloomington, IN 47405, USA
[g] Theory Center, Thomas Jefferson National Accelerator Facility, Newport News, VA 23606, USA

ARTICLE INFO

Editor: B. Grinstein

ABSTRACT

In view of the recent high-statistic KLOE data for the $\eta \to \pi^+\pi^-\pi^0$ decay, a new determination of the quark mass double ratio has been done. Our approach relies on a dispersive model that takes into account rescattering effects between three pions via subenergy unitarity. The latter is essential to reproduce the Dalitz plot distribution. A simultaneous description of the KLOE and WASA-at-COSY data is achieved in terms of just two real parameters. From a global fit, we determine $Q = 21.6 \pm 1.1$. The predicted slope parameter for the neutral channel $\alpha = -0.025 \pm 0.004$ is in reasonable agreement with the PDG average value.

1. Introduction

Three meson systems play an important role in studies of hadron reaction dynamics and hadron spectroscopy. For example, in three-particle decays of heavy quarkonia several candidates for non-quark model resonances have recently been observed [1–3]. Three-body decays of B and D mesons are a promising laboratory for studies of CP-violation [4,5]. In the light meson sector the limited phase space makes three-particle decays an ideal testing ground of effective theories of strong interactions. Detailed amplitude analysis of three meson production becomes even more important in light of the current and forthcoming high precision data from various hadron facilities [6–9].

The isospin breaking $\eta \to 3\pi$ decay, which we consider here, is of great importance as it allows to measure the light quark mass difference. The electromagnetic effects are known to be small [10–12], and the decay is driven by strong interactions through the $\Delta I = 1$ isospin breaking transition that appears directly in the QCD Lagrangian. The decay amplitude is proportional to the light quark mass difference, $(m_u - m_d)$, and it is conventionally expressed in terms of the parameter Q^2 defined by

$$\frac{1}{Q^2} = \frac{m_d^2 - m_u^2}{m_s^2 - \hat{m}^2}, \quad \hat{m} = \frac{(m_u + m_d)}{2}, \tag{1}$$

with m_s being the strange quark mass. Note, that this double ratio (1) is protected to strong high order corrections. Given the small breakup momenta the distribution of pions in the Dalitz plot of the $\eta \to \pi^+\pi^-\pi^0$ decay can be analyzed in terms of a small number of parameters that determine deviations from a uniform distribution. These parameters are referred to as Dalitz plot parameters. Early analyses [13–15] could determine only a few, leading Dalitz plot parameters. A few more parameters were determined using the 2008 KLOE measurement [16]. These analyses were further improved thanks to the high-quality WASA-at-COSY [17] and new KLOE [18] data. The statistics of the most resent measurements is high enough to allow for binned, data-driven analysis.

On the theoretical side there has been significant progress in chiral, effective field theory analysis of η decays. Chiral perturbation theory (χPT) seems to converge poorly, yielding $\Gamma_{\eta\to\pi^+\pi^-\pi^0}=$ 66, 167 ± 50, ~ 300 eV at leading (LO), next-to-leading (NLO) and next-to-next-to-leading (NNLO) order, respectively [19–22]. This indicates importance of pion-pion interactions and various approaches have been used to implement these effects to all orders [23–27].

In our recent study [28] we implemented the S-matrix constraints of unitarity, analyticity and crossing symmetry via a set of dispersion relations [29–31]. Connection with QCD is achieved by

* Corresponding author.
E-mail address: danilkin@uni-mainz.de (I.V. Danilkin).

matching the dispersive amplitudes with χPT at the point where the latter converges best *i.e.* below the threshold. In [28] we used WASA-at-COSY data [17] to determine the free parameters, *i.e.* subtraction constants of the dispersive integrals. As a result, we achieved a simultaneous description of the Dalitz plot distributions of the charged and neutral η decay modes. In order to extract the parameter Q we matched the dispersive amplitude with the next-to-leading order (NLO) χPT result near the Adler zero and we obtained $Q = 21.4 \pm 0.4$ [28]. The purpose of this letter is to revisit the result of [28] in a view of the new high statistic data from the KLOE experiment [18].

2. The method

In this section, we briefly review the η decay amplitudes that were developed in [28]. For $\eta \to 3\pi$, the transition amplitude $A(s,t,u)$ is a function of three Mandelstam variables $s = (p_{\pi^+} + p_{\pi^-})^2$, $t = (p_{\pi^-} + p_{\pi^0})^2$, and $u = (p_{\pi^+} + p_{\pi^0})^2$ which are related by $s + t + u = m_\eta^2 + 3m_\pi^2$. Except for the phase-space boundary, we work in the isospin limit and take $m_\pi = (2m_{\pi^+} + m_{\pi^0})/3$. At low energies, one can perform a partial wave (p.w.) decomposition while crossing symmetry implies unitarity cuts in all three Mandelstam variables. Therefore we symmetrize the p.w. expansion in all three channels [29,32–37], which for the charged decay, $\eta \to \pi^+\pi^-\pi^0$, implies the following representation,

$$A^C(s,t,u) = \sum_{L=0}^{L_{max}=1} \frac{(2L+1)}{2} \left[\frac{2}{3} P_L(z_s)\,(a_{0L}(s) - a_{2L}(s)) \right.$$
$$\left. + P_L(z_t)\,(a_{1L}(t) + a_{2L}(t)) - P_L(z_u)\,(a_{1L}(u) - a_{2L}(u)) \right]. \quad (2)$$

The amplitudes a_{IL} have only the right-hand, unitary cuts. In Eq. (2) $z_i \equiv \cos\theta_i$ and $\theta_{s,t,u}$ are the center-of-mass scattering angles in the s, t and u-channels, respectively. The subscript (I, L) labels isospin and orbital angular momentum, with $I + L = even$ due to Bose symmetry. The latter implies that for $L_{max} = 1$ there are three unknown isospin amplitudes with $(I, L) = (0, 0)$, $(2, 0)$, $(1, 1)$. An amplitude of the neutral decay, $\eta \to 3\pi^0$ can be easily reconstructed using $\Delta I = 1$ relation,

$$A^N(s,t,u) = A^C(s,t,u) + A^C(t,u,s) + A^C(u,s,t). \quad (3)$$

We emphasize, that the decomposition in Eq. (2) has the same analytical properties as the amplitude in NNLO chiral expansion [38, 39]. However, in contrast to χPT, we can impose unitarity to all orders on the a_{IL} amplitudes. The discontinuity along the right-hand cut can be expressed through the elastic $\pi\pi$ partial wave amplitudes f_{IL}, which leads to

$$\Delta a_{IL}(s) = \frac{1}{2i}\,(a_{IL}(s+i\epsilon) - a_{IL}(s-i\epsilon))$$
$$= f^*_{IL}(s)\rho(s)\Bigg(a_{IL}(s) + 2\sum_{L'=0}^{L_{max}}\sum_{I'}(2L'+1)$$
$$\times \int_{-1}^{+1} \frac{dz_s}{2}\, P_L(z_s)\, P_{L'}(z_t) C^{II'}_{st}\, a_{I'L'}(t) \Bigg). \quad (4)$$

Note, that the amplitudes $a_{IL}(s)$ and $f_{IL}(s)$ for $L > 0$ are subject to kinematical constraints [40] which have to be removed before application of dispersion relations. This is done by introducing the reduced amplitudes $\tilde{a}_{IL}(s) = a_{IL}(s)/Z_L(s)$ where the factor $Z_L(s)$ is proportional to the product of the c.m. momenta of $\pi\pi$ and $\pi\eta$ [28]. The normalization is fixed by the phase space factor $\rho(s) = \sqrt{1 - 4m_\pi^2/s}$, with $\mathrm{Im}(1/f_{IL}(s)) = -\rho(s)$. The explicit form of the crossing matrices $C^{II'}_{st,su}$ can be found in [28]. The contribution from the first term on the right-hand side of Eq. (4) reproduces the direct s-channel unitarity, while the second term contains the left-hand cuts from the t and u-channels. While calculating the latter, special care has to be taken for $4m_\pi^2 \le s < (m_\eta + m_\pi)^2$, i.e. one has to deform the contour to avoid the cut along the real axis [41,42,32]. The kinematical singularity free amplitudes $\tilde{a}_{IL}(s)$ satisfy the Cauchy representation, which up to subtraction constants yields,

$$\tilde{a}_{IL}(s) = \frac{1}{\pi} \int_{4m_\pi^2}^{\infty} ds' \frac{\Delta \tilde{a}_{IL}(s')}{s' - s}. \quad (5)$$

The combination of (5) and (4) sets the so-called Khuri–Treiman (KT) framework that can be solved using several techniques [43]. The most popular method is to write a set of dispersion relation for the ratio $\tilde{a}_{IL}(s)/\Omega_{IL}(s)$ with $\Omega_{IL}(s)$ being the Omnès function [23,27]. In this case it is necessary to make further assumptions about the unknown high-energy region which is typically done by introducing subtractions. This procedure is relatively easy for $\omega \to 3\pi$ decay which depends dominantly on the pion-pion P-wave scattering input [44,45]. However, this is more challenging for $\eta \to 3\pi$ decay, where the dominant contribution comes from the S-wave and the form of the $\pi\pi$ isoscalar Omnès function is very sensitive to the asymptotic behavior of the phase shift $\delta_{IL=00}(s \to \infty)$. In [46] (Figs. 4 and 7) different scenarios for the $\pi\pi$ phase shift inputs were investigated and the corresponding Omnès functions were produced. In order to minimize these differences, the subtraction polynomial of the sufficient order is required in the dispersion representation.

A complementary approach is the Pasquier inversion [36,47] that we applied to analyze WASA-at-COSY data in [28] and use here as well. This method uses contour deformation to exchange the order of double integral appearing on the right hand side of Eq. (5). As it was shown in [43], once the two-body amplitudes $f_{IL}(s)$ are known, different methods for solving the dispersive integral provide the same result. However, when the Pasquier inversion is applied, the input of $f_{IL}(s)$ is required in a different energy region.

To proceed we write $\tilde{a}_{IL}(s)$ in the form

$$\tilde{a}_{IL}(s) = \mathcal{F}_{IL}(s)\, f_{IL}(s)\, g_{IL}(s), \quad (6)$$

where the function $\mathcal{F}_{IL}(s)$ is introduced to remove Adler zeros specific to the elastic amplitude $f_{IL}(s)$ and introduce zeros in the decay amplitude as required by chiral symmetry. From Eq. (4) and Eq. (6) one can derive the discontinuity of g_{IL} and write the dispersive representation for $g_{IL}(s)$. As a result we obtain a double integral equations for $g_{IL}(s)$, which can be reduced to a single integral equation using the Pasquier inversion,

$$g_{IL}(s) = -\frac{1}{\pi} \int_{-\infty}^{0} ds' \frac{1}{s'-s} \frac{\Delta f_{IL}(s')}{f^*_{IL}(s')}\, g_{IL}(s')$$
$$+ \frac{1}{\pi} \int_{-\infty}^{(M-m_\pi)^2} dt \sum_{L'=0}^{L_{max}} \sum_{I'} \mathcal{K}_{IL,I'L'}(s,t)$$
$$\times C^{II'}_{st}\, f_{I'L'}(t)\, g_{I'L'}(t). \quad (7)$$

The explicit form of the kernel functions, $\mathcal{K}_{IL,I'L'}(s,t)$ can be found in [28]. Currently, they are only calculated in the region $s \in (0, (M - m_\pi)^2)$, which includes physical region and therefore cover the whole Dalitz plot region. In order to compute the amplitudes beyond that region, a proper analytical continuation is required within Pasquier inversion technique. Important steps in that direction were already elaborated in [48]. We will come this issue later in the Q-value determination.

The first term and the part of the second term on the right-hand side have the left-hand cut and can be expanded in the Taylor series in the physical region. Retaining only a single term in the expansion we arrive at the following relation [43,49]

$$g_{IL}(s) = g_{IL}(s_0) + \frac{1}{\pi} \int_0^{(M-m_\pi)^2} dt \sum_{L'=0}^{L_{max}} \sum_{I'} C_{st}^{II'} \times \left(\mathcal{K}_{IL,I'L'}(s,t) - \mathcal{K}_{IL,I'L'}(s_0,t)\right) f_{I'L'}(t)\, g_{I'L'}(t)\,, \quad (8)$$

which is solved by discretizing the integral and inverting the kernel matrix. The subtraction point, s_0 is chosen to be near the Adler zero at leading order of χPT, $s_0 \simeq 4/3\, m_\pi^2$, and the subtraction constants $g_{IL}(s_0)$, which absorb the left hand cut contribution are the free parameters that are to be determined by fitting to the data.

3. Numerical results

The $\eta \to 3\pi$ Dalitz plot distribution is conventionally expressed in terms of the variables x, y which are defined by

$$x = \frac{\sqrt{3}}{2 m_\eta Q_c} (t-u)\,,$$
$$y = \frac{3}{2 m_\eta Q_c} \left((m_\eta - m_{\pi^0})^2 - s\right) - 1\,. \quad (9)$$

For charge decay $Q_c = m_\eta - 2m_\pi^{\pm} - m_\pi^0$ and for the neutral decay $Q_n = m_\eta - 3m_\pi^0$. Kinematics restrict the events to be contained within the unit disk $x^2 + y^2 \leq 1$. The KLOE [18] and WASA-at-COSY [17] data were binned into 371 and 59 sectors of the unit disk, respectively (only bins that lie completely inside the physical region are included). We determine the unknown parameters, $g_{IL}(s_0)$ by minimizing

$$\chi^2 = \sum_{bins}^{N} \left(\frac{|A|^2_{\rm data} - |A^C(\{g_{IL}(s_0)\})|^2}{\Delta |A|^2_{\rm data}} \right)^2, \quad (10)$$

where $|A|^2_{\rm data}$ is the acceptance corrected data in each bin and $\Delta |A|_{\rm data}$ is the uncertainty (assumed to be only statistical).

In our earlier work [28] we performed two different fits to the WASA-at-COSY data. The first fit was done using only two resonant p.w. amplitudes, i.e. $(I,L) = (0,0), (1,1)$ and the second fit included all isospin amplitudes for the S and the P waves, i.e. $(I,L) = (0,0), (2,0), (1,1)$. Here we follow the same procedure. The resulting parameters are collected in Table 1, where we show fits with and without three-particle rescattering effects (so called "two-body" and "three-body" scenarios), i.e using (8) to determine $g_{IL}(s)$, or just setting it to a constant $g_{IL}(s) = g_{IL}(s_0)$.

In the first step we fit the KLOE data alone. When only $(I,L) = (0,0), (1,1)$ amplitudes are taken into account, we observe a significant reduction of $\chi^2/d.o.f$ while moving from the "two-body" to the "three body" case. At the same time, when a complete set of S and P waves is incorporated, the $\chi^2/d.o.f$ stabilizes at around 1.2-1.3 in both cases. In the second step, we combine the KLOE and WASA-at-COSY data. The results are in general very similar, showing the consistency of two different data sets. The results of the fit are shown in Fig. 1.

The Dalitz plot parameters are defined as an effective range expansion around the center of the Dalitz plot $x = y = 0$,

Table 1
Results of two-body (2b) and three-body (3b) fits to WASA-at-COSY [17] data, KLOE data [18] and the combined fit. For two-body fits we quote $(g_{IL}^{2b}(s_0) \pm \Delta g_{IL}^{2b}(s_0))/g_{00}^{2b}(s_0)$, while when presenting results of three-body fit we quote $(g_{IL}^{3b}(s_0) \pm \Delta g_{IL}^{3b}(s_0))/g_{00}^{2b}(s_0)$, where $g_{00}^{2b}(s_0)$ is the central value obtained in the two-body fit with the same number of partial waves. We do the latter to illustrate the relative change in normalization between two- and three-body fits.

$(I,L) = (0,0), (1,1)$				
	$g_{00}/g_{00}^{(2b)}$	$g_{20}/g_{00}^{(2b)}$	$g_{11}/g_{00}^{(2b)}$	$\chi^2/d.o.f.$
(2b)				
COSY	1.000 ± 0.002	-	0.058 ± 0.009	1.45
KLOE	1.000 ± 0.005	-	0.019 ± 0.025	10.4
Comb	1.000 ± 0.005	-	0.020 ± 0.026	9.5
(3b)				
COSY	1.043 ± 0.005	-	0.233 ± 0.009	0.95
KLOE	1.046 ± 0.006	-	0.194 ± 0.024	2.61
Comb	1.046 ± 0.006	-	0.195 ± 0.026	(Set 1) 1.64
$(I,L) = (0,0), (2,0), (1,1)$				
	$g_{00}/g_{00}^{(2b)}$	$g_{20}/g_{00}^{(2b)}$	$g_{11}/g_{00}^{(2b)}$	$\chi^2/d.o.f.$
(2b)				
COSY	1.00 ± 0.02	-0.26 ± 0.05	0.38 ± 0.07	0.94
KLOE	1.00 ± 0.06	-0.44 ± 0.17	0.56 ± 0.21	1.21
Comb	1.00 ± 0.07	-0.44 ± 0.18	0.56 ± 0.23	1.54
(3b)				
COSY	1.19 ± 0.01	0.14 ± 0.003	0.28 ± 0.04	0.90
KLOE	1.215 ± 0.002	0.015 ± 0.005	0.427 ± 0.008	1.29
Comb	1.214 ± 0.002	0.018 ± 0.005	0.423 ± 0.008	(Set 2) 1.61

$$\frac{|A^C(x,y)|^2}{|A^C(0,0)|^2} = 1 + a\,y + b\,y^2 + d x^2 + f\,y^3 + g\,x^2 y + \cdots$$
$$\frac{|A^N(z,\phi)|^2}{|A^N(0,0)|^2} = 1 + 2\alpha\, z + 2\beta\, z^{3/2} \sin 3\phi + \cdots\,, \quad (11)$$

where $x = \sqrt{z}\cos\phi$ and $y = \sqrt{z}\sin\phi$. In Table 2 we show the averaged Dalitz Plot parameters between three-body fits with $(I,L) = (0,0), (1,1)$ and $(I,L) = (0,0), (2,0), (1,1)$ wave sets. We also predict the slope parameter α for the neutral decay mode to be

$$\alpha = -0.024 \pm 0.004, \quad \beta = -0.000 \pm 0.002\,, \quad (12)$$
$$\alpha = -0.025 \pm 0.004, \quad \beta = -0.000 \pm 0.002\,,$$

from the KLOE and combined KLOE & WASA-at-COSY fits, respectively. Note, that without three body effects $\alpha^{2b} = -0.021 \pm 0.004$ for both sets of data. The new results (12) compares favorably with the most recent PDG value $\alpha^{PDG} = -0.0315 \pm 0.0015$ [1]. This difference is expected to get even smaller once electromagnetic corrections are fully considered (not only in kinematic factors).

3.1. Matching to χPT and the Q-value

We note, that NLO χPT result depends on four low energy constants (LECs). These can be reduced to a single one $L_3 = (-2.35 \pm 0.37) \times 10^{-3}$ [50] if one employs Gell-Mann–Okubo constraint between meson masses and meson decay constants. This is not the case at NNLO where one has to deal with several unknown LEC's. Therefore, in our analysis we match the dispersive amplitudes with NLO χPT near the Adler zero. Note, that we match single variable partial wave amplitudes $a_{IL}(s)$ to χPT and not the full amplitude $A^C(s,t,u)$ along the lines $s = t$ or $t = u$. This procedure should be equivalent, since $a_{IL}^{\chi PT}(s)$ possess Adler zeros as well. In order to perform the matching at $t = u$ we would need to make an additional analytic continuation of our results. In Fig. 2 we show our

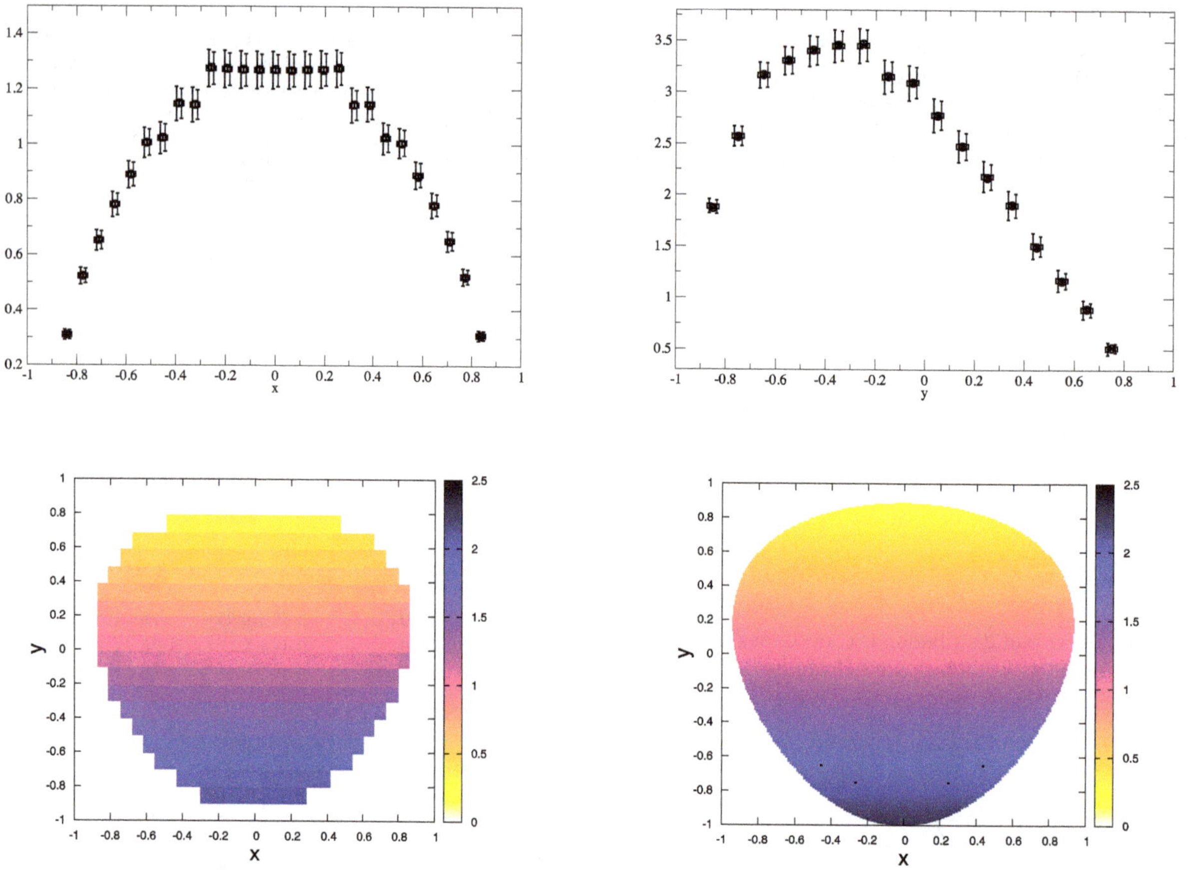

Fig. 1. Upper panels are the x- and y-projection plots. Black circles are the data. Red squares and blue squares represent results of the two-body and three-body fits, respectively. The fits are performed on the Dalitz distribution shown in the bottom left panel using three waves, $(I, L) = (0, 0), (2, 0), (1, 1)$. For better visualization the experimental points are shifted horizontally from the fit results. The bottom right panel is the Dalitz distribution from the three-body fit with $(I, L) = (0, 0), (2, 0), (1, 1)$ waves, while the bottom left panel is the new KLOE [18] data. (For interpretation of the references to color in this figure legend, the reader is referred to the web version of this article.)

Table 2
Dalitz plot parameters for $\eta \to \pi^+\pi^-\pi^0$. We present the results of the separate fits to WASA-at-COSY [17] and KLOE [18] data and the combined fit. In both cases we average Dalitz Plot parameters between $(I, L) = (0, 0), (1, 1)$ and $(I, L) = (0, 0), (2, 0), (1, 1)$ scenarios (see Table 1).

	a	b	d	f	g
WASA-at-COSY [17]	-1.144 ± 0.018	$0.219 \pm 0.019 \pm 0.037$	$0.086 \pm 0.018 \pm 0.018$	0.115 ± 0.037	–
KLOE [18]	$-1.095 \pm 0.003^{+0.003}_{-0.002}$	$0.145 \pm 0.003 \pm 0.005$	$0.081 \pm 0.003^{+0.006}_{-0.005}$	$0.141 \pm 0.007^{+0.007}_{-0.008}$	$-0.044 \pm 0.009^{+0.012}_{-0.013}$
Theory (fit to COSY)	-1.116 ± 0.032	0.188 ± 0.012	0.063 ± 0.004	0.091 ± 0.003	-0.042 ± 0.009
Theory (fit to KLOE)	-1.077 ± 0.029	0.170 ± 0.008	0.060 ± 0.002	0.091 ± 0.003	-0.044 ± 0.003
Theory (combined fit)	-1.075 ± 0.028	0.155 ± 0.006	0.084 ± 0.002	0.101 ± 0.003	-0.074 ± 0.003

results of a combined fit to KLOE and WASA-at-COSY with a fixed overall normalization to NLO χPT near the Adler zero in the region $s = (0, 10m_\pi^2)$ along the lines $s = t$ or $t = u$. The updated Q-value is

$$Q = 21.6 \pm 1.1\,, \tag{13}$$

which should be compared to the result of [28] $Q = 21.4 \pm 1.1$ (the fit to WASA-at-COSY data only) and $Q = 21.7 \pm 1.1$ (the fit to KLOE data only). Note, that the obtained Q-value is consistent with the latest ($N_f = 2 + 1 + 1$) lattice computations $Q = 22.2 \pm 1.8$ [51].

There are several challenges in the accurate determination of the Q-value. The first one comes from the elastic $\pi\pi$ scattering amplitudes, which are available from the Roy equation analysis [52] and implies the error $\Delta Q_{\pi\pi} = 0.25$. Second uncertainty is due to experimental $\eta \to \pi^+\pi^-\pi^0$ decay width, which serves as an input in our analysis. Its value increased by more than 3σ over the last thirty years, resulting in the current PDG value $\Gamma_{\eta\to\pi^+\pi^-\pi^0} = 296 \pm 16$ eV [1]. This error propagates to $\Delta Q_\Gamma = 0.29$. Third source of uncertainty is the experimental data on Dalitz plot itself, which thanks to the recent high-statistical analyses has improved significantly. Its contribution to the Q-value is $< 10\%$ of the size of error bars coming from the $\pi\pi$ amplitudes and therefore we included it in $\Delta Q_{\pi\pi}$. Another uncertainty comes from matching to χPT amplitude. The error associated with L_3 LEC is very small and therefore the resulting error bar in our previous analysis was $\Delta Q_{\text{total}} = 0.4$ [28]. That error was dominated by the experimental error bars and therefore should be viewed as a lower bound of the full error. We note, however, that the Q-value determination is very sensitive to the matching to NLO amplitude. Though the region around the Adler zero is supposed to be stable against contributions from higher orders in the chiral

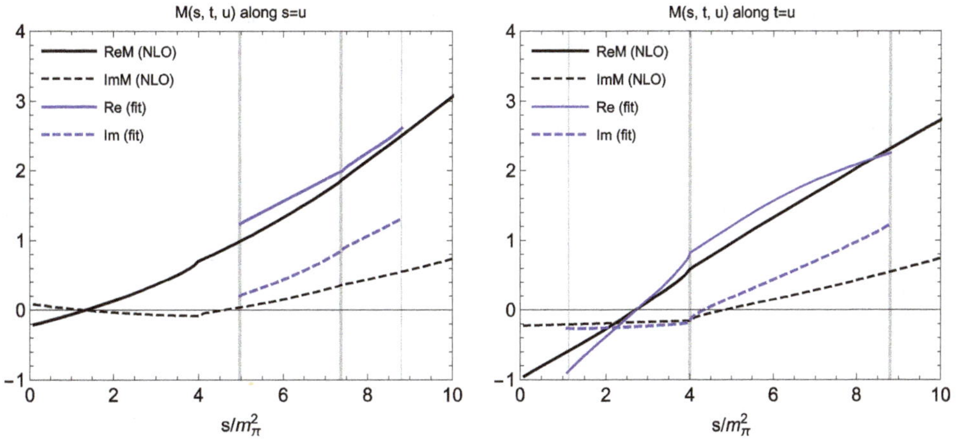

Fig. 2. The amplitude along the lines $s=u$ and $s=t$ with a comparison to NLO ChPT. The relation between $M(s,t,u)$ and $A^C(s,t,u)$ is given in [28]. The solid thick and thin vertical lines correspond to the physical region and the region where we calculated our amplitudes, respectively. As explained in the text, in order to compute the amplitudes beyond that region, a proper analytical continuation is required within Pasquier inversion technique.

expansion, we cannot completely exclude them. Assuming conservatively an additional error of 10% on NLO χPT amplitude, gives $\Delta Q_{match} = 1.08$ and the total $\Delta Q_{\text{total}} = 1.1$ quoted in Eq. (13).

4. Conclusions

In this work we revisited our previous dispersive analysis [28] of the $\eta \to 3\pi$ decay in light of the new KLOE [18] data. Within our unitary model we established a unified description of charged and neutral decay modes. The method is based on Khuri–Treiman equation which is consistent with elastic unitarity, analyticity and crossing symmetry. Using the input from the $\pi\pi$ amplitude, the Khuri–Treiman equation was solved using Pasquier inversion technique. This allowed to establish a significant reduction of the unknown parameters compared to a more straightforward Omnès solution. However, the price is the treatment of the left-hand cuts, which is in general not known. We assume, that the unitarity in the physical region, where it can be constrained by the data, plays the key role and does not depend on an accurate form of the unphysical left-hand cuts. The latter we absorbed in the subtraction constants [43]. With these model assumptions we were able to describe the data from KLOE [18] and WASA-at-COSY [17] with a minimal number fitting parameters.

The new results are $\alpha = -0.025 \pm 0.004$ and $Q = 21.6 \pm 1.1$. Since the experimental data on $\eta \to 3\pi$ Dalitz plot is very precise now, the main experimental uncertainties come from $I=0$ two-pion scattering amplitudes and the decay width $\Gamma_{\eta\to\pi^+\pi^-\pi^0}$. Improving them are relevant for further Q-value and α determinations.

After submission of our manuscript an improved dispersive analysis based on Omnès functions was announced in [53]. The new Q-value is $Q = 22.0 \pm 0.7$ which is consistent with our estimate.

The codes employed to compute the partial wave amplitudes and the Dalitz plot distribution are available for downloading as well as in an interactive form online at the Joint Physics Analysis Center (JPAC) webpage [54].

Acknowledgements

We thank Astrid Blin for the valuable comments on this manuscript. This material is based upon work supported in part by the U.S. Department of Energy, Office of Science, Office of Nuclear Physics under contracts DE-AC05-06OR23177, DE-FG0287ER40365, National Science Foundation under Grants PHY-1415459 and PHY-1205019. The work of I.V.D. is supported by the Deutsche Forschungsgemeinschaft (DFG) through the Collaborative Research Center SFB 1044. P.G. acknowledges support from Department of Physics and Engineering, California State University, Bakersfield, CA. C.F.-R. work is supported in part by CONACYT (Mexico) under grant No. 251817.

References

[1] K.A. Olive, et al., Chin. Phys. C 38 (2014) 090001.
[2] R. Aaij, et al., Phys. Rev. Lett. 112 (2014) 222002.
[3] E.S. Swanson, Phys. Rep. 429 (2006) 243.
[4] R. Aaij, et al., Phys. Rev. Lett. 111 (2013) 101801.
[5] R. Aaij, et al., Phys. Rev. Lett. 112 (2014) 011801.
[6] M. Battaglieri, Int. J. Mod. Phys. E 19 (2010) 837.
[7] P. Eugenio, JLAB-E04-005, 2003.
[8] C. Adolph, et al., Phys. Lett. B 740 (2015) 303.
[9] M. Ablikim, et al., Phys. Rev. D 92 (2015) 012014.
[10] D.G. Sutherland, Phys. Lett. 23 (1966) 384.
[11] J.S. Bell, D.G. Sutherland, Nucl. Phys. B 4 (1968) 315.
[12] C. Ditsche, B. Kubis, U.-G. Meissner, Eur. Phys. J. C 60 (2009) 83.
[13] M. Gormley, et al., Phys. Rev. D 2 (1970) 501.
[14] J.G. Layter, et al., Phys. Rev. D 7 (1973) 2565.
[15] A. Abele, et al., Phys. Lett. B 417 (1998) 197.
[16] F. Ambrosino, et al., J. High Energy Phys. 0805 (2008) 006.
[17] P. Adlarson, et al., Phys. Rev. C 90 (2014) 045207.
[18] A. Anastasi, et al., J. High Energy Phys. 05 (2016) 019.
[19] J.A. Cronin, Phys. Rev. 161 (1967) 1483.
[20] H. Osborn, D.J. Wallace, Nucl. Phys. B 20 (1970) 23.
[21] J. Gasser, H. Leutwyler, Nucl. Phys. B 250 (1985) 539.
[22] J. Bijnens, K. Ghorbani, J. High Energy Phys. 0711 (2007) 030.
[23] G. Colangelo, S. Lanz, E. Passemar, PoS CD09 (2009) 047.
[24] S. Lanz, PoS CD12 (2013) 007.
[25] S.P. Schneider, B. Kubis, C. Ditsche, J. High Energy Phys. 1102 (2011) 028.
[26] K. Kampf, M. Knecht, J. Novotny, M. Zdrahal, Phys. Rev. D 84 (2011) 114015.
[27] S. Descotes-Genon, B. Moussallam, Eur. Phys. J. C 74 (2014) 2946.
[28] P. Guo, et al., Phys. Rev. D 92 (2015) 054016.
[29] N.N. Khuri, S.B. Treiman, Phys. Rev. 119 (1960) 1115.
[30] J. Kambor, C. Wiesendanger, D. Wyler, Nucl. Phys. B 465 (1996) 215.
[31] A.V. Anisovich, H. Leutwyler, Phys. Lett. B 375 (1996) 335.
[32] J.B. Bronzan, C. Kacser, Phys. Rev. 132 (1963) 2703.
[33] I. Aitchison, Il Nuovo Cimento 35 (1965) 434.
[34] I. Aitchison, Phys. Rev. 137 (1965) B1070.
[35] I.J.R. Aitchison, R. Pasquier, Phys. Rev. 152 (1966) 1274.
[36] R. Pasquier, J.Y. Pasquier, Phys. Rev. 170 (1968) 1294.
[37] R. Pasquier, J.Y. Pasquier, Phys. Rev. 177 (1969) 2482.
[38] J. Stern, H. Sazdjian, N.H. Fuchs, Phys. Rev. D 47 (1993) 3814.
[39] M. Knecht, B. Moussallam, J. Stern, N.H. Fuchs, Nucl. Phys. B 457 (1995) 513.
[40] M.F.M. Lutz, I. Vidana, Eur. Phys. J. A 48 (2012) 124.
[41] V. Gribov, V. Anisovich, A. Anselm, Sov. Phys. JETP 15 (1962) 159.
[42] C. Kacser, Phys. Rev. 132 (1963) 2712.
[43] P. Guo, I.V. Danilkin, A.P. Szczepaniak, Eur. Phys. J. A 51 (2015) 135.
[44] F. Niecknig, B. Kubis, S.P. Schneider, Eur. Phys. J. C 72 (2012) 2014.

[45] I.V. Danilkin, et al., Phys. Rev. D 91 (2015) 094029.
[46] G. Colangelo, E. Passemar, P. Stoffer, Eur. Phys. J. C 75 (2015) 172.
[47] I.J.R. Aitchison, J.J. Brehm, Phys. Rev. D 17 (1978) 3072.
[48] P. Guo, Phys. Rev. D 91 (2015) 076012.
[49] P. Guo, Mod. Phys. Lett. A 31 (2015) 1650058.
[50] G. Amoros, J. Bijnens, P. Talavera, Nucl. Phys. B 602 (2001) 87.
[51] S. Aoki, et al., Eur. Phys. J. C 77 (2017) 112.
[52] R. Garcia-Martin, et al., Phys. Rev. D 83 (2011) 074004.
[53] G. Colangelo, S. Lanz, H. Leutwyler, E. Passemar, Phys. Rev. Lett. 118 (2017) 022001.
[54] http://www.indiana.edu/~jpac/index.html.

Universal seesaw and $0\nu\beta\beta$ in new 3331 left-right symmetric model

Debasish Borah [a], Sudhanwa Patra [b,*]

[a] Department of Physics, Indian Institute of Technology, Guwahati, Assam 781039, India
[b] Center of Excellence in Theoretical and Mathematical Sciences, Siksha 'O' Anusandhan University, Bhubaneswar, 751030, India

ARTICLE INFO

Editor: A. Ringwald

ABSTRACT

We consider a class of left-right symmetric model with enlarged gauge group $SU(3)_c \times SU(3)_L \times SU(3)_R \times U(1)_X$ without having scalar bitriplet. In the absence of scalar bitriplet, there is no Dirac mass term for fermions including usual quarks and leptons. We introduce new isosinglet vector-like fermions so that all the fermions get their masses through a universal seesaw mechanism. We extend our discussion to neutrino mass and its implications in neutrinoless double beta decay ($0\nu\beta\beta$). We show that for TeV scale $SU(3)_R$ gauge bosons, the heavy-light neutrino mixing contributes dominantly to $0\nu\beta\beta$ that can be observed at ongoing experiments. The new physics contributions arising from purely left-handed currents via exchange of keV scale right-handed neutrinos and the so called mixed helicity λ-diagram can saturate the KamLANDZen bound. We show that the right handed neutrinos in this model can have mass in the sub keV range and can be long lived compared to the age of the Universe. The contributions of these right handed neutrinos to flavour physics observables like $\mu \to e\gamma$ and muon $g-2$ is also discussed. Towards the end we also comment on different possible symmetry breaking patterns of this enlarged gauge symmetry to that of the standard model.

1. Introduction

The Standard Model (SM) of particle physics has been the most successful phenomenological theory specially after the discovery of its last missing piece, the Higgs boson at the Large Hadron Collider (LHC) back in 2012 with subsequent null results for Beyond Standard Model (BSM) searches. However, the SM fails to address several observed phenomena as well as theoretical questions. For example, it fails to explain the sub-eV neutrino mass [1–6], the origin of parity violation in weak interactions and the origin of three fermion families. The first two questions can be naturally addressed within the framework of the Left-right symmetric model (LRSM) [7,8], one of the most widely studied BSM frameworks. These models not only explain tiny neutrino masses naturally through seesaw mechanism but also give rise to an effective parity violating SM at low energy through spontaneous breaking of a parity preserving symmetry at high scale. The conventional LRSM based on the gauge group $SU(3)_c \times SU(2)_L \times SU(2)_R \times U(1)_{B-L}$ can be enhanced to a more general LRSM based on the gauge group $SU(3)_c \times SU(3)_L \times SU(3)_R \times U(1)_X$ (or in short 3331). The advantage of such an up gradation of the gauge symmetry is the ability of the latter in providing an explanation to the origin of three fermion families of the SM in addition to having other generic features of the LRSM. In such a model, the number of three fermion generations is no longer a choice, but a necessity in order to cancel chiral anomalies. In such models, where the usual lepton and quark representations are enlarged from a fundamental of $SU(2)_L$ in the SM to a fundamental of $SU(3)_L$, the number of generations must be equal to the number of colours in order to cancel the anomalies [9]. This is in contrast with the SM or the usual LRSM where the gauge anomalies are canceled within each fermion generation separately. One can also build such a sequential 3331 model by including additional chiral fermions. But since such a model does not explain the origin of three families from the anomaly cancellation point of view and contain non-minimal chiral fermion content, we stick to discussing a special type of non-sequential 3331 model here.

There have been a few works [10–14] recently done within the framework of such 3331 models with different motivations. Particularly from the origin of neutrino mass point of view, the work [10] considered a scalar sector comprising of bitriplets plus sextets which gives rise to tiny neutrino masses through canonical type I [15] and type II [16,17] seesaw. Another recent work [12] studied a specific 3331 model with bitriplet and triplet scalar

* Corresponding author.
E-mail addresses: dborah@iitg.ernet.in (D. Borah), sudha.astro@gmail.com (S. Patra).

fields that can explain tiny neutrino masses through inverse [18, 19] and linear seesaw mechanism [19]. The earlier work [13] considered effective higher dimensional operators to explain fermion masses in 3331 models while the recent work [14] studied the model and several of its variants from LHC phenomenology point of view. Here, we simply consider another possible way of generating fermion masses in 3331 models through the universal seesaw mechanism [20–22] where all fermions acquire their masses through a common seesaw mechanism.[1] Incorporating additional vector like fermion pairs corresponding to each fermion generation, we show that the correct fermion mass spectrum can be generated in such a model with a scalar sector where all of them transform as fundamentals under $SU(3)_{L,R}$ without the need of bi-fundamental and sextet scalars shown in [10,12] for the implementation of different seesaw mechanism for neutrino masses. We also discuss the possibilities of light neutral fermions apart from sub-eV active neutrinos, their role in neutrinoless double beta decay ($0\nu\beta\beta$) and different possible symmetry breaking chains of the gauge symmetry $SU(3)_c \times SU(3)_L \times SU(3)_R \times U(1)_X$ to that of the SM. We show that for TeV scale $SU(3)_R$ gauge bosons, the right handed neutrinos are constrained to lie around the sub keV mass regime having interesting consequences for $0\nu\beta\beta$. We find that although the pure heavy neutrino contribution to $0\nu\beta\beta$ remains suppressed compared to the one from light neutrinos, the heavy-light neutrino mixing which can be quite large in this model without any fine-tuning, gives a large contribution to $0\nu\beta\beta$ keeping it within experimental reach. We also show that the relatively light right handed neutrinos in this model can be long lived compared to the age of the Universe. The contribution of these right handed neutrinos to lepton flavour violating decay and anomalous magnetic moment of the muon is found to be suppressed.

This letter is organized as follows. In section 2 we briefly discuss the model with the details of the particle spectrum, fermion masses via universal seesaw and gauge boson masses. In section 3 we discuss possible contributions to $0\nu\beta\beta$ from purely light (heavy) neutrinos and heavy-light neutrino mixing respectively. We then briefly discuss other interesting implications for cosmology and flavour physics in section 4. Finally we discuss about different possible symmetry breaking chains in section 5 and then conclude in section 6.

2. The model framework

2.1. Particle spectrum

The usual fermions transform under $SU(3)_c \times SU(3)_L \times SU(3)_R \times U(1)_X$ as

$$\Psi_{aL} = \begin{pmatrix} \nu_{aL} \\ \ell^-_{aL} \\ \xi^q_{aL} \end{pmatrix}, \quad \Psi_{aR} = \begin{pmatrix} \nu_{aR} \\ \ell^-_{aR} \\ \xi^q_{aR} \end{pmatrix},$$

$$Q_{mL} = \begin{pmatrix} d_{\alpha L} \\ u_{\alpha L} \\ J^{-q-1/3}_{\alpha L} \end{pmatrix}, \quad Q_{mR} = \begin{pmatrix} d_{\alpha R} \\ u_{\alpha R} \\ J^{-q-1/3}_{\alpha R} \end{pmatrix},$$

$$Q_{3L} = \begin{pmatrix} u_{3L} \\ d_{3L} \\ J^{q+2/3}_{3L} \end{pmatrix}, \quad Q_{3R} = \begin{pmatrix} u_{3R} \\ u_{3R} \\ J^{q+2/3}_{3R} \end{pmatrix}, \tag{1}$$

with a = 1, 2, 3 whereas m = 1, 2.

[1] See Refs. [23–26] for implementation of universal seesaw mechanism for fermion mass generation within left-right symmetric model.

Table 1
Particle content of the model.

Particle	$SU(3)_c \times SU(2)_L \times SU(3)_R \times U(1)_X$
Q_{mL}	$(3, 3^*, 1, -\frac{q}{3})$
Q_{mR}	$(3, 1, 3^*, -\frac{q}{3})$
Q_{3L}	$(3, 3, 1, \frac{q+1}{3})$
Q_{3R}	$(3, 1, 3, \frac{q+1}{3})$
Ψ_{aL}	$(1, 3, 1, \frac{q-1}{3})$
Ψ_{aR}	$(1, 1, 3, \frac{q-1}{3})$
$U_{L,R}$	$(3, 1, 1, \frac{2}{3})$
$D_{L,R}$	$(3, 1, 1, \frac{2}{3})$
$E_{L,R}$	$(1, 1, 1, -1)$
$N_{L,R}$	$(1, 1, 1, 0)$
χ_L	$(1, 3, 1, -\frac{q+2}{3})$
χ_R	$(1, 1, 3, -\frac{q+2}{3})$
ϕ_L	$(1, 3, 1, \frac{1-q}{3})$
ϕ_R	$(1, 1, 3, \frac{1-q}{3})$

The transformation of the fields under the gauge symmetry $SU(3)_c \times SU(3)_L \times SU(3)_R \times U(1)_X$ are given in Table 1.

Assuming $q = 0$, if the neutral components of the scalar fields acquire their vacuum expectation value (vev) as

$$\langle\chi_L\rangle \equiv \frac{1}{\sqrt{2}}\begin{pmatrix} 0 \\ v_{1L} \\ 0 \end{pmatrix}, \quad \langle\phi_L\rangle \equiv \frac{1}{\sqrt{2}}\begin{pmatrix} v_{2L} \\ 0 \\ \omega_L \end{pmatrix}$$
$$\langle\chi_R\rangle \equiv \frac{1}{\sqrt{2}}\begin{pmatrix} 0 \\ v_{1R} \\ 0 \end{pmatrix}, \quad \langle\phi_R\rangle \equiv \frac{1}{\sqrt{2}}\begin{pmatrix} v_{2R} \\ 0 \\ \omega_R \end{pmatrix} \tag{2}$$

2.2. Fermion mass

The Yukawa Lagrangian can be written as

$$\begin{aligned}
\mathcal{L}_Y = &-[Y_D]_{ma}\left(\overline{Q}_{mL}\phi_L D_{aR} + \overline{Q}_{mR}\phi_R D_{aL}\right) - [M_D]_{ab}\,\overline{D}_{aL}D_{bR} \\
&- [Y_U]_{ma}\left(\overline{Q}_{mL}\chi_L U_{aR} + \overline{Q}_{mR}\chi_R U_{aL}\right) - [M_U]_{ab}\,\overline{U}_{aL}U_{bR} \\
&- \left[Y'_D\right]_{3a}\left(\overline{Q}_{3L}\chi^*_L D_{aR} + \overline{Q}_{3R}\chi^*_R D_{aL}\right) \\
&- \left[Y'_U\right]_{3a}\left(\overline{Q}_{3L}\phi^*_L U_{aR} + \overline{Q}_{3R}\phi^*_R U_{aL}\right) \\
&- [Y_E]_{ab}\left(\overline{\Psi}_{aL}\chi^*_L E_{bR} + \overline{\Psi}_{mR}\chi^*_R E_{bL}\right) - [M_E]_{ab}\,\overline{E}_{aL}E_{bR} \\
&- [Y_N]_{ab}\left(\overline{\Psi}_{aL}\phi^*_L N_{bR} + \overline{\Psi}_{mR}\phi^*_R N_{bL}\right) - [M_{LR}]_{ab}\,\overline{N}_{aL}N_{bR} \\
&- [M_{LL}]_{ab}\,N_{aL}N_{bL} - [M_{RR}]_{ab}\,N_{aR}N_{bR} + \text{h.c.}
\end{aligned} \tag{3}$$

After integrating out the heavy fermions, we can write down the effective Yukawa terms for charged fermions of the standard model as follows

$$\begin{aligned}
y_u &= Y_U \frac{v_{1R}}{M_U} Y_U^T, \\
y_d &= Y_D \frac{v_{2R}}{M_D} Y_D^T, \\
y_e &= Y_E \frac{v_{1R}}{M_E} Y_E^T
\end{aligned} \tag{4}$$

Similarly, the heavy neutral singlet fields $N_{L,R}$ can be integrated out to generate the effective mass matrix of neutrinos ν_L, ν_R which contains a Dirac mass term and two Majorana mass terms. The effective Dirac mass as well as Majorana mass terms are given by

$$M_D = Y_N \frac{1}{M_{RR}} M_{LR}^T \frac{1}{M_{RR}} Y_N^T v_{2L} v_{2R},$$

$$M_L = Y_N \frac{1}{M_{RR}} Y_N^T v_{2L}^2,$$

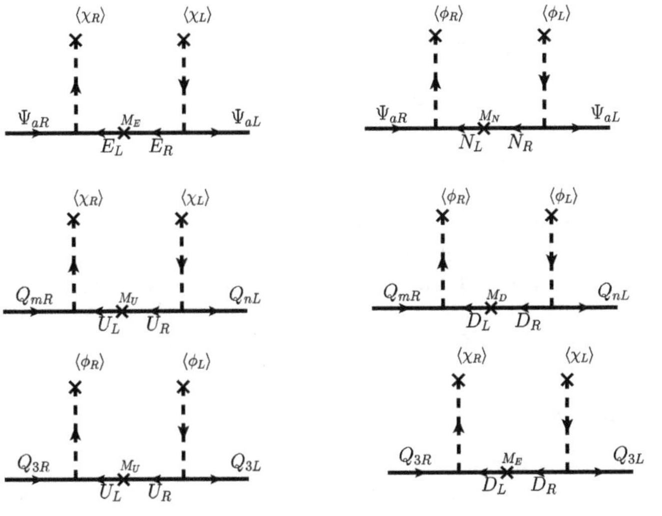

Fig. 1. Feynman diagram for Dirac mass of fermions within $SU(3)_c \times SU(3)_L \times SU(3)_R \times U(1)_X$ model.

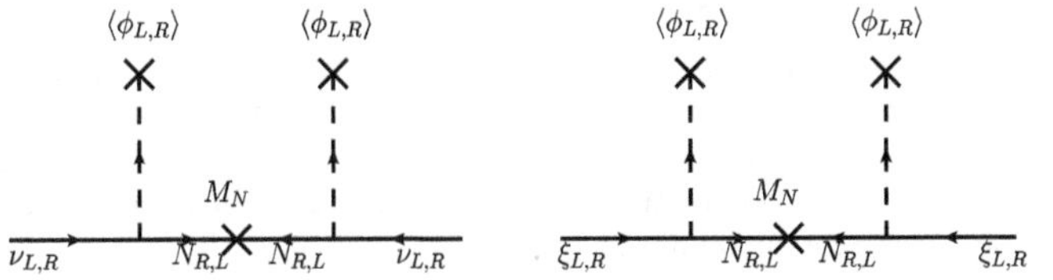

Fig. 2. Feynman diagram for Majorana mass of neutral fermions within $SU(3)_c \times SU(3)_L \times SU(3)_R \times U(1)_X$ model.

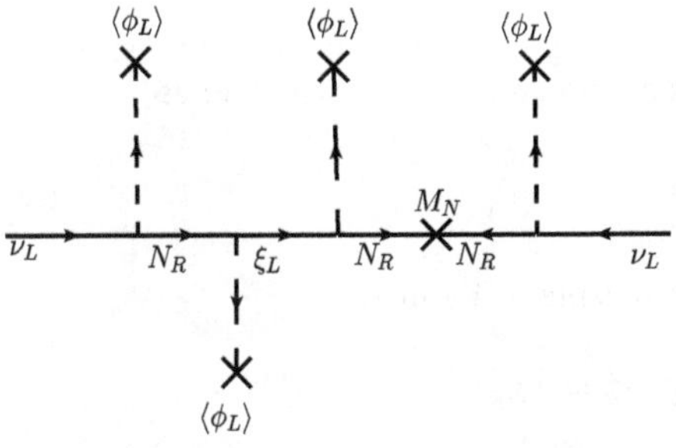

Fig. 3. Feynman diagram for new contribution to the Majorana mass of light neutrinos within $SU(3)_c \times SU(3)_L \times SU(3)_R \times U(1)_X$ model.

$$M_R = Y_N \frac{1}{M_{RR}} Y_N^T v_{2R}^2 . \tag{5}$$

There are additional neutrino leptons ξ_L, ξ_R which acquire Dirac and Majorana masses similar to $\nu_{L,R}$ shown above. They are given by

$$M_{\xi_D} = Y_N \frac{1}{M_{RR}} M_{LR}^T \frac{1}{M_{RR}} Y_N^T \omega_L \omega_R ,$$
$$M_{\xi_L} = Y_N \frac{1}{M_{RR}} Y_N^T \omega_L^2 ,$$
$$M_{\xi_R} = Y_N \frac{1}{M_{RR}} Y_N^T \omega_R^2 . \tag{6}$$

The origin of the Dirac masses can be understood from the mass diagrams shown in Fig. 1 whereas the Majorana mass diagrams are given in Fig. 2. Apart from these, the light neutrinos also receive non-leading contribution to their masses from the diagram shown in Fig. 3. The contribution of this diagram can be written as

$$M_L' = Y_N \frac{1}{M_{RR}} Y_N^T \frac{1}{M_{\xi_L}} Y_N \frac{1}{M_{RR}} Y_N^T v_{2L}^2 \omega_L^2 \tag{7}$$

Here we are assuming equality of left and right sector Yukawa couplings as well as masses $M_{LL} = M_{RR} = M_N$. The approximate scale of light neutrino mass matrix M_L has to be less than 0.1 eV which puts limit on model parameters Y_N, M_{RR}. For example, if $Y_N \simeq 0.01$ then $M_{RR} \geq 10^{10}$ GeV to keep $M_L \leq 0.1$ eV. Further lowering the scale of M_{RR} will involve more fine-tuning in the Yukawa coupling Y_N. In the limit of tiny M_D, the light neutrino mass solely originates from M_L given in equation (5) and the mixing between heavy and light neutrinos can be neglected. In such a case, the light neutral lepton mass matrix can be written in the basis (ν_L, ν_R) as

$$M_\nu = \begin{pmatrix} Y_N \frac{1}{M_{RR}} Y_N^T v_{2L}^2 & 0 \\ 0 & Y_N \frac{1}{M_{RR}} Y_N^T v_{2R}^2 \end{pmatrix} . \tag{8}$$

As a result of this particular structure of the mass matrix, both the mass eigenvalues for left-handed and right-handed neutrinos are proportional to each other. The two mixing matrices are related as

$$V_{\nu_R} = V_{\nu_L} \equiv U , \tag{9}$$

where U is the Pontecorvo–Maki–Nakagawa–Sakata (PMNS) leptonic mixing matrix. For a representative sets of input model parameters like $v_L \approx 174$ GeV, $v_R \approx 10$ TeV, and $m_i \approx 0.1$ eV, the right-handed neutrino masses lie in the range of keV scale. Such light keV scale right handed neutrinos can also have very interesting implications for cosmology [31]. We leave such a detailed study of this model from cosmology point of view to an upcoming work.

2.3. Gauge boson mass

The relevant kinetic terms leading to gauge boson masses are given by

$$\begin{aligned} \mathcal{L}_{\text{G.B.}} \supset & \left| \left(\frac{g_L}{2} \mathbf{W}_\mu^L + g_X(-2/3) B_\mu \right) \chi_L \right|^2 \\ & + \left| \left(\frac{g_R}{2} \mathbf{W}_\mu^R + g_X(-2/3) B_\mu \right) \chi_R \right|^2 \\ & + \left| \left(\frac{g_L}{2} \mathbf{W}_\mu^L + g_X(1/3) B_\mu \right) \phi_L \right|^2 \\ & + \left| \left(\frac{g_R}{2} \mathbf{W}_\mu^R + g_X(1/3) B_\mu \right) \phi_R \right|^2 \end{aligned} \tag{10}$$

where the factor $\mathbf{W}_\mu^{L,R}$ is defined as

$$\begin{aligned} \mathbf{W}_\mu^{L,R} &= \sum_{i=1}^{8} W_{L,R\,\mu}^i \Lambda_i \\ &= \begin{pmatrix} W^3 + \frac{1}{\sqrt{3}} W^8 & W^+ & V^{-q} \\ W^- & -W^3 + \frac{1}{\sqrt{3}} W^8 & V'^{-q-1} \\ V^q & V'^{q+1} & -\frac{2}{\sqrt{3}} W^8 \end{pmatrix}_{L,R} \end{aligned}$$

Using respective vev's for scalar fields shown in equation (2) we can derive the gauge boson masses for the present model. In the gauge boson spectrum, we have

- One massless photon A,
- Four neutral gauge bosons $Z_{L,R}, Z'_{L,R}$,
- Four charged gauge bosons $W^\pm_{L,R}$,
- Four gauge bosons with charge $q+1$, $X^{\pm(1+q)}_{L,R}$,
- Four gauge bosons with charge q, $Y^\pm_{L,R}$.

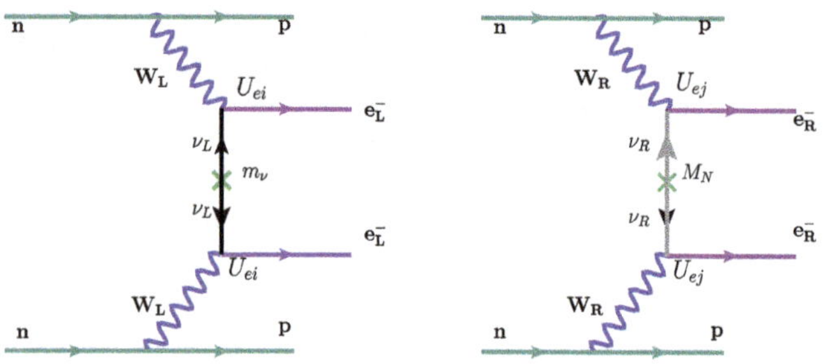

Fig. 4. $0\nu\beta\beta$ decay diagrams due to exchange of light left-handed neutrinos with left-handed charged currents and right-handed neutrinos with right-handed charged currents.

Table 2
Dimensionless particle physics parameters due to exchange of left-handed and right-handed neutrinos and the corresponding effective mass parameters.

η_i	**Effective mass parameter**
$\eta_{\nu_L} \approx \frac{1}{m_e}\sum_{i=1}^{3} U_{ei}^2 m_i$	$m_{ee}^{\nu_L} \approx \sum_{i=1}^{3} U_{ei}^2 m_i$
$\eta_{\nu_R} \approx \frac{1}{m_e}\left(\frac{M_{W_L}}{M_{W_R}}\right)^4 \sum_{i=1}^{3} U_{ei}^2 M_i$	$m_{ee}^{\nu_R} \approx \left(\frac{M_{W_L}}{M_{W_R}}\right)^4 \sum_{i=1}^{3} U_{ei}^2 M_i$

3. Contributions to $0\nu\beta\beta$

We find that the light sub-eV scale left-handed neutrinos (ν_L) and the heavy right-handed neutrinos (ν_R) with keV scale masses can give sizable contributions to neutrinoless double beta decay. Since the bitriplet scalar is absent in the present left-right symmetric 3331 model, there are no Dirac mass term for light neutrinos at tree level. However one can eventually generate Majorana masses for ν_L and $\nu_R \equiv N_R$ through universal seesaw, see Fig. 2. Such Majorana nature of neutrinos violate lepton number by two units and thus, contributes to $0\nu\beta\beta$ decay (see Fig. 4). The Feynman diagram for $0\nu\beta\beta$ decay is depicted in Fig. 6 due to exchange of left-handed as well as right-handed neutrinos. The corresponding Feynman amplitudes due to exchange of left-handed and right-handed neutrinos are given by

$$\mathcal{A}_{\nu_L} \propto G_F^2 \frac{U_{ei}^2 m_i}{p^2}$$
$$\mathcal{A}_{\nu_R} \propto G_F^2 \left(\frac{M_{W_L}^2}{M_{W_R}^2}\right) U_{ei}^2 \frac{M_i}{p^2} \tag{11}$$

The inverse half-life for a given isotope for $0\nu\beta\beta$ decay - due to exchange of left-handed light neutrinos via left-handed currents, and right-handed neutrinos via right-handed currents is given by

$$[T_{1/2}^{0\nu}]^{-1} = G_{01}\left(|\mathcal{M}_\nu \eta_\nu|^2 + |\mathcal{M}_N \eta_N|^2\right), \tag{12}$$

where G_{01} is $0\nu\beta\beta$ phase space factor, $\mathcal{M}_i$ correspond to the nuclear matrix elements (NME) and η_i is the corresponding dimensionless particle physics parameter. Since we have $M_i \approx 1$–10 MeV masses of right-handed neutrinos in the present model and satisfying $|M_i^2| \ll p^2$ where p being the neutrino virtually momentum around 100 MeV, the NMEs for right-handed neutrinos and left-handed neutrinos are same i.e., $\mathcal{M}_\nu = \mathcal{M}_N$ (see Table 2).

3.1. *Standard mechanism via left-handed neutrinos ν_L*

The standard mechanism for neutrinoless double beta decay due to exchange of light left-handed neutrinos via left-handed currents gives dimensionless particle physics parameter as,

$$\eta_{\nu_L} = \frac{1}{m_e}\sum_{i=1}^{3} U_{ei}^2 m_i = \frac{m_{ee}^{\nu_L}}{m_e}. \tag{13}$$

Here, m_e is the electron mass. The effective mass parameter for standard mechanism is explicitly given by

$$m_{ee}^{\nu_L} = \left|c_{12}^2 c_{13}^2 m_1 + s_{12}^2 c_{13}^2 m_2 e^{i\alpha} + s_{13}^2 m_3 e^{i\beta}\right|. \tag{14}$$

3.2. *New contribution from right-handed neutrinos ν_R*

In the present left-right symmetric 3331 model, we found that the right-handed neutrino mass lies around a few keV (for TeV scale W_R) which is much less than its momentum, $M_i \ll |p|$. Under this condition, the propagator simplifies in a similar way as for the light neutrino exchange,

$$P_R \frac{\not{p} + M_i}{p^2 - M_i^2} P_R \approx \frac{M_i}{p^2}. \tag{15}$$

This results dimensionless particle physics parameter η_{ν_R} due to exchange of right-handed neutrinos via right-handed currents as,

$$\eta_{\nu_R} \approx \frac{1}{m_e}\left(\frac{M_{W_L}}{M_{W_R}}\right)^4 \sum_{i=1}^{3} U_{ei}^2 M_i \propto \eta_{\nu_L}, \tag{16}$$

where the proportionality relation between η_{ν_R} and η_{ν_L} appears at the last step due to the proportionality between heavy and light neutrino mass matrix discussed above. After a little simplification, the effective mass parameter due to exchange of right-handed neutrinos can be expressed as,

$$m_{ee}^{\nu_R} \approx \left(\frac{M_{W_L}}{M_{W_R}}\right)^4 \sum_{i=1}^{3} U_{ei}^2 M_i \propto m_{ee}^{\nu_L}. \tag{17}$$

It is clear from eq. (6) that both light and heavy neutrino mass eigenvalues are proportional to each other as

$$M_i \approx \frac{v_R^2}{v_L^2} m_i. \tag{18}$$

Thus, one can express W_R mass as

$$M_{W_R}^2 \approx \frac{1}{4} g_R^2 v_R^2 = \frac{1}{4} g_R^2 v_L^2 \frac{M_i}{m_i}. \tag{19}$$

As a result, the effective Majorana mass parameter - with $g_L \approx g_R$ and $M_W \approx \frac{1}{2} g_L v_L$ - is modified to

$$m_{ee}^{\nu_R} \approx \left(\frac{m_1}{M_1}\right)^2 \sum_{i=1}^{3} U_{ei}^2 M_i. \tag{20}$$

Comparing this with the light neutrino contribution $m_{ee}^{\nu_L} = \sum_{i=1}^{3} U_{ei}^2 m_i$, it is straightforward to estimate that the heavy neutrino contribution is suppressed by a factor of $m_i/M_i = (v_L/v_R)^2$ compared to the light neutrino contribution.

3.2.1. *For NH pattern*

For normal hierarchical (NH) pattern of light neutrinos we consider the following mass structures for left-handed and right-handed neutrinos,

$$m_1 = m_{\text{lightest}}, \qquad m_2 = \sqrt{m_1^2 + \Delta m_{\text{sol}}^2},$$
$$m_3 = \sqrt{m_1^2 + \Delta m_{\text{sol}}^2 + +\Delta m_{\text{atm}}^2},$$
$$M_> = M_3,$$
$$M_1 = \frac{m_1}{m_3} M_3, \quad M_2 = \frac{m_2}{m_3} M_3, \tag{21}$$

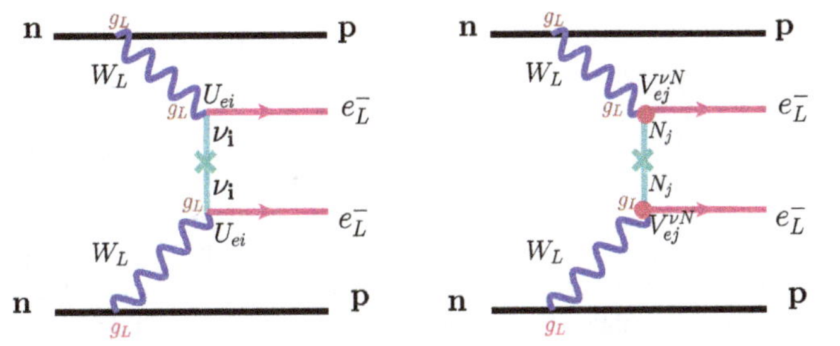

Fig. 5. $0\nu\beta\beta$ decay diagrams due to purely left-handed charge current interaction and with the exchange of ν_L and N_R.

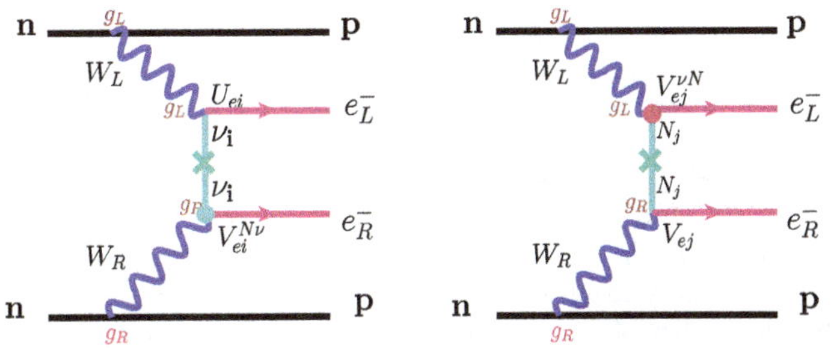

Fig. 6. $0\nu\beta\beta$ decay diagrams due to mediation of one W_L and one W_R which also involves light-heavy neutrino mixing.

where M_3 is fixed around few keV range, as a result of choosing the W_R mass scale at a few TeV. The analytic form for effective mass parameters due to exchange of right-handed neutrinos is given by

$$m_{\rm ee}^{\nu_L} = \left|c_{12}^2 c_{13}^2 m_1 + s_{12}^2 c_{13}^2 m_2 e^{i\alpha} + s_{13}^2 m_3 e^{i\beta}\right| ,$$
$$|m_{\rm ee}^{\nu_R}|_{\rm NH} = \left(\frac{m_1}{M_1}\right)^2 M_3 \left|\frac{m_1}{m_3} c_{12}^2 c_{13}^2 + \frac{m_2}{m_3} s_{12}^2 c_{13}^2 e^{i\alpha} + s_{13}^2 e^{i\beta}\right| \tag{22}$$

3.2.2. For IH pattern

Similarly for inverse hierarchical (IH) pattern of light neutrinos, the masses for light left-handed and right-handed neutrinos are fixed as,

$$m_3 = m_{\rm lightest}\,, \qquad m_2 = \sqrt{m_1^2 + \Delta m_{\rm sol}^2}\,,$$
$$m_3 = \sqrt{m_1^2 + \Delta m_{\rm sol}^2 + +\Delta m_{\rm atm}^2}\,,$$
$$M_> = M_2\,,$$
$$M_1 = \frac{m_1}{m_2} M_2, \quad M_3 = \frac{m_3}{m_2} M_2\,, \tag{23}$$

where we fixed the heaviest right-handed neutrino mass M_2 around few keV. The effective mass parameters due to exchange of right-handed neutrinos is given by

$$m_{\rm ee}^{\nu_L} = \left|c_{12}^2 c_{13}^2 m_1 + s_{12}^2 c_{13}^2 m_2 e^{i\alpha} + s_{13}^2 m_3 e^{i\beta}\right| ,$$
$$|m_{\rm ee}^{\nu_R}|_{\rm IH} = \left(\frac{m_1}{M_1}\right)^2 M_2 \left|\frac{m_1}{m_2} c_{12}^2 c_{13}^2 + s_{12}^2 c_{13}^2 e^{i\alpha} + \frac{m_3}{m_2} s_{13}^2 e^{i\beta}\right| \tag{24}$$

3.3. $0\nu\beta\beta$ with heavy-light neutrino mixing

Though the purely heavy neutrino contribution to $0\nu\beta\beta$ remains suppressed, there can be sizable contributions from heavy-light neutrino mixing diagrams (see Fig. 5). This heavy-light neutrino mixing can also contribute to light neutrino masses through a type I seesaw formula which was ignored in the above discussion for simplicity. Such an assumption is valid for negligible heavy-light neutrino mixing. However, if we go beyond this simple assumption or equivalently consider $M_{LR} \approx M_{RR}$, then the Dirac as well as Majorana mass matrices for ν_L and $\nu_R \equiv N_R$ can be written as

$$M_D = Y_N \frac{1}{M_{RR}} Y_N^T v_{2L} v_{2R}\,,$$
$$M_L = Y_N \frac{1}{M_{RR}} Y_N^T v_{2L}^2\,,$$
$$M_R = Y_N \frac{1}{M_{RR}} Y_N^T v_{2R}^2\,. \tag{25}$$

Thus, the neutral lepton mass matrix in the basis (ν_L, N_R) as

$$M_\nu = \begin{pmatrix} Y_N \frac{1}{M_{RR}} Y_N^T v_{2L}^2 & Y_N \frac{1}{M_{RR}} Y_N^T v_{2L} v_{2R} \\ Y_N^T \frac{1}{M_{RR}} Y_N v_{2L} v_{2R} & Y_N \frac{1}{M_{RR}} Y_N^T v_{2R}^2 \end{pmatrix}.$$
$$= \begin{pmatrix} M_L & M_D \\ M_D^T & M_R \end{pmatrix} \tag{26}$$

In the limit $M_L \ll M_D \ll M_R$, the type-I seesaw contribution to light neutrino mass is given by

$$M_\nu^{\rm I} = -M_D^T \frac{1}{M_R} M_D\,, \tag{27}$$

and the light-heavy neutrino mixing is proportional to $M_D/M_R \approx v_{2L}/v_{2R}$. With $v_{2L} \approx 174$ GeV and v_{2R} around few TeV, we find that light-heavy neutrino mixing is large of the order of $\lesssim 0.1$. This large value of light-heavy neutrino mixing where heavy neutrinos are fixed at few keV scale, can contribute to $0\nu\beta\beta$ significantly.

3.3.1. Purely left-handed current effects

The new physics contributions to $0\nu\beta\beta$ decay arising from purely left-handed currents due to exchange of keV scale right-handed neutrinos results in the following effective mass parameter,

$$\mathbf{m}_{\rm ee,LL}^N = \sum_{i=1}^{3} {V_{ei}^{\nu N}}^2 M_i \tag{28}$$

here M_i is in keV range and $V^{\nu N}$ is the light-heavy neutrino mixing. Since $V^{\nu N} \propto v_{2L}/v_{2R} \approx 0.01$, the effective mass parameter for $0\nu\beta\beta$ is estimated to be $\mathbf{m}_{\rm ee,LL}^N = (0.01)^2 \cdot 10^3$ eV which is of the order of 0.1 eV, saturating the KamLAND-Zen [30] bound.

3.3.2. From λ-diagram

The effective mass parameters due to the $W_L - W_R$ mediated diagrams (known as λ diagrams) shown in Fig. 6 are given by

$$\mathbf{m}_{\rm ee,\lambda}^{\nu} = 10^{-2} \left(\frac{M_{W_L}}{M_{W_R}}\right)^2 \sum_{i=1}^{3} U_{ei} V_{ei}^{N\nu} |p|$$
$$\mathbf{m}_{\rm ee,\lambda}^{N} = 10^{-2} \left(\frac{M_{W_L}}{M_{W_R}}\right)^2 \sum_{j=1}^{3} V_{ej} V_{ej}^{\nu N} |p| \tag{29}$$

With $M_{W_L} \approx 80.4$ GeV, $M_{W_R} \approx 4$ TeV and $V^{\nu N} \simeq 0.01$, the effective mass parameters due to these λ diagrams are found to be around sub-eV which can translated to a life-time of 10^{26} yrs. This value is very close to experimental bound for Xe isotope. It is interesting that, such large observable heavy-light neutrino mixing arises naturally in the model without any fine-tuning of the Yukawa couplings involved. Although we consider one contribution at a time in the above discussions, in general one has to include all the contributions and at the same time keeping the light neutrino mass and mixing in the allowed range. We intend to perform a detailed study, considering the most general neutrino mass formula $M_\nu = M_L + M_\nu^{I}$ to an upcoming work.

3.4. Total contribution to $0\nu\beta\beta$ transition

The total contribution to inverse half-life for neutrinoless double beta decay for a given isotope in the present left-right symmetric 3331 model is given by

$$[T_{1/2}^{0\nu}]^{-1} = G_{01} \left| \frac{\mathcal{M}_\nu}{m_e} \right|^2 |m_{\rm eff}^{\rm ee}|^2, \tag{30}$$

where,

$$|m_{\rm ee}^{\rm eff}|^2 = |m_{\rm ee}^{\nu}|^2 + |m_{\rm ee,L}^{N}|^2 + |m_{\rm ee,R}^{N}|^2 + |m_{\rm ee,\lambda}^{\nu/N}|^2 \tag{31}$$

We have generated effective mass with the variation of lightest neutrino mass, m_{ν_1} for NH and m_{ν_3} for IH as shown in Fig. 7. The standard mechanism for $0\nu\beta\beta$ decay due to exchange of light neutrinos is displayed in green band for NH (m_{ee}^{ν} − NH) and in red band for IH (m_{ee}^{ν} − IH). The present bound on half-life is $T_{1/2}^{0\nu}(^{136}{\rm Xe}) > 1.07 \times 10^{26}$ yr at 90% C.L. from KamLAND-Zen [30]

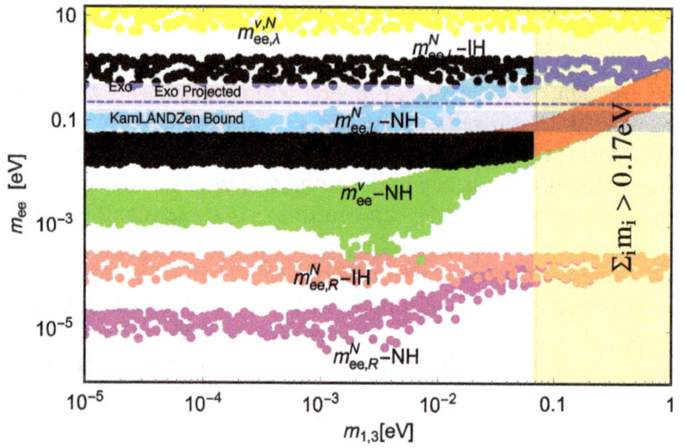

Fig. 7. Effective Majorana mass parameter m_{ee} as a function of the lightest neutrino mass m_1 for NH (m_3 for IH). We have used best-fit oscillation data [27] and the Majorana phases are varied between $[0, 2\pi]$. The vertical shaded regions are excluded from cosmology [28,29] while the dashed horizontal line is for from KamLAND-Zen [30] bound. The standard mechanism for $0\nu\beta\beta$ decay due to exchange of light neutrinos is presented in green band for NH (m_{ee}^{ν} − NH) and in red band for IH (m_{ee}^{ν} − IH). The sub-dominant contribution due to exchange of right-handed neutrinos via $W_R - W_R$ mediation is displayed in magenta points for NH ($m_{ee,R}^{N}$ − NH) and in purple points for IH ($m_{ee,R}^{N}$ − IH). Similarly, the new physics contributions due to large light-heavy neutrino mixing: i) arising from purely left-handed current through exchange of keV masses for right-handed neutrinos saturating the present KamLANDZen bound is displayed in cyan points for NH ($m_{ee,L}^{N}$ − NH) and in blue points for IH ($m_{ee,L}^{N}$ − IH), ii) from mixed helicity λ-diagram represented by yellow points ($m_{ee,\lambda}^{\nu/N}$) same for both NH and IH case. (For interpretation of the references to colour in this figure legend, the reader is referred to the web version of this article.)

and thus, the corresponding bound on effective mass is represented in horizontal dashed line while the cyan region is excluded due to the 95% C.L. limit on $\sum_i m_i < 0.17$ eV obtained from Planck+WMAP low multipole polarisation+high resolution CMB+BAO data.

As shown in Fig. 7, it is clear that if only the light Majorana neutrinos contribute to the $0\nu\beta\beta$ decay process, it is very difficult to see the signal even at next generation experiments. It is found that the new physics contributions to $0\nu\beta\beta$ decay due to large light-heavy neutrino mixing–arising: i) from purely left-handed currents through exchange of keV scale right-handed neutrinos and ii) from mixed helicity λ-diagram – can saturate the experimental bound which can be observed in current ongoing experiments. These contributions i) due to purely left-handed currents are displayed in cyan points for NH ($m_{ee,L}^{N}$ − NH) and in blue points for IH ($m_{ee,L}^{N}$ − IH), ii) from mixed helicity λ-diagram represented by yellow points ($m_{ee,\lambda}^{\nu/N}$) same for both NH and IH case. The sub-dominant contribution due to exchange of right-handed neutrinos via $W_R - W_R$ mediation is also displayed in magenta points for NH ($m_{ee,R}^{N}$ − NH) and in purple points for IH ($m_{ee,R}^{N}$ − IH).

4. Other implications

As noted above, the right handed neutrinos in the model can be as light as a keV or below depending on the scale of symmetry breaking. If long lived enough, such keV scale right handed neutrinos can have interesting implications in cosmology as they can contribute to dark matter or relativistic degrees of freedom for keV or eV scale masses. Such a light right handed neutrino can decay into a standard model neutrino and a photon at one loop having the following decay width [32]

$$\Gamma_{N \to \nu\gamma} = \frac{9\alpha G_F^2}{256(4\pi^4)} \sin^2(2\theta_{\nu N}) m_N^2 \tag{32}$$

where α is the fine structure constant and G_F is the Fermi coupling constant. The heavy-light neutrino mixing angle $\theta_{\nu N} \approx \frac{v_{2L}}{v_{2R}} = \frac{v_L}{v_R}$ where in the last step we have identified the symmetry breaking scales $v_{2L,2R}$ with $v_{L,R}$ for simplicity. The lightest right handed neutrino mass is related to the lightest active neutrino mass by the proportionality factor $(v_L/v_R)^2$ where v_L is the electroweak symmetry breaking scale and v_R is the left-right symmetry breaking scale, taken to be heavier than 6 TeV. Since the lightest active neutrino mass is still a free parameter from light neutrino oscillation data, we vary it in the range $(10^{-6}, 0.1)$ eV and accordingly calculate the lightest right handed neutrino mass using the proportionality factor. The resulting mass and life time of the lightest right handed neutrino are shown in Fig. 8. It can be seen that the lightest right handed neutrino can have masses in the range of a few eV to a few keV for symmetry breaking scale v_R in the 6–100 TeV range. The lifetime of such a right handed neutrino can

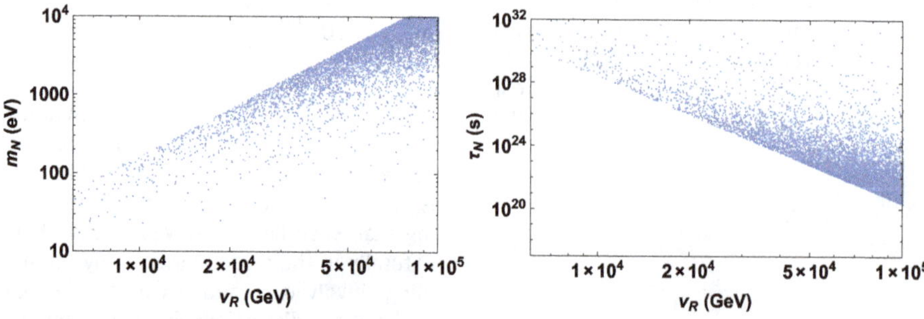

Fig. 8. The left panel shows the variation of the mass of the lightest right handed neutrino with v_R while the right panel shows its lifetime.

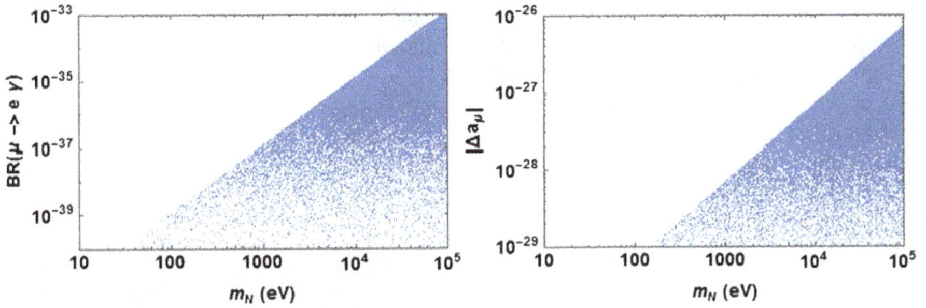

Fig. 9. The left panel shows the contribution to $\mu \to e\gamma$ from right handed neutrino and the right panel shows the corresponding contribution to muon $g-2$.

be significantly greater than the age of the Universe $\approx 10^{17}$ s as seen from the right panel of Fig. 8.

We also check the possible contribution of these right handed neutrinos to lepton flavour violating decay $\mu \to e\gamma$ as well as anomalous magnetic moment of the muon. In the SM, such LFV decays occur at loop level and heavily suppressed due to the smallness of neutrino masses, far beyond the current experimental sensitivity [33] around $\text{BR}(\mu \to e\gamma) < 4.2 \times 10^{-13}$ at 90% confidence level. The anomalous magnetic moment is the discrepancy between the experimentally observed and the predicted value of muon $(g-2)$ [34]

$$\Delta a_\mu = a_\mu^{\text{exp}} - a_\mu^{\text{pred}} = (29.0 \pm 9.0) \times 10^{-10} \tag{33}$$

The right handed neutrinos can give rise to additional contributions to these flavour observables due to heavy light neutrino mixing. There can be other diagrams due to right handed gauge bosons as well, but their contributions will not be more than the left counterparts we consider here. The contribution of right handed neutrinos to $\mu \to e\gamma$ can be written as [35]

$$\text{BR}(\mu \to e\gamma) = \frac{\sqrt{2}G_F^2 m_\mu^5}{\Gamma_\mu} \Big|\sum_i V_{\mu i}^{\nu N}(V_{ei}^{\nu N})^* G_\gamma\left(\frac{M_i^2}{M_{W_L}^2}\right)\Big|^2 \tag{34}$$

where

$$G_\gamma(x) = \frac{x - 6x^2 + 3x^3 + 2x^4 - 6x^3 \log x}{4(1-x)^4}$$

and Γ_μ is the total decay of the muon. Also, $M_i = m_N$ correspond to the right handed neutrino masses. Similarly, the right handed neutrino contribution to the muon $g-2$ is given by [35]

$$\Delta a_\mu = -\frac{4\sqrt{2}G_F m_\mu^2}{(4\pi)^2} \sum_i |V_{\mu i}^{\nu N}|^2 G_\gamma\left(\frac{M_i^2}{M_{W_L}^2}\right) \tag{35}$$

For simplicity, we consider the heavy light mixing to be identical for all generations and approximately equal that is $V^{\nu N} \approx \frac{v_L}{v_R}$ as considered before. We find that the right handed neutrino contribution to both these flavour observables remain suppressed as can be seen from Fig. 9. Therefore, some additional new physics is required in order to saturate the experimental bounds on these observables.

5. Symmetry breaking pattern

Depending on the scale of different vev's mentioned above, the gauge symmetry of the model can be broken down to that of the Standard Model through several possible symmetry breaking chains. They are summarised pictorially in Fig. 10. The relevant scalar potential for the model can be written as

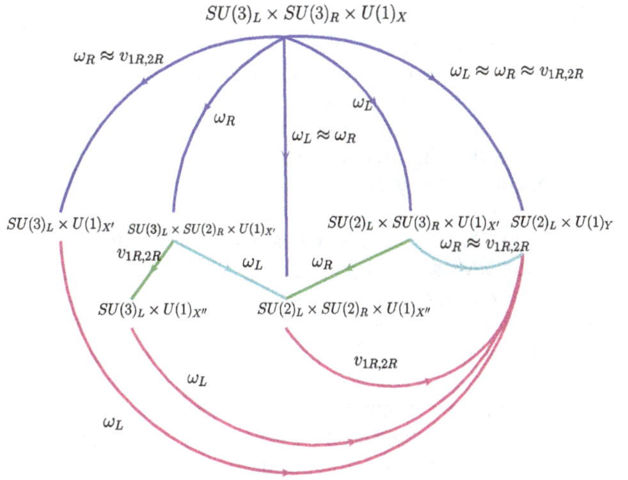

Fig. 10. Symmetry breaking patterns of $SU(3)_C \times SU(3)_L \times SU(3)_R \times U(1)_X$ gauge symmetry to the Standard Model gauge symmetry $SU(3)_C \times SU(2)_L \times U(1)_Y$.

$$\begin{aligned}
V(\phi,\chi) &= V_\chi + V_\phi + V_{\chi\phi} \\
V_\chi &= \mu_\chi^2\left((\chi_L^\dagger\chi_L) + (\chi_R^\dagger\chi_R)\right) + \lambda_\chi\left((\chi_L^\dagger\chi_L)^2 + (\chi_R^\dagger\chi_R)^2\right) \\
&\quad + \lambda'_\chi(\chi_L^\dagger\chi_L)(\chi_R^\dagger\chi_R) \\
V_\phi &= \mu_\phi^2\left((\phi_L^\dagger\phi_L) + (\phi_R^\dagger\phi_R)\right) + \lambda_\phi\left((\phi_L^\dagger\phi_L)^2 + (\phi_R^\dagger\phi_R)^2\right) \\
&\quad + \lambda'_\phi(\phi_L^\dagger\phi_L)(\phi_R^\dagger\phi_R) \\
V_{\chi\phi} &= \rho_{\phi\chi}\left((\chi_L^\dagger\chi_L) + (\chi_R^\dagger\chi_R)\right)\left((\phi_L^\dagger\phi_L) + (\phi_R^\dagger\phi_R)\right) \\
&\quad + \rho'_{\phi\chi}\left(\epsilon_{ijk}\phi_L^i\phi_L^j\chi_L^k + \epsilon_{ijk}\phi_R^i\phi_R^j\chi_R^k + \text{h.c.}\right)
\end{aligned}$$

In the scalar potential written above, the discrete left-right symmetry is assumed which ensures the equality of left and right sector couplings. However, as shown in earlier works [21] in the context of usual LRSM with universal seesaw that the scalar potential of such a model with exact discrete left-right symmetry is too restrictive and gives to either parity preserving $(v_L = v_R)$ solution or a solution with $(v_R \neq 0, v_L = 0)$ at tree level. While the first one is not phenomenologically acceptable the latter solution can be acceptable if a non-zero vev $v_L \neq 0$ can be generated through radiative corrections [36]. While it may naturally explain the smallness of v_L compared to v_R, it will constrain the parameter space significantly [36]. Another way of achieving a parity breaking vacuum is to consider softly broken discrete left-right symmetry by considering different mass terms for the left and right sector scalars [8,21]. As it was pointed out by the authors of [8], such a model which respects the discrete left-right symmetry everywhere except in the scalar mass terms, preserve the *naturalness* of the left-right symmetry in spite of radiative corrections.

Another interesting way is to achieve parity breaking vacuum is to decouple the scale of parity breaking and gauge symmetry breaking by introducing a parity odd singlet scalar [37]. While we do not perform a detailed analysis of different possible symmetry breaking chains and their constraints on the parameter space of the model, we outline them pictorially in the cartoon shown in Fig. 10. As can be seen from Fig. 10, there are seven different symmetry breaking chains through which the gauge symmetry of the model $SU(3)_c \times SU(3)_L \times SU(3)_R \times U(1)_X$ can be broken down to that of the SM as summarised below.

- One step breaking: The vev's satisfy $v_{1L,2L} \ll v_{1R,2R} \approx \omega_L \approx \omega_R$ in this case.
- Two step breaking: The vev's satisfy either $v_{1L,2L} \ll \omega_L \ll v_{1R,2R} \approx \omega_R$ or $v_{1L,2L} \ll v_{1R,2R} \ll \omega \approx \omega_R$ or $v_{1L,2L} \ll \omega_R \approx v_{1R,2R} \ll \omega_L$. The usual 331 model presumes an intermediate stage in the first case while the usual LRSM or 3221 symmetry arises an intermediate symmetry in the second case. In the third case, the 3231 symmetry assumes an intermediate stage. The phenomenology of such asymmetric LRSM was discussed recently by [14].
- Three step breaking: This is possible in three different ways when the vev's satisfy $v_{1L,2L} \ll v_{1R,2R} \ll \omega_R \ll \omega_L$ or $v_{1L,2L} \ll v_{1R,2R} \ll \omega_R \ll \omega_L$ or $v_{1L,2L} \ll \omega_L \ll v_{1R,2R} \ll \omega_R$. One can have both the usual LRSM or 3221 or asymmetric LRSM (3321 or 3231) or the usual 331 model as an intermediate stage.

All these different symmetry breaking chains can not only provide a different phase transition history in cosmology but also give rise to different particle spectra including gauge bosons as well as neutral fermions which could be tested in different experiments.

6. Conclusion

We have demonstrated a class of left-right symmetric model with extended gauge group $SU(3)_c \times SU(3)_L \times SU(3)_R \times U(1)_X$ with a universal seesaw mechanism for fermion masses and mixing and the implications for neutrinoless double beta ($0\nu\beta\beta$) decay. The novel feature of the model is that masses and mixing for left-handed and right-handed neutrinos are exactly determined by oscillation parameters and lightest neutrino mass. This forces the heavy neutrino masses to lie in keV regime if the W_R mass is fixed at a few TeV. We show that for such a TeV scale model, the heavy-light neutrino mixing can be quite large and can contribute substantially to $0\nu\beta\beta$ transition keeping it within experimental reach. Our numerical estimation showed that the new physics contributions due to purely left-handed currents via exchange of keV scale right-handed neutrinos and the λ-diagram can saturate the experimental KamLANDZen bound. We also show that the right handed neutrinos in this model can be long lived and can have interesting cosmological implications the details of which we leave for future studies. We also check the contributions of such light right handed neutrinos to lepton flavour violating decay and anomalous magnetic moment of the muon and find them to be suppressed with respect to the experimental bounds. In the end we have discussed the scalar potential and possible symmetry breaking patterns that can be allowed for spontaneous breaking of the 3331 gauge symmetry to that of the standard model.

Acknowledgements

The authors would like to thank the organisers of the *Indo-US Bilateral Workshop on Understanding the Origin of the Invisible Sector: From Neutrinos to Dark Matter and Dark Energy* during November 16–18, 2016 at the School of Physics, University of Hyderabad, India where this work was initiated.

References

[1] S. Fukuda, et al., Super-Kamiokande, Phys. Rev. Lett. 86 (2001) 5656, arXiv:hep-ex/0103033;
Q.R. Ahmad, et al., SNO, Phys. Rev. Lett. 89 (2002) 011301, arXiv:nucl-ex/0204008;
Phys. Rev. Lett. 89 (2002) 011302, arXiv:nucl-ex/0204009;
J.N. Bahcall, C. Pena-Garay, New J. Phys. 6 (2004) 63, arXiv:hep-ph/0404061;
K. Nakamura, et al., J. Phys. G 37 (2010) 075021.
[2] K. Abe, et al., T2K Collaboration, Phys. Rev. Lett. 107 (2011) 041801, arXiv:1106.2822 [hep-ex].
[3] Y. Abe, et al., Phys. Rev. Lett. 108 (2012) 131801, arXiv:1112.6353 [hep-ex].
[4] F.P. An, et al., DAYA-BAY Collaboration, Phys. Rev. Lett. 108 (2012) 171803, arXiv:1203.1669 [hep-ex].
[5] J.K. Ahn, et al., RENO Collaboration, Phys. Rev. Lett. 108 (2012) 191802, arXiv:1204.0626.
[6] P. Adamson, et al., MINOS, Phys. Rev. Lett. 110 (2013) 171801.
[7] J.C. Pati, A. Salam, Phys. Rev. D 10 (1974) 275;
G. Senjanovic, R.N. Mohapatra, Phys. Rev. D 12 (1975) 1502;
R.N. Mohapatra, R.E. Marshak, Phys. Rev. Lett. 44 (1980) 1316;
J.F. Gunion, J. Grifols, A. Mendez, B. Kayser, F.I. Olness, Phys. Rev. D 40 (1989) 1546;
N.G. Deshpande, J.F. Gunion, B. Kayser, F.I. Olness, Phys. Rev. D 44 (1991) 837.
[8] R.N. Mohapatra, J.C. Pati, Phys. Rev. D 11 (1975) 2558.
[9] M. Singer, J.W.F. Valle, J. Schechter, Phys. Rev. D 22 (1980) 738;
F. Pisano, V. Pleitez, Phys. Rev. D 46 (1992) 410;
P.H. Frampton, Phys. Rev. Lett. 69 (1992) 2889;
R. Foot, O.F. Hernandez, F. Pisano, V. Pleitez, Phys. Rev. D 47 (1993) 4158.
[10] M. Reig, J.W.F. Valle, C.A. Vaquera-Araujo, arXiv:1611.02066 [hep-ph].
[11] E.T. Franco, V. Pleitez, arXiv:1611.06568 [hep-ph].
[12] M. Reig, J.W.F. Valle, C.A. Vaquera-Araujo, arXiv:1611.04571 [hep-ph].
[13] A.G. Dias, C.A. de S. Pires, P.S. Rodrigues da Silva, Phys. Rev. D 82 (2010) 035013, arXiv:1003.3260 [hep-ph].
[14] D.T. Huong, P.V. Dong, Phys. Rev. D 93 (2016) 095019.
[15] P. Minkowski, Phys. Lett. B 67 (1977) 421;
M. Gell-Mann, P. Ramond, R. Slansky, print-80-0576, (CERN), 1980;
T. Yanagida, in: Proceedings of the Workshop on the Baryon Number of the Universe and Unified Theories, Tsukuba, Japan, 13–14 Feb. 1979, 1979;
R.N. Mohapatra, G. Senjanovic, Phys. Rev. Lett. 44 (1980) 912;
J. Schechter, J.W.F. Valle, Phys. Rev. D 22 (1980) 2227.
[16] R.N. Mohapatra, G. Senjanovic, Phys. Rev. D 23 (1981) 165.
[17] G. Lazarides, Q. Shafi, C. Wetterich, Nucl. Phys. B 181 (1981) 287;
C. Wetterich, Nucl. Phys. B 187 (1981) 343;
J. Schechter, J.W.F. Valle, Phys. Rev. D 25 (1982) 774;
B. Brahmachari, R.N. Mohapatra, Phys. Rev. D 58 (1998) 015001;
R.N. Mohapatra, Nucl. Phys. Proc. Suppl. 138 (2005) 257;
S. Antusch, S.F. King, Phys. Lett. B 597 (2) (2004) 199.
[18] R.N. Mohapatra, J.W.F. Valle, Phys. Rev. D 34 (1986) 1642;
M. Gonzalez-Garcia, J.W.F. Valle, Phys. Lett. B 216 (1989) 360;
M.E. Catano, R. Martinez, F. Ochoa, Phys. Rev. D 86 (2012) 073015.
[19] E.K. Akhmedov, M. Lindner, E. Schnapka, J.W.F. Valle, Phys. Lett. B 368 (1996) 270;
E.K. Akhmedov, M. Lindner, E. Schnapka, J.W.F. Valle, Phys. Rev. D 53 (1996) 2752;
M. Malinsky, J. Romao, J.W.F. Valle, Phys. Rev. Lett. 95 (2005) 161801.
[20] B. Brahmachari, E. Ma, U. Sarkar, Phys. Rev. Lett. 91 (2003) 011801.
[21] A. Davidson, K.C. Wali, Phys. Rev. Lett. 59 (1987) 393;
R.N. Mohapatra, Y. Zhang, JHEP 1406 (2014) 072.
[22] K.S. Babu, R.N. Mohapatra, Phys. Rev. Lett. 62 (1989) 1079;
P.-H. Gu, M. Lindner, Phys. Lett. B 698 (2011) 40.
[23] S. Patra, Phys. Rev. D 87 (1) (2013) 015002, http://dx.doi.org/10.1103/PhysRevD.87.015002, arXiv:1212.0612 [hep-ph].
[24] P.S.B. Dev, R.N. Mohapatra, Y. Zhang, JHEP 1602 (2016) 186, http://dx.doi.org/10.1007/JHEP02(2016)186, arXiv:1512.08507 [hep-ph].
[25] F.F. Deppisch, C. Hati, S. Patra, P. Pritimita, U. Sarkar, Phys. Lett. B 757 (2016) 223, http://dx.doi.org/10.1016/j.physletb.2016.03.081, arXiv:1601.00952 [hep-ph].
[26] F.F. Deppisch, C. Hati, S. Patra, P. Pritimita, U. Sarkar, arXiv:1701.02107 [hep-ph].
[27] M.C. Gonzalez-Garcia, M. Maltoni, J. Salvado, T. Schwetz, JHEP 1212 (2012) 123, arXiv:1209.3023 [hep-ph].
[28] P.A.R. Ade, et al., Planck Collaboration, Astron. Astrophys. 594 (2016) A13, arXiv:1502.01589 [astro-ph.CO].
[29] S. Mertens, KATRIN Collaboration, Phys. Proc. 61 (2015) 267.

[30] A. Gando, et al., KamLAND-Zen Collaboration, Addendum: Phys. Rev. Lett. 117 (8) (2016) 082503, http://dx.doi.org/10.1103/PhysRevLett.117.082503, Phys. Rev. Lett. 117 (10) (2016) 109903, arXiv:1605.02889 [hep-ex].
[31] D. Borah, Phys. Rev. D 94 (2016) 075024.
[32] R. Adhikari, et al., JCAP 1701 (2017) 025.
[33] A.M. Baldini, et al., MEG Collaboration, Eur. Phys. J. C 76 (2016) 434.
[34] G.W. Bennett, et al., Muon g-2 Collaboration, Phys. Rev. D 73 (2006) 072003.
[35] A. Abada, T. Toma, JHEP 1602 (2016) 174.
[36] A. Kobakhidze, A. Spencer-Smith, JHEP 1308 (2013) 036.
[37] D. Chang, R.N. Mohapatra, M.K. Parida, Phys. Rev. Lett. 52 (1984) 1072.

Constraining dark photon model with dark matter from CMB spectral distortions

Ki-Young Choi [a], Kenji Kadota [b], Inwoo Park [c,b,*]

[a] Institute for Universe and Elementary Particles and Department of Physics, Chonnam National University, 77 Yongbong-ro, Buk-gu, Gwangju, 61186, Republic of Korea
[b] Center for Theoretical Physics of the Universe, Institute for Basic Science (IBS), Daejeon, 34051, Republic of Korea
[c] Department of Physics, Korea Advanced Institute of Science and Technology, Daejeon, 34141, Republic of Korea

ARTICLE INFO

Editor: M. Trodden

Keywords:
Dark matter
Dark photon
CMB distortion

ABSTRACT

Many extensions of Standard Model (SM) include a dark sector which can interact with the SM sector via a light mediator. We explore the possibilities to probe such a dark sector by studying the distortion of the CMB spectrum from the blackbody shape due to the elastic scatterings between the dark matter and baryons through a hidden light mediator. We in particular focus on the model where the dark sector gauge boson kinetically mixes with the SM and present the future experimental prospect for a PIXIE-like experiment along with its comparison to the existing bounds from complementary terrestrial experiments.

1. Introduction

The energy spectrum of the cosmic microwave background (CMB) follows the most perfect blackbody spectrum ever observed. There yet can exist a minuscule deviation from the blackbody when the CMB photons are not in a perfect equilibrium. The number-changing interactions such as Bremsstrahlung and double Compton scatterings are not efficient enough for the redshift $z \lesssim 2 \times 10^6$ and the energy injection/extraction can result in the Bose–Einstein distribution with a non-vanishing μ parameter (rather than the blackbody distribution with $\mu = 0$) [1]. For $z \lesssim 5 \times 10^4$, even the kinetic equilibrium cannot be maintained due to the inefficient Compton scatterings and the spectrum distortion can be characterized by the Compton y-parameter which is given by the line of sight integral of electron pressure [2].

The attempt to measure potential CMB spectral distortion has been made by the Far Infrared Absolute Spectrophotometer (FIRAS) instrument aboard the COBE satellite [3] two decades ago, leading to the upper bounds $|\mu| \lesssim 10^{-4}$ and $|y| \lesssim 10^{-5}$. The next generation space-telescope PIXIE [4] is expected to improve the sensitivity to $|\mu| \sim 5 \times 10^{-8}$ and $|y| \sim 10^{-8}$.

The CMB spectral distortion can, for instance, be induced by the energy injection into the background plasma in many non-standard cosmological scenarios [5]. The examples include the energy release from decaying heavy relics [6,7], evaporating primordial black holes [8], the annihilating dark matter (DM) [9,10] and the dissipation of acoustic waves [11–13].

Even in the standard cosmology, however, the CMB distortion can occur due to the energy transfer between the photons and the "baryons" (protons and electrons) [5,14,15]. The Coulomb interactions of non-relativistic plasma consisting of baryons with photons can extract energy from the CMB and maintain the kinetic equilibrium. The temperature of baryons follows that of photons and decreases inversely proportional to the scale factor of the Universe, $T_b \simeq T_\gamma \sim 1/a$, instead of $1/a^2$ for the decoupled non-relativistic matter. This extraction of energy from the CMB results in the μ-distortion of the order of $\mu \simeq -3 \times 10^{-9}$.

The analogous effects can be induced when the DM is thermally coupled to the photon-baryon plasma by the elastic scatterings, and such effects on the CMB spectral distortions were first discussed in [16] and elaborated on in [17]. The additional energy extraction from CMB into DM enhances the spectral distortion of CMB with a negative μ. Since the DM number density is inversely proportional to its mass, for a given DM mass density, the FIRAS can constrain the DM mass up to $m_\chi \sim 0.1$ GeV and a fu-

* Corresponding author at: Department of Physics, Korea Advanced Institute of Science and Technology, Daejeon, 34141, Republic of Korea.
E-mail addresses: kiyoungchoi@jnu.ac.kr (K.-Y. Choi), kadota@ibs.re.kr (K. Kadota), inwpark@kaist.ac.kr (I. Park).

ture experiment such as PIXIE can further extend its sensitivity to $m_\chi \sim 1$ GeV. The CMB distortion measurements would complement the other heavy DM searches such as the direct detection experiments which rapidly lose the sensitivity to sub-GeV DM due to the small recoil energy of the nuclear target.

One of the intriguing models which can realize the coupling of the DM to the SM particles is a "dark photon" scenario where there exists a dark sector with a broken U(1) gauge symmetry [18,19]. The phenomenology associated with such a novel dark sector has received considerable attention in recent years and a wide range of experimental searches have been performed in the collider and beam dump experiments such as BarBar, PHENIX, E137 and Charm [20–25]. The constraints on the dark photon model from the cosmological and astrophysical observations have also been discussed recently [26,27].

In this paper, we study the spectral distortion of CMB in the dark photon model, where the DM and baryons can interact via a dark photon, caused by the momentum transfer between CMB and DM via the elastic scatterings. We also illustrate the comparison with the existing constraints on the dark photon model in the laboratory and astrophysical observations. We first review the model in §2 followed by the estimation of CMB distortions in §3. §4 gives our results, followed by the conclusion in §5.

2. Dark photon and DM

We consider the dark sector consisting of the dark photon and DM. We assume that $U(1)_d$ gauge symmetry in the dark sector has a kinetic mixing with $U(1)_Y$ in the SM of $SU(3)_C \times SU(2)_L \times U(1)_Y$ [18,19]. The mixing is parametrized by a small parameter ε as

$$\mathcal{L}_{mixing} = \frac{\varepsilon}{2}\hat{B}_{\mu\nu}\hat{Z}_d^{\mu\nu} \tag{1}$$

where $\hat{B}_{\mu\nu}$ and $\hat{Z}_{d\mu\nu}$ are the field strengths of $U(1)_Y$ and $U(1)_d$ respectively. We also assume that the fermion DM χ has the $U(1)_d$ gauge interaction with the gauge coupling g_d as

$$\mathcal{L}_{int} = -g_d \hat{Z}_{d\mu}\overline{\chi}\gamma^\mu\chi. \tag{2}$$

After the electroweak symmetry breaking, we replace $\hat{B}_{\mu\nu} = -s_W\hat{Z}_{\mu\nu} + c_W\hat{A}_{\mu\nu}$ with $s_W = \sin\theta_W$ and $c_W = \cos\theta_W$ and the mass of $\hat{Z}_{\mu\nu}$, m_Z^0, is generated from the Higgs mechanism. Similarly we assume that the hidden gauge boson has a mass $m_{Z_d}^0$ by $U(1)_d$ symmetry breaking through the hidden sector Higgs mechanism.

The kinetic mixings between the gauge fields can be removed and the kinetic terms can be canonically normalized by the following field re-definition

$$\begin{pmatrix} A_{SM\mu} \\ Z_\mu^0 \\ Z_{d\mu}^0 \end{pmatrix} = \begin{pmatrix} 1 & 0 & -\varepsilon c_W \\ 0 & 1 & \varepsilon s_W \\ 0 & 0 & \sqrt{1-\varepsilon^2} \end{pmatrix} \begin{pmatrix} \hat{A}_\mu \\ \hat{Z}_\mu \\ \hat{Z}_{d\mu} \end{pmatrix} \tag{3}$$

leading to

$$\begin{aligned}\mathcal{L} = & -\frac{1}{4}A_{SM\mu\nu}A_{SM}^{\mu\nu} - \frac{1}{4}Z_{\mu\nu}^0 Z^{0\mu\nu} - \frac{1}{4}Z_{d\mu\nu}^0 Z_d^{0\mu\nu} \\ & + \frac{1}{2}m_Z^{0\,2} Z_\mu^0 Z^{0\mu} - m_Z^{0\,2}\frac{\varepsilon s_W}{\sqrt{1-\varepsilon^2}} Z_\mu^0 Z_d^{0\mu} \\ & + \frac{1}{2}\left(m_Z^{0\,2}\frac{\varepsilon^2 s_W^2}{1-\varepsilon^2} + m_{Z_d}^{0\,2}\frac{1}{1-\varepsilon^2}\right) Z_{d\mu}^0 Z_d^{0\mu}\end{aligned} \tag{4}$$

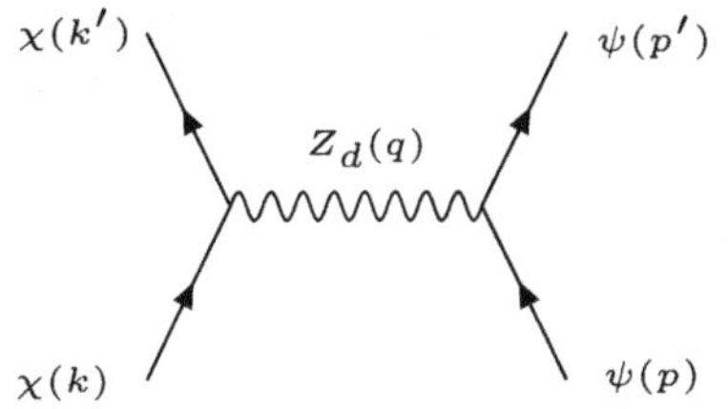

Fig. 1. Elastic scattering between baryon (ψ) and DM (χ) through a dark photon (Z_d) exchange.

The mass matrix of Z_μ^0 and $Z_{d\mu}^0$ can be diagonalized by a mixing parameter θ_X,

$$\begin{pmatrix} Z_{SM\mu} \\ Z_{d\mu} \end{pmatrix} = \begin{pmatrix} \cos\theta_X & -\sin\theta_X \\ \sin\theta_X & \cos\theta_X \end{pmatrix} \begin{pmatrix} Z_\mu^0 \\ Z_{d\mu}^0 \end{pmatrix} \tag{5}$$

where

$$\tan 2\theta_X = \frac{2m_Z^{0\,2}\varepsilon s_W/\sqrt{1-\varepsilon^2}}{m_Z^{0\,2} - m_Z^{0\,2}\{\varepsilon^2 s_W^2/(1-\varepsilon^2)\} - m_{Z_d}^{0\,2}\frac{1}{1-\varepsilon^2}}. \tag{6}$$

The bare gauge fields are consequently related to the mass eigenstates as

$$\begin{aligned}\hat{A}_\mu &= A_{SM\mu} - \frac{\varepsilon c_W s_X}{\sqrt{1-\varepsilon^2}}Z_{SM\mu} + \frac{\varepsilon c_W c_X}{\sqrt{1-\varepsilon^2}}Z_{d\mu}, \\ \hat{Z}_{d\mu} &= -\frac{s_X}{\sqrt{1-\varepsilon^2}}Z_{SM\mu} + \frac{c_X}{\sqrt{1-\varepsilon^2}}Z_{d\mu}, \\ \hat{Z}_\mu &= \left(c_X + \frac{\varepsilon s_W s_X}{\sqrt{1-\varepsilon^2}}\right)Z_{SM\mu} + \left(s_X - \frac{\varepsilon s_W c_X}{\sqrt{1-\varepsilon^2}}\right)Z_{d\mu},\end{aligned} \tag{7}$$

where $s_X = \sin\theta_X$ and $c_X = \cos\theta_X$.

The electromagnetic current hence has the interaction

$$\mathcal{L}_{int} = -eJ_{em}^\mu\left(A_{SM\mu} - \frac{\varepsilon c_W s_X}{\sqrt{1-\varepsilon^2}}Z_{SM\mu} + \frac{\varepsilon c_W c_X}{\sqrt{1-\varepsilon^2}}Z_{d\mu}\right), \tag{8}$$

and the DM interacts with $Z_{d\mu}$ and Z_μ as

$$\mathcal{L}_{int} = -g_d\overline{\chi}\gamma^\mu\chi\left(\frac{c_X}{\sqrt{1-\varepsilon^2}}Z_{d\mu} - \frac{s_X}{\sqrt{1-\varepsilon^2}}Z_{SM\mu}\right). \tag{9}$$

We can therefore see that the electromagnetic current in the SM which couples to $\hat{A}_\mu$ can interact with the dark photon Z_d suppressed by ε. Since we are interested in the parameter range $m_{Z_d} \sim$ GeV $\ll m_Z$, we can represent our dark sector model with two free parameters ε and m_{Z_d} in the following sections. We hence discuss the CMB spectral distortions when the DM interactions with the SM fields ψ_{SM} are mediated by the dark photon, represented by the Lagrangian

$$\mathcal{L}_{int} = -e\varepsilon c_W\overline{\psi}_{SM}\gamma^\mu\psi_{SM}Z_{d\mu} - g_d\overline{\chi}\gamma^\mu\chi Z_{d\mu}. \tag{10}$$

The corresponding Feynman diagram is shown in Fig. 1. We note here that the DM does not interact with the SM photon and only couples to the SM particles by mediating Z_d gauge boson.[1]

3. CMB spectrum distortion from DM-baryon scattering

For the decoupled non-relativistic DM, the temperature decreases as $T_\chi \sim a^{-2}$ (a is the scale factor). When DM is kinetically

[1] The DM coupling to the SM Z is suppressed by $\tan\theta_X$ compared with that to dark photon and hence negligible in the limit of $m_{Z_d} \ll m_Z$ and a small ε.

coupled to the background baryons ($z \gtrsim 10^4$), however, T_χ evolves along with baryon temperature T_b obeying the Boltzmann equation [26,17]

$$\dot{T}_\chi = -2HT_\chi + \Gamma_{\chi b}(T_b - T_\chi), \tag{11}$$

with

$$\Gamma_{\chi b} = \frac{2c_n N_b \sigma_n m_b m_\chi}{(m_b + m_\chi)^2}\left(\frac{T_b}{m_b} + \frac{T_\chi}{m_\chi}\right)^{(n+1)/2}, \tag{12}$$

where m_b, $N_b = N_b^0 a^{-3}$ are the baryon mass and number density. c_n is a constant of the order of unity depending on the power n of the DM-baryon elastic scattering cross section $\sigma_{tr}(v) = \sigma_n v^n$ with v being the DM-baryon relative velocity. We use the conventional cross section for the momentum-transfer

$$\sigma_{tr} \equiv \int d\Omega(1 - \cos\theta)\frac{d\sigma}{d\Omega}, \tag{13}$$

where the weight factor $(1 - \cos\theta)$ represents the longitudinal momentum transfer and regulates the spurious infrared divergence for the forward scattering (corresponding to no momentum transfer with $\cos\theta \to 1$) [28].

The DM-baryon scatterings can cause the distortion of the photon spectra and the rate of the photon energy extraction from these elastic scatterings becomes [5,17]

$$\rho_\gamma \frac{d}{dt}\left(\frac{\Delta\rho_\gamma}{\rho_\gamma}\right) = -\frac{3}{2}\left(N_b^{tot} + r_{\chi b}N_\chi\right)HT_\gamma, \tag{14}$$

where $r_{\chi b} \equiv \Gamma_{\chi b}(T_b - T_\chi)/(HT_b)$ parametrizes the efficiency of the momentum transfer from photons to DM, while the first term on RHS represents the energy transfer from the photons to baryons due to Compton scattering. The baryon number density $N_b^{tot} = \rho_b/m_H(2 - \frac{5}{4}Y_{He})$, with m_H the mass of the hydrogen, and Y_{He} helium fraction by mass. Its integration can give the estimation for the amplitude of the spectral distortion $\Delta \equiv \Delta\rho_\gamma/\rho_\gamma$. The observational bound from the FIRAS is $|\Delta| \lesssim 6 \times 10^{-5}$, and this bound is expected to be improved for the PIXIE to the level of $\Delta \approx 10^{-8}$.

For a simple power law form of the DM-baryon elastic scattering cross section $\sigma_{tr}(v) = \sigma_n v^n$, the FIRAS gives the upper bound on the cross section as [17]

$$\sigma_n \leq \sigma_n^{max} \equiv C_n \frac{m_\chi}{m_b}\left(1 + \frac{m_b}{m_\chi}\right)^{\frac{3-n}{2}}\left(\frac{a_{\max}}{a_\mu}\right)^{\frac{n+3}{2}\frac{m_\chi}{m_\chi^{\max}}}, \tag{15}$$

$a_{\max} = 10^{-4}$, $a_\mu = 0.5 \times 10^{-7}$ with $m_\chi^{\max} = 0.18$ MeV (the same formulae are applicable for the future sensitivity of PIXIE with the replacement $m_\chi^{\max} = 1.3$ GeV). For the DM-proton scattering, $C_n = (1.4 \times 10^{-30}, 1.1 \times 10^{-27}, 8.2 \times 10^{-25}, 5.5 \times 10^{-22})$ cm^2 for $n = (-1, 0, 1, 2)$ respectively and m_b with the proton mass m_p [26].

The analogous bounds can be obtained for the scatterings between DM and electrons by replacing the coefficients C_n in Eq. (15) with $C_n = (1.4 \times 10^{-30}, 2.6 \times 10^{-29}, 4.5 \times 10^{-28}, 7.0 \times 10^{-27})$ cm^2 for $n = (-1, 0, 1, 2)$ respectively and m_b with the electron mass m_e.

4. CMB spectral distortion in dark photon model

We now consider new constraints on the dark photon model from the CMB spectral distortions due to the elastic scatterings between DM and baryons. CMB distortions can probe the DM mass smaller than GeV and complement the existing bounds from other experiments as we shall discuss in the following.

In the dark photon model with a kinetic mixing outlined in §2, the momentum transfer between DM and the baryon is mediated by the dark photon as in Fig. 1. The corresponding matrix element is

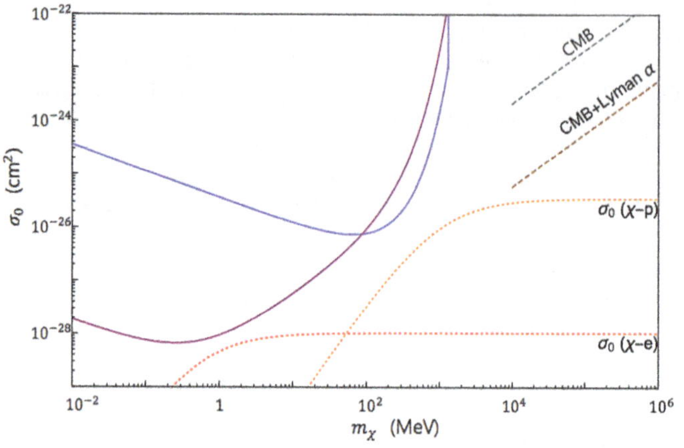

Fig. 2. The expected upper bound on the cross section from the PIXIE-like CMB spectral distortion experiment is shown with the solid lines: $\sigma_0^{max}(\chi - p)$ for DM-proton scattering (blue) and $\sigma_0^{max}(\chi - e)$ for DM-electron scattering (purple) respectively [17]. We also show the constraints from Planck CMB data, and CMB+ SDSS Lyman α data [26] with dashed lines for comparison. The cross sections in the dark photon model are shown with dotted lines: $\sigma_0(\chi - p)$ for the interaction of DM with protons while $\sigma_0(\chi - e)$ with electrons. Here we used $\alpha_D = 0.1$, $m_{Z_d} = 1$ MeV and $\varepsilon = 10^{-5}$ (for DM-proton) and 10^{-3} (for DM-electron). (For interpretation of the references to color in this figure legend, the reader is referred to the web version of this article.)

$$|\mathcal{M}|^2 = \frac{64\pi^2 c_W^2 \varepsilon^2 \alpha\alpha_D}{(q^2 - m_{Z_d}^2)^2}\left[4(k \cdot p)(k' \cdot p) + m_b^2 q^2 + k \cdot k' q^2 + q^4\right], \tag{16}$$

where $\alpha \equiv e^2/4\pi \simeq 1/137$ and $\alpha_D \equiv g_d^2/4\pi$. Here DM momentum and the relative velocity of baryon-DM in the CM frame are related as $|\vec{k}| = v m_\chi m_b/(m_\chi + m_b)$ assuming both the baryon and DM are non-relativistic. The corresponding momentum transfer cross section for $m_{Z_d} \gg |\vec{k}|$ is given by

$$\sigma_{tr} = \frac{16\pi c_W^2 \varepsilon^2 \alpha\alpha_D}{(m_\chi + m_b)^2 m_{Z_d}^4} m_\chi^2 m_b^2 + O(v^2). \tag{17}$$

Note that the leading term is independent of the velocity for the non-relativistic hidden gauge boson.

Fig. 2 shows how the momentum-transfer cross section varies in terms of m_χ (dotted lines) along with the expected upper bounds from the CMB distortion with the PIXIE-like sensitivity $\Delta \simeq 10^{-8}$, (solid lines). The region above $\sigma_0^{\max}$ is disfavored due to the large spectral distortion. For the PIXIE experiment, the constraint can be applied for the DM mass $m_\chi \leq 1.3$ GeV, since, for a larger DM mass, the distortion is too small due to the smaller DM abundance as $N_\chi/N_b^{tot} \sim 3(\text{GeV}/m_\chi)$ [17]. The dotted lines represent the constraints from the Planck CMB and SDSS Lyα forest data obtained in Ref. [26] whose analysis are applicable only to heavier DM $m_\chi \geq 10$ GeV for comparison.

Figs. 3 and 4 show the bounds from the CMB distortion on the dark photon mass (m_{Z_d}) and the kinetic mixing (ε^2) for different DM masses. We show the constraints from the DM-proton interaction with $m_\chi = 1$ MeV, 300 MeV, 1 GeV in Fig. 3, and those from the DM-electron interaction with $m_\chi = 0.1$ MeV, 1 MeV, 100 MeV in Fig. 4. We here used $\alpha_D = 0.1$ and $m_\chi^{\max} = 1.3$ GeV corresponding to the PIXIE sensitivity and the colored regions are excluded. The parameter sets producing the distortion of the order $|\Delta| \approx 3 \times 10^{-9}$ (corresponding to the expected magnitude in the conventional standard cosmology as discussed in the introduction section) are also shown to indicate the ultimate precision limit for

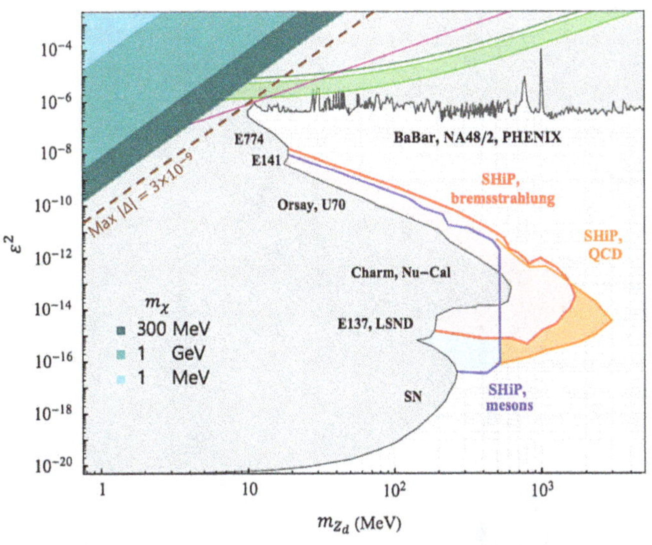

Fig. 3. The expected bounds from the CMB spectral distortion by PIXIE (colored regions are excluded) when $m_{Z_d} \gg$ keV for a few representative DM masses ($m_\chi =$ 1 MeV, 300 MeV, 1 GeV), due to the elastic scattering between DM and protons. $\alpha_d = 0.1$ is used for concreteness and the parameter sets producing the CMB distortion of the order $|\Delta| \approx 3 \times 10^{-9}$ expected in the conventional standard cosmology are indicated in a dashed line (brown). The other experimental constraints are adopted from [25]. (For interpretation of the references to color in this figure legend, the reader is referred to the web version of this article.)

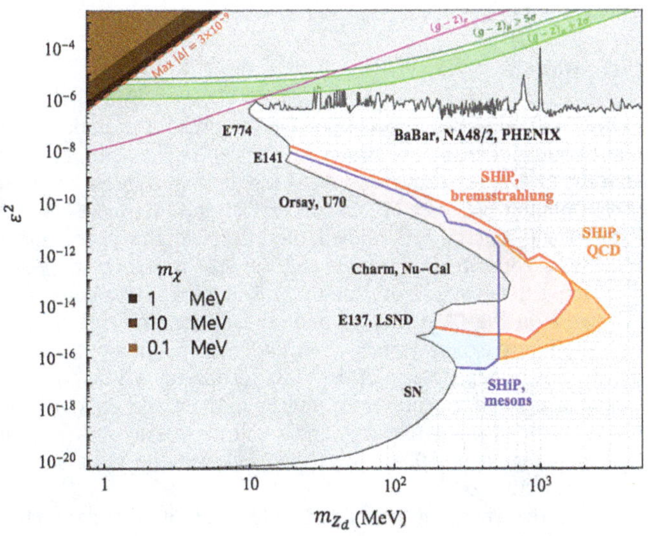

Fig. 4. The bounds due to the elastic scattering between DM and electrons, to be compared with the bounds from the DM-proton scattering in Fig. 3. (For interpretation of the colors in this figure, the reader is referred to the web version of this article.)

the CMB spectral distortion measurements. The other experimental constraints are adopted from [25].

Fig. 5 shows the exclusion plots on the plane of the DM mass (m_χ) and the kinetic mixing (ε^2). The expected excluded regions from the CMB spectral distortion with a PIXIE-like sensitivity due to the elastic scattering between DM-proton (solid line) and those for the DM-electron (dashed line) scattering are shown with different colors representing different dark photon masses m_{Z_d} ($\alpha_d = 0.1$ is used for concreteness). We expect the momentum transfer is most efficient when two scattering particles are of the same mass and our figure indeed confirms that the bound from the spectral distortion becomes tightest when the DM mass is around the pro-

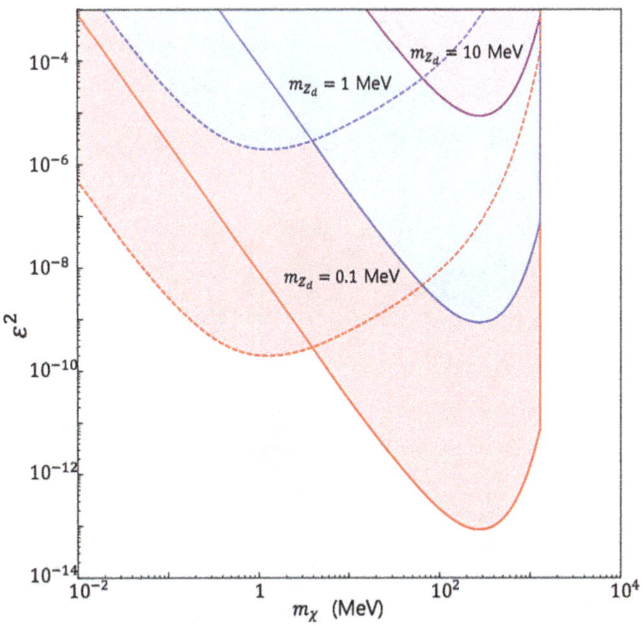

Fig. 5. The expected upper bounds from PIXIE (colored regions are excluded) in terms of the DM mass (m_χ) and the kinetic mixing (ε^2) for $m_{Z_d} \gg$ keV. The bounds from the DM-proton (DM-electron) scattering are shown with solid (dashed) lines. Different colors are for a few representative dark photon masses ($m_{Z_d} =$ 0.1 MeV, 1 MeV, 10 MeV) and $\alpha_d = 0.1$ is used for concreteness. (For interpretation of the colors in this figure, the reader is referred to the web version of this article.)

ton mass for the DM-protons scattering and around the electron mass for the DM-electrons scattering.

An interesting feature is that the constraints due to the DM-proton interaction is stronger at $m_\chi \sim 100$ MeV than those due to the DM-electron interaction even though $\sigma_0^{max}(\chi - p)$ is approximately 100 times larger than $\sigma_0^{max}(\chi - e)$. This is because the cross section for DM-proton interaction is larger than that for DM-electron by m_p^2/m_e^2 as seen in Fig. 2, thus the constraint becomes stronger compensating for the larger upper bound.

Our discussions so far focused on the dark photon mass larger than the scale of the exchanged momentum $m_{Z_d} \gg |\vec{k}|$, where the velocity dependence in the momentum-transfer cross section disappears at the leading order. We briefly discuss, before concluding our study, the opposite limit for a small dark photon mass, where the cross section behaves as $\sigma \sim v^{-4}$.

For $m_{Z_d} \ll |\vec{k}|$, the differential cross section becomes

$$\frac{d\sigma}{d\Omega} \simeq \frac{4c_W^2 \varepsilon^2 \alpha \alpha_D m_\chi^2 m_b^2}{(m_\chi + m_b)^2} \frac{1}{(2\vec{k}^2(1-\cos\theta) + m_{Z_d}^2)^2}, \tag{18}$$

and the corresponding momentum transfer cross section is

$$\begin{aligned} \sigma_{tr} &\simeq 2\pi c_W^2 \varepsilon^2 \alpha \alpha_D \frac{m_b^2 m_\chi^2}{(m_b + m_\chi)^2 \vec{k}^4} \left[\ln\left(\frac{4\vec{k}^2}{m_{Z_d}^2}\right) - 1 \right], \\ &\simeq 2\pi c_W^2 \varepsilon^2 \alpha \alpha_D \frac{(m_b + m_\chi)^2}{m_b^2 m_\chi^2 v^4} \left[\ln\left(\frac{4 \times 10^4\ \text{eV}^2}{m_{Z_d}^2}\right) - 1 \right], \\ &\equiv \sigma_{-4} v^{-4}. \end{aligned} \tag{19}$$

In the second line we used the relation between $|\vec{k}|$ and v, and used the approximation that the logarithmic term does not change much during the epoch of our interest for $10^6 \lesssim z \lesssim 10^4$ (we thus used $|\vec{k}| = 100$ eV, a typical momentum scale around $z \sim 10^6$).

While the DM decoupling epoch can be approximated by the step function for $n = 0$, the DM kinetic decoupling is far from

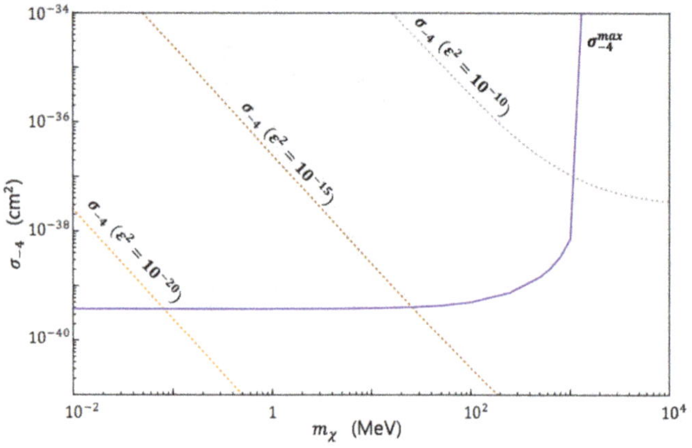

(a) DM-proton scattering

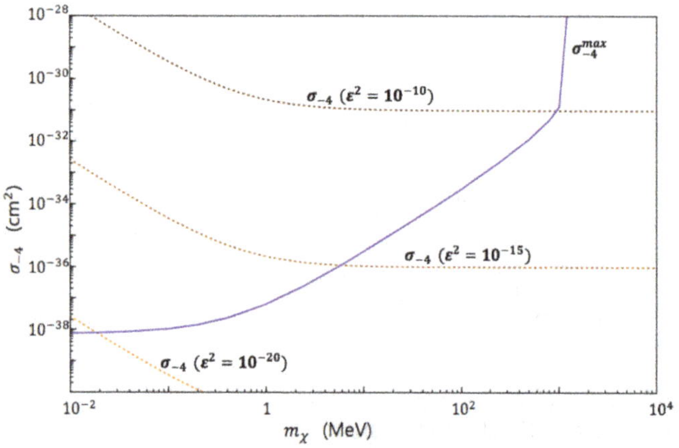

(b) DM-electron scattering

Fig. 6. Expected upper bound on the momentum-transfer cross section σ_{-4} with $\sigma = \sigma_{-4} v^{-4}$ (blue solid) for DM-protons (a), DM-electrons scattering (b). Three dotted lines are the predictions from the light dark photon model with $m_{Z_d} \ll |\vec{k}|$ with $\varepsilon^2 = 10^{-20}, 10^{-15}$, and 10^{-10} respectively. Here we used $\alpha_d = 0.1$ and $m_{Z_d} = 1$ eV. (For interpretation of the references to color in this figure legend, the reader is referred to the web version of this article.)

instantaneous transition for a light m_{Z_d} where $n = -4$. Therefore instead of using the step-function approximation as done in [17], we here solve Eqs. (11) and (14) numerically to obtain the upper bound on the momentum transfer cross section. The corresponding bound is shown in Fig. 6.[2] We can see the bound has little dependence on the DM mass, which can be expected from Eq. (14) characterizing the magnitude of the spectral distortion. For a light DM, $r_{\chi b} N_\chi$ in Eq. (14) is independent of DM mass because $\Gamma_{\chi b} \propto m_\chi$ and $N_\chi / N_b^{tot} \sim 3(\text{GeV}/m_\chi)$. The mass dependence shows up for a larger DM mass $m_\chi \gtrsim m_b$ where $\Gamma_{\chi b} \propto 1/m_\chi$ and thus $r_{\chi b} N_\chi \propto 1/m_\chi^2$, before the distortion signals become too small to be detected for $m_\chi \gtrsim 1.3$ GeV. Also note the bounds have little dependence on the dark photon mass m_{Z_d} because the cross section only depends logarithmically on m_{Z_d}. This is reasonable because the dark photon propagator $1/(k^2 - m_{Z_d}^2)$ has a small dependence on m_{Z_d} when $m_{Z_d} \ll k$. Fig. 7 shows the expected constraints on m_χ and ε^2 from the DM-proton (solid lines) scattering and the DM-electron scattering (dashed lines) for $m_{Z_d} = 1$ eV and $\alpha_d = 0.1$.

[2] For $n \leq -2$, the thermal decoupling is gradual and the Maxwell–Boltzmann distribution would not be a good approximation [17]. We need, in this case, a more rigorous treatment by solving the Boltzmann equation in the phase space and defer it to our future work.

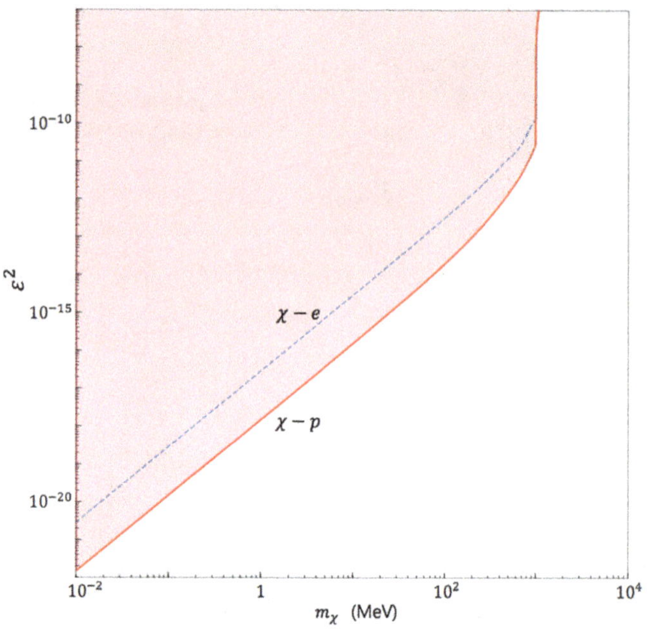

Fig. 7. The experimental bounds (colored regions are excluded) in terms of the DM mass (m_χ) and the kinetic mixing (ε) for $m_{Z_d} \ll$ keV. The expected excluded regions from the CMB spectral distortion by PIXIE due to elastic scatterings between DM-proton (solid line) and DM-electron (dashed line) are shown with the dark photon mass $m_{Z_d} = 1$ eV and $\alpha_d = 0.1$ for concreteness. (For interpretation of the colors in this figure, the reader is referred to the web version of this article.)

5. Conclusion

We have explored the possibilities to probe the dark sector where the hidden gauge boson kinetically mixes with the SM from the CMB spectral distortion. The momentum transfer between baryon-photon plasma and DM can extract energy from CMB and distort their spectra. We studied the effects in the dark photon model as a concrete example beyond the SM. In particular, we focused on a relatively light (sub-GeV) dark photon for detectable distortions in the CMB spectra, and studied the expected bounds from the future experiments such as PIXIE. We pointed out the different velocity dependence of the cross section for a different dark photon mass and we presented the bounds on the dark photon model in the regimes for large and small masses of dark photon corresponding to $n = 0$ and -4 (the power of the cross section $\propto v^n$) respectively.

While the stringent bounds already exist on the dark photon model, in particular, from the collider experiments, we illustrated that the astrophysical observables can also give the compelling limits on the dark photon parameters totally independent from those coming from the particle physics experiments. Our new constraints from the CMB spectral distortion are comparable with those already existing constraints at $m_{Z_d} = 10$ MeV. More specifically, we found the CMB spectral distortion observables can give the tight bounds, for $m_{Z_d} \gg$ keV (which corresponds to $n = 0$), when $m_\chi \sim m_p$(GeV) for $\chi - p$ scattering and when $m_\chi \sim m_e$(MeV) for $\chi - e$ scattering. It can be understood by the fact that momentum transfer is maximized when the scattering particles have comparable masses. The DM-electron scattering can give the tighter bounds than that from DM-proton scattering for a lighter dark matter mass range as illustrated in Fig. 5. For $m_{Z_d} \ll$ keV (which corresponds to $n = -4$), in contrast, $\chi - p$ scattering gives stronger constraints than $\chi - e$ scattering for the dark matter mass range considered in our analysis. This is because, as Fig. 6 illustrates, the upper bound on the momentum-transfer cross section of $\chi - p$ scattering is always stronger than $\chi - e$ scattering.

We leave the study for a more general dark photon mass range taking account of the collisional Boltzmann equations without assuming the Maxwell–Boltzmann distribution for our future work.

Acknowledgements

K.-Y.C. was supported by the National Research Foundation of Korea (NRF) grant funded by the Korea government (MEST) (NRF-2016R1A2B4012302). KK and I. Park were supported by Institute for Basic Science (IBS-R018-D1). K.-Y.C. appreciates Asia Pacific Center for Theoretical Physics for the support to the Focus Research Program.

References

[1] R.A. Sunyaev, Y.B. Zeldovich, Astrophys. Space Sci. 7 (1970) 20.
[2] Y.B. Zeldovich, R.A. Sunyaev, Astrophys. Space Sci. 4 (1969) 301, http://dx.doi.org/10.1007/BF00661821.
[3] D.J. Fixsen, E.S. Cheng, J.M. Gales, J.C. Mather, R.A. Shafer, E.L. Wright, Astrophys. J. 473 (1996) 576, http://dx.doi.org/10.1086/178173, arXiv:astro-ph/9605054.
[4] A. Kogut, et al., JCAP 1107 (2011) 025, http://dx.doi.org/10.1088/1475-7516/2011/07/025, arXiv:1105.2044 [astro-ph.CO].
[5] J. Chluba, R.A. Sunyaev, Mon. Not. R. Astron. Soc. 419 (2012) 1294, http://dx.doi.org/10.1111/j.1365-2966.2011.19786.x, arXiv:1109.6552 [astro-ph.CO].
[6] W. Hu, J. Silk, Phys. Rev. Lett. 70 (1993) 2661, http://dx.doi.org/10.1103/PhysRevLett.70.2661.
[7] S. Sarkar, A.M. Cooper-Sarkar, Phys. Lett. B 148 (1984) 347, http://dx.doi.org/10.1016/0370-2693(84)90101-1.
[8] B.J. Carr, K. Kohri, Y. Sendouda, J. Yokoyama, Phys. Rev. D 81 (2010) 104019, http://dx.doi.org/10.1103/PhysRevD.81.104019, arXiv:0912.5297 [astro-ph.CO].
[9] P. McDonald, R.J. Scherrer, T.P. Walker, Phys. Rev. D 63 (2001) 023001, http://dx.doi.org/10.1103/PhysRevD.63.023001, arXiv:astro-ph/0008134.
[10] N. Padmanabhan, D.P. Finkbeiner, Phys. Rev. D 72 (2005) 023508, http://dx.doi.org/10.1103/PhysRevD.72.023508, arXiv:astro-ph/0503486.
[11] R.A. Sunyaev, Y.B. Zeldovich, Astrophys. Space Sci. 9 (1970) 368.
[12] J. Chluba, R. Khatri, R.A. Sunyaev, Mon. Not. R. Astron. Soc. 425 (2012) 1129, http://dx.doi.org/10.1111/j.1365-2966.2012.21474.x, arXiv:1202.0057 [astro-ph.CO].
[13] J. Silk, Astrophys. J. 151 (1968) 459, http://dx.doi.org/10.1086/149449.
[14] R. Khatri, R.A. Sunyaev, J. Chluba, Astron. Astrophys. 540 (2012) A124, http://dx.doi.org/10.1051/0004-6361/201118194, arXiv:1110.0475 [astro-ph.CO].
[15] E. Pajer, M. Zaldarriaga, JCAP 1302 (2013) 036, http://dx.doi.org/10.1088/1475-7516/2013/02/036, arXiv:1206.4479 [astro-ph.CO].
[16] H. Tashiro, K. Kadota, J. Silk, Phys. Rev. D 90 (8) (2014) 083522, http://dx.doi.org/10.1103/PhysRevD.90.083522, arXiv:1408.2571 [astro-ph.CO].
[17] Y. Ali-Haïmoud, J. Chluba, M. Kamionkowski, Phys. Rev. Lett. 115 (7) (2015) 071304, http://dx.doi.org/10.1103/PhysRevLett.115.071304, arXiv:1506.04745 [astro-ph.CO].
[18] L.B. Okun, Sov. Phys. JETP 56 (1982) 502, Zh. Eksp. Teor. Fiz. 83 (1982) 892.
[19] B. Holdom, Phys. Lett. B 166 (1986) 196, http://dx.doi.org/10.1016/0370-2693(86)91377-8.
[20] H. Davoudiasl, H.S. Lee, W.J. Marciano, Phys. Rev. D 85 (2012) 115019, http://dx.doi.org/10.1103/PhysRevD.85.115019, arXiv:1203.2947 [hep-ph].
[21] R. Essig, et al., arXiv:1311.0029 [hep-ph].
[22] A. Adare, et al., PHENIX Collaboration, Phys. Rev. C 91 (3) (2015) 031901, http://dx.doi.org/10.1103/PhysRevC.91.031901, arXiv:1409.0851 [nucl-ex].
[23] E. Goudzovski, NA48/2 Collaboration, EPJ Web Conf. 96 (2015) 01017, http://dx.doi.org/10.1051/epjconf/20159601017, arXiv:1412.8053 [hep-ex].
[24] J.P. Lees, et al., BaBar Collaboration, Phys. Rev. Lett. 114 (17) (2015) 171801, http://dx.doi.org/10.1103/PhysRevLett.114.171801, arXiv:1502.02580 [hep-ex].
[25] S. Alekhin, et al., Rep. Prog. Phys. 79 (12) (2016) 124201, http://dx.doi.org/10.1088/0034-4885/79/12/124201, arXiv:1504.04855 [hep-ph].
[26] C. Dvorkin, K. Blum, M. Kamionkowski, Phys. Rev. D 89 (2) (2014) 023519, http://dx.doi.org/10.1103/PhysRevD.89.023519, arXiv:1311.2937 [astro-ph.CO].
[27] J. Berger, K. Jedamzik, D.G.E. Walker, arXiv:1605.07195 [hep-ph].
[28] S.A. Raby, G. West, Nucl. Phys. B 292 (1987) 793, http://dx.doi.org/10.1016/0550-3213(87)90671-7.

29

Complementarity and stability conditions

Howard Georgi

Center for the Fundamental Laws of Nature, ThePhysics Laboratories, Harvard University, Cambridge, MA 02138, United States

ARTICLE INFO

Editor: B. Grinstein

ABSTRACT

We discuss the issue of complementarity between the confining phase and the Higgs phase for gauge theories in which there are no light particles below the scale of confinement or spontaneous symmetry breaking. We show with a number of examples that even though the low energy effective theories are the same (and trivial), discontinuous changes in the structure of heavy stable particles can signal a phase transition and thus we can sometimes argue that two phases which have different structures of heavy particles that cannot be continuously connected and thus the phases cannot be complementary. We discuss what this means and suggest that such "stability conditions" can be a useful physical check for complementarity.

1. Introduction

This note is an attempt to understand better the classic papers by Fradkin and Shenker [1], Banks and Rabinovici [2], 't Hooft [3] and Dimopoulos, Raby and Susskind [4,5] related to complementarity between the Higgs and confining phases in gauge theories.[1] In model building, this is important because it sometimes happens that one takes a Higgsed theory that is perturbatively calculable for small couplings and pushes it into regions in which perturbation theory is questionable. If the Higgs phase and confining phase are complementary, that is if there is no phase transition separating the Higgs phase and confining phase, then one may hope that this will give a picture of the physics that is qualitatively correct even if it is not quantitatively reliable. But if the two phases are genuinely different, then you have no right to expect that this procedure will make any sense at all.

A recent example is an $SU(N+3) \times SU(3) \times U(1)$ model that was suggested as a possible explanation of the di-photon excess at 750 GeV [8]. The model has $(N+3, \overline{3})$ scalar field ξ that is trying to break the symmetry down to $SU(N) \times SU(3) \times U(1)$.[2] In the limit in which only one of the couplings gets strong, we can think of the strong non-Abelian group as the gauge symmetry and treat the other approximately as a global symmetry.

If $SU(3)$ gets strong and $SU(N+3)$ is global, the issue is easy. Here, I think that there is no hope of complementarity. Because in this case, in the Higgs phase, we have the $SU(N+3) \times U(1)$ global symmetry broken down to $SU(N) \times SU(3) \times U(1)$. There is a coset space

$$\frac{SU(N+3)}{SU(N) \times SU(3)} \tag{1.1}$$

describing an $(N, \overline{3})$ of massless Goldstone bosons in the Higgs phase and there is no unbroken gauge symmetry And even if the $SU(N+3)$ is weakly gauged, the heavy vectors are light and still present in the low energy theory.

In the confining $SU(3)$ theory, there is no reason for the global $SU(N+3)$ to break and no reason for anything to be light. So in this situation, the phases are distinguished by different symmetries and different massless particles in the low energy theory.

What happens if $SU(N+3)$ gets strong? Then presumably the $SU(3)$ is unbroken both in the confining phase and in the Higgs phase. So this could perhaps be complementary. In the Higgs phase we have massless $SU(N)$ gauge bosons, and the rest of the $SU(N+3)$ gauge bosons have mass of order gv. And Λ_N is of the same order of magnitude times the exponential factor that goes to 1 as the coupling gets large. Thus in the gauge invariant spectrum there are glueballs and bound states of heavy vectors. As the coupling increases, all of these things get heavy! Likewise, in the confining phase of the full $SU(N+3)$ theory, we expect that all the particle states will have mass of the order of the $SU(N+3)$ confinement scale or greater.

Thus in both the confining phase and the Higgs phase, the low energy theories are trivial. This is consistent with complementarity, and in this case, we believe that the phases are in fact complementary. However, in general, the equivalence of the effective low energy theories in the confining and Higgs phases [3] is not a suf-

E-mail address: hgeorgi@fas.harvard.edu.

[1] See also [6]. One other reference that might be useful is [7].

[2] There are no other matter fields that carry the $SU(N+3)$.

ficient condition for complementarity.[3] And we suggest another diagnostic for complementarity that can be useful.

It may be that even when the low energy particles and symmetries acting on them are identical, there are sectors describing heavy particles in the two phases with different properties that distinguish the two phases. The property that we will focus on is stability. In a sense, a heavy stable particle is part of the effective low energy theory because if something puts one in the low-energy world, it stays there and its interactions do not involve any high-energies. [10] Stability conditions can be an easy and very physical way of identifying this situation.

It is important to note that stability for a particular set of parameters is not enough because complementarity is about how the physics changes as parameters change. We are interested in the situation in which stability is guaranteed independent of the phase space. An example of this is a theory with a conserved quantized charge. A conserved charge divides the space of physical states up into sectors with definite charge, separated by superselection rules. In a theory with a single conserved charge, the sector with the lowest non-zero positive charge must contain stable states — either a single particle with the minimum charge or a collection of stable particles with total charge equal to the minimum. There is stability here, but it is not a property of the particle. We can certainly imagine changing the parameters in the theory continuously to make some a different particle carrying the conserved charge (not necessarily the same value of the charge) the lightest particle. And indeed, no single particle with the lowest charge has to exist at all. But at least some particles carrying the charge will always be stable so long as the charge is conserved. We might say that each sector of charged states is unconditionally stable, because there is always some combination of particles that is the lightest state with the appropriate charge.

As a very explicit (and fairly silly) example imagine a world with a conserved charge and three types of charged particle, A, B and C with charges 2, 3, and 5 respectively. The lowest positive charge is 1, and the stable states in the charge 1 sector could be $\bar{A}B$, $\bar{A}\bar{A}C$ or $\bar{C}BB$, depending on the particle masses. Charge conservation guarantees that two of the particle types are stable, and which two are actually stable depends on the masses, but the charge 1 sector is stable independent of the details of the masses..

If in a phase transition, the lowest positive charge changes, then even if the light particles in the two phases are qualitatively similar, the possible structures of stable particles in the effective low energy theory must be different in the two phases. There is then no way to get continuously from one effective theory to the other, and the two phases cannot be complementary.

In the remainder of this note, we will give a series of examples based on familiar $SU(N)$ groups. We hope they will convince the reader that this is an interesting approach.

2. SU(5) with a scalar 10

As a warm-up, and to get the reader used to the style of analysis, consider an $SU(5)$ theory with a single 10 of scalars, $\xi^{jk} = -\xi^{kj}$. The most general renormalizable Lagrangian has a global $U(1)$ symmetry, and for a range of parameters, ξ develops a VEV that can be put in the form[4]

$$\langle\xi\rangle = \begin{pmatrix} 0 & 0 & 0 & 0 & 0 \\ 0 & 0 & 0 & 0 & 0 \\ 0 & 0 & 0 & 0 & 0 \\ 0 & 0 & 0 & 0 & -v \\ 0 & 0 & 0 & v & 0 \end{pmatrix} \tag{2.2}$$

This breaks the $SU(5)$ gauge symmetry down to $SU(3) \times SU(2)$, under which ξ transforms as

$$(\bar{3}, 1) + (3, 2) + (1, 1) \tag{2.3}$$

with the VEV in the $(1, 1)$. The $(3, 2)$ and the imaginary part of the $(1, 1)$ are eaten by the Higgs mechanism producing a $(3, 2)$ and $(1, 1)$ of massive vector bosons.

There is also a global $U(1)$ symmetry that is a combination of the original global $U(1)$ and the $U(1)$ generator of the $SU(5)$ that commutes with $SU(3) \times SU(2)$. The $(1, 1)$ in (2.3) must be neutral under the unbroken symmetry, so the charges must look like

$$(\bar{3}, 1)_2 + (3, 2)_1 + (1, 1)_0 \tag{2.4}$$

in some arbitrary normalization, and because the $U(1)$ charge of the multiplet must be the average charge of the multiplet after symmetry breaking, we know that ξ is a $10_{6/5}$. The condensate also breaks the global 5-ality of the $SU(5)$ theory. down to triality×duality for the $SU(3) \times SU(2)$ In the Higgsed theory, the uneaten $(\bar{3}, 1)$ of scalars has triality 2 and charge 2, the $(3, 2)$ massive gauge boson has triality 1, duality 1 and charge 1.

In both the Higgs phase and the confining phase, heavy particles carry a quantized conserved charge. Now we can examine the stable sectors in the Higgs phase and the confining phase. In this case, they match up perfectly. In the Higgs phase, all the triality and duality zero gauge singlet combinations like 3 $(\bar{3}, 1)_2$ scalars or 6 $(3, 2)_1$ massive vector bosons all have $U(1)$ charges which are multiples of 6. In the confining theory the 5-ality zero states are combinations of 5 $10_{6/5}$ scalars, which have the same property. The lowest positive charge is 6 in both cases.

Thus the stability conditions do not distinguish between this Higgs phase and the confining phase, and this is consistent with complementarity.

3. SU(5) with a scalar 15

Contrast the model discussed in section 2 with an $SU(5)$ theory with a single 15 of scalars, $\xi^{jk} = \xi^{kj}$. The most general renormalizable Lagrangian again has a global $U(1)$ symmetry, and for a range of parameters, ξ develops a VEV that can be put in the form[5]

$$\langle\xi\rangle = \begin{pmatrix} 0 & 0 & 0 & 0 & 0 \\ 0 & 0 & 0 & 0 & 0 \\ 0 & 0 & 0 & 0 & 0 \\ 0 & 0 & 0 & 0 & 0 \\ 0 & 0 & 0 & 0 & v \end{pmatrix} \tag{3.5}$$

This breaks the $SU(5)$ gauge symmetry down to an $SU(4)$, under which ξ transforms as

$$10 + 4 + 1 \tag{3.6}$$

with the VEV in the 1. The 4 and the imaginary part of the 1 are eaten by the Higgs mechanism producing a 4 and 1 of massive vector bosons.

There is also a global $U(1)$ symmetry that is a combination of the original global $U(1)$ and the $U(1)$ generator of the $SU(5)$ that

[3] This has been emphasized in a very different context in [9].

[4] See section A.2. Note that this statement is not trivial, and such details are too often ignored in treatments of Higgs theories. However, here, we want to focus on other things, so in this and subsequent sections, we will relegate the discussion of the potentials to Appendix A.

[5] See section A.3.

commutes with $SU(4)$. The 1 in (3.6) must be neutral under the unbroken symmetry, so the charges must look like (again in an arbitrary normalization)

$$10_2 + 4_1 + 1_0 \tag{3.7}$$

And because the $U(1)$ charge of the multiplet must be the average charge of the multiplet after symmetry breaking, we know that ξ is a $15_{8/5}$.

This time the heavy stable particle sectors in the Higgs phase and the confining phase have different $U(1)$ quantum numbers. In the Higgs phase, a bound state of 4 4_1 massive vector bosons confined by the $SU(4)$ has charge 4. The sector with charge 4 has the smallest non-zero value of the conserved $U(1)$ charge, and thus it is unconditionally stable.

In the confining phase, there are no states with charge 4. The lowest nonzero charged state is a bound state of 5 $15_{8/5}$ scalars, with charge 8, and the lightest charge 8 particle is unconditionally stable. Thus the Higgs and confining theories have different unconditionally stable sectors and cannot be complementary.

The Higgs phase and the confining phase are distinguished in spite of the fact that there is nothing in the low energy theory in either case, because the stable heavy particle sectors have different global $U(1)$ charges. There is no complementarity.

It is interesting to compare this with a model with ξ being a single 5 of scalars, where we know that complementarity is preserved. In this case, again the gauged $SU(5)$ is broken to $SU(4)$ preserving a global, but now ξ breaks up into

$$4_1 + 1_0 \tag{3.8}$$

and again the 4_1 is eaten by the Higgs mechanism to become the longitudinal component of the massive gauge boson. The Higgs phase in this case is missing the 10_2 of scalars, but otherwise looks remarkably similar to the 15 case. The 4-ality zero states have charges that are multiples of 4. But now the confining phase is not qualitatively different, because the 5 has global charge 4/5 (the average charge of the multiplet in (3.8)), so the 5-ality zero states also have charges that are multiples of 4.

One of the issues in the difference between $\xi = 15$ and $\xi = 5$ is that the charge structure of the Higgs phase is determined in part by the charges of the eaten Goldstone bosons which depend on the symmetry breaking but are independent of the details of the rest of the ξ multiplet. But in the composite phase, the full multiplet is involved in everything.

4. SU(5) with 3 scalar 10s

Next consider an $SU(5)$ gauge group with three 10s of scalars. We can write the scalar fields as

$$\xi^{ajk} = \xi^{a[jk]} \tag{4.9}$$

where a is the $SU(3)$ flavor index and j, k are $SU(5)$ indices. We show below that we can find a potential with a global $SU(3) \times U(1)$ symmetry that produces the vev[6]

$$\langle \xi^{ajk} \rangle = v\, \epsilon^{ajk} \tag{4.10}$$

where ϵ_{ajk} is the 3-dimensional Levi-Civita tensor.

The VEV (4.10) preserves a global $SU(3)$ symmetry generated by the sum of the global $SU(3)_G$ symmetry generator and the generator of an $SU(3)_g$ subgroup of the gauged $SU(5)$ acting on the first 3 of the $SU(5)$ indices. And it preserves a gauged $SU(2)$ acting on $SU(5)$ indices 4 and 5. Under $SU(3)_G \times SU(5)_g \to SU(3)_G \times SU(3)_g \times SU(2)_g \to SU(3)_{G+g} \times SU(2)_g$, the $SU(5)$ generators break up into

$$\begin{aligned}(1, 24) &\to (1, 1, 3) + (1, 8, 1) + (1, 3, 2) + (1, \overline{3}, 2) + (1, 1, 1)\\ &\to (1, 3) + (8, 1) + (3, 2) + (\overline{3}, 2) + (1, 1)\end{aligned} \tag{4.11}$$

and the (complex) ξs transform like

$$\begin{aligned}(3, 10) &\to (3, \overline{3}, 1) + (3, 3, 2) + (3, 1, 1)\\ &\to (8, 1) + (1, 1) + (6, 2) + (\overline{3}, 2) + (3, 1)\end{aligned} \tag{4.12}$$

The vev (4.10) is in the real part of the singlet. The imaginary part of the (8, 1) and (1, 1) in (4.12) and the $(\overline{3}, 2)$ in (4.12) are eaten by the Higgs mechanism giving massive gauge bosons, producing an $SU(3)$ adjoint, a complex (3, 2) and a singlet. If the gauge coupling is small, their masses are in the ratio $1 : 1 : \sqrt{8/5}$. But the details here don't really matter if the coupling is strong. They just all get heavy.

There is also a global $U(1)$ symmetry that is a combination of the original global $U(1)$ and the $U(1)$ generator of the $SU(5)$ that commutes with $SU(3) \times SU(2)$. The $SU(3)_{G+g}$ singlet in ξ must be neutral under the unbroken $U(1)$, so the charges must look like

$$\begin{aligned}&(3, \overline{3}, 1)_0 + (3, 3, 2)_q + (3, 1, 1)_{2q}\\ &\to (8, 1)_0 + (1, 1)_0 + (6, 2)_q + (\overline{3}, 2)_q + (3, 1)_{2q}\end{aligned} \tag{4.13}$$

for some q. The global charge of the ξ field in the theory before symmetry breaking must then be $4q/5$, because this is the average charge of the multiplet after symmetry.

In the Higgs phase, there are no states with charge q, so the sector with charge $2q$ must contain stable particles with total charge $2q$. For weak coupling, there are both "fundamental" charge $2q$ states, like the scalar $(3, 1)_{2q}$ in (4.13), and composite charge $2q$ states, like the bound states of two $(\overline{3}, 2)_q$ vector bosons, confined when the unbroken $SU(2)$ gauge interaction gets strong.

In the confining phase, on the other hand, physical states confined by the strong $SU(5)$ gauge interactions must have 5-ality 0. They therefore contain a multiple of 5 ξs, and thus have charges which are a multiple of $4q$ and there is no stable sector with charge $2q$. Thus the Higgs phase defined by (4.10) cannot not complementary to the confining phase.

Note that the global $SU(3)$ symmetry here is almost certainly not necessary. It makes the analysis of the potential much easier, but if it is explicity broken, a phase with the same unbroken $U(1)$ and the same charges will very likely exist in some region of the parameter space.

5. SU(5) with 4 scalar $\overline{10}$s

The examples in sections 3 and 4 have confining unbroken gauge symmetries in the Higgs phase. Again, this is not necessary. Here is an example similar to that in section 4, but slightly more complicated in which the non-Abelian gauge symmetry is completely broken. Consider an $SU(5)$ gauge group with four $\overline{10}$s of scalars. We can write the scalar fields as

$$\xi^{ajk\ell} = \xi^{a[jk\ell]} \tag{5.14}$$

where a is the $SU(4)$ flavor index and j, k, ℓ are $SU(5)$ indices. Here we can find a potential with a global $SU(4) \times U(1)$ symmetry that produces the vev[7]

$$\langle \xi^{ajk\ell} \rangle = v\, \epsilon^{ajk\ell} \tag{5.15}$$

[6] See section A.4.

[7] See section A.5.

where $\epsilon_{ajk\ell}$ is the 4-dimensional Levi-Civita tensor. As usual, we first discuss the non-Abelian structure and then go back and discuss the $U(1)$s.

The VEV (5.15) preserves a global $SU(4)$ symmetry generated by the sum of the global $SU(4)_G$ symmetry generator and the generator of an $SU(4)_g$ subgroup of the gauged $SU(5)$ acting on the first 4 of the $SU(5)$ indices. Under $SU(4)_G \times SU(4)_g \to SU(4)_{G+g}$, the (complex) ξs transform like

$$(4,6)+(4,\overline{4}) \to 20+\overline{4}+15+1 \tag{5.16}$$

When the singlet gets a vev corresponding to (4.10), the $SU(5)$ symmetry breaks completely and the $SU(5)$ generators break up into

$$24 \to 15+4+\overline{4}+1 \tag{5.17}$$

all of which eat parts of the ξ field giving rise to massive gauge bosons. At tree level, this gives mass to all the gauge bosons, producing an $SU(4)$ adjoint, a $4+\overline{4}$ and 1 with masses in the ratio $1:\sqrt{3/2}:3/\sqrt{5}$. Again the details here don't really matter if the coupling is strong.

As in the example in section 4, there is also a global $U(1)$ symmetry that is a combination of the original global $U(1)$ and the $U(1)$ generator of the $SU(5)$. The $SU(4)_{G+g}$ singlet in ξ must be neutral under the unbroken $U(1)$, so the charges must look like

$$(4,6)_q+(4,\overline{4})_0 \to 20_q+\overline{4}_q+15_0+1_0 \tag{5.18}$$

for some q. The global charge of the ξ field in the theory before symmetry breaking is the average charge of the multiplet which is $3q/5$.

Now the Higgs phase at small coupling has particles with charge are the q — for example the $\overline{4}$ state. In the confining phase, however, the physical states are all built out of multiples of 5 ξs and thus have charges which are multiples of $3q$.

So again, in this case, this Higgs phase and the confining phase are distinguished in spite of the fact that there is nothing in the low energy theory in either case, because there are different stable sectors of heavy particles. As in section 4, the $SU(4)$ global symmetry makes it easy to analyze the more general potential, but it is probably not necessary for the stability analysis, which depends only on the global $U(1)$.

6. Conclusion

The examples in this note should convince the reader that in constructing an effective theory, it is important to consider heavy stable particles as well as light particles. This can contain important information about the structure of the quantum field theory. In particular, we have shown that discontinuous changes in the structure of heavy stable particles can signal a phase transition. While this can show conclusively that two phases are not continuously related, we do not know of any way to sharpen these argument to determine conclusively that two phases are complementary. For this we still need "theorems" like those of reference [1] and [2].

Acknowledgements

Savas Dimopoulos, Yuichiro Nakai, Stuart Raby, Matt Reece, Matt Schwartz, Steve Shenker and Lenny Susskind have contributed with important remarks. HG is supported in part by the National Science Foundation under grant PHY-1418114.

Appendix A. Potentials and VEVs

A.2. SU(5) with a scalar 10

For an $SU(5)$ theory with a single 10 of scalars, $\xi^{jk}=-\xi^{kj}$, we want to show that the most general renormalizable Lagrangian has a global $U(1)$ symmetry, and for a range of parameters, ξ develops a VEV that can be put in the form

$$\langle\xi\rangle = \begin{pmatrix} 0&0&0&0&0\\ 0&0&0&0&0\\ 0&0&0&0&0\\ 0&0&0&0&-v\\ 0&0&0&v&0 \end{pmatrix} \tag{A.19}$$

which breaks the symmetry down to $SU(3)\times SU(2)$. This is easy because we can treat the ξ field as a 2×2 matrix and write the most general renormalizable potential as

$$\lambda_1\left(\left(\mathrm{Tr}(\xi\xi^\dagger)\right)^2 - 4v^2\,\mathrm{Tr}(\xi\xi^\dagger)\right) - \lambda_2\left(\mathrm{Tr}(\xi\xi^\dagger\xi\xi^\dagger) - 2v^2\,\mathrm{Tr}(\xi\xi^\dagger)\right) \tag{A.20}$$

This evidently has a global $U(1)$ and it is extremized for the VEV (A.19). If

$$2\lambda_1 > \lambda_2 > 0 \tag{A.21}$$

then (A.19) is a local minimum. The massive scalars are a $(1,1)$ with mass squared $8(2\lambda_1-\lambda_2)v^2$ and a complex $(\overline{3},1)$ with mass squared $4\lambda_2 v^2$.

A.3. SU(5) with a scalar 15

For an $SU(5)$ theory with a single 15 of scalars, $\xi^{jk}=\xi^{kj}$, we want to show that the most general renormalizable Lagrangian has a global $U(1)$ symmetry, and for a range of parameters, ξ develops a VEV that can be put in the form

$$\langle\xi\rangle = \begin{pmatrix} 0&0&0&0&0\\ 0&0&0&0&0\\ 0&0&0&0&0\\ 0&0&0&0&0\\ 0&0&0&0&v \end{pmatrix} \tag{A.22}$$

which breaks the symmetry down to $SU(4)$. Again we can treat the ξ field as a 2×2 matrix and this time we will write the most general renormalizable potential as

$$\lambda_1\left(\left(\mathrm{Tr}(\xi\xi^\dagger)\right)^2 - 2v^2\,\mathrm{Tr}(\xi\xi^\dagger)\right) - \lambda_2\left(\mathrm{Tr}(\xi\xi^\dagger\xi\xi^\dagger) - 2v^2\,\mathrm{Tr}(\xi\xi^\dagger)\right) \tag{A.23}$$

This again has a global $U(1)$ and it is extremized for the VEV (A.22). If

$$\lambda_1 > \lambda_2 > 0 \tag{A.24}$$

then (A.22) is a local minimum. The massive scalars are a real singlet with mass squared $8(\lambda_1-\lambda_2)v^2$ and a complex 10 with mass squared $4\lambda_2 v^2$.

A.4. SU(5) with 3 scalar 10s

Here we are interested an $SU(5)$ gauge group with three 10s of scalars which we write as

$$\xi^{ajk} = -\xi^{akj} \tag{A.25}$$

where a is the $SU(3)$ flavor index and j, k are $SU(5)$ indices. We show below that we can find a potential with a global $SU(3) \times U(1)$ symmetry that produces the vev

$$\langle \xi^{ajk} \rangle = v\, \epsilon^{ajk} \tag{A.26}$$

where ϵ_{ajk} is the 3-dimensional Levi-Civita tensor.

The VEV (A.26) preserves a global $SU(3)$ symmetry generated by the sum of the global $SU(3)_G$ symmetry generator and the generator of an $SU(3)_g$ subgroup of the gauged $SU(5)$ acting on the first 3 of the $SU(5)$ indices. And it preserves a gauged $SU(2)$ acting on $SU(5)$ indices 4 and 5.

To see that this Higgs phase actually exists, consider the most general potential. The potential must involve 2 ξs and 2 $\xi^\dagger$s. Bose symmetry implies that the 2 ξ transform like

$$(3, 10) \times (3, 10)_{\text{symmetric}} = (6, 5) + (6, 50) + (\overline{3}, 45) \tag{A.27}$$

so there are three independent quartic terms in the potential which we can take to be

$$\kappa_1 = \kappa_0^2 \quad \text{where } \kappa_0 \text{ is the invariant mass term} \quad \kappa_0 = \xi^{aj_1k_1}\, \overline{\xi}_{aj_1k_1} \tag{A.28}$$

$$\kappa_2 = \xi^{bj_1k_1}\, \overline{\xi}_{aj_1k_1}\, \xi^{aj_2k_2}\, \overline{\xi}_{bj_2k_2} \tag{A.29}$$

$$\kappa_3 = \xi^{aj_2k_1}\, \overline{\xi}_{aj_1k_1}\, \xi^{bj_1k_2}\, \overline{\xi}_{bj_2k_2} \tag{A.30}$$

If we then write the most general potential as

$$V = \lambda_1(\kappa_1 - 12v^2\kappa_0) + \lambda_2(\kappa_2 - 4v^2\kappa_0) - \lambda_3(\kappa_3 - 4v^2\kappa_0) \tag{A.31}$$

V is extremized for the vev (4.10), and if the λs satisfy

$$3\lambda_1 + \lambda_2 > \lambda_3\,, \quad 4\lambda_2 > \lambda_3\,, \quad \lambda_3 > 0 \tag{A.32}$$

then (A.26) is a local minimum so the example works. The squared masses of the massive scalars are

$$\begin{array}{lc} \text{a real } (1,1) & 16v^2\,(3\lambda_1 + \lambda_2 - \lambda_3) \\ \text{a real } (8,1) & 4v^2\,(4\lambda_2 - \lambda_3) \\ \text{a complex } (6,2) & 4v^2\lambda_3 \\ \text{a complex } (3,1) & 8v^2\lambda_3 \end{array} \tag{A.33}$$

A.5. SU(5) with 4 scalar $\overline{10}$s

The examples in sections 3 and 4 have confining unbroken gauge symmetries in the Higgs phase. Again, this is not necessary. Here is an example similar to that in section 4, but slightly more complicated example in which the non-Abelian gauge symmetry is completely broken. Again consider an $SU(5)$ gauge group with four $\overline{10}$s of scalars. We can write the scalar fields as

$$\xi^{ajk\ell} = \xi^{a[jk\ell]} \tag{A.34}$$

where a is the $SU(4)$ flavor index and j, k, ℓ are $SU(5)$ indices. Here we can find a potential with a global $SU(4) \times U(1)$ symmetry that produces the vev

$$\langle \xi^{ajk\ell} \rangle = v\, \epsilon^{ajk\ell} \tag{A.35}$$

where $\epsilon_{ajk\ell}$ is the 4-dimensional Levi-Civita tensor.

The VEV (A.35) preserves a global $SU(4)$ symmetry generated by the sum of the global $SU(4)_G$ symmetry generator and the generator of an $SU(4)_g$ subgroup of the gauged $SU(5)$ acting on the first 4 of the $SU(5)$ indices. Under $SU(4)_G \times SU(4)_g \to SU(4)_{G+g}$, the (complex) ξs transform like

We can analyze the potential as we did in the previous example. The potential must involve two ξs and 2 $\xi^\dagger$s. Bose symmetry implies that the two ξ transform like

$$(4, \overline{10}) \times (4, \overline{10})_{\text{symmetric}} = (10, 5) + (10, \overline{50}) + (6, \overline{45}) \tag{A.36}$$

so there are three independent quartic terms in the potential which we can take to be

$$\kappa_1 = \kappa_0^2 \quad \text{where } \kappa_0 \text{ is the invariant mass term}$$

$$\kappa_0 = \xi^{aj_1k_1\ell_1}\, \overline{\xi}_{aj_1k_1\ell_1} \tag{A.37}$$

$$\kappa_2 = \xi^{bj_1k_1\ell_1}\, \overline{\xi}_{aj_1k_1\ell_1}\, \xi^{aj_2k_2\ell_2}\, \overline{\xi}_{bj_2k_2\ell_2} \tag{A.38}$$

$$\kappa_3 = \xi^{aj_2k_1\ell_1}\, \overline{\xi}_{aj_1k_1\ell_1}\, \xi^{bj_1k_2\ell_2}\, \overline{\xi}_{bj_2k_2\ell_2} \tag{A.39}$$

We could write down a 4th along the same lines,

$$\xi^{bj_2k_1\ell_1}\, \overline{\xi}_{aj_1k_1\ell_1}\, \xi^{aj_1k_2\ell_2}\, \overline{\xi}_{bj_2k_2\ell_2} \tag{A.40}$$

but we know from (A.36) that it is not independent. If we then write the most general potential as

$$V = \lambda_1(\kappa_1 - 48v^2\kappa_0) + \lambda_2(\kappa_2 - 12v^2\kappa_0) - \lambda_3(\kappa_3 - 12v^2\kappa_0) \tag{A.41}$$

Then V is extremized for the vev (A.35), and if the λs satisfy

$$4\lambda_1 + \lambda_2 > \lambda_3\,, \quad 9\lambda_2 > \lambda_3\,, \quad \lambda_3 > 0 \tag{A.42}$$

then (A.35) is a local minimum. The squared masses of the massive scalars are

$$\begin{array}{lc} \text{a real singlet} & 48v^2\,(4\lambda_1 + \lambda_2 - \lambda_3) \\ \text{a real 15} & \frac{16}{3}v^2\,(9\lambda_2 - \lambda_3) \\ \text{a complex 20} & 8v^2\lambda_3 \end{array} \tag{A.43}$$

References

[1] E.H. Fradkin, S.H. Shenker, Phase diagrams of lattice gauge theories with Higgs fields, Phys. Rev. D 19 (1979) 3682–3697.
[2] T. Banks, E. Rabinovici, Finite temperature behavior of the lattice Abelian Higgs model, Nucl. Phys. B 160 (1979) 349–379.
[3] G. 't Hooft, Naturalness, chiral symmetry, and spontaneous chiral symmetry breaking, NATO Sci. Ser. B 59 (1980) 135.
[4] L. Susskind, Lattice models of quark confinement at high temperature, Phys. Rev. D 20 (1979) 2610–2618.
[5] S. Raby, S. Dimopoulos, L. Susskind, Tumbling gauge theories, Nucl. Phys. B 169 (1980) 373–383.
[6] F.J. Wegner, Duality in generalized Ising models and phase transitions without local order parameters, J. Math. Phys. 12 (1971) 2259–2272.
[7] W. Caudy, J. Greensite, On the ambiguity of spontaneously broken gauge symmetry, Phys. Rev. D 78 (2008) 025018, arXiv:0712.0999 [hep-lat].
[8] H. Georgi, Y. Nakai, Diboson excess from a new strong force, arXiv:1606.05865 [hep-ph].
[9] J. Terning, 't Hooft anomaly matching for QCD, Phys. Rev. Lett. 80 (1998) 2517–2520, arXiv:hep-th/9706074.
[10] H. Georgi, An effective field theory for heavy quarks at low-energies, Phys. Lett. B 240 (1990) 447–450.

A "gauged" $U(1)$ Peccei–Quinn symmetry

Hajime Fukuda [a], Masahiro Ibe [a,b], Motoo Suzuki [a,b], Tsutomu T. Yanagida [a]

[a] Kavli IPMU (WPI), UTIAS, University of Tokyo, Kashiwa, Chiba 277-8583, Japan
[b] ICRR, University of Tokyo, Kashiwa, Chiba 277-8582, Japan

ARTICLE INFO

Editor: A. Ringwald

ABSTRACT

The Peccei–Quinn (PQ) solution to the strong CP problem requires an anomalous global $U(1)$ symmetry, the PQ symmetry. The origin of such a convenient global symmetry is quite puzzling from the theoretical point of view in many aspects. In this paper, we propose a simple prescription which provides an origin of the PQ symmetry. There, the global $U(1)$ PQ symmetry is virtually embedded in a gauged $U(1)$ PQ symmetry. Due to its simplicity, this mechanism can be implemented in many conventional models with the PQ symmetry.

1. Introduction

The Peccei–Quinn (PQ) mechanism [1–4] is the most successful solution to the strong CP problem. There, a global $U(1)$ symmetry (the PQ symmetry) which is almost exact but broken by the axial anomaly of QCD plays a crucial role. After spontaneous breaking, the effective θ-angle of QCD is canceled by the vacuum expectation value (VEV) of the associated pseudo Nambu-Goldstone boson, the axion a.

The origin of such a convenient global symmetry is, however, quite puzzling from the theoretical point of view in many aspects. By definition, the PQ symmetry is not an exact symmetry. Besides, the postulation of global symmetries is not comfortable in the sense of general relativity. It is also argued that all global symmetries are broken by quantum gravity effects [5–10].

In this paper, we address a question in which circumstances a theory admits the global PQ symmetry. If we could regard the PQ symmetry as a $U(1)$ gauge symmetry, there would be no suspicion about the exactness and the consistency with quantum gravity. The PQ symmetry is, however, broken by the QCD anomaly, and hence, it cannot be a consistent gauge symmetry as it is.

To circumvent the dilemma, let us recall that, for example, the $U(1)_Y$ gauge symmetry of the Standard Model would be anomalous if it coupled only to the lepton sector. The anomalies of the $U(1)_Y$ gauge symmetry in the lepton sector are canceled only when it also couples to the quark sector. In a similar manner, it seems conceivable that the anomalies of the gauged PQ symmetry, $U(1)_{gPQ}$, are canceled between the contributions from two (or more) PQ charged sectors.

To make one step forward, let us assume that the PQ charged sectors are completely decoupled with each other except for gauge interactions. In this limit, an additional accidental $U(1)$ symmetry appears, whose charge assignment coincides with the $U(1)_{gPQ}$ symmetry in each sector up to relative normalizations. There, the accidental symmetry is broken only by the QCD anomaly, and hence, it plays the role of the global PQ symmetry for the PQ mechanism.

The interactions between the PQ charged sectors inevitably break the accidental symmetry. Thus, the original question about the plausibility of the global PQ symmetry is reduced to the question how well such cross-sector symmetry breaking operators are suppressed. To this question, the gauged PQ symmetry again provides an answer. The cross-sector symmetry breaking operators can be suppressed by an appropriate charge assignment of $U(1)_{gPQ}$. Therefore, the origin of the anomalous global PQ symmetry can be attributed to a gauged $U(1)$ PQ symmetry.

In the literature, there have been many attempts to achieve the PQ symmetry as an accidental symmetry resulting from (discrete) gauge symmetries [11–25].[1] There have also been arguments of the origin of the axion in string theory [26–28] and in extra dimensional setups [29–35].

In this context, our prescription adds a simple field theoretical explanation of the origin of the PQ symmetry. There, the PQ symmetry is virtually embedded in a gauged $U(1)$ PQ symmetry.[2]

E-mail address: m0t@icrr.u-tokyo.ac.jp (M. Suzuki).

1 Discrete gauge symmetries are immune to quantum gravity effects [52–55].

2 A model discussed in [29] also achieve an virtual embedment of the PQ symmetry in a gauged $U(1)$ PQ symmetry in an extra dimensional setup.

Due to its simplicity, this mechanism can be implemented in many conventional models with the PQ symmetry. We also emphatically refer [16,36,37] which discuss the domain wall problems of axion models with similar structures we consider in the following.

2. General prescription

Let us recall invisible axion models such as the KSVZ model [38, 39] or the DSFZ model [40,41]. There, the postulated anomalous global PQ symmetry is spontaneously broken with which the axion field associates. The non-perturbative effects of QCD generate the axion potential through the axial anomalies.

Now let us bring two sectors of the invisible axion models. The two PQ symmetries in each sector, $U(1)_{PQ}$ and $U(1)_{PQ'}$, are explicitly broken by the QCD anomalies, and the corresponding Noether currents j^{μ}_{PQ} and $j^{\mu}_{PQ'}$ satisfy the anomalous ward identities,

$$\partial j_{PQ} = \frac{g_s^2}{32\pi^2} N_1 G\tilde{G}\,, \quad \partial j_{PQ'} = \frac{g_s^2}{32\pi^2} N_2 G\tilde{G}\,. \tag{1}$$

Here, G the gauge field strength of QCD, g_s the QCD coupling constant. The Lorentz indices and the color indices are suppressed. The coefficients N_1 and N_2 depend on each invisible axion model.

In the two anomalous symmetries, there is a linear combination which is free from the QCD anomaly. Hereafter, we consider that the anomaly free combination is a gauge symmetry, which we name the $U(1)_{gPQ}$ symmetry. Here, we assume that the $U(1)_{gPQ}$ is free from all anomalies.[3]

In each sector, breaking operators of the global PQ symmetries are forbidden by the $U(1)_{gPQ}$ symmetry. Therefore, the $U(1)_{gPQ}$ symmetry provides protection of the PQ symmetries in each sector.

Let us further assume that there are no interactions between the two sectors except for the gauge interactions. In this limit, the PQ symmetries in each sector are broken only by the anomalies. It should be noted that the radiative corrections generate interactions between the two sectors. Those corrections, however, do not break the PQ symmetries in each sector since they are broken only by the $U(1)_{gPQ}$ and the QCD anomalies. Therefore, in this limit, the theory possesses an accidental $U(1)$ symmetry in addition to the $U(1)_{gPQ}$ gauge symmetry. In the following, we call this anomalous accidental symmetry, $U(1)_{aPQ}$. As it has been noted, the $U(1)_{aPQ}$ symmetry plays the role of the PQ symmetry for the PQ mechanism.

In reality, there are interaction terms between the two sectors. In particular, there are terms which are invariant under the $U(1)_{gPQ}$ gauge symmetry but break the $U(1)_{aPQ}$ symmetry. For example, let us consider operators $\mathcal{O}_1$ and $\mathcal{O}_2$ which consist of fields in each sector, respectively. When these two operators have non-vanishing and opposite $U(1)_{gPQ}$ charges, the interaction terms

$$\mathcal{L}_{aPQ} = \frac{1}{M_{PL}^{d_{\mathcal{O}_1}+d_{\mathcal{O}_2}-4}} \mathcal{O}_1\mathcal{O}_2 + h.c.\,, \tag{2}$$

explicitly break the $U(1)_{aPQ}$ symmetry. Here, $d_{\mathcal{O}_{1,2}}$ denote the mass dimensions of the corresponding operators, and $M_{\rm PL}$ denotes the reduced Planck scale. Given the general discussion that all global symmetries are broken by quantum gravity effects, there is no principle to suppress these terms since it is consistent with gauge symmetries.

Such explicit breaking terms of the $U(1)_{aPQ}$ symmetry are, however, acceptable as long as the breaking effects are small enough not to spoil the PQ mechanism. In practice, the current experimental upper limit on the θ angle, $\theta \lesssim 10^{-10}$ [42], can be satisfied for $d_{\mathcal{O}_1} + d_{\mathcal{O}_2} > 10$ when the PQ symmetries are spontaneously broken at 10^{10-12} GeV [17–19].

The mass dimensions of the lowest dimensional symmetry breaking operator depends on the charge assignment of $U(1)_{gPQ}$. In fact, as we exemplify later, there are many possible charge assignments which suppress the $U(1)_{aPQ}$ breaking effects down to an acceptable level.

3. Decomposition of $U(1)_{gPQ}$ and $U(1)_{aPQ}$

Before moving to explicit examples, let us discuss how to decompose the $U(1)_{gPQ}$ and the $U(1)_{aPQ}$ symmetries. For that purpose, let us consider a simple example where the invisible axion candidates in the two sectors correspond to the axial components of complex SM gauge singlet scalar fields ϕ and ϕ',

$$\phi = \frac{1}{\sqrt{2}} f_a e^{i\tilde{a}/f_a}\,, \quad \phi' = \frac{1}{\sqrt{2}} f_b e^{i\tilde{b}/f_b}\,. \tag{3}$$

Here, $f_{a,b}$ are the decay constants of each sector and we keep only the axial components, $\tilde{a}$ and $\tilde{b}$. The domains of them are given

$$\tilde{a}/f_a = [0, 2\pi)\,, \quad \tilde{b}/f_b = [0, 2\pi)\,, \tag{4}$$

respectively.

Let us assume that the $U(1)_{gPQ}$ gauge charges of the complex scalars are q and q', respectively. In this case, the axial components are shifted by,

$$\tilde{a}/f_a \to \tilde{a}/f_a + q\alpha\,, \quad \tilde{b}/f_b \to \tilde{b}/f_b + q'\alpha\,, \tag{5}$$

under the $U(1)_{gPQ}$ symmetry. Hereafter, we take the normalization of α such that q and q' are relatively prime integers without loosing generality.

From the covariant kinetic terms of ϕ and ϕ', we obtain

$$\begin{aligned}\mathcal{L} &= |D_\mu\phi|^2 + |D_\mu\phi'|^2 \\ &= \frac{1}{2}(\partial\tilde{a})^2 + \frac{1}{2}(\partial\tilde{b})^2 - gA_\mu(qf_a\partial^\mu\tilde{a} + q' f_b\partial^\mu\tilde{b}) \\ &\quad + \frac{g^2}{2}(q^2 f_a^2 + q'^2 f_b^2)A_\mu A^\mu \\ &= \frac{1}{2}(\partial a)^2 + \frac{1}{2}m_A^2\left(A_\mu - \frac{1}{m_A}\partial_\mu b\right)^2\,,\end{aligned} \tag{6}$$

where, g is the gauge coupling constant of $U(1)_{gPQ}$. The mass of the $U(1)_{gPQ}$ gauge boson, A_μ, is given by,

$$m_A^2 = g^2(q^2 f_a^2 + q'^2 f_b^2)\,. \tag{7}$$

In the final expression, we redefine the axial fields by

$$\begin{pmatrix} a \\ b \end{pmatrix} = \frac{1}{\sqrt{q^2 f_a^2 + q'^2 f_b^2}} \begin{pmatrix} q' f_b & -qf_a \\ qf_a & q' f_b \end{pmatrix} \begin{pmatrix} \tilde{a} \\ \tilde{b} \end{pmatrix}\,. \tag{8}$$

The field b is the would-be Nambu–Goldstone boson, while the gauge invariant field a corresponds to the PQ axion.

To extract an gauge invariant $U(1)_{aPQ}$ global symmetry, let us remember that a gauge orbit of $U(1)_{gPQ}$ winds the domain of $(\tilde{a}/f_a, \tilde{b}/f_b)$ more than once for $q \neq q'$ (see Fig. 1). Then, the domain of a is given by the interval of the gauge orbit in the domain since the field points connected by a gauge orbit is physically equivalent. When we take that q and q' are relatively prime integers, we find the axion interval in the figure is given by,

[3] The $[U(1)_{gPQ}]^3$ anomaly and the gravitational anomaly of $U(1)_{gPQ}$ can be canceled by adding fermions which are singlet under the Standard Model gauge groups.

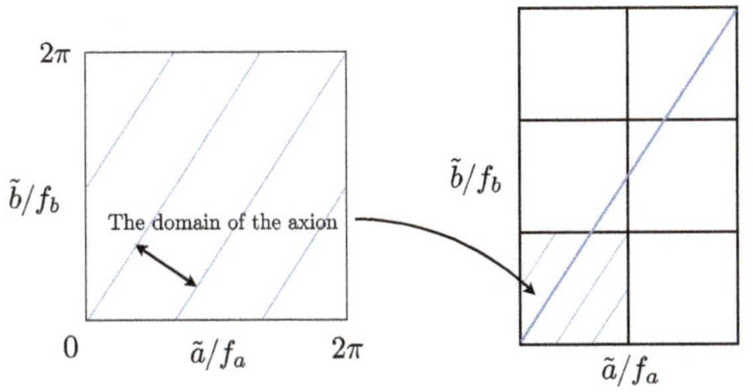

Fig. 1. (Left) A gauge orbit in the domain of $(\tilde{a}/f_a, \tilde{b}/f_b)$ for $q=2$, $q'=3$. The domain of a is given by the interval between the orbits. (Right) The unwind gauge orbits.

$$a = \left[0, \frac{2\pi f_a f_b}{\sqrt{q^2 f_a^2 + q'^2 f_b^2}}\right) . \tag{9}$$

Thus, with a decay constant,

$$F_a = \frac{f_a f_b}{\sqrt{q^2 f_a^2 + q'^2 f_b^2}} , \tag{10}$$

the $U(1)_{aPQ}$ symmetry is realized by the shift of the axion,

$$\frac{a}{F_a} \to \frac{a}{F_a} + \delta_{PQ} , \tag{11}$$

with δ_{PQ} ranging from 0 to 2π.[4]

The anomalous coupling of the axial components depends on models of the invisible axion models. In order for the $U(1)_{gPQ}$ symmetry is free from the anomaly, the anomalous coupling should appear in the form of

$$\mathcal{L}_{\rm QCD} = \frac{g_s^2}{32\pi^2} N \left(\frac{q'\tilde{a}}{f_a} - \frac{q\tilde{b}}{f_b}\right) G\tilde{G} , \tag{12}$$

$$= \frac{g_s^2}{32\pi^2} N \frac{a}{F_a} G\tilde{G} . \tag{13}$$

Here, N is a model dependent integer.

4. Examples

4.1. Barr–Seckel model

As the simplest example, let us discuss a model based on two KSVZ axion models [38,39]. This example corresponds to the model discussed in [17].

In each KSVZ sector, the PQ symmetry is spontaneously broken by the VEVs of complex scalars ϕ and ϕ' whose PQ charges are unity. In each sector, the scalars couple to extra vector-like quarks via

$$\mathcal{L} = y\phi Q \bar{Q} + h.c. , \tag{14}$$

and

$$\mathcal{L} = y'\phi' Q' \bar{Q}' + h.c. . \tag{15}$$

The PQ charges of the extra quarks are taken to be $Q(0)$ and $\bar{Q}(-1)$ in the first KSVZ sector and $Q'(0)$ and $\bar{Q}'(-1)$ in the second sector. We assume that there are N_f and N'_f flavors of the extra quarks in each sector.

Due to the QCD anomaly, the axion candidates in each sector have anomalous coupling,

$$\mathcal{L} = \frac{g_s^2}{32\pi^2}\left(\frac{N_f \tilde{a}}{f_a} + \frac{N'_f \tilde{b}}{f_b}\right) G\tilde{G} . \tag{16}$$

Here, we define the axial components of the KSVZ scalars as in Eq. (3). From this expression, we find that a linear combination of the two PQ symmetries with the charge assignments $\phi(q)$ and $\phi(q')$ is free from the anomaly for

$$q'/q = -N_f/N'_f . \tag{17}$$

As discussed in the previous section, we regard the anomaly free PQ symmetry as the $U(1)_{gPQ}$ gauge symmetry, where q and q' are normalized so that they are relatively prime integers.

Under the $U(1)_{gPQ}$ symmetry, no explicit PQ breaking operators appear in each sector. The interaction terms between the two KSVZ sectors, on the other hand, generically break $U(1)_{aPQ}$. In fact, the lowest dimensional operator which breaks the $U(1)_{aPQ}$ symmetry is given by,

$$\mathcal{L}_{\cancel{aPQ}} = \frac{1}{M_{\rm PL}^{|q|+|q'|-4}} \phi^{|q'|}\phi'^{|q|} + h.c. \tag{18}$$

As we have seen in the previous section, the explicit breaking of the PQ symmetry is acceptable when $|q|+|q'| > 10$. Once this condition is satisfied, the anomalous $U(1)_{aPQ}$ of an acceptable quality appears as a result of the $U(1)_{gPQ}$ gauge symmetry.

Let us comment here that q and q' in our normalization are given by,

$$q = N'_f/n_{gcd} , \quad q' = -N_f/n_{gcd} , \tag{19}$$

when N_f and N'_f has common divisors, $n_{gcd} > 1$. In this case, the anomalous coupling of the axion is given by,

$$\mathcal{L}_{\rm QCD} = \frac{g_s^2}{32\pi^2} n_{gcd} \frac{a}{F_a} G\tilde{G} , \tag{20}$$

which means $N = n_{gcd}$ in Eq. (13).

4.2. Composite axion model

As a second example, let us apply our prescription to the so-called composite axion model [43,44].[5]

There, we consider an $SU(N_c)$ gauge theory with vector-like fermions of $SU(N_c) \times$ QCD quantum numbers,

$$Q(N_c, 3), \quad \bar{Q}(\bar{N}_c, \bar{3}), \quad q(N_c, 1), \quad \bar{q}(\bar{N}_c, 1) . \tag{21}$$

This model possesses an axial $U(1)$ symmetry with the charge assignments,

$$Q(1), \quad \bar{Q}(1), \quad q(-3), \quad \bar{q}(-3) . \tag{22}$$

This symmetry is free from the anomaly of $SU(N_c)$ but broken by the QCD anomaly. We identify this symmetry with the anomalous PQ symmetry in the first sector. The anomalous PQ symmetry is spontaneously broken at the dynamical scale of $SU(N_c)$, where the axion appears as an composite field.[6]

[4] We may extend our analysis where there is a kinetic mixing between $\tilde{a}$ and $\tilde{b}$, although the kinetic mixing does not change our discussion.

[5] For other attempts to obtain a high-quality PQ symmetry in the composite axion model, see e.g. [25,56].

[6] There are 15 light pseudo-goldstone modes associated with the chiral symmetry breaking at Λ_N, which are color charged except for the axion candidate. The colored pseudo Nambu–Goldstone bosons obtain masses of $\mathcal{O}(\alpha_s \Lambda_N)$ where $\alpha_s = g_s^2/4\pi$. See e.g. [30].

According to the general prescription, we further introduce another sector of the composite axion where N_c is replaced by N_c'. The PQ symmetry in this sector is also broken spontaneously at the dynamical scale of $SU(N_c')$.

In this model, the anomalous couplings of the axion candidates are given by

$$\mathcal{L} = \frac{g_s^2}{32\pi^2}\left(\frac{N_c \tilde{a}}{f_a} + \frac{N_c' \tilde{b}}{f_b}\right) G^a \tilde{G}^a \,. \tag{23}$$

Here, the decay constants are taken so that the domains of $\tilde{a}/f_a = [0, 2\pi)$ and $\tilde{b}/f_b = [0, 2\pi)$ coincide with the domains of the axial components of the quark bilinears, $Q\bar{Q}$ and $Q'\bar{Q}'$, respectively. From Eq. (23), we find an anomaly free combination is given by taking

$$q'/q = -N_c/N_c', \tag{24}$$

with which we identify the $U(1)_{gPQ}$ gauge symmetry in our general prescription. The anomalous $U(1)_{aPQ}$ symmetry is, on the other hand, given by Eq. (11). The axion domain wall number corresponds to the greatest common devisor of N_c and N_c'.

Under the $U(1)_{gPQ}$ symmetry, there are explicit breaking terms of the $U(1)_{aPQ}$ symmetry,

$$\mathcal{L} \sim \frac{1}{M_{\rm PL}^{3|q|+3|q'|-4}} (Q\bar{Q})^{|q'|} (Q'\bar{Q}')^{|q|} \,. \tag{25}$$

These operators does not spoil the PQ solution for $3(|q| + |q'|) > 10$. Thus, for example, a model with $N_c = 2$ and $N_c' = 5$ provides the origin of the anomalous PQ symmetry for the successful PQ mechanism.

For $q = 3k$ or $q' = 3k'(k, k' \in \mathbb{Z}\backslash\{0\})$, there are additional lower dimensional operators which break the $U(1)_{aPQ}$ symmetry,

$$\mathcal{L} \sim \frac{1}{M_{\rm PL}^{|q|+3|q'|-4}} (Q\bar{Q})^{|q'|} (q'\bar{q}')^{|q|/3} \,, \tag{26}$$

or

$$\mathcal{L} \sim \frac{1}{M_{\rm PL}^{3|q|+|q'|-4}} (q\bar{q})^{|q'|/3} (Q'\bar{Q}')^{|q|} \,. \tag{27}$$

Those operators are harmless for $|q| + 3|q'| > 10$ or $3|q| + |q'| > 10$, which can be satisfied for $N_c = 3$ and $N_c' = 4$ for example.

5. Discussions

In this paper, we made an attempt to explain an origin of the anomalous global PQ symmetry. In our prescription, the anomalous global PQ symmetry originates from the gauged $U(1)$ symmetry where the PQ symmetry is virtually embedded in a gauged $U(1)$ symmetry. Due to its simplicity, this mechanism can be implemented in many conventional models with the PQ symmetry.

In this prescription, the anomalous PQ symmetry appears as an approximate symmetry. Thus, it is expected that the PQ symmetry is broken not only by the QCD anomaly but by some very higher dimensional operators to some extent. Thus, the effective θ-angle at the vacuum of the axion field is expected to be non-vanishing completely, though its numerical value highly depends on models.

As we have seen, our prescription allows models with either $N = 1$ or $N > 1$. For $N > 1$, the axion potential generated by the non-perturbative QCD effects has a $\mathbb{Z}_N(\subset U(1)_{aPQ})$ symmetry. When the $\mathbb{Z}_N$ symmetry is an exact symmetry, models with $N > 1$ causes a serious domain wall problem if spontaneous breaking of the PQ symmetry takes place after inflation.[7] On top of the above arguments, there can also be a serious domain wall problem even for $N = 1$ [45].

A trivial solution to the domain wall problems is to assume that the PQ symmetry breaking takes place before the end of inflation. In this case, the Hubble constant during inflation is limited from above to avoid the so-called isocurvature problem [46–51].

Acknowledgements

This work is supported in part by Grants-in-Aid for Scientific Research from the Ministry of Education, Culture, Sports, Science and Technology (MEXT) KAKENHI, Japan, No. 25105011 and No. 15H05889 (M. I.) as well as No. 26104009 (T. T. Y.); Grant-in-Aid No. 26287039 (M. I. and T. T. Y.) and No. 16H02176 (T. T. Y.) from the Japan Society for the Promotion of Science (JSPS) KAKENHI; and by the World Premier International Research Center Initiative (WPI), MEXT, Japan (M. I., and T. T. Y.). The work of H.F. is supported in part by a Research Fellowship for Young Scientists from the Japan Society for the Promotion of Science (JSPS).

References

[1] R.D. Peccei, H.R. Quinn, Phys. Rev. Lett. 38 (1977) 1440.
[2] R.D. Peccei, H.R. Quinn, Phys. Rev. D 16 (1977) 1791.
[3] S. Weinberg, Phys. Rev. Lett. 40 (1978) 223.
[4] F. Wilczek, Phys. Rev. Lett. 40 (1978) 279.
[5] S.W. Hawking, Phys. Lett. B 195 (1987) 337.
[6] G.V. Lavrelashvili, V.A. Rubakov, P.G. Tinyakov, JETP Lett. 46 (1987) 167; Pis'ma Zh. Eksp. Teor. Fiz. 46 (1987) 134.
[7] S.B. Giddings, A. Strominger, Nucl. Phys. B 307 (1988) 854.
[8] S.R. Coleman, Nucl. Phys. B 310 (1988) 643.
[9] G. Gilbert, Nucl. Phys. B 328 (1989) 159.
[10] T. Banks, N. Seiberg, Phys. Rev. D 83 (2011) 084019, arXiv:1011.5120 [hep-th].
[11] J.E. Kim, Phys. Rev. D 24 (1981) 3007.
[12] H.M. Georgi, L.J. Hall, M.B. Wise, Nucl. Phys. B 192 (1981) 409.
[13] S. Dimopoulos, P.H. Frampton, H. Georgi, M.B. Wise, Phys. Lett. B 117 (1982) 185.
[14] P.H. Frampton, Phys. Rev. D 25 (1982) 294.
[15] K. Kang, I.-G. Koh, S. Ouvry, Phys. Lett. B 119 (1982) 361.
[16] G. Lazarides, Q. Shafi, Phys. Lett. B 115 (1982) 21.
[17] S.M. Barr, D. Seckel, Phys. Rev. D 46 (1992) 539.
[18] M. Kamionkowski, J. March-Russell, Phys. Lett. B 282 (1992) 137, arXiv:hep-th/9202003.
[19] R. Holman, S.D.H. Hsu, T.W. Kephart, E.W. Kolb, R. Watkins, L.M. Widrow, Phys. Lett. B 282 (1992) 132, arXiv:hep-ph/9203206.
[20] M. Dine, in: Conference on Topics in Quantum Gravity, Cincinnati, Ohio, April 3–4, 1992, 1992, pp. 157–169, arXiv:hep-th/9207045.
[21] A.G. Dias, V. Pleitez, M.D. Tonasse, Phys. Rev. D 67 (2003) 095008, arXiv:hep-ph/0211107.
[22] L.M. Carpenter, M. Dine, G. Festuccia, Phys. Rev. D 80 (2009) 125017, arXiv:0906.1273 [hep-th].
[23] K. Harigaya, M. Ibe, K. Schmitz, T.T. Yanagida, Phys. Rev. D 88 (2013) 075022, arXiv:1308.1227 [hep-ph].
[24] K. Harigaya, M. Ibe, K. Schmitz, T.T. Yanagida, Phys. Rev. D 92 (2015) 075003, arXiv:1505.07388 [hep-ph].
[25] M. Redi, R. Sato, J. High Energy Phys. 05 (2016) 104, arXiv:1602.05427 [hep-ph].
[26] E. Witten, Phys. Lett. B 149 (1984) 351.
[27] R. Kallosh, A.D. Linde, D.A. Linde, L. Susskind, Phys. Rev. D 52 (1995) 912, arXiv:hep-th/9502069.
[28] P. Svrcek, E. Witten, J. High Energy Phys. 06 (2006) 051, arXiv:hep-th/0605206.
[29] H.-C. Cheng, D.E. Kaplan, arXiv:hep-ph/0103346, 2001.
[30] K.I. Izawa, T. Watari, T. Yanagida, Phys. Lett. B 534 (2002) 93, arXiv:hep-ph/0202171.
[31] C.T. Hill, A.K. Leibovich, Phys. Rev. D 66 (2002) 075010, arXiv:hep-ph/0205237.
[32] A. Fukunaga, K.I. Izawa, Phys. Lett. B 562 (2003) 251, arXiv:hep-ph/0301273.

[7] In our prescription, however, $\mathbb{Z}_N$ is not an exact symmetry, and hence, the domain wall problem associated for $N > 1$ might be avoidable by the effects of the explicit $U(1)_{aPQ}$ symmetry breaking [57].

[33] K.I. Izawa, T. Watari, T. Yanagida, Phys. Lett. B 589 (2004) 141, arXiv:hep-ph/0403090.
[34] K.-w. Choi, Phys. Rev. Lett. 92 (2004) 101602, arXiv:hep-ph/0308024.
[35] B. Grzadkowski, J. Wudka, Phys. Rev. D 77 (2008) 096004, arXiv:0705.4307 [hep-ph].
[36] S.M. Barr, X.C. Gao, D. Reiss, Phys. Rev. D 26 (1982) 2176.
[37] K. Choi, J.E. Kim, Phys. Rev. Lett. 55 (1985) 2637.
[38] J.E. Kim, Phys. Rev. Lett. 43 (1979) 103.
[39] M.A. Shifman, A.I. Vainshtein, V.I. Zakharov, Nucl. Phys. B 166 (1980) 493.
[40] M. Dine, W. Fischler, M. Srednicki, Phys. Lett. B 104 (1981) 199.
[41] A.R. Zhitnitsky, Sov. J. Nucl. Phys. 31 (1980) 260; Yad. Fiz. 31 (1980) 497.
[42] C.A. Baker, et al., Phys. Rev. Lett. 97 (2006) 131801, arXiv:hep-ex/0602020.
[43] J.E. Kim, Phys. Rev. D 31 (1985) 1733.
[44] K. Choi, J.E. Kim, Phys. Rev. D 32 (1985) 1828.
[45] S.M. Barr, K. Choi, J.E. Kim, Nucl. Phys. B 283 (1987) 591.
[46] M. Axenides, R.H. Brandenberger, M.S. Turner, Phys. Lett. B 126 (1983) 178.
[47] D. Seckel, M.S. Turner, Phys. Rev. D 32 (1985) 3178.
[48] A.D. Linde, Phys. Lett. B 158 (1985) 375.
[49] A.D. Linde, D.H. Lyth, Phys. Lett. B 246 (1990) 353.
[50] M.S. Turner, F. Wilczek, Phys. Rev. Lett. 66 (1991) 5.
[51] D.H. Lyth, Phys. Rev. D 45 (1992) 3394.
[52] L.M. Krauss, F. Wilczek, Phys. Rev. Lett. 62 (1989) 1221.
[53] J. Preskill, L.M. Krauss, Nucl. Phys. B 341 (1990) 50.
[54] J. Preskill, S.P. Trivedi, F. Wilczek, M.B. Wise, Nucl. Phys. B 363 (1991) 207.
[55] T. Banks, M. Dine, Phys. Rev. D 45 (1992) 1424, arXiv:hep-th/9109045.
[56] L. Randall, Phys. Lett. B 284 (1992) 77.
[57] T. Hiramatsu, M. Kawasaki, K. Saikawa, J. Cosmol. Astropart. Phys. 1108 (2011) 030, arXiv:1012.4558 [astro-ph.CO].

Permissions

All chapters in this book were first published in PLB, by Elsevier; hereby published with permission under the Creative Commons Attribution License or equivalent. Every chapter published in this book has been scrutinized by our experts. Their significance has been extensively debated. The topics covered herein carry significant findings which will fuel the growth of the discipline. They may even be implemented as practical applications or may be referred to as a beginning point for another development.

The contributors of this book come from diverse backgrounds, making this book a truly international effort. This book will bring forth new frontiers with its revolutionizing research information and detailed analysis of the nascent developments around the world.

We would like to thank all the contributing authors for lending their expertise to make the book truly unique. They have played a crucial role in the development of this book. Without their invaluable contributions this book wouldn't have been possible. They have made vital efforts to compile up to date information on the varied aspects of this subject to make this book a valuable addition to the collection of many professionals and students.

This book was conceptualized with the vision of imparting up-to-date information and advanced data in this field. To ensure the same, a matchless editorial board was set up. Every individual on the board went through rigorous rounds of assessment to prove their worth. After which they invested a large part of their time researching and compiling the most relevant data for our readers.

The editorial board has been involved in producing this book since its inception. They have spent rigorous hours researching and exploring the diverse topics which have resulted in the successful publishing of this book. They have passed on their knowledge of decades through this book. To expedite this challenging task, the publisher supported the team at every step. A small team of assistant editors was also appointed to further simplify the editing procedure and attain best results for the readers.

Apart from the editorial board, the designing team has also invested a significant amount of their time in understanding the subject and creating the most relevant covers. They scrutinized every image to scout for the most suitable representation of the subject and create an appropriate cover for the book.

The publishing team has been an ardent support to the editorial, designing and production team. Their endless efforts to recruit the best for this project, has resulted in the accomplishment of this book. They are a veteran in the field of academics and their pool of knowledge is as vast as their experience in printing. Their expertise and guidance has proved useful at every step. Their uncompromising quality standards have made this book an exceptional effort. Their encouragement from time to time has been an inspiration for everyone.

The publisher and the editorial board hope that this book will prove to be a valuable piece of knowledge for researchers, students, practitioners and scholars across the globe.

List of Contributors

Hai Tao Li and Peter Skands
ARC Centre of Excellence for Particle Physics at the Terascale, School of Physics and Astronomy, Monash University, VIC-3800,Australia

Tomohiro Abe
Institute for Advanced Research, Nagoya University, Furo-cho Chikusa-ku, Nagoya, Aichi, 464-8602, Japan
Kobayashi-Maskawa Institute for the Origin of Particles and the Universe, Nagoya University, Furo-cho Chikusa-ku, Nagoya, Aichi, 464-8602, Japan

Yoshikazu Hagiwara
Department of Physics, Kyoto University, Kyoto 606-8502, Japan

Yoshitaka Hatta
Yukawa Institute for Theoretical Physics, Kyoto University, Kyoto 606-8502, Japan

Bo-Wen Xiao
Key Laboratory of Quark and Lepton Physics (MOE) and Institute of Particle Physics, Central China Normal University, Wuhan 430079, China

Feng Yuan
Nuclear Science Division, Lawrence Berkeley National Laboratory, Berkeley, CA 94720, USA

HayatoIto, Takeo Moroia and Natsumi Nagataa,
Department of Physics, University of Tokyo, Tokyo 113-0033, Japan

Osamu Jinnouchi
Department of Physics, Tokyo Institute of Technology, Tokyo 152-8551, Japan

Hidetoshi Otono
Research Center for Advanced Particle Physics, Kyushu University, Fukuoka 819-0395, Japan

Shintaro Eijima and Mikhail Shaposhnikov
Institute of Physics, Laboratory for Particle Physics and Cosmology, École Polytechnique Fédérale de Lausanne, CH-1015 Lausanne, Switzerland

Shohini Bhattacharyaa, and Andreas Metza
Department of Physics, SERC, Temple University, Philadelphia, PA19122, USA

Jian Zhou
School of Physics and Key Laboratory of Particle Physics and Particle Irradiation (MOE), Shandong University, Jinan, Shandong250100, China

Dingli Hu
George P. and Cynthia W. Mitchell Institute for Fundamental Physics and Astronomy, Texas A&M University, College Station, TX 77843, USA

Tianjun Li
Key Laboratory of Theoretical Physics, Institute of Theoretical Physics, Chinese Academy of Sciences, Beijing 100190, China
School of Physical Sciences, University of Chinese Academy of Sciences, No.19A Yuquan Road, Beijing 100049, China

Adam Lux
Department of Physics and Engineering Physics, The University of Tulsa, Tulsa, OK 74104, USA

James A. Maxin
Department of Physics and Engineering Physics, The University of Tulsa, Tulsa, OK 74104, USA
Department of Chemistry and Physics, Louisiana State University, Shreveport, LA71115, USA

Dimitri V. Nanopoulos
George P. and Cynthia W. Mitchell Institute for Fundamental Physics and Astronomy, Texas A&M University, College Station, TX 77843, USA
Astroparticle Physics Group, Houston Advanced Research Center (HARC), Mitchell Campus, Woodlands, TX 77381, USA
GAcademy of Athens, Division of Natural Sciences, 28 Panepistimiou Avenue, Athens 10679, Greece

Matteo Rinaldi
Instituto de Fisica Corpuscular, CSIC-Universitat de Valencia, Parc Cientific UV, C/ Catedratico Jose Beltran2, E-46980 Paterna, Valencia, Spain

Giorgio Arcadi, Manfred Lindner and Farinaldo S. Queiroz
Max-Planck-Institut für Kernphysik, Saupfercheckweg 1, 69117 Heidelberg, Germany

Yann Mambrini and Mathias Pierre
Laboratoire de Physique Théorique, CNRS, Univ. Paris-Sud, Université Paris-Saclay, 91405 Orsay, France

Triparno Bandyopadhyay and Amitava Raychaudhuri
Department of Physics, University of Calcutta, 92 Acharya Prafulla Chandra Road, Kolkata 700009, India

Debasish Borah and Shibananda Sahoo
Department of Physics, Indian Institute of Technology Guwahati, Assam 781039, India

Soumya Sadhukhan
Physical Research Laboratory, Ahmedabad 380009, India

M.J. Tannenbaum
Brookhaven National Laboratory, Upton, NY, 11973, USA

J.J. Cobos-Martínez
Laboratório de Física Teórica e Computacional- LFTC, Universidade Cruzeiro do Sul, 01506-000, São Paulo, SP, Brazil
Instituto de Física y Matemáticas, Universidad Michoacana de San Nicolás de Hidalgo, Edificio C-3, Ciudad Universitaria, Morelia, Michoacán 58040, México

K. Tsushima
Laboratório de Física Teórica e Computacional- LFTC, Universidade Cruzeiro do Sul, 01506-000, São Paulo, SP, Brazil

G. Krein
Instituto de Física Teórica, Universidade Estadual Paulista, Rua Dr. Bento Teobaldo Ferraz, 271- Bloco II, 01140-070, São Paulo, SP, Brazil

A.W. Thomas
CSSM and ARC Centre of Excellence for Particle Physics at the Terascale, Department of Physics, University of Adelaide, Adelaide, SA 5005, Australia

Benjamín Grinstein, and Andrew Kobach
Physics Department, University of California, San Diego, La Jolla, CA 92093, USA

Izabela Babiarz
Faculty of Mathematics and Natural Sciences, University of Rzeszów, ul. Pigonia 1, 35-310 Rzeszów, Poland

Rafał Staszewski and Antoni Szczurek
Institute of Nuclear Physics, Polish Academy of Sciences, Radzikowskiego 152, 31-342Kraków, Poland Also at University of Rzeszów, PL-35-959 Rzeszów, Poland.

Jiajun Liao and Danny Marfatia
Department of Physics and Astronomy, University of Hawaii at Manoa, Honolulu, HI 96822, USA

Kerry Whisnant
Department of Physics and Astronomy, Iowa State University, Ames, IA 50011, USA

D.E. López-Fogliani
IFIBA, UBA & CONICET, Departamento de Física, FCEyN, Universidad de Buenos Aires, 1428 Buenos Aires, Argentina
Pontificia Universidad Católica Argentina, Buenos Aires, Argentina

C. Muñoz
Departamento de Física Teórica, Universidad Autónoma de Madrid, Campus de Cantoblanco, E-28049 Madrid, Spain
Instituto de Física Teórica UAM-CSIC, Campus de Cantoblanco, E-28049 Madrid, Spain

Hosein Bagheri, Mohammadmahdi Ettefaghi and Reza Moazzemi
Department of Physics, University of Qom, Ghadir Blvd., Qom 371614-6611, I.R. Iran

C. Beskidta, W. deBoer and S. Wayand
Institut für Experimentelle Kernphysik, Karlsruhe Institute of Technology, 76128 Karlsruhe, Germany

D.I. Kazakov
Institut für Experimentelle Kernphysik, Karlsruhe Institute of Technology, 76128 Karlsruhe, Germany
Bogoliubov Laboratory of Theoretical Physics, Joint Institute for Nuclear Research, 141980, 6 Joliot-Curie, Dubna, Moscow Region, Russia

Francesco D'Eramo
Department of Physics, University of California Santa Cruz, 1156 High St., Santa Cruz, CA 95064, USA
Santa Cruz Institute for Particle Physics, 1156 High St., Santa Cruz, CA 95064, USA

Bradley J. Kavanagh
Laboratoire de Physique Théorique et Hautes Energies, CNRS, UMR 7589, 4 Place Jussieu, F-75252 Paris, France

Paolo Panci
CERN Theoretical Physics Department, CERN, Case C01600, CH-1211 Genève, Switzerland
Institut d'Astrophysique de Paris, UMR 7095 CNRS, Université Pierre et Marie Curie, 98 bis Boulevard Arago, F-75014 Paris, France

Satoshi Mishima and Kei Yamamoto
Theory Center, IPNS, KEK, Tsukuba, Ibaraki 305-0801, Japan

Motoi Endo
Theory Center, IPNS, KEK, Tsukuba, Ibaraki 305-0801, Japan

The Graduate University of Advanced Studies (Sokendai), Tsukuba, Ibaraki 305-0801, Japan

Teppei Kitahara
Institute for Theoretical Particle Physics (TTP), Karlsruhe Institute of Technology, Engesserstraße 7, D-76128 Karlsruhe, Germany
Institute for Nuclear Physics (IKP), Karlsruhe Institute of Technology, Hermann-von-Helmholtz-Platz 1, D-76344 Eggenstein-Leopoldshafen, Germany

Benjamin Fuks
Sorbonne Universités, Université Pierre et Marie Curie (Paris 06), UMR 7589, LPTHE, F-75005 Paris, France
CNRS, UMR 7589, LPTHE, F-75005 Paris, France
Institut Universitaire de France, 103 boulevard Saint-Michel, 75005 Paris, France

Jeong Han Kim
Department of Physics and Astronomy, University of Kansas, Lawrence, KS, 66045, USA

Seung J. Lee
Department of Physics, Korea University, Seoul 136-713, Republic of Korea
School of Physics, Korea Institute for Advanced Study, Seoul 130-722, Republic of Korea

Lin Hana, and Yan-Ju Zhanga,
School of Biomedical Engineering, Xinxiang Medical University, Xinxiang 453003, China

Yao-Bei Liu
Henan Institute of Science and Technology, Xinxiang 453003, PR China

Vincenzo Cirigliano
Theoretical Division, Los Alamos National Laboratory, Los Alamos, NM 87545, USA

Sacha Davidson
IPNL, CNRS/IN2P3, Université Lyon 1, Univ. Lyon, 69622 Villeurbanne, France

Yoshitaka Kuno
Department of Physics, Osaka University, 1-1 Machikaneyama, Toyonaka, Osaka 560-0043, Japan

Johannes Blümlein
Deutsches Elektronen-Synchrotron, DESY, Platanenallee 6, D-15738 Zeuthen, Germany

Carsten Schneider
Research Institute for Symbolic Computation (RISC), Johannes Kepler University Linz, Altenbergerstraße 69, A-4040 Linz, Austria

P. Guo
Department of Physics and Engineering, California State University, Bakersfield, CA 93311, USA

I.V. Danilkin
Institut für Kernphysik and PRISMA Cluster of Excellence, Johannes Gutenberg Universität, D-55099 Mainz, Germany
SSC RF ITEP, Bolshaya Cheremushkinskaya 25, 117218 Moscow, Russia

C. Fernández-Ramírez
Instituto de Ciencias Nucleares, Universidad Nacional Autónoma de México, Ciudad de México 04510, Mexico

V. Mathieu
Center for Exploration of Energy and Matter, Indiana University, Bloomington, IN 47403, USA
Physics Department, Indiana University, Bloomington, IN 47405, USA

A.P. Szczepaniak
Center for Exploration of Energy and Matter, Indiana University, Bloomington, IN 47403, USA
Physics Department, Indiana University, Bloomington, IN 47405, USA
Theory Center, Thomas Jefferson National Accelerator Facility, Newport News, VA 23606, USA

Debasish Borah
Department of Physics, Indian Institute of Technology, Guwahati, Assam 781039, India

Sudhanwa Patra
Center of Excellence in Theoretical and Mathematical Sciences, Siksha 'O' Anusandhan University, Bhubaneswar, 751030, India

Ki-Young Choi
Institute for Universe and Elementary Particles and Department of Physics, Chonnam National University, 77 Yongbong-ro, Buk-gu, Gwangju, 61186, RepublicofKorea

Kenji Kadota
Center for Theoretical Physics of the Universe, Institute for Basic Science (IBS), Daejeon, 34051, Republic of Korea

Inwoo Park
Center for Theoretical Physics of the Universe, Institute for Basic Science (IBS), Daejeon, 34051, Republic of Korea
Department of Physics, Korea Advanced Institute of Science and Technology, Daejeon, 34141, Republic of Korea

Howard Georgi
Center for the Fundamental Laws of Nature, ThePhysics Laboratories, Harvard University, Cambridge, MA 02138, United States

Hajime Fukuda and Tsutomu T.Yanagida
Kavli IPMU (WPI), UTIAS, University of Tokyo, Kashiwa, Chiba 277-8583, Japan

Masahiro Ibe and Motoo Suzuki
Kavli IPMU (WPI), UTIAS, University of Tokyo, Kashiwa, Chiba 277-8583, Japan
ICRR, University of Tokyo, Kashiwa, Chiba 277-8582, Japan

Index

www.ingramcontent.com/pod-product-compliance
Lightning Source LLC
Chambersburg PA
CBHW080658200326
41458CB00013B/4898
* 9 7 8 1 6 8 2 8 5 6 4 5 1 *